SALVAGE

Proceedings

10th IEEE Symposium on
COMPUTER ARITHMETIC

Editors:
Peter Kornerup
David W. Matula

PROCEEDINGS

June 26-28, 1991
Grenoble, France

10th SYMPOSIUM on

COMPUTER ARITHMETIC

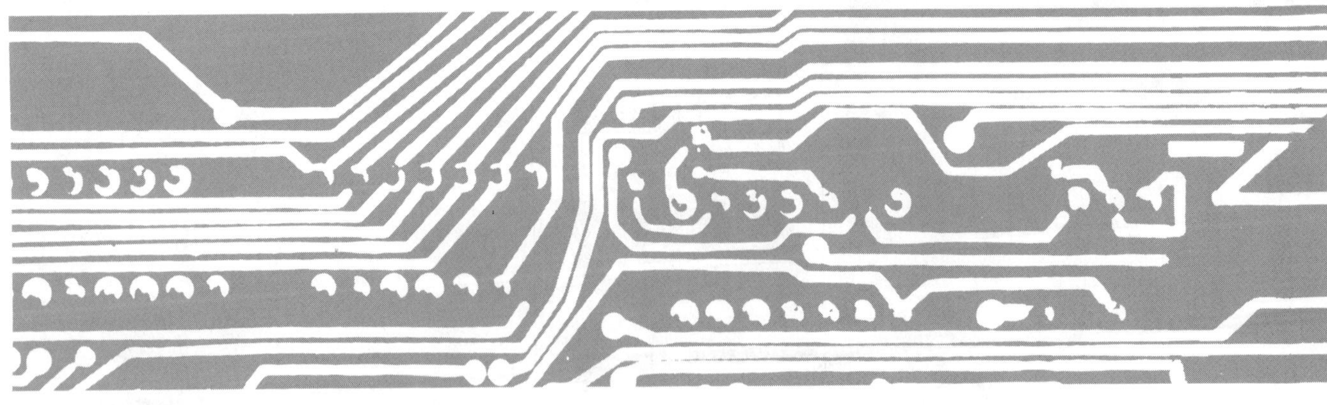

1951-1991

IEEE COMPUTER SOCIETY PRESS

THE INSTITUTE OF ELECTRICAL AND ELECTRONICS ENGINEERS, INC.

Proceedings

10th IEEE Symposium on
COMPUTER ARITHMETIC

June 26-28, 1991   Grenoble, France

Edited by:
Peter Kornerup
David W. Matula

Sponsored by
IEEE Computer Society
IEEE Technical Committee on VLSI

In cooperation with
Centre National de la Recherche Scientifique (CNRS)
Institut d'Informatique et de
Mathématiques Appliquées de Grenoble (IMAG)

1951-1991

IEEE Computer Society Press
Los Alamitos, California

Washington   ●   Brussels   ●   Tokyo

Published by the
IEEE Computer Society Press
10662 Los Vaqueros Circle
PO Box 3014
Los Alamitos, CA 90720-1264

IEEE Computer Society Press Order Number 2151
Library of Congress Number 88-641433
IEEE Catalog Number 91CH3015-5
ISBN 0-8186-6151-8 (microfiche)
ISBN 0-8186-9151-4 (case)

Additional copies can be ordered from

IEEE Computer Society Press
Customer Service Center
10662 Los Vaqueros Circle
PO Box 3014
Los Alamitos, CA 90720-1264

IEEE Service Center
445 Hoes Lane
PO Box 1331
Piscataway, NJ 08855-1331

IEEE Computer Society
13, avenue de l'Aquilon
B-1200 Brussels
BELGIUM

IEEE Computer Society
Ooshima Building
2-19-1 Minami-Aoyama
Minato-ku, Tokyo 107
JAPAN

Editorial production: Wally Hutchins
Printed in the United States of America by Braun-Brumfield, Inc.

 THE INSTITUTE OF ELECTRICAL AND ELECTRONICS ENGINEERS, INC.

Foreword

The 10th Symposium on Computer Arithmetic was held in Grenoble, France, on June 26-28, 1991. The Symposia on Computer Arithmetic are held every other year, and have alternated between the United States and Europe since the 6th (1983). They involve researchers from around the world. A total of 36 papers were presented in 10 technical sessions. The contributions to the present symposium and these proceedings represent a wide spectrum of interests, ranging from number systems to VLSI-oriented algorithms for arithmetic operations.

A total of 89 papers were submitted and it was a difficult task for the Program Committee to select the 36 papers finally accepted for presentation. Each paper was reviewed by at least three referees before the meeting of the Program Committee on January 25-26, 1991. Further reviews were obtained from among the 13 program committee members present at that meeting. Decisions on papers authored or coauthored by program committee members present were made in two subcommittees, chosen such that no member could unduly influence the decision concerning his/her own paper. Some papers were conditionally accepted, subject to a revision and another reviewing cycle. The authors of such conditionally accepted papers were very cooperative in expediting revisions satisfying the referees. We want to express our thanks to the referees and the members of the Program Committee for a thorough and timely job in the evaluation and selection process.

Thanks are also due to the IEEE Computer Society, the IEEE Technical Committee on VLSI, the Centre National de la Recherche Scientifique (CNRS) and the Institut d'Informatique et de Mathématiques Appliquées de Grenoble (IMAG) for their sponsorship and support.

General Chairman

Jean-Michel Muller
CNRS, LIP-IMAG
Lyon, France

Program Co-Chairmen

Peter Kornerup
Odense University, Denmark

David W. Matula
Southern Methodist University
Dallas, Texas, USA

Symposium Committee

General Chairman

Jean-Michel Muller
LIP-IMAG, École Normale Supérieure de Lyon

Program Co-Chairmen

Peter Kornerup
Odense University, Denmark
David W. Matula
Southern Methodist University, Dallas, TX

Program Committee

M.A. Bayoumi, *Univ. of Southwestern Louisiana, LA*
T.M. Carter, *University of Utah, UT*
L. Ciminiera, *Politecnico di Torino, Italy*
W.J. Cody, *Argonne Nat. Laboratory, IL*
M.D. Ercegovac, *UCLA, CA*
J. Fandrianto, *Integrated Information Technology, CA*
W.E. Ferguson, *Cyrix Corp., Dallas, TX*
T.E. Hull, *University of Toronto, Canada*
M. Iri, *University of Tokyo, Japan*
M-J. Irwin, *Penn. State, PA*
G.A. Jullien, *University of Windsor, Canada*
S. Knowles, *INMOS, Great Britain*
U. Kulisch, *Universität Karlsruhe, Germany*
J-M. Muller, *LIP-IMAG, Lyon, France*
S.M. Rump, *Tech. Univ. Hamburg-Harburg, Germany*
H.J. Sips, *Delft Tech. Univ., The Netherlands*
R. Stefanelli, *Politecnico di Milano, Italy*
E.E. Swartzlander, *University of Texas at Austin, TX*
N. Takagi, *Kyoto University, Japan*
G. Taylor, *MIPS, CA*
P.R. Turner, *U.S. Naval Academy, MD*
J. Vignes, *Univ. Pierre et Marie Curie, France*
J. Vuillemin, *DEC-PRL, France*

Local Arrangements

Jean-Michel Muller
LIP-IMAG, École Normale Supérieure de Lyon

List of Referees

Alt, R.	Lang, T.
Bayoumi, M.A.	Lewis
Birman, M.	Lin, H.
Bischof	Lozier, D.
Bohlender, G.	M-Nielsen, P.
Briggs, W.S.	Matula, D.W.
Buchanani, J.L.	Montuschi, P.
Burgess, N.	Muller, J-M.
Carter, T.M.	Murota, K.
Chen, H.	Neff, R.
Chow, E.T.	Okabe, Y.
Ciminiera, L.	Olver, F.W.J.
Cody, W.J.	Overbeek, R.
Coonen, J.T.	Owens, R.M.
Cosnard, M.	Parikh, S.
Dadda, L.	Piuri, V.
Deprettere, E.F.	Privat, G.
Ercegovac, M.D.	Quach, N.
Fandrianto, J.	Rump, S.M.
Ferguson, W.	Santoro, M.
Fraigniaud, P.	Seetharaman, G.
Frandsen, G.	Silverman, R.
Franklin, C.	Sips, H.
Gavrielov, M.	Skavantzos, A.
Gladwell, I.	Srinivasan, P.
Gonnella, J.	Stefanelli, R.
Guyot, A.	Sturges, A.
Hannon, G.	Swartzlander, E.
Hekstra, H.	Takagi, N.
Hinton	Tang, P.T.P.
Hull, T.E.	Taylor, G.
Hutchings, B.	Teufel, T.
Iri, M.	Turner, P.R.
Irwin, M-J.	van Westrheuen, S.C.
Jansson, C.	Vignes, J.
Johansen, S.P.	Vishwanath, V.
Jullien, G.	Vuillemin, J.
Karageorgis, A.	Vuong, I.
Kelliher	Wigley, N.W.
Knowles, S.	Woods, R.
Kornerup, P.	Yamahata, H.
Kulisch, U.	Yves, R.

Previous IEEE Symposia on Computer Arithmetic

1. June 16, 1969

   One-day workshop
   preceding the IEEE Computer Group Conference
   Minneapolis, MN
   Organized by R.R. Shively

2. May 15-16, 1972

   One-and-a-half-day symposium (SCA-2)
   University of Maryland, College Park, MD
   Organized by H.L. Garner and D.E. Atkins

3. Nov. 19-20, 1975

   Two-day symposium (SCA-3)
   Southern Methodist University, Dallas, TX
   Organized by T.R.N. Rao and D.W. Matula

4. Oct. 25-27, 1978

   Two-and-a-half-day symposium (SCA-4)
   UCLA, Los Angeles, CA
   Organized by A. Avizienis and M.D. Ercegovac

5. May 18-19, 1981

   Two-day symposium plus tutorial (SCA-5)
   University of Michigan, Ann Arbor, MI
   Organized by K.S. Trivedi and D.E. Atkins

6. June 20-22, 1983

   Three-day symposium (ARITH6)
   Aarhus University, Aarhus, Denmark
   Organized by T.R.N. Rao and P. Kornerup

7. June 4-6, 1985

   Two-and-a-half-day symposium (ARITH7)
   University of Illinois, Urbana, IL
   Organized by K. Hwang, D.D. Gajski, and A. Sameh

8. May 19-21, 1987

   Two-and-a-half-day symposium (ARITH8)
   Villa Olmo, Como, Italy
   Organized by Luigi Dadda, Mary-Jane Irwin,
   and R. Stefanelli

9. Sept. 6-9, 1989

   Two-and-a-half-day symposium (ARITH9)
   Santa Monica, CA
   Organized by A. Avizienis, M.D. Ercegovac,
   and E.E. Swartzlander

Keynote Speaker

W. J. Cody
Mathematics and Computer Science Division
Argonne National Laboratory
Argonne, IL 60439-4801

Arithmetic Standards: The Long Road*

Abstract

This is an informal discussion of the events leading to the IEEE floating-point standards, the practical implications of those standards, and the barriers remaining to full realization of their potential.

W. J. Cody retired from Argonne National Laboratory in March 1991 after almost 32 years of service. At that time he was a Senior Mathematician in the Mathematics and Computer Science Division. He received the BS degree in mathematics from Elmhurst College, Elmhurst, Ill., in 1951; the MA degree in mathematics from the University of Oklahoma, Norman, in 1956; and an honorary ScD degree from Elmhurst in 1977.

His research interests include the approximation and evaluation of elementary and special functions, the design and evaluation of numerical software, and the interaction between computer arithmetic design and numerical algorithms. Among his contributions are the book "Software Manual for the Elementary Functions," coauthored with W. Waite, the program MACHAR for dynamically determining the characteristics of a floating-point system, and the widely used software packages ELEFUNT and CELEFUNT for testing real and complex elementary functions routines, and FUNPACK and SPECFUN containing special functions programs. The 1972 paper "A Statistical Study of the Accuracy of Floating-point Number Systems," coauthored with H. Kuki, was selected in 1983 as one of the most influential papers published in the first 25 years of CACM.

He is currently a member of IFIP Working Group 2.5 on Mathematical Software and of the ACM SIGAda Numerics Working Group now drafting Ada standards for primitive and elementary functions. He was a member of the committee that drafted ANSI/IEEE Std 754-1985 and chaired the committee that drafted ANSI/IEEE Std 854-1987. He received an IEEE Computer Society Outstanding Contribution Award in 1988 for the latter effort.

* This work was supported by the Applied Mathematical Sciences subprogram of the Office of Energy Research, U.S. Department of Energy, under Contract W-31-109-Eng-38.

Table of Contents

Session 1: Number Systems

(Chair: Michel Scott, Dublin City University)

Session 2: Multiplication

(Chair: Simon Knowles, INMOS, GB)

Session 3: Inner Products

(Chair: Svetoslav Markov, Bulgarian Academy of Science)

Session 1:

Number Systems

Chair:
Michel Scott
Dublin City University

New Redundant Representations of Complex Numbers and Vectors

Jean Duprat   Yvan Herreros   Sylvanus Kla

Laboratoire LIP-IMAG
Ecole Normale Supérieure de Lyon, 46 Allée d'Italie
69364 Lyon Cedex 07 FRANCE

Abstract

*In this paper, we present a new redundant representation for complex numbers, called **polygonal representation**. This representation enables fast carry-free addition (in a way quite similar to the carry-free addition in signed-digits number systems), and is convenient for multiplication. Then we generalize our technique in order to handle n-dimensional vectors.*

Introduction

Complex numbers and vectors are used in various fields of Computer Science. We need to represent these objects as efficiently as possible. The most common way is of course to represent them as arrays of real numbers: here we shall study an other way, using complex digit sets (in the real case, digit sets have been widely studied, for instance by Matula [1], Petkovsek [2], Carter and Robertson [3]).

Our goal is to generalize the carry-free addition technique of Avizienis [4] for signed-digit arithmetic to complex and vectorial arithmetic. As in Avizienis' number systems, we need *redundancy* in order to enable carry free addition. In the first part of this paper, we deal with manipulation of Complex Numbers. Then we shall generalize our technique to the manipulation of vectors.

1 Polygonal representation of complex numbers

1.1 Usual representation

The most common way is of course to represent the complex number $a + i.b$ by the couple (a, b) of real numbers. This representation has some drawbacks: the sets of numbers of the form $a + ib$ with a and b in real intervals (rectangles) are not stable by complex multiplication: for example $E = \left\{ a + i.b / (a, b) \in [-1, 1]^2 \right\}$ is represented in *Fig. 1* with the set $F = \left\{ xy / (x, y) \in E^2 \right\}$

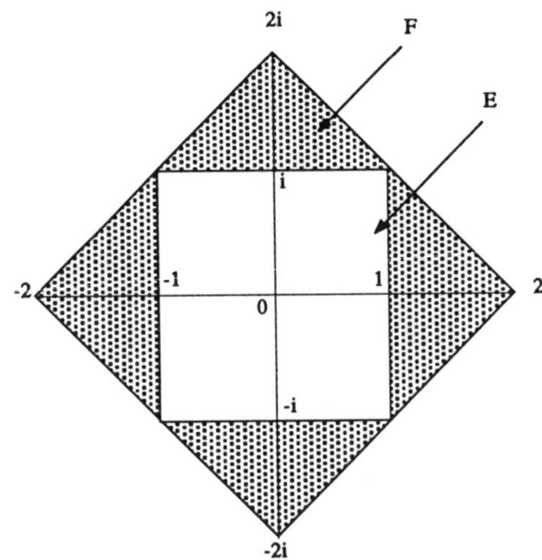

Fig. 1: Sets E and F

Some other representations have been previously proposed:

- Radix $i\sqrt{2}$ with digits in $\{0, 1\}$
- Radix $i - 1$ with digits in $\{0, 1\}$ [5].

These two representations use as radix a 8th root of 16 (of modulus $\sqrt{2}$). Both representations use a complex radix and integer digits: in a dual way, we shall use here an integer radix and complex digits.

1.2 Use of pth roots of unity and zero as digits

Assume that we are in radix 2. We shall consider the case where the digits are chosen in the set containing the pth roots of the unity and zero:

$$D_p = \{0, 1, \omega, \omega^2, \cdots, \omega^{p-1}\} \text{ where } \omega = e^{\frac{2i\pi}{p}}$$

As in conventional number systems, a number x is represented by a digit sequence $(d_i), d_i \in D_p$ which satisfies

$$x = \sum_{i=-\infty}^{\infty} d_i 2^i.$$

i. p=1: It is the usual representation of real numbers in radix 2 with digits in $\{0, 1\}$. Each positive real number is representable.

ii. p=2: We obtain the binary signed-digit representation of real numbers, with digits in $\{-1, 0, 1\}$ [4]. Each real number is representable.

iii. p=3: The digits are taken in $\{0, 1, j, j^2\}$, with $j = e^{2i\pi/3}$. The set of representable numbers has a fractal structure. *Fig 2* presents the set of numbers representable only with "fractional" digits, i.e. the set of numbers of the form $x = \sum_{i=0}^{\infty} d_i 2^{-i}$.

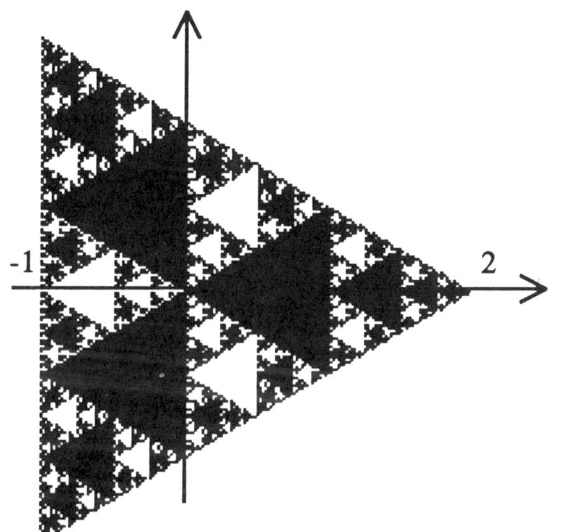

Fig. 2: Numbers representable in radix 2 with digits in $\{0, 1, \mathbf{j}, \mathbf{j^2}\}$

iv. p≥4: If $p \geq 4$, every complex number can be represented in radix 2 with digits in D_p. As a proof, we give for $p = 4$ an algorithm which computes a representation of a given number. This algorithm may be easily extended to higher values of p (see [6]).

Assume that we want to compute a representation of a number x. Since from a representation of x, a representation of $2^k x$ may be obviously deduced, we assume without loss of generality, that x lies into the square S delimited by numbers $2, 2i, -2, -2i$ (see *Fig. 3*). We divide this square into the 5 areas labelled $1, -1, i, -i$ and 0 depicted in *Fig. 3*.

Let us denote $x^{(0)} = x$. The sequence d_i of digits of a representation of x is defined by induction as:

- $d_i = k$ if $x^{(i)}$ lies in the area labelled k
- $x^{(i+1)} = 2(x^{(i)} - k)$

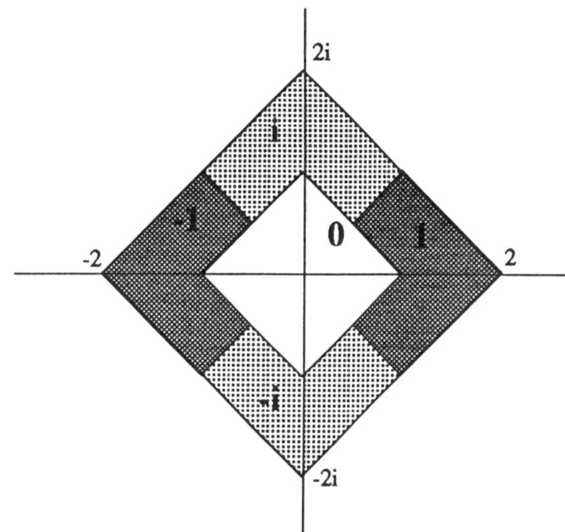

Fig. 3: The square S

1.3 Hexagonal binary representation

i. Definitions: Now let us consider the representation of complex numbers in radix 2 with digits in D_6. This representation will give us a compact encoding of complex numbers. Moreover, it offers the ability of performing fully parallel addition in constant time (i.e. independent from the length of the operands). Let us denote $H(1) = D_6$. By extension, $H(n)$ is the set of the points which are sum of n elements of $H(1)$ (examples are shown *Fig. 4*). Then $H(m) + H(n) = \{x + y / x \in H(m), y \in H(n)\}$ satisfies $H(m) + H(n) = H(m + n)$, and $H(m) * H(n) = \{xy / x \in H(m), y \in H(n)\} \subset H(m * n)$. Finally, we define $H(\infty) = H$.

Let us denote $h(a) = \{0, 1, 2, \cdots, a\}$. Since $D_6 = H(1) = \{0, 1\} + j.\{0, 1\} + j^2\{0, 1\}$, we deduce that $H(a) = h(a) + j.h(a) + j^2.h(a)$. Therefore H may be written as $\mathbf{N} + j\mathbf{N} + j^2\mathbf{N}$. Therefore any element of H may be represented in radix 2 with digits in $H(1)$. (Notice that H is a lattice of the complex field $\mathbf{C}$.)

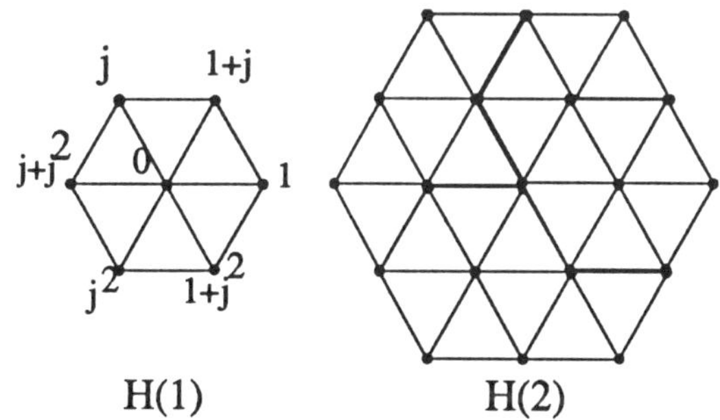

Fig. 4: Sets H(1) and H(2).

ii. Representation of the elements of H: In a similar fashion, we can represent any element x of H in radix b with digits in $H(b-1)$:

$$x = \sum_{i=0}^{\infty} d_i b^i, \quad d_i \in H(b-1)$$

In the following, we deal with representations of H in radix b with digits in $H(a)$, $a \leq b-1$. We shall study what conditions must be satisfied by a and b in order to represent H and to perform fully parallel additions without carry propagation. Before doing that, let us propose a convenient way to represent the "hexagonal digits" of $H(a)$: an "hexagonal digit" d of $H(a)$ is represented by 3 elements d^1, d^2 and d^3 of $h(a) = \{0, 1, 2, \cdots, a\}$, satisfying $d = d^1 + d^2 j + d^3 j^2$. This representation has some advantages, including:

- If $d = d^1 + d^2 j + d^3 j^2$ then $-d = (a - d^1) + (a - d^2)j + (a - d^3)j^2$
- If $d = d^1 + d^2 j + d^3 j^2$ then $\bar{d} = d^1 + d^3 j + d^2 j^2$ ($\bar{d}$ is the complex conjugate of d)

Theorem 1.
H can be represented in radix b with digits in $H(a)$, $a \leq b-1$, if $3(a+1) > 2b$. ($a \in \mathbf{N}$, $b \in \mathbf{N}$)

Proof:
We just need to prove that each element of $H(b)$ can be represented, since a representation of a number $x.b^k$ is deducible immediately from a representation of x. It suffices to show that $H(b)$ is covered by the seven sets $H(a)$, $H(a) + b$, $H(a) + \omega b$, $H(a) + \omega^2 b$, $H(a) + \omega^3 b$, $H(a) + \omega^4 b$ and $H(a) + \omega^5 b$ (see *Fig. 5*).

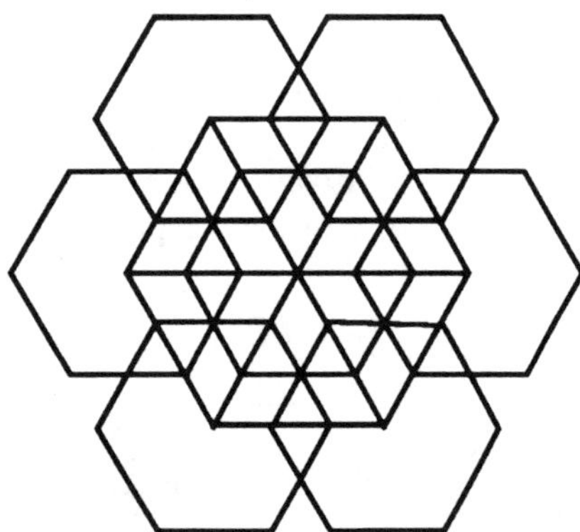

Fig. 5: Covering of H(b)

Therefore, from geometrical reasons, it suffices that $a \geq 2b/3$, as depicted *Fig. 6*. Since the sets $H(b)$ and $H(a)$, $H(a) + b$, $H(a) + \omega b$, ... are *discrete* sets, the condition $a \geq 2b/3$ is equivalent to the condition $a > 2b/3 - 1$.

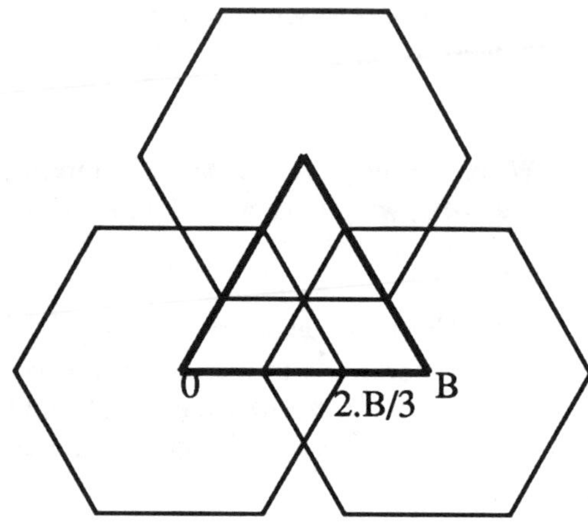

Fig. 6: Majoration of a

iii. Fully parallel addition in radix b with digits in H(a): We give an algorithm derived from the technique of Avizienis [4], in order to add elements of H written in radix b with digits in $H(a)$.

Let X and Y be in H, $X = \sum_{i=0}^{N} x_i b^i$ and $Y = \sum_{i=0}^{N} y_i b^i$. For each integer i, $x_i + y_i$ belongs to $H(2a)$: we find $c_{i+1} \in H(2)$ and $s_i \in H(a-2)$ such that $x_i + y_i = b.c_{i+1} + s_i$. The value $t_i = c_i + s_i$ belongs to $H(a)$, and $X + Y$ is obviously equal to $\sum t_i b^i$, therefore, if we are able to compute the values c_{i+1} and s_i (i.e. if $H(2a) \subset bH(2) + H(a-2)$), then we are able to perform an addition.

Theorem 2.
If $3((a-2)+1) > 2b$, then $H(2a) \subset bH(2) + H(a-2)$, therefore a fully parallel addition is possible in radix 2 with digits in $H(a)$.

Proof:
Since $a \leq b$ then $H(2a) \subset H(2b)$. From geometrical considerations, $H(2b)$ is equal to $bH(1) + H(b)$ (see *Fig. 7*).

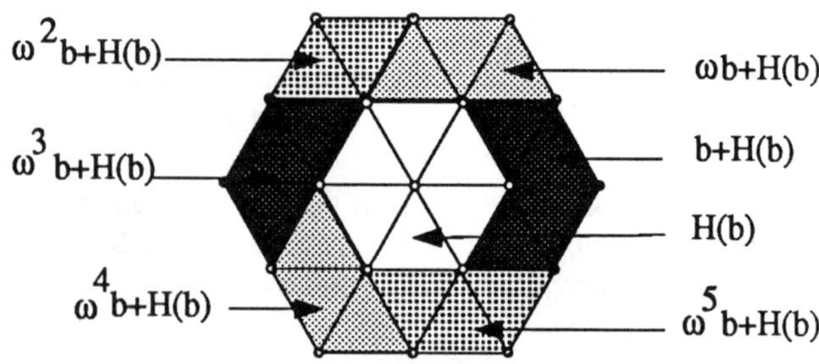

Fig. 7: H(2b)=bH(1)+H(b)

If $3((a-2)+1) > 2b$ then from theorem 1, $H(b) \subset bH(1) + H(a-2)$. Therefore $H(2a) \subseteq bH(2) + H(a-2)$. From $a \leq b-1$ and $3((a-2)+1) > 2b$, we deduce that if $b > 6$ there exists a fully parallel addition algorithm.

1.4 Hadwired fully parallel adder in H

We present in *Fig. 8* an hadwired fully parallel adder for hexagonal binary representation, divided in slices. We can notice that in fact, this adder works in radix 8. We perform the addition $Z = X + Y$ with $x_i = a_i + b_i j + c_i j^2$, $y_i = a'_i + b'_i j + c'_i j^2$, $z_i = A_i + B_i j + C_i j^2$ In *Fig. 8*, the terms a_i, b_i, c_i are permuted in order to obtain 3 identical slices.

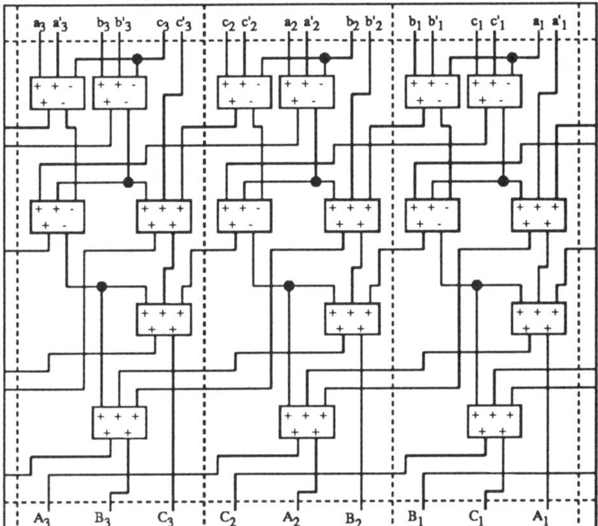

Fig. 8: A 3–digit (or 1–radix
8 digit) slice of the adder.

This adder uses the elementary cells described below. One of these cells is well known (it is a *Full Adder*), and the other cell is quite similar.

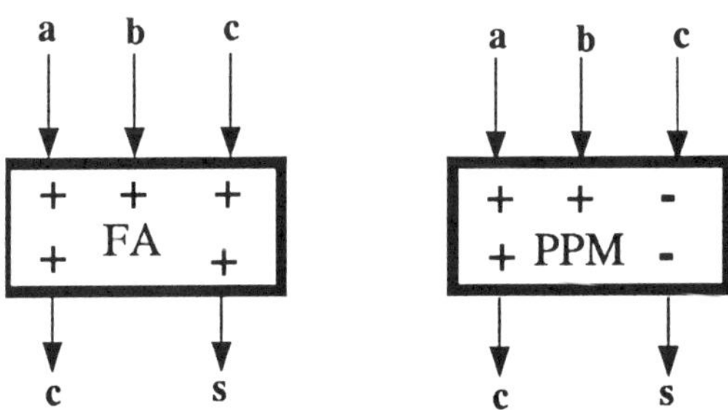

Full Adder cell   Plus Plus Minus cell

Fig. 9: The elementary cells of the adder.
Those cells compute a + b ± c = 2c ± s

2 Addition of vectors

In this part, we define techniques for adding n-dimentional vectors in time independent from the number of digits used to represent each component of the vectors.

We define redundant representation of vectors and carry free addition algorithm.

2.1 Definitions

We define a redundant representation of an n-dimentional vector as a decomposition of this vector in a system $e = (e_1, e_2, ..., e_{n+1})$ satisfying two conditions:

1. $e_1 + e_2 + ... + e_{n+1} = 0$
2. n vectors of the system e are always a basis.

Each vector of R^n is defined by $n+1$ non unique positive real numbers $x^1, ..., x^{n+1}$ such as $x = \sum_{i=1}^{n+1} x^i e_i$. If $x^i = \sum_{j=-\infty}^{+\infty} x_j^i b^j$, then $x = \sum_{j=-\infty}^{+\infty} X_j b^j$, where $X_j = \sum_{i=1}^{n+1} x_j^i e_i$ is called a "**vectorial digit**".

We denote H the subset of vectors of R^n whose coordinates are integers:

$$H = \left\{ x = \sum_{i=1}^{n+1} x^i e_i / \forall i \; x^i \in N \right\}$$

These notions generalize those presented in part A, where $n = 2$ and $e = (1, j, j^2)$.

The norm of a given $x = \sum_{i=1}^{n+1} x^i e_i$, where x^i is positive or negative, is defined as

$$|x| = \max_i x^i - \min_i x^i$$

We define a distance d as $d(x, y) = |x - y|$.

If each real is written in radix b with digits in $\mathbf{h}(\omega) = \{0, 1, ..., \omega\}$, then each vector is written in radix b with vectorial digits in $\mathbf{H}(\omega) = \{\mathbf{x} \in \mathbf{H} / |\mathbf{x}| \leq \omega\}$. The set $H(a)$ satisfies the property $H(a) + H(b) = H(a + b)$. $'a'$ is called radius of $H(a)$

Our addition algorithm is a generalization of Avizienis' algorithm [4]: the sum of two vectorial digits X_i and Y_i in $H(\omega)$ must be decomposed as a sum $bC_{i+1} + S_i$, where C_{i+1} in $H(c)$ is a carry and S_i in $H(s)$ is a partial sum, and $c + s = \omega$. We have to solve two problems: first, we need to find c and s such that $H(2\omega) \subset bH(c) + H(s)$ and $c + s = \omega$; second, we need to find an algorithm to decompose each vectorial digit of $H(2\omega)$ in $bH(c)$ and $H(s)$.

If we call "**interior radius**" of a set X the highest integer a such that X contains $H(a)$,

$$r_{int}(X) = max\{a / H(a) \subset X\}$$

then $H(2\omega) \subset bH(c) + H(s)$ iff
$2\omega \leq r_{int}(bH(c) + H(s)) = r_{int}$.

We call **"recover radius"** of a set $X = \{x^1, ..., x^q\}$ the value:

$$r_{rec}(X) = min\{a/[X] \subset X + H(a)\}$$
$$= \max_{x \in [X]} d(x, X)$$

where $[X]$ is defined by

$$[X] = \{x \in H / \exists \lambda_1, ..., \lambda_q \geq 0, \sum_{i=1}^{q} \lambda_i = 1 \text{ and } x = \sum_{i=1}^{q} \lambda_i x^i\}.$$

2.2 Preliminary results

We have the following results:

Theorem 3.

(1) $r_{rec} = r_{rec}(bH(c)) = \left\lfloor \frac{nb}{n+1} \right\rfloor$

(2) $r_{int} = s$ if $s < r_{rec}$

(3) $r_{int} \geq bc$ if $s \geq r_{rec}$

(4) if $n \geq 2$ then
$\quad H(2\omega) \subset bH(c) + H(s) \Rightarrow c \geq 2$

(5) if $n = 1$ then
$\quad s \geq r_{rec} \Rightarrow r_{int} = bc + s$

Proof

We have an immediate result: $[bH(c)] = H(bc)$ and $r_{int} \geq s$

(1) is proved if we show:

$i \quad \forall z \in [bH(c)] \; d(z, bH(c)) \leq \left\lfloor \frac{nb}{n+1} \right\rfloor$

$ii \quad \exists z \in [bH(c)], \; d(z, bH(c)) = \left\lfloor \frac{nb}{n+1} \right\rfloor$

To this purpose, we first show the following lemma:

Lemma

$Let \quad z \in [bH(c)]$

$\exists P = \{P_0, ..., P_n\} \subset bH(c)$ such that

$$z \in [P]$$
$$d(z, bH(c)) = d(z, P)$$
$$\sum_{i=0}^{n} d(z, P_i) = nb$$

Let $z = \sum_{i=1}^{n+1} z^i e_i$ with $z^i \geq 0$

and $z^i = b\alpha_i + \beta_i$ where $0 \leq \alpha_i \leq c-1, \; 0 \leq \beta_i \leq b$

then, reordering the β_i such that $b \geq \beta_{i_1} \geq \beta_{i_2} \geq ... \geq \beta_{i_{n+1}}$, we obtain

$$z = P_0 + \sum_{j=1}^{n+1} \beta_{i_j} e_{i_j} \quad \text{where } P_0 = \sum_{i=1}^{n+1} b\alpha_i e_i$$
$$= P_0 + \sum_{j=1}^{n} (\beta_{i_j} - \beta_{i_{n+1}}) e_{i_j} \text{ since } \sum_{j=1}^{n+1} e_{i_j} = 0.$$

Let $\beta_{i_n} - \beta_{i_{n+1}} = b\lambda_n$,

$\beta_{i_j} - \beta_{i_{n+1}} = (\beta_{i_{j+1}} - \beta_{i_{n+1}}) + b\lambda_j$

$\beta_{i_j} - \beta_{i_{j+1}} \geq 0 \; \Rightarrow \; \lambda_j \geq 0$

$\beta_{i_1} - \beta_{i_{n+1}} \leq b \; \Rightarrow \; \lambda_1 + \lambda_2 + ... + \lambda_n \leq 1$

Let $\lambda_0 = 1 - (\lambda_1 + \lambda_2 + ... + \lambda_n)$

then

$$z = \sum_{j=0}^{n} \lambda_j \left(P_0 + b \sum_{k=1}^{j} e_{i_k} \right) = \sum_{j=0}^{n} \lambda_j P_j$$
$$\text{where } P_j = P_0 + b \sum_{k=1}^{j} e_{i_k} = P_{j-1} + b e_{i_j}.$$

By induction we can show that:

$$z - P_j = b \sum_{k=1, j \geq 2}^{j-1} \left(\sum_{l=k}^{j-1} \lambda_l \right) e_{i_k} + b \sum_{k=j+1}^{n+1} \left(\sum_{l=0}^{j-1} \lambda_l + \sum_{l=k, k \leq n}^{n} \lambda_l \right) e_{i_k}.$$

If we set $d_i = d(z, P_i)$ then, using the previous formula,
$d_i = b \sum_{j \neq i} \lambda_j = b(1 - \lambda_i)$ and $\sum_{i=0}^{n} \lambda_i = 1 \Rightarrow$
$\sum_{i=0}^{n} d_i = nb$. The point $z = \sum_{i=0}^{n} \lambda_i P_i = \sum_{i=0}^{n} \left(1 - \frac{d_i}{b}\right) P_i$ is
completely defined by the set P and a choice of $d_0, ..., d_n$
such that $\sum_{i=0}^{n} d_i = nb$

An other point Q in $bH(c)$ can be written:
$Q = P_0 + \sum_{j=1}^{n} \gamma_j (P_j - P_0)$ where the γ_j are all integers.
We always have $d(z, Q) \geq min(d_i) = d(z, P)$, and the
result $d(z, bH(c)) = d(z, P)$ holds.

We denote $P(P_0, b; e_{i_1}, e_{i_2}, ..., e_{i_n}) = \{P_0, ..., P_n\}$

Let $z \in H(bc)$. Applying the lemma
$\exists P = \{P_0, ..., P_n\} \subset bH(c), \; d(z, bH(c)) = d(z, P)$.

Let $d_i = d(z, P_i)$

$d(z, bH(c)) = d(z, P) = min(d_i) \leq \left\lfloor \frac{nb}{n+1} \right\rfloor$ and the point z defined by

$$d_0 = ... = d_{q-1} = \left\lceil \frac{nb}{n+1} \right\rceil$$

$$d_q = ... = d_n = \left\lfloor \frac{nb}{n+1} \right\rfloor \quad \text{where } nb \equiv q \mod(n+1)$$

satisfies $d(z, bH(c)) = min(d_i) = \left\lfloor \frac{nb}{n+1} \right\rfloor$. So (1) is proved.

To prove (2) we show that if $s \leq r_{rec} - 1$ then $\exists z \in H(s+1), \; d(z, bH(c)) > s$.

$z \in [P(0, b; e_1, ..., e_n)]$ defined by

$$d_0 = s + 1$$

$$d_1 = ... = d_{\dot{q}} = \left\lceil \frac{nb - (s+1)}{n} \right\rceil$$

$$d_{q+1} = ... = d_n = \left\lfloor \frac{nb - (s+1)}{n} \right\rfloor$$

$$\text{where } nb - (s+1) \equiv q \mod(n)$$

is such that $d(z, bH(c)) > s$ since $d(z, bH(c)) = \min_i (d_i)$.

(3) is a consequence of the result $[bH(c)] = H(bc)$

To prove (4) we show that $H(2\omega) \not\subset bH(1) + H(s)$. In order to do that, let us consider the point $z = (b+s)e_1 + (s+1)e_2$, which belongs obviously to $H(2\omega)$. Let us show that for any $y \in bH(1), \; d(z, y) > s$. y may be written $b \sum_{i=1}^{n+1} \epsilon_i e_i$ with $\epsilon_i = 0, 1$.

If $\epsilon_2 = 1$ then $d(y, z) \geq 2b - b\epsilon_1 - 1 \geq b - s > s$

If $\epsilon_2 = 0$ then $d(y, z) \geq b\epsilon_3 + s + 1 \geq s + 1 > s$

(5) is obvious.

2.3 Determination of b, ω and s

Now we can determine b, ω, s such that $H(2\omega) \subset bH(c) + H(s)$.

The relation $H(2\omega) \subset bH(c) + H(s)$ is equivalent to $2\omega \leq r_{int}$. Since $2\omega = 2(b-1) > s$, r_{int} may be greater than s, so s may be greater than r_{rec}:

$$H(2\omega) \subset bH(c) + H(s) \Leftrightarrow \begin{cases} s \geq r_{rec} \\ 2\omega \leq r_{int} \end{cases}$$

If $n \geq 2$, $2\omega = 2b - 2 \leq 2b \leq bc$: the relation $2\omega \leq r_{int}$ always holds. The relation $s \geq r_{rec}$ gives $(s+1)(n+1) > nb$. With $c+s = \omega = b-1$ and $c \geq 2$, we obtain $s+1 \leq b-2$ and then $b > 2n+2$. This result shows again that in the complex field the basis should be greater than 6 (theorem 2).

We can choose

$$b > 2n + 2$$

$$s = \left\lfloor \frac{nb}{n+1} \right\rfloor \quad c = b - s$$

If $n = 1$, we have a second relation $2\omega \leq r_{int} = bc + s$ and we deduce $b \geq 3$.

2.4 Circuit building method

Now let us deal with a circuit able to compute the sum of two vectors. We generalize the circuit used to add complex numbers presented in figure 8.

We want to add two n-dimensional vectors x and y, written in radix $b = 2^{n+1}$ with p vectorial digits in $H(b-1)$:

$$z = x + y = \sum_{j=0}^{p(n+1)-1} \left(\sum_{i=1}^{n+1} (x_j^i + y_j^i) e_i \right) 2^j$$

We use a matrix Z whose term Z_{ij} counts the number of terms in $e_i 2^j$ we need to add. So $Z_{ij} = 2 \; \forall i, j$.

For instance:

$$n = 2 \quad p = 3 \quad b = 8$$

$$Z = \begin{pmatrix} \overbrace{2 \quad 2 \quad 2}^{b^2} & \overbrace{2 \quad 2 \quad 2}^{b^1} & \overbrace{2 \quad 2 \quad 2}^{b^0} \\ 2 \quad 2 \quad 2 & 2 \quad 2 \quad 2 & 2 \quad 2 \quad 2 \\ 2 \quad 2 \quad 2 & 2 \quad 2 \quad 2 & 2 \quad 2 \quad 2 \end{pmatrix} \begin{matrix} e_3 \\ e_2 \\ e_1 \end{matrix}$$

We split this matrix into p blocks: The ith blocks corresponds to the terms in b^i. Each block is equal to a $(n+1)(n+1)$ matrix Z^0

$$2^2 \quad 2^1 \quad 2^0$$

$$Z^0 = \begin{pmatrix} 2 & 2 & 2 \\ 2 & 2 & 2 \\ 2 & 2 & 2 \end{pmatrix} \begin{matrix} e_3 \\ e_2 \\ e_1 \end{matrix}$$

We can apply the following transformations:

1. Vertical transformation:
   In a column, one term is redistributed from its place to the other places of the same column. This transformation is based upon the relation $\sum_{i=1}^{n+1} e_i = 0$

2. Horizontal transformation:
   A term equal to 3, may be transformed in a term equal to 1 in the same place, and a carry transferred to the same place in next column.

For instance the vertical transformation transforms Z^0 into $W^0 = \begin{pmatrix} 1 & 3 & 3 \\ 3 & 1 & 3 \\ 3 & 3 & 1 \end{pmatrix}$, while the horizontal transformation transforms W^0 into

$$Z^1 = 1 \leftarrow \begin{pmatrix} 1+1 \leftarrow & 1+1 \leftarrow & 1 \\ 1 & 1+1 \leftarrow & 1+1 \\ 1 \leftarrow & 1+1 \leftarrow & 1 & 1+1 \end{pmatrix} \leftarrow = \begin{pmatrix} 2 & 2 & 1 \\ 1 & 2 & 2 \\ 2 & 1 & 2 \end{pmatrix}$$

These transformations are implemented using PPM or FA Cell (see figure 9). It leads to the circuit *Fig. 10*.

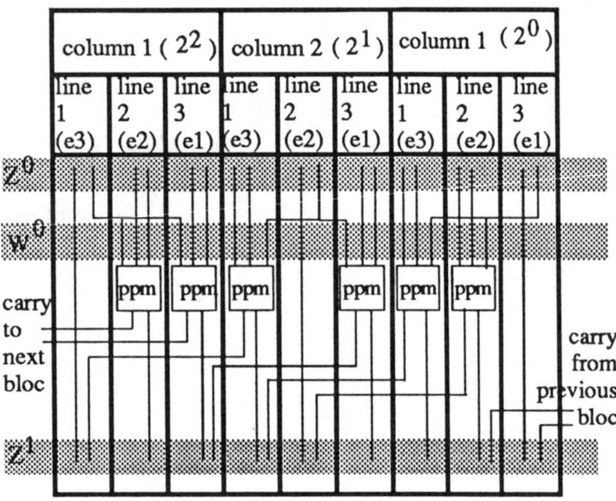

Fig 10: Circuit implementing horizontal and vertical transformations

We use horizontal and vertical transformations in order to build a sequence Z^j which converges to the matrix whose all terms equal 1.

2.5 Examples

i. A parallel adder of real numbers: $n = 1$, $b = 2^{1+1} = 4$

$$\begin{array}{ccc} & Z^j & W^j \\ j = 0 & \begin{pmatrix} 2 & 2 \\ 2 & 2 \end{pmatrix} \equiv & \begin{pmatrix} 1 & 3 \\ 3 & 1 \end{pmatrix} \\ j = 1 & \begin{pmatrix} 2 & 1 \\ 1 & 2 \end{pmatrix} \equiv & \begin{pmatrix} 3 & 0 \\ 0 & 3 \end{pmatrix} \\ j = 2 & \begin{pmatrix} 1 & 1 \\ 1 & 1 \end{pmatrix} \end{array}$$

We obtain the circuit shown *Fig. 11*

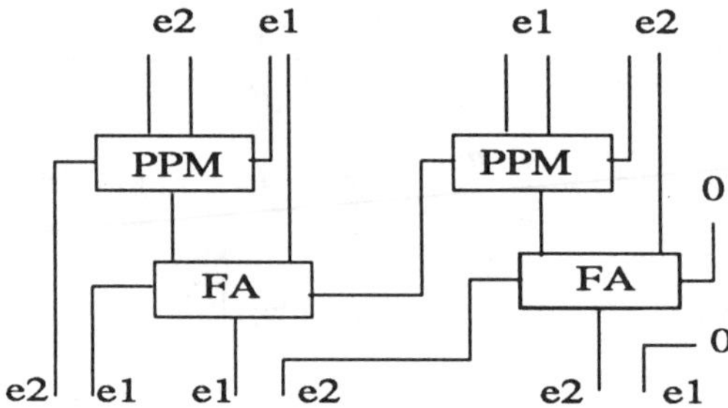

Fig 11: A real redundant adder

ii. A parallel adder of two-dimensional vectors: $n = 2$, $b = 2^3 = 8$

The sequences Z^j and W^j are:

$$j = 0 \quad \begin{pmatrix} 2 & 2 & 2 \\ 2 & 2 & 2 \\ 2 & 2 & 2 \end{pmatrix} \equiv \begin{pmatrix} 1 & 3 & 3 \\ 3 & 1 & 3 \\ 3 & 3 & 1 \end{pmatrix}$$

$$j = 1 \quad \begin{pmatrix} 2 & 2 & 1 \\ 1 & 2 & 2 \\ 2 & 1 & 2 \end{pmatrix} \equiv \begin{pmatrix} 3 & 3 & 0 \\ 0 & 3 & 3 \\ 3 & 0 & 3 \end{pmatrix}$$

$$j = 2 \quad \begin{pmatrix} 2 & 1 & 1 \\ 1 & 2 & 1 \\ 1 & 1 & 2 \end{pmatrix} \equiv \begin{pmatrix} 3 & 0 & 2 \\ 2 & 3 & 0 \\ 0 & 2 & 3 \end{pmatrix}$$

$$j = 3 \quad \begin{pmatrix} 1 & 0 & 3 \\ 3 & 1 & 0 \\ 0 & 3 & 1 \end{pmatrix}$$

$$j = 4 \quad \begin{pmatrix} 1 & 1 & 1 \\ 1 & 1 & 1 \\ 1 & 1 & 1 \end{pmatrix}$$

We obtain the circuit presented in figure 8.

iii. A parallel adder of three-dimensional vectors: $n = 3$, $b = 2^4$. The sequences Z^j and W^j are:

$$j = 0 \quad \begin{pmatrix} 2 & 2 & 2 & 2 \\ 2 & 2 & 2 & 2 \\ 2 & 2 & 2 & 2 \\ 2 & 2 & 2 & 2 \end{pmatrix} \equiv \begin{pmatrix} 1 & 3 & 3 & 3 \\ 3 & 1 & 3 & 3 \\ 3 & 3 & 1 & 3 \\ 3 & 3 & 3 & 1 \end{pmatrix}$$

$$j = 1 \quad \begin{pmatrix} 2 & 2 & 2 & 1 \\ 1 & 2 & 2 & 2 \\ 2 & 1 & 2 & 2 \\ 2 & 2 & 1 & 2 \end{pmatrix} \equiv \begin{pmatrix} 3 & 3 & 3 & 0 \\ 0 & 3 & 3 & 3 \\ 3 & 0 & 3 & 3 \\ 3 & 3 & 0 & 3 \end{pmatrix}$$

$$j = 2 \quad \begin{pmatrix} 2 & 2 & 1 & 1 \\ 1 & 2 & 2 & 1 \\ 1 & 1 & 2 & 2 \\ 2 & 1 & 1 & 2 \end{pmatrix} \equiv \begin{pmatrix} 3 & 3 & 0 & 2 \\ 2 & 3 & 3 & 0 \\ 0 & 2 & 3 & 3 \\ 3 & 0 & 2 & 3 \end{pmatrix}$$

$$j = 3 \quad \begin{pmatrix} 2 & 1 & 0 & 3 \\ 3 & 2 & 1 & 0 \\ 0 & 3 & 2 & 1 \\ 1 & 0 & 3 & 2 \end{pmatrix} \equiv \begin{pmatrix} 3 & 0 & 1 & 4 \\ 4 & 3 & 0 & 1 \\ 1 & 4 & 3 & 0 \\ 0 & 1 & 4 & 3 \end{pmatrix}$$

$$
j = 4 \quad
\begin{pmatrix}
1 & 0 & 2 & 3 \\
3 & 1 & 0 & 2 \\
2 & 3 & 1 & 0 \\
0 & 2 & 3 & 1
\end{pmatrix}
$$

$$
j = 5 \quad
\begin{pmatrix}
1 & 0 & 3 & 1 \\
1 & 1 & 0 & 3 \\
3 & 1 & 1 & 0 \\
0 & 3 & 1 & 1
\end{pmatrix}
$$

$$
j = 6 \quad
\begin{pmatrix}
1 & 1 & 1 & 1 \\
1 & 1 & 1 & 1 \\
1 & 1 & 1 & 1 \\
1 & 1 & 1 & 1
\end{pmatrix}
$$

We obtain the circuit presented *fig. 12.*

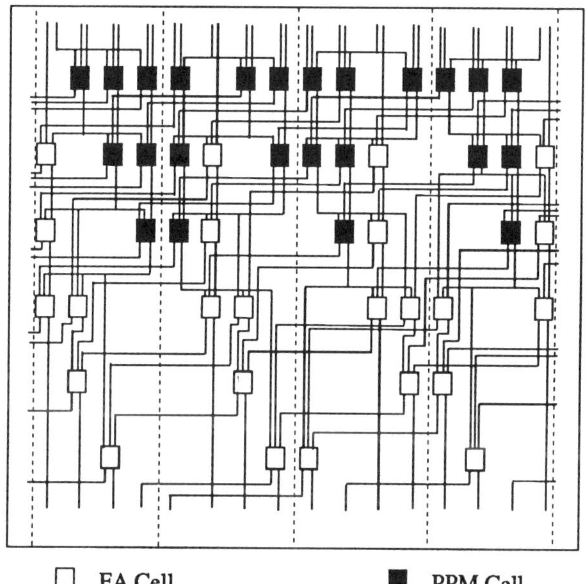

Fig. 12: A 1 radix-16 vectorial digit slice of the adder.

References

[1] D. Matula, "Basic digit sets for radix representation," *J. ACM*, vol. 29, pp. 1131–1143, 1982.

[2] M. Petkovsek, "Contiguous digit sets and local rounding," in 9th *Symposium on Computer Arithmetic*, (Santa Monica USA), Sept. 1989.

[3] T. Carter and J. Robertson, "The set theory of arithmetic decomposition," *IEEE Trans. on Computers*, vol. 39, Aug. 1990.

[4] A. Avizienis, "Signed-digit number representations for fast parallel arithmetic," *IRE Transactions on electronic computers*, vol. 10, pp. 389–400, 1961.

[5] D. Knuth, *The art of computer programming*, vol. 2. Addison Wesley, 1973.

[6] Y. Herreros, *Contribution à l'arithmétique des ordinateurs*. Ph.d. dissertation, INPG Grenoble, 1991.

Analysis of Arithmetic Algorithms:
A Statistical Study

F. Chatelin
IBM-FRANCE
5 Pl. Vendôme
75021 Paris Cedex 01, France

V. Frayssé
IRIT-CERFACS
42 av. Coriolis
31057 Toulouse Cedex, France

Abstract

In order to get an insight on the perturbations generated by running algorithms on a computer, one may simulate them by random perturbations on the data. For linear systems, we find that such a statistical estimation gives results which compare favorably with those given by the backward analysis of Wilkinson and Skeel. We intend to use such a technique mainly for nonlinear problems when no theoretical analysis is available.

1 Stability, backward error and condition number

Consider in the finite dimensional space $\Re^n$ the linear system $Ax = b$. In order to solve this problem, we choose a direct method implemented on a computer. Because of the finite precision arithmetic, the algorithm generates a perturbation, so that the computed solution is not x but x_ϵ, where ϵ is a parameter associated with the arithmetic (ϵ tends to zero means the precision tends to infinity). In exact arithmetic, x_ϵ is the exact solution of the linear system $A_\epsilon x_\epsilon = b_\epsilon$. The algorithm generates a perturbation $(A - A_\epsilon, b - b_\epsilon)$ on the initial problem. It is usually the case that this perturbation is neither unique nor easily computable.

In order to estimate the arithmetic error $x - x_\epsilon$, one needs to have a model for the perturbations generated by the algorithm. This can be done by means of a backward error analysis.

1.1 Backward error

In the context of computational stability analysis, we consider that the computed x_ϵ is the exact solution of a perturbed problem :

$$(A + \Delta A)x_\epsilon = b + \Delta b.$$

where ΔA and Δb belong to the class of perturbations generated by the algorithm. The **backward error** is the minimal relative amplitude ω of the perturbations $(\Delta A, \Delta b)$ of a given class, for which x_ϵ is still a solution of $(A + \Delta A)x_\epsilon = b + \Delta b$.

We recall here the two well-known models of perturbations used to study Gaussian elimination. First, Wilkinson [7] used a global model based on matrix norms and a global backward error of the type :

$$\frac{\|\Delta A\|}{\|A\|} + \frac{\|\Delta b\|}{\|b\|}.$$

which leads to the **normwise perturbations** $\{\|\Delta A\| \leq \omega\|A\|, \|\Delta b\| \leq \omega\|b\|\}$. Then, Oettli and Prager [4] introduced a more local model, focused on matrix elements. Skeel [6] studied it in detail. He defines the class of **componentwise perturbations** i.e.

$$\{\Delta A, \Delta b; |\Delta A| \leq \omega E, |\Delta b| \leq \omega f\},$$

where E and f are given as respectively a matrix and a vector of positive elements and where the inequalities are componentwise. For this class of perturbations, the relative backward error is

$$\eta = \min\{\omega; (A + \Delta A)x_\epsilon = b + \Delta b$$

$$with \ |\Delta A| \leq \omega E, |\Delta b| \leq \omega f\}.$$

E and f are respectively a matrix and a vector to be defined by the user. As noted by Skeel, they may be seen as the maximal uncertainty on the data (tolerance), the special structure of external perturbations ... The choice $(E = |A|, f = |b|)$ seems to be a good model for Gaussian elimination.

Once the backward error is computed, one may want to estimate the error $x - x_\epsilon$ of the solution. This is where the condition number plays a role.

1.2 Condition number

A problem is said to be stable if a "small" variation of the data induces a "small" variation of the solution. Consider a problem (P) with data ξ and solution $\Phi(\xi)$. The stability of this problem (P), when subjected to a certain type of perturbations, can be quantified by means of the relative condition number C:

$$C = \lim_{\bar{\xi} \to \xi} \frac{\text{relative distance from } \Phi\left(\bar{\xi}\right) \text{ to } \Phi(\xi)}{\text{relative distance from } \bar{\xi} \text{ to } \xi},$$

if this limit exists, which is always the case for a nonsingular linear system.

The definition of C majorizes, **to the first order**, the relative error:

$$\frac{\|\Delta\Phi(\xi)\|}{\|\Phi(\xi)\|} \leq C\eta,$$

where η is the relative backward error associated to the chosen type of perturbations. Of course, C depends on the model of the perturbations and on the metric. The condition number also depends on the choice of the data to be perturbed.

Below is a table of the condition numbers (or upper bounds) of a linear system for Wilkinson's global model and Skeel's structured model.

data	global	structured
A, b	$\leq 2\|A\|\|A^{-1}\|$	$\dfrac{\||A^{-1}|E|x| + \||A^{-1}|f|\|_\infty}{\|x\|_\infty}$
A	$\leq \|A\|\|A^{-1}\|$	$\dfrac{\||A^{-1}|E|x|\|_\infty}{\|x\|_\infty}$
b	$\|A^{-1}\|_\infty \dfrac{\|b\|_\infty}{\|x\|_\infty}$	$\dfrac{\||A^{-1}|f|\|_\infty}{\|x\|_\infty}$

Table 1: Some condition numbers or upper bounds for a linear system

2 Statistical estimation

A simple way to get a statistical estimate of a condition number is to perturb the data with random perturbations taken inside the class of perturbations one wants to study. Then, one measures the induced variation on the solution. One can estimate the condition number by computing the ratio of the size of the induced variation and the size of the perturbation [1].

Of course, this method is very costly for linear systems where theoretical tools are already available (explicit formulations of condition numbers). But it allows one to estimate a condition number even when its mathematical formulation is unknown. Therefore, it will be very useful for nonlinear problems where theory has not provided yet such formulations. Then our aim is to test the reliability of the method in the "simpler" linear case first.

2.1 The method and its implementation

$$\text{data } A, b \; \xrightarrow{\; G_\epsilon \;} \; X \; \xrightarrow{\; F \;} \; Y = AX - b \; \cdot$$

Perturbing randomly the data of the algorithm G_ϵ, we estimate:

- the relative condition number K_ϵ of F^{-1} by measuring $\frac{\Delta_r(X)}{\Delta_r(Y)}$,

- the relative condition number L_ϵ of G_ϵ by measuring $\frac{\Delta_r(X)}{\Delta_r(data)}$,

- the relative condition number I_ϵ of $\mathcal{I}_\epsilon = F_\epsilon \circ G_\epsilon$ by measuring $\frac{\Delta_r(Y)}{\Delta_r(data)}$

where $\Delta_r(z)$ is a measure of a relative variation around z, with an appropriate norm. One can show that $I_\epsilon \sim \frac{L_\epsilon}{K_\epsilon}$. We call I_ϵ the partial arithmetic stability. If I_ϵ is close to one, then $\mathcal{I}_\epsilon$ is a good approximation of the identity. If I_ϵ is too large, it means that the algorithm is more ill-conditioned than the mathematical problem. I_ϵ is useful to distinguish between the contribution of the algorithm and of the problem itself, to an instability. We introduce two kinds of perturbations:

- type-1 perturbations are global. We define them by $(A_{ij})_{per} = A_{ij} + \alpha\|A\|t$ and $(b_i)_{per} = b_i + \alpha\|b\|t$,

- type-2 perturbations are structured. We define them by $(A_{ij})_{per} = A_{ij}(1 + \alpha t)$ and $(b_i)_{per} = b_i(1 + \alpha t)$,

where t controls the amplitude of the perturbations and α is a discrete random variable such as, for example $pb(\alpha = 1) = pb(\alpha = -1) = 1/4$, $pb(\alpha = 0) = 1/2$. From now, when not written explicitly, $\|.\|$ is $\|.\|_\infty$

For each type of perturbation, we vary the amplitude of this perturbation from machine precision to 10^{-1} (or more if it is relevant for the problem): this is a way to simulate perturbations of different origins (arithmetic, numerical approximations, physical measurements ...). After applying the algorithm G_ϵ to the randomised data, we collect a sample of computed solutions X and their associated sample of residuals Y. Let x_ϵ be the computed solution of the linear system $Ax = b$.

Let m (resp. ρ) be the mean and σ (resp. v) be the standard deviation of the sample X (resp. Y). Let σ_π be the norm of the relative variation of the data. We define the estimators for the condition numbers following Wilkinson's definition for type-1 perturbations, and Skeel's definition for type-2 perturbations. Tables 2 and 3 present formulae which estimate the condition number and the relative error $\frac{\|x - x_\epsilon\|}{\|x_\epsilon\|}$. β takes the value

- $\|A\|\|x_\epsilon\| + \|b\|$ when both A and b are perturbed,

- $\|A\|\|x_\epsilon\|$ when only A is perturbed,

- $\|b\|$ when only b is perturbed.

$I_{1\epsilon}(t)$	$\dfrac{\sqrt{\|v\|^2 + \|\rho\|^2}}{\beta\sigma_\pi}$
$L_{1\epsilon}(t)$	$\dfrac{\|\sigma\|}{\|x_\epsilon\|\sigma_\pi}$
$K_{1\epsilon}(t)$	$\dfrac{\|\sigma\|\beta}{\|x_\epsilon\|\|v\|}$
error estimation	$K_{1\epsilon}\dfrac{\|Ax_\epsilon - b\|}{\beta}$

Table 2: Statistical estimators for type-1 perturbations.

perturbation	E, f
$I_{2\epsilon}(t)$	$\dfrac{1}{\sigma_\pi}\max_{1\le i\le n}\dfrac{\sqrt{v_i^2 + \rho_i^2}}{(E|x_\epsilon| + f)_i}$
$L_{2\epsilon}(t)$	$\dfrac{\|\sigma\|}{\|x_\epsilon\|\sigma_\pi}$
$K_{2\epsilon}(t)$	$\dfrac{\|\sigma\|}{\|x_\epsilon\|}\dfrac{1}{\max_{1\le i\le n}\dfrac{|v_i|}{(E|x_\epsilon|+f)_i}}$
error estimation	$K_{2\epsilon}\max_{1\le i\le n}\dfrac{|r_i|}{(E|x_\epsilon| + f)_i}$

Table 3: Statistical estimators for type-2 perturbations.

For more details on the method and for proofs, see the appendix and [2].

2.2 General behaviour of the condition number

Figure 1 shows the standard behaviour of the condition number estimate for the problem $Ax = b$ when the matrix A is perturbed with our type-1 or type-2 perturbations, and when only the right-hand side is perturbed. t controls the amplitude of the perturbations. For perturbations of A and possibly b, and for $t \le t_0$, the condition number K is constant, smaller than the classical condition number for type-1 perturbations, and close to Skeel's condition number for type-2 perturbations. For $t > t_0$, K varies like $1/t$ (which appears as a line with slope -1 on the log-log scale of figure 1). For $t > t_0$, the perturbed system is seen as singular by the computer and no estimate is reliably available.

For perturbations of b only, K is constant for all t. This experimentation is very useful because in many cases it allows to make an estimation of the error, by definition of the condition number.

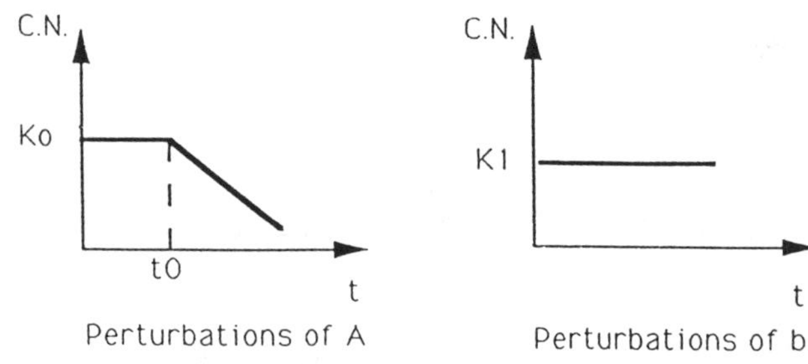

Figure 1: Behaviour of the condition of a linear system with the size of the perturbation (log-log scale).

3 Results and comments

We present here the results when Gaussian elimination, as provided by the LAPACK library, is applied to the linear system $DDx = b$ whose coefficient matrix and exact solution x are given by:

$$DD(i,j) = \frac{i-1}{i+j-1}, i \ne j$$
$$DD(i,i) = n$$
$$x(i) = \sqrt{(i)}, i = 1, n$$

For each plot, the value of the parameter t, which controls the amplitude of the perturbation, is on the horizontal axis. On the vertical axis are plotted the relative condition number of the identity I, the relative condition number of the algorithm L, the relative condition number of the mathematical problem K, the exact relative error Eex and the estimated relative error $Eest$. Classical and componentwise condition numbers of the unscaled matrix are all close to 1 and lead to very good estimations of the error which is around 10^{-16}. All statistical experiments show constant condition numbers (and thus stability) for type-1 and type-2 perturbations. We "descale" this diagonal dominant matrix by multiplying each even row by 10^6 and each odd row by 10^{-6}. We modify the right-hand side in the same way, so that the descaled system has the same solution as the original one. The exact arithmetic error is now around 10^{-14}. This has no influence on row-scaling independant condition numbers (i.e Skeel's condition numbers) but the classical condition numbers are multiplied by 10^{12}. Nevertheless, Skeel's condition numbers still provide a very good error estimation where the classical condition numbers are far too large.

We observe on figures 2 and 3 that type-2 perturbations do not generate any instability. When the condition number of the identity is one, our estimate of the condition number of the mathematical problem is close to that defined by Skeel and yields a very good estimation of the error (figures 2 and 3).

On the contrary, with type-1 perturbations, the problem is unstable for values of t smaller than 10^{-13} when both A and b are perturbed (figure 4). When only b is perturbed, the condition number of the mathematical problem is constant as expected. But the estimation of the arithmetical error is not good at all: it is largely overestimated. This means that type-1 perturbations are not a good model for the perturbations generated by the LU algorithm: they are too "large" (figure 5). We are actually measuring the error for a different problem which would be a type-1 perturbation of the initial problem. Figure 6 illustrates this point. Let (A, b) be the original "descaled" matrix and vector, and let (A', b') a given type-1 perturbation of (A, b). Let x and x' such that : $Ax = b$ and $A'x' = b'$. If K_0 is the condition number of the linear system $Ax = b$ subjected to type-1 perturbations, then we can make the following estimation:

$$\frac{\|x - x'\|}{\|x\|} \sim K_0 \frac{\|Ax' - b\|}{\|A\|\|x'\| + \|b\|}.$$

The results shown on figure 6 are now very good. The conclusion of this example is in threefold:

1. it is very important to have a good model of the perturbations generated by an algorithm if one wants to estimate reliably the arithmetical error,

2. given a model of the perturbation, the statistical method allows one to measure the condition number of a linear system subject to these perturbations and estimate the error generated by this kind of perturbations,

3. the perturbation generated by the LU algorithm seems to be row scaling independant. This is in agreement with the insistence in the literature for building condition numbers which are independent of row scaling.

More extended results, using different matrices are presented in [2].

Appendix

Type-1 perturbations

We perturb A and/or b in the following way:

$$\begin{aligned}
(A_{ij})_{per} &= A_{ij} + \alpha \|A\| t, \\
(b_i)_{per} &= b_i + \alpha \|b\| t,
\end{aligned}$$

where t controls the size of the perturbations. We have $\|\Delta A\| \le nt\|A\|$ and $\|\Delta b\| \le t\|b\|$ where n is the size of the matrix.

We are then simulating a discrete version of the global perturbations of Wilkinson. Therefore we use the appropriate formulation of the relative backward error stated by Rigal and Gaches [5, 3]: if Δy is an **absolute** variation of the residual, $\frac{\Delta y}{\|A\|\|x\| + \|b\|}$ is the associated **relative** variation when both A and b are perturbed, $\frac{\Delta y}{\|A\|\|x\|}$ (resp. $\frac{\Delta y}{\|b\|}$) when only A (resp. b) is perturbed.

We would like to estimate $\Delta_r(x_\epsilon)$ by $\frac{\sqrt{\|\sigma\|^2 + \|m - x\|^2}}{\|x_\epsilon\|}$ but we dot not know x, the exact solution. That is why we have to take away the term $\|m - x\|$ called the bias and use $\Delta_r(x_\epsilon) \sim \frac{\|\sigma\|}{\|x_\epsilon\|}$. The estimation will be justified when $\|\sigma\| \le \|m - x\|$, which will happen as the size of the perturbation grows.

Nevertheless, we know the exact residual which is $y = Ax - b = 0$. We can then estimate $\Delta(y)$ by $\sqrt{\|v\|^2 + \|\rho\|^2}$. That is what we do for computing the condition number of the identity $I_{1\epsilon}(t)$. But since the condition number of the mathematical problem involves both $\Delta_r(x)$ and $\Delta_r(y)$ and that we have to take away the bias for $\Delta_r(x)$, we decide to use $\Delta(y_\epsilon) = \|v\|$ in the computation of $K_{1\epsilon}(t)$. This choice is heuristic but we obtained our best results with it. We can also consider that the absolute condition number of the mathematical problem is estimated by $\frac{\|\sigma\|}{\|v\|}$.

$I_{1\epsilon}(t)$ is the ratio of $\Delta_r(y_\epsilon)$ and σ_π, $L_{1\epsilon}(t)$ is the ratio of $\Delta_r(x_\epsilon)$ and σ_π, and $K_{1\epsilon}(t)$ is the ratio of $\Delta_r(x_\epsilon)$ and $\Delta_r(y_\epsilon)$. All the estimates were given in table 2.

Type-2 perturbations

We perturb A and/or b in the following way:

$$\begin{aligned}
(A_{ij})_{per} &= A_{ij}(1 + \alpha t), \\
(b_i)_{per} &= b_i(1 + \alpha t),
\end{aligned}$$

where t controls the size of the perturbations.

We have: $|\Delta A| \le t|A|$ and $|\Delta b| \le t|b|$.

In this case, we are simulating a discrete version of the structured perturbations of Skeel [6]. Therefore we use the appropriate formulation of the relative backward error: if Δy is an **absolute** variation of the residual, $\max_{i=1,n} \frac{\Delta y_i}{(|A||x| + |b|)_i}$ is the associated **relative** variation when both A and b are perturbed, and $\max_{i=1,n} \frac{\Delta y_i}{(|A||x|)_i}$ (resp. $\max_{i=1,n} \frac{\Delta y_i}{|b|_i}$) when only A (resp. b) is perturbed.

Like in the previous paragraph, we will estimate $\Delta_r(x_\epsilon)$ by $\frac{\|\sigma\|}{\|x_\epsilon\|}$.

When computing the condition number of the identity, we will use $\Delta_r(y) = \max_{1 \le i \le n} \frac{\sqrt{v_i^2 + \rho_i^2}}{(A|x_\epsilon| + b)_i}$. When computing the condition number of the mathematical problem, we will use $\max_{1 \le i \le n} \frac{|v_i|}{(A|x_\epsilon| + b)_i}$.

$I_{2\epsilon}(t)$ is the ratio of $\Delta_r(y_\epsilon)$ and σ_π, $L_{2\epsilon}(t)$ is the ratio of $\Delta_r(x_\epsilon)$ and σ_π, and $K_{2\epsilon}(t)$ is the ratio of $\Delta_r(x_\epsilon)$ and $\Delta_r(y_\epsilon)$. The estimates were given in table 3.

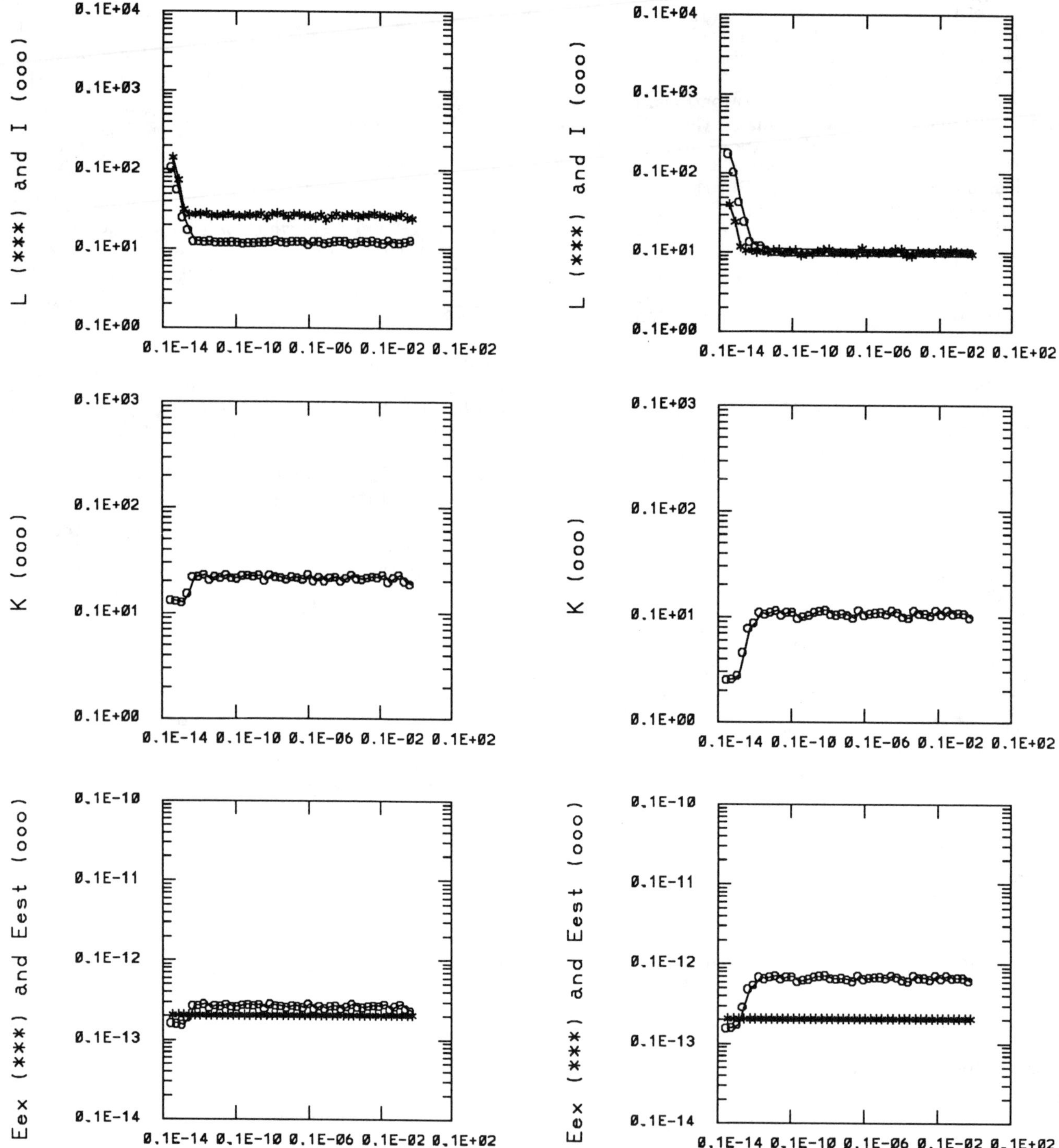

Figure 2: Gaussian elimination. Type-2 perturbations of A and b.

Figure 3: Gaussian elimination. Type-2 perturbations of b.

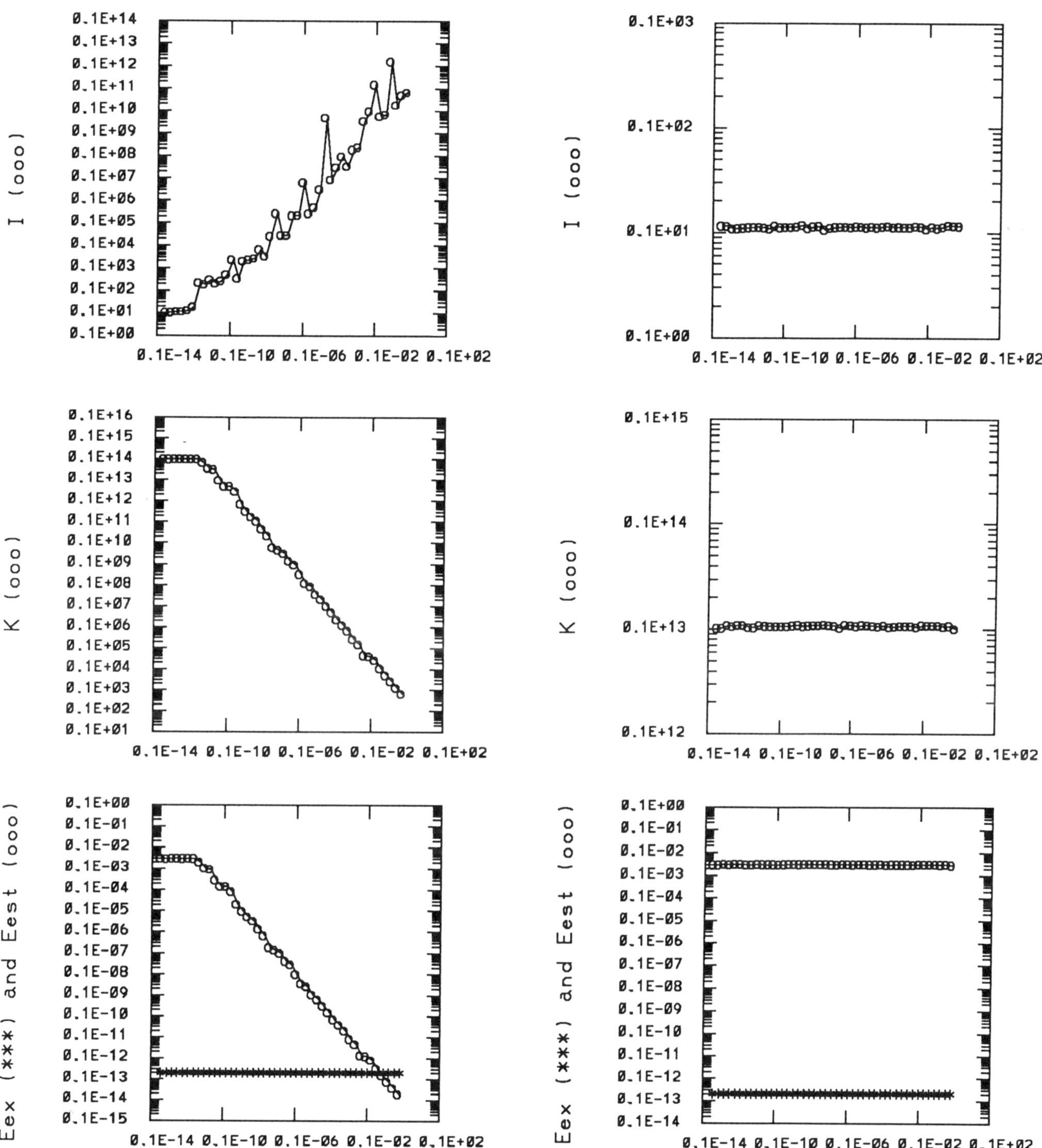

Figure 4: Gaussian elimination. Type-1 perturbations of A and b.

Figure 5: Gaussian elimination. Type-1 perturbations of b.

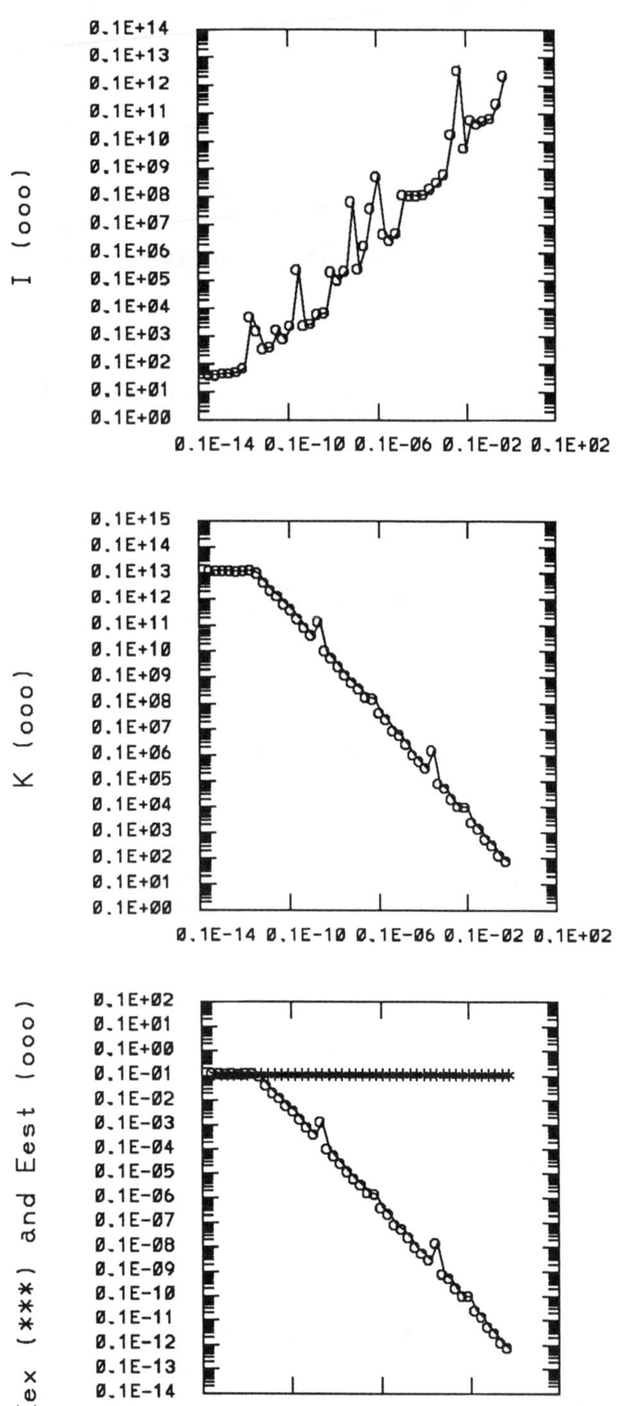

Figure 6: Study of a given type-1 perturbations of A and b. Comparison with the exact error $x - x'$.

References

[1] F. Chatelin. Résolution approchée d'équations sur ordinateur, 1989. notes de DEA, Université Paris Dauphine.

[2] F. Chatelin and V. Fraysse. A statistical study of the stability of linear systems. Technical Report TR/PA/90/43, CERFACS, 1990.

[3] N. J. Higham. How accurate is gaussian elimination? Technical Report TR 89-1024, Cornell University, Ithaca NY 14853-7501, July 1989.

[4] W. Oettli and W. Prager. Compatibily of approximate solution of linear equations with given error bounds for coefficients and right-hand sides. *Numer. Math.*, 6:405–409, 1964.

[5] J.L. Rigal and J. Gaches. On the compatiblity of a given solution with the data of a linear system. *J. Assoc. Comput. Mach.*, 14(3):543–526, July 1967.

[6] R. D. Skeel. Scaling for numerical stability in gaussian elimination. *J. Assoc. Comput. Mach.*, 26(3):494–526, July 1979.

[7] J.H. Wilkinson. Error analysis of direct methods of matrix inversion. *J. Assoc. Comput. Mach.*, 8(3):281–330, July 1961.

Representation of Numbers in Non-Classical Numeration Systems

Christiane Frougny

Université Paris 8
and
Litp, Institut Blaise Pascal
4 place Jussieu, 75252 Paris Cedex 05

Abstract

Numeration systems the basis of which is defined by a linear recurrence with integer coefficients are considered. We give conditions on the recurrence under which the function of normalization which transforms any representation of an integer into the normal one — obtained by the usual algorithm — can be realized by a finite automaton. Addition is a particular case of normalization. The same questions are discussed for the representation of real numbers in basis θ, where θ is a real number > 1. In particular it is shown that if θ is a Pisot number, then the normalization and the addition in basis θ are computable by a finite automaton.

1 Introduction

Numbers are used through a symbolic expression and the way they are represented plays an important role in computer science, in arithmetic and in coding theory. The research of numeration systems adequate to specific problems, and in which the arithmetical operations can be accelerated is far from being achieved. The interest for parallel architectures has led to algorithms like the "weak addition" ([1], [12]) where an integer has several representations.

We present here some theoretical results about the possibility of realizing the addition of numbers represented in some non-classical numeration system (extending the usual ones) by means of finite automata.

Finite automata are a "simple" model of computation, since only a finite memory is required. It is known that in the standard k-ary numeration system, where k is an integer ≥ 2, the addition is computable by a finite automaton (*cf* [4]).

In this paper we study numeration systems the basis of which is not a geometric progression but a sequence of integers given by a linear recurrence relation, which paradigm is the sequence of Fibonacci numbers. These numeration systems have also been considered in [5] and [13]. In the Fibonacci numeration system every integer can be represented using digits 0 and 1. The representation is not unique, but one of them is distinguished, the one which does not contain two consecutive 1's (*cf* [15], [11]).

More generally, let U be a strictly increasing sequence of integers such that $1 \in U$. By the greedy algorithm every integer has a representation in basis U, that we call the *normal* representation. The *normalization* is the function which transforms any representation on any alphabet onto the normal one. The addition of two integers represented in basis U can be performed that way: just add the two representations digit by digit, without carry, which gives a word on the double alphabet. Then normalize this word to obtain the normal representation of the sum. Thus addition can be viewed as a particular case of normalization.

Our purpose is to study the process of normalization in numeration systems where the basis is defined by a linear recurrence relation with integer coefficients. We call these numeration systems *linear numeration systems*.

In previous works we considered particular cases of linear numeration systems which generalize the Fibonacci numeration system and we showed that normalization is computable by a finite automaton which is obtained by the composition of two sequential machines, one processing words from left to right and the other one from right to left ([6], [7] and [9]). Here we first prove that if the set of normal representations is recognizable by a finite automaton, then the normalization is computable by a finite automaton if and only if the set of words having value 0 in basis U is recognizable by a finite automaton (Proposition 2.1). To every word one associates a polynomial. Then we consider words which can be associated to polynomials belonging to the ideal generated by the characteristic polynomial P of the linear recurrence. Obviously every word of this set is equal to 0 in basis U. We give a construction which links recognizability by a finite automaton and division of polynomials by P. We prove that the set of words associated to the ideal (P), on any alphabet, is recognizable by a finite automaton if and only if P has no root of modulus 1 (Theorem 2.1). If P has one root of modulus 1, then there exist alphabets on which the normalization is not computable by a finite automaton.

In a similar manner we discuss the representation of real numbers in basis θ where θ is a real number > 1. The normal θ-representation of a real number is called the *θ-development* or the *θ-expansion* in the literature [14]. The θ-developments of real numbers, when θ is not necessarily an integer, have been used for fast computation of elementary functions (*cf* [12]).

The notion of normalization is defined for the θ-representation as for the integers. If θ is an algebraic integer then a construction similar to the one given for the integers links the recognizability of the set of infinite words equal to 0 to the property of the minimal polynomial of θ of having no root of modulus 1 (Theorem 3.1).

We prove that the normalization is computable by a finite automaton if and only if the set of infinite words equal to 0 in basis θ is recognizable by a finite automaton (Proposition 3.2). Thus the normalization in basis θ is computable by a finite automaton on any alphabet if and only if the minimal polynomial of θ has no root of modulus 1 and if $\sum_{n \geq 0} s_n \theta^{-n} = 0$ implies $\sum_{n \geq 0} s_n \alpha^{-n} = 0$ for every conjugate α of modulus > 1 (Theorem 3.2).

Let θ be an algebraic integer > 1; θ is a *Pisot* number if its conjugates have modulus < 1; θ is a *Salem* number if its conjugates have modulus ≤ 1, and it is not a Pisot number. Thus, if θ is a Pisot number, then the normalization in basis θ is computable by a finite automaton on any alphabet — and addition also. If θ is a Salem number, there exist alphabets on which normalization is not computable by a finite automaton (Corollary 3.1). These results have strong connexion with symbolic dynamics, that we do not discuss here.

The integers and the golden mean $\frac{1+\sqrt{5}}{2}$ being Pisot numbers, our results cover the most standard numeration systems. All proofs can be found in [8].

2   The integers

Representation of integers

Only positive numbers are considered. Let $U = (u_n)_{n \geq 0}$ be a strictly increasing sequence of integers with $u_0 = 1$. Every positive integer N can be written with respect to the basis U, *i.e.* it is possible to find $n \geq 0$ and integers $d_0, \cdots, d_n$ such that $N = d_0 u_n + \cdots + d_n u_0$ by the following algorithm (folklore):
Given integers x and y let us denote by $q(x, y)$ and $r(x, y)$ the quotient and the remainder of the Euclidean division of x by y.
Let $n \geq 0$ such that $u_n \leq N < u_{n+1}$ and let $d_0 = q(N, u_n)$ and $r_0 = r(N, u_n)$, $d_i = q(r_{i-1}, u_{n-i})$ and $r_i = r(r_{i-1}, u_{n-i})$ for $i = 1, \cdots, n$. Then $N = d_0 u_n + \cdots + d_n u_0$.

For $0 \leq i \leq n$, $d_i < \frac{u_{n-i+1}}{u_{n-i}}$; thus if the ratio $\frac{u_{n+1}}{u_n}$ is bounded by a positive constant K for all $n \geq 0$ (K minimal), then $0 \leq d_i \leq K-1$. The set $A = \{0, 1, \cdots, K-1\}$ is called the *canonical alphabet* of digits associated to the basis U, and (U, A) is the *canonical numeration system* associated to U.

The word $d_0 \cdots d_n$ of A^* obtained by this algorithm is called the *normal representation* of the integer N in basis U. It is denoted by $< N > = d_0 \cdots d_n$. The normal representation of 0 is the empty word ε.

More generally, a *numeration system* is given by a strictly increasing sequence $U = (u_n)_{n \geq 0}$ of positive integers, with $u_0 = 1$, called the *basis*, and a finite subset C of **N**, the alphabet of *digits*. A *representation* of an integer N in the system (U, C) is a word $d_0 \cdots d_n$ of the free monoid C^* such that $N = d_0 u_n + \cdots + d_n u_0$.

The normal representation of an integer N has maximal length among the representations of N not beginning by a 0. It is also the greatest (for the lexicographical ordering) of all the representations of N of this same length in basis U. Given (U, C), the mapping $\pi : C^* \to \mathbf{N}$ is defined by $\pi(d_0 \cdots d_n) = d_0 u_n + \cdots + d_n u_0$. The *normalization* ν_C is the mapping which associates to a word f of C^* the normal representation of the integer represented by f.

The normalization is linked to the problem of addition of two integers written in basis U. To add two integers N et P of respective representation $f = f_0 \cdots f_k$ and $g = g_0 \cdots g_j$ in (U, A) we add f and g digit by digit from the right and without carry. Let $f \oplus g = f_0 \cdots f_{k-j-1}(f_{k-j} + g_0) \cdots (f_k + g_j)$ (if $k \geq j$). Then $f \oplus g$ is a word written on the alphabet $\{0, \cdots, 2K-2\}$. The addition of N and P reduces to the normalization of $f \oplus g$.

In this paper we study numeration systems where the basis is defined by

$$u_{n+m} = a_1 u_{n+m-1} + \cdots + a_m u_n$$

$$a_i \in \mathbf{Z}, \ 1 \leq i \leq m, \ a_m \neq 0.$$

These systems are called *linear numeration systems*. The ratio $\frac{u_{n+1}}{u_n}$ is bounded for all $n \geq 0$ and the canonical alphabet is included in $\{0, \cdots, K-1\}$ with $K < \max(a_1 + \cdots + a_m, \max\{\frac{u_{i+1}}{u_i} \mid 0 \leq i \leq m-2\})$.
If $m = 1$ and $a_1 \geq 2$ the system is the standard a_1-ary numeration system with $A = \{0, \cdots, a_1 - 1\}$ for canonical alphabet.

EXAMPLE 2.1 . — The Fibonacci numeration system $\mathcal{F}$ is defined by the sequence of Fibonacci numbers generated by the linear recurrence $u_{n+2} = u_{n+1} + u_n$, with $u_0 = 1$ and $u_1 = 2$. The canonical alphabet is $\{0, 1\}$. The representations of the integer 24 in $\mathcal{F}$ on $\{0, 1\}$ are the following : 101111, 110011, 110100, 1000011, 1000100. The normal representation of 24 is 1000100. The normal representation of an integer in $\mathcal{F}$ is the one that does not contain two consecutive 1's (*cf* [15]).   □

Normalization of finite words

First let us give some definitions. More details can be found in [4] and [2]. Let M be a monoid. The family Rat$M$ of *rational* subsets of M is the least family of subsets of M containing the finite subsets and closed under product, union and the star operation.
A *finite automaton* $\mathcal{A} = (E, Q, I, T)$ is a directed graph labelled by letters of the alphabet E, with a finite set Q of vertices called *states*. $I \subset Q$ is the set of *initial* states, and $T \subset Q$ is the set of *terminal* states. A path in $\mathcal{A}$

is said to be *successful* if it starts in I and terminates in T. The set of successful pathes is the *behavior* of $\mathcal{A}$. A word w of E^* is *recognized* by $\mathcal{A}$ if it is the label of a successful path of $\mathcal{A}$. A subset of E^* is *recognizable* if it is the behavior of a finite automaton on E. The recognizable subsets of E^* are exactly the rational subsets of E^* by Kleene Theorem (cf [4]), and we shall use both denominations.

Let E and F be two alphabets. A *transducer* $\mathcal{T}$ is a finite automaton with edges labelled by couples of $E^* \times F^*$. A relation $R \subset E^* \times F^*$ is *rational* if and only if it is the behavior of a transducer. From now on we shall use the denomination *rational*.

We assume that the characteristic polynomial $P(X) = X^m - a_1 X^{m-1} - \cdots - a_m$ of U has a real root $\theta > 1$ which dominates strictly the modulus of its conjugates. A is the canonical alphabet and $L(U) \subset A^*$ is the set of normal representations of the integers in basis U.

If $c > 0$ is an integer, let $C = \{0, \cdots, c\}$, $\tilde{C} = \{-c, \cdots, c\}$ and

$$Z(U,c) = \{f = f_0 \cdots f_n \in \tilde{C}^* \mid f_0 u_n + \cdots + f_n u_0 = 0\}$$

be the set of words on $\tilde{C}$ equal to 0 in basis U.

PROPOSITION **2.1** . — *If the set of normal representations $L(U)$ is rational, then the normalization $\nu_C : C^* \to A^*$ is a rational function if and only if the set $Z(U,c)$ of words of $\tilde{C}^*$ equal to 0 in basis U is rational.*

To prove that if ν_C is rational then $Z(U,c)$ is rational, it is necessary to give a precise characterization of the normalization.

Let E and F be two alphabets. The length of a word f is denoted by $|f|$. The set of words on E of length $\leq k$ is denoted by $E^{\leq k}$. Recall that a relation $R \subseteq E^* \times F^*$ is *length-preserving* if, for every $(f,g) \in R$, $|\bar{f}| = |g|$ (cf [4]). This is equivalent to $R \subseteq (E \times F)^*$.

DEFINITION **2.1** . — *A relation $R \subseteq E^* \times F^*$ is said to have* bounded length differences *if there exists $k \in \mathbf{N}$ such that, for every $(f,g) \in R$, $\mid |f| - |g| \mid \leq k$.*

PROPOSITION **2.2** . — [8], [10] *A rational relation of $E^* \times F^*$ which has length differences bounded by k is equal to the behavior of a transducer $T = (E \times F, Q, \alpha, T)$ with edges labelled by elements of $E \times F$, equipped with an initial partial function $\alpha : Q \to (E^{\leq k} \times \varepsilon) \cup (\varepsilon \times F^{\leq k})$.*

The behavior of a transducer of this kind is defined as follows. A couple $(f,g) \in E^* \times F^*$ is recognized by $\mathcal{T}$ if there exist $i \in Q$ and $t \in T$, such that $\alpha(i) = (u,v)$ is defined, $f = uf'$, $g = vg'$ and (f',g') is the label of a path from i to t.

Coming back to the linear numeration systems we have

PROPOSITION **2.3** . — *The normalization in basis U, restricted to words not beginning by 0, has bounded length differences.*

Define a mapping between words of $\tilde{C}^*$ and polynomials of $\mathbf{Z}[X]$ by :
$$f = f_0 \cdots f_n \in \tilde{C}^* \mapsto F(X) = f_0 X^n + \cdots + f_n, \ f_i \in \tilde{C}.$$
The *Gaussian norm* of F is $\|F\| = \max_{i=0 \cdots n} |f_i|$. This gives a correspondence between words of $\tilde{C}^*$ and polynomials of $\mathbf{Z}[X]$ of norm at most c.

Let us denote by (P) the ideal of $\mathbf{Z}[X]$ generated by P, and by $I(P,c)$ the trace on $\tilde{C}^*$ of (P), that is $I(P,c) = \{f = f_0 \cdots f_n \in \tilde{C}^* \mid F(X) = f_0 X^n + \cdots + f_n \in (P)\}$. This set is strictly included in $Z(U,c)$.

Let $f = uvw$. Then u is a *left factor*, v is a *factor* and w is a *right factor* of f. The set of left factors of elements of a language L is denoted by $LF(L)$.

PROPOSITION **2.4** . — *The set $I(P,c)$ is recognizable by a finite automaton if and only if the number of remainders of the Euclidean division by P of polynomials associated to words of $LF(I(P,c))$ is finite.*

Denote by $[f]$ the remainder of the division by P of the polynomial associated to the word f. When the number of remainders by P of the words of $LF(I(P,c))$ is finite, the explicit construction of the minimal finite automaton $\mathcal{A} = (\tilde{C}, Q, i, i)$ which recognizes $I(P,c)$ is the following.
(i) the (finite) set of states Q is equal to the set of remainders by P of the elements of $LF(I(P,c))$
(ii) the initial state i is equal to $\{[\varepsilon]\}$
(iii) the terminal state is defined by : $\{[v] \mid v \in I(P,c)\} = i$
(iv) the transitions are of the form $[f] \xrightarrow{a} [fa]$ where $a \in \tilde{C}$.

EXAMPLE **2.2** . — Let $P(X) = X^2 - X - 1$ be the characteristic polynomial of the Fibonacci sequence. The following finite automaton recognizes $I(P,1)$.

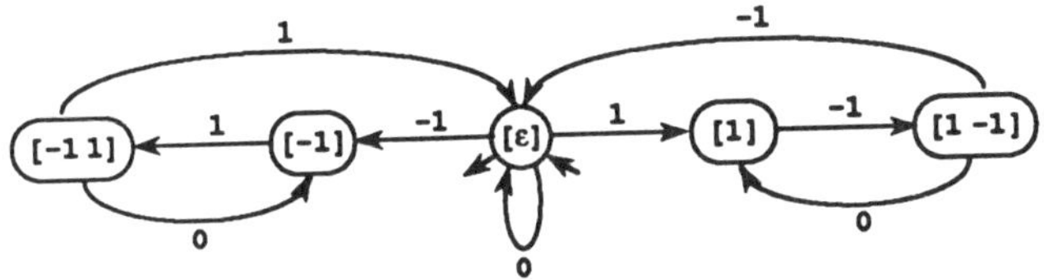

□

Since the polynomials considered belong to $\mathbf{Z}[\mathbf{X}]$ the number of remainders is finite if and only if the coefficients of the quotient by P are bounded. We thus set the

DEFINITION **2.2** . — *A polynomial P of $\mathbf{C}[X]$ satisfies the* bounded division property *(in short (BD)) if, for* every $c > 0$, there exists a constant $\beta(P,c)$ such that for every polynomial F of $\mathbf{C}[X]$, $F = PQ$, $Q \in \mathbf{C}[X]$, $||F|| \leq c$, implies that $||Q|| \leq \beta(P,c)$.

PROPOSITION 2.5 . — [3] *The polynomials satisfying the bounded division property are exactly the polynomials having no root of modulus 1.*

From the characterization *supra* we deduce

THEOREM 2.1 . — *The set of words of $\tilde{C}^*$ the associated polynomial of which belongs to (P) is recognizable by a finite automaton for every positive integer c if and only if P has no root of modulus 1.*

EXAMPLE 2.3 . — The Fibonacci polynomial $P(X) = X^2 - X - 1$ has no root of modulus 1, thus $I(P,c)$ is recognizable for every $c \geq 1$. $\qquad\square$

EXAMPLE 2.4 . — Let $u_{n+2} = u_{n+1} + 2u_n$ and $P(X) = X^2 - X - 2 = (X+1)(X-2)$ be its characteristic polynomial. One can verify that $I(P,3) \cap (-1)(3(-3))^*1(3(-3))^*2 = \{(-1)(3(-3))^p1(3(-3))^p2 \mid p \geq 0\}$. Since this set is not rational, $I(P,3)$ is not rational either. $\qquad\square$

We give now a necessary condition for the rationality of the normalization in basis U.

THEOREM 2.2 . — *If P has one root of modulus 1, then there exists $c_0 > 0$ such that for every $c \geq c_0$, the normalization ν_C is not rational.*

The question whether P has no root of modulus 1 implies that the normalization in basis U is rational on any alphabet is still open.

3   The real numbers

Representation of real numbers

Let $\theta > 1$ and $x \geq 0$ be two real numbers. Every infinite sequence of positive integers $(z_n)_{n \geq 0}$ such that $x = \sum_{n \geq 0} z_n \theta^{-n}$ is a θ-*representation* of x. A particular θ-representation called the θ-*development* or the θ-*expansion* can be computed by the following algorithm (cf [14]).
Denote by $[y]$ and by $\{y\}$ the integer part and the fractional part of a number y.
Let $x_0 = [x]$ and $r_0 = \{x\}$, and, for $i \geq 1 : x_i = [\theta r_{i-1}]$ and $r_i = \{\theta r_{i-1}\}$. Then $x = \sum_{k \geq 0} x_k \theta^{-k}$.

For $i \geq 1, x_i < \theta$. If $\theta \in \mathbf{N}$, the canonical alphabet is $A = \{0, \cdots, \theta - 1\}$ and if $\theta \notin \mathbf{N}, A = \{0, \cdots, [\theta]\}$. We write $x = x_0.x_1x_2 \cdots$ where x_0 is the integer and $.x_1x_2 \cdots$ is the fractional part of x. The θ-development of x is the normal θ-representation of x and it is greater for the lexicographic ordering than any θ-representation of x.

It is clear that if $\theta = t_0.t_1t_2 \cdots$ is the θ-development of θ, then $1 = 0.t_0t_1 \cdots$. The sequence $t_0t_1 \cdots$ is denoted by $d(1)$ and called by extension the θ-development of 1. Let $x \in [0, 1[$ of θ-development $0.x_1x_2 \cdots$. The sequence $x_1x_2 \cdots \in A^{\mathbf{N}}$ is also said to be the θ-development of x.

EXAMPLE 3.1 . — Let $\theta = (1+\sqrt{5})/2$. Then $d(1) = 11$. Let $\theta = (3+\sqrt{5})/2$. Then $d(1) = 21^\omega$. $\qquad\square$

Normalization of infinite words

Let C be any finite subset of integers. As for the integers the *normalization* function $\nu_C : C^{\mathbf{N}} \to A^{\mathbf{N}}$, where A is the canonical alphabet, maps a sequence $(y_n)_n$ of numerical value x in basis θ onto the θ-development of x. We characterize the numbers θ such that the normalization in basis θ is rational on any alphabet.

Let us fix some definitions. An infinite path in a finite automaton $\mathcal{A} = (E, Q, I, T)$ is *successful* if it starts in I and goes infinitely often through T. The *infinite behavior* of an automaton is the set of all its successful pathes. A subset of $E^{\mathbf{N}}$ is said to be *recognizable* if it is the infinite behavior of a finite automaton, that is if it is Büchi-recognizable (cf [4]).
A relation $R \subset E^{\mathbf{N}} \times F^{\mathbf{N}}$ is *rational* if it is the infinite behavior of a transducer.

As for the integers we first consider the set of infinite words on $\tilde{C}^{\mathbf{N}}$ equal to 0 in basis θ, $Z(\theta, c) = \{s = (s_n)_{n \geq 0} \in \tilde{C}^{\mathbf{N}} \mid \sum_{n \geq 0} s_n \theta^{-n} = 0\}$.

To every infinite word $s = (s_n)_{n \geq 0}$ of $\tilde{C}^{\mathbf{N}}$ is associated a formal power series $S(X) = \sum_{n \geq 0} s_n X^n$ in $\mathbf{Z}[[X]]$ which *Gaussian norm* is $||S|| = \sup_{n \geq 0} |s_n| \leq c$.

One can show that it is not a restriction to suppose that θ is an algebraic integer. A construction similar to the one given in Section 2 links the recognizability of $Z(\theta, c)$ and the division of polynomials by the minimal polynomial M of θ. Let us denote by $LF(Z(\theta, c))$ the set $\{w \in \tilde{C}^* \mid \exists s \in \tilde{C}^{\mathbf{N}}, ws \in Z(\theta, c)\}$.

PROPOSITION 3.1 . — *Let θ be an algebraic integer > 1. The set $Z(\theta, c)$ is recognizable by a finite automaton if and only if the number of remainders of the division by the minimal polynomial M of θ of polynomials associated to words of $LF(Z(\theta, c))$ is finite.*

EXAMPLE 3.2 . — The Fibonacci polynomial $P(X) = X^2 - X - 1$ is the minimal polynomial of $\theta = (1+\sqrt{5})/2$. The finite automaton constructed in Example 2.2, with every state terminal, has for infinite behavior the set of infinite words on $\{-1, 0, 1\}$ equal to 0 in Fibonacci basis $(1 + \sqrt{5})/2$. $\qquad\square$

As above, the number of remainders is finite if and only if the coefficients of the quotient of the division are bounded since the polynomials belong to $\mathbf{Z}[X]$. With a result similar to the one expressed in Proposition 2.5, we prove that.

THEOREM 3.1 . — *Let θ be an algebraic integer > 1, M its minimal polynomial. The set $Z(\theta, c)$ is recognizable for every c if and only if M has no root of modulus 1, and if for every infinite word $s = (s_n)_{n \geq 0}$ of $Z(\theta, c)$, one has $\sum_{n \geq 0} s_n \alpha^{-n} = 0$ for every root α of modulus > 1 of M.*

Using the same tools as in Proposition 2.1 we are able to show

PROPOSITION 3.2 . — *The normalization $\nu_C : C^{\mathbf{N}} \to A^{\mathbf{N}}$ is rational if and only if $Z(\theta, c)$ is recognizable.*

The proof uses the following property of the normalization (cf [10]).

PROPOSITION 3.3 . — *If the normalization in basis θ is rational, it has a bounded delay.*

The previous results can be put together into the following statement.

THEOREM 3.2 . — *The normalization ν_C in basis θ is rational on any alphabet C if and only if the minimal polynomial of θ has no root of modulus 1 and if $|s_n| \leq c$, $\sum_{n \geq 0} s_n \theta^{-n} = 0$ implies $\sum_{n \geq 0} s_n \alpha^{-n} = 0$ for every conjugate α of modulus > 1.*

COROLLARY 3.1 . — *Let θ be a Pisot number. For every alphabet C, the normalization ν_C in basis θ is rational (and in particular the addition also).*
Let θ be a Salem number. There exists an integer c_0 such that for every integer $c \geq c_0$ the normalization ν_C in basis θ is not rational.

EXAMPLE 3.3 . — Let $\theta = (1 + \sqrt{5})/2$. Then θ is a Pisot number, and the normalization is rational on any alphabet. □

EXAMPLE 3.4 . — Let $\theta = (3 + \sqrt{5})/2$. The minimal polynomial of θ is $X^2 - 3X + 1$ and θ is a Pisot number. The normalization is rational on any alphabet. □

EXAMPLE 3.5 . — Let θ be the dominant root of the polynomial $X^4 - 2X^3 - 2X^2 - 2X + 1$. θ is a Salem number and $d(1) = 2(211)^\omega$. There exists c_0 such that for every $c \geq c_0$ the normalization on C is not rational. □

Acknowledgements. I thank J.P. Bézivin, G. Rauzy and J. Sakarovitch for fruitful discussions.

References

[1] A. Avizienis, Signed-digit number representations for fast parallel arithmetic. *IRE Transactions on Electronic Computers* **10**, 1961, 389-400.

[2] J. Berstel, *Transductions and Context-Free Languages.* Teubner, 1979.

[3] J.P. Bézivin, personal communication, 1989.

[4] S. Eilenberg, *Automata, Languages and Machines*, vol. A, Academic Press, 1974.

[5] A.S. Fraenkel, Systems of numeration. *Amer. Math. Monthly* **92(2)**, 1985, 105-114.

[6] Ch. Frougny, Linear numeration systems of order two. *Information and Computation* **77**, 1988, 233-259.

[7] Ch. Frougny, Linear numeration systems, θ-developments and finite automata. *Proceedings of STACS* 89, L.N.C.S. **349**, 1989, 144-155.

[8] Ch. Frougny, Systèmes de numération linéaires et automates finis. Thèse d'Etat, Université Paris 7, Rapport LITP 89-69, 1989.

[9] Ch. Frougny, Fibonacci representations and finite automata. *I.E.E.E. Information Theory*, 1991. To appear.

[10] Ch. Frougny and J. Sakarovitch, Rational relations with bounded delay. *Proceedings of STACS* 91, L.N.C.S. **480**, 1991, 50-63.

[11] D. E. Knuth, *The Art of Computer Programming.* Vol. 1, 2 and 3, Addison-Wesley, 1975.

[12] J.M. Muller, *Arithmétique des ordinateurs.* Masson, 1989.

[13] A. Pethö and R. Tichy, On digit expansions with respect to linear recurrences. *J. of Number Theory* **33**, 1989, 243-256.

[14] A. Rényi, Representations for real numbers and their ergodic properties. *Acta Math. Acad. Sci. Hungar.* **8**, 1957, 477-493.

[15] E. Zeckendorf, Représentation des nombres naturels par une somme de nombres de Fibonacci ou de nombres de Lucas. *Bull. Soc. Royale des Sciences de Liège.* **3-4**, 1972, 179-182.

Semantics for Exact Floating Point Operations

G. Bohlender, W. Walter
Institut für Angewandte Mathematik
Universität Karlsruhe
Kaiserstr. 12, D-7500 Karlsruhe
W. Germany

P. Kornerup
Dept. of Math and Computer Sci.
Odense University
DK - 5230 Odense
Denmark

D.W. Matula
Dept. of Computer Science and Eng.
Southern Methodist University
Dallas, TX 75275
USA

Abstract

Semantics are given for the four elementary arithmetic operations and the square root, to characterize what we term exact floating point operations. The operands of the arithmetic operations and the argument of the square root are all floating point numbers in one format. In every case, the result is a pair of floating point numbers in the same format with no accuracy lost in the computation. These semantics allow us to realize the following principle: it shall be a user option to discard any information in the result of a floating point arithmetic operation. The reliability and portability previously associated only with mathematical software implementations in integer arithmetic can thus be attained exploiting the generally higher efficiency of floating point hardware. Currently, these operations are partially supported in hardware on many existing architectures. This proposal is intended to influence the design of future floating point processors to include the full functionality with hardware efficiency.

1 Introduction and Summary

Hardware implementations of floating point arithmetic operations have been prone to discard supplemental information necessarily generated in the process of computing and returning a rounded result. Most notably the low order bits of floating point addition, subtraction and multiplication are often available at intermediate stages of the hardware algorithm, as also is the remainder for divide and square root, yet these results are generally unavailable as a secondary supplemental result. It is our contention that for many purposes the user would be well served to have this supplemental information provided in a convenient form, particularly in view of its availability at very modest additional hardware cost.

In this paper we first state a goal for hardware floating point design in the appealing sense of "exact floating point operations," designed to dispel the popular misconception that floating point operations are inherently inexact. We support our goal with prescribed semantics characterizing the notion of exact floating point operations, where the prescribed results can be readily obtained and provided to the user by traditional hardware procedures. Of the utmost significance in this approach is then that any compromise in floating point computation by truncation of either the accuracy of a number representation, or the limiting value of a process, is shifted to *user control*. With floating point accuracy compromises then an artifact of a mathematical software system, any ill-contrived and/or ill-defined approach may be appropriately corrected at that level.

Our motivation is exposed in the following principle.

Exact Floating Point Operation Principle:
It shall be a user option to discard any information in the result of a floating point arithmetic operation.

This dictate is effectively directed to hardware arithmetic unit designers with the term "user" denoting either a person or program having access to and controlling the operations. By "floating point arithmetic operation" we mean the five operations: addition, subtraction, multiplication, division, and square root; explicitly described in the IEEE/ANSI floating point arithmetic standards[1,2]. We shall herein describe semantics for exact floating point arithmetic operations satisfying this principle.

Our proposed semantics may briefly be summarized as follows. The "exact" finite precision result of each of these floating point operations shall be expressed by two component numbers, the first of which is a rounded value of the infinitely precise result rounded by an appropriate one of the IEEE defined controlled roundings: to-nearest, to-positive-infinity, to-negative-infinity; and designated the *high order part*. Let x and y be floating point numbers of the same format. Letting h denote the high order part for addition, subtraction or multiplication of x, y, we define the second component number to be the *low order part* ℓ given, respectively, by $\ell = x + y - h$, $\ell = x - y - h$, and $\ell = x \times y - h$. Letting q denote the high order part for division of x by y or the square root of x, we define the second component number to be the *remainder* r given by $r = x - q \times y$ for division and $r = x - q \times q$ for square root. We note as a principal result that if the high order part is chosen by round to-nearest, the second component number can itself always be represented exactly by another floating point number of the same format (in the absence of overflow/underflow or invalid operation exceptions). It is possible for some, but not all, of the five operations that the high order part may be chosen by an alternative rounding with the second component number still representable by a floating point number of the same format.

Our semantics is simply to prescribe as allowed roundings for the high order part those roundings which guarantee that the second component number will be exactly representable as a second floating point number of the same format. The exact floating point result is then the floating point number pair in each instance, in the absence of any exception condition.

The fact that the low order part and/or remainder so defined can be represented by a floating point number of the same format in each operation instance is straightforward to verify and is not claimed to be a significant result in and of itself. The importance we ascribe here is to the consistent package of results, namely, that barring exceptions, the result of addition, subtraction, multiplication, division, and square root can each be expressed by a pair of floating point numbers in the same floating point format. Furthermore, these value pairs satisfy well known mathematical identities, allowing the user total flexibility to build more complex expressions of arbitrarily high precision by well known techniques [3,4,5,6].

In section 2 we give a brief formal account of the semantics and prove their attainability. In section 3 we consider alternative roundings avilable to prescribe the high order part in conformance with the low order part/remainder guaranteed to be representable in a second floating point word. In section 4 we note exception conditions that can arise, and refer to a recommended known approach to the handling of underflow/overflow.

2 Semantics for Exact Floating Point Operations

We consider a floating point system $F = F(b, p, emin, emax)$ characterized by its base b, the length p of a mantissa, and the minimal and maximal exponents, *emin* and *emax*. A floating point number x in this system consists of a sign, a fixed-length mantissa with digits $d_1, d_2, \ldots, d_p$, where $0 \leq d_i \leq b - 1$ for all i, and an exponent e with $emin \leq e \leq emax$, both given to the base b:

$$x = \pm b^e \sum_{i=1}^{p} d_i\, b^{-i}.$$

For simplicity, we will restrict our discussion to normalized floating point numbers, that is, numbers with $d_1 \neq 0$, and we do not define the representation of 0.

For floating point numbers, we define $e(x)$ to mean the exponent of x in the preceding sense. Also, we define the notation

$$ulp(x) := b^{e(x) - p}$$

to mean one unit in the last place relative to the floating point number x. Note that for all floating point numbers x, $x + ulp(x)$ and $x - ulp(x)$ are the adjacent floating point numbers above and below x, unless one of them overflows or underflows, or unless $x = \pm b^n$ for some integer n. If $x = b^n$ for some integer n, then there are $b - 1$ floating point numbers between $x - ulp(x)$ and x. By symmetry to zero, an analogous statement holds for $x = -b^n$.

For the remainder of this paper, we will assume the above definitions. In particular, we will assume a floating point system F with an arbitrary base $b \geq 2$ and with a number of mantissa digits $p \geq 2$. All floating point numbers are assumed to be members of this floating point system F.

The mathematical specification of the proposed exact floating point operations is summarized in the following table. Note that all arguments are of the same floating point format.

operation	arguments		mathematical specification
	in	out	
exact_add	(x, y,	h, l)	$x + y = h + l$ with $e(l) \le e(h) - p$ unless $l = 0$
exact_sub	(x, y,	h, l)	$x - y = h + l$ with $e(l) \le e(h) - p$ unless $l = 0$
exact_mul	(x, y,	h, l)	$x * y = h + l$ with $e(l) \le e(h) - p$ unless $l = 0$
exact_div	(x, y,	q, r)	$x = qy + r$ with $|r| < |y| ulp(q)$ unless $q = 0$
exact_sqrt	(x,	q, r)	$x = q^2 + r$ with $-2q \cdot ulp(q) + ulp^2(q) < r$ $r \le 2q \cdot ulp(q)$

We will demonstrate that the above specifications are reasonable: they can be fulfilled, and they require the first part of the result to be of one ulp accuracy.

Lemma: For every pair of floating point operands x, y in the four arithmetic operations and for every floating point argument x in the square root operation, it is possible to find a pair of floating point numbers h, l or q, r, respectively, such that the above specifications are fulfilled — unless an exception occurs.

Proof:

Addition, Subtraction:

Without loss of generality, let us assume that $e(x) \ge e(y)$. Two cases have to be distinguished. If $e(x) - e(y) < p$, their sum (difference) can be represented with $2p$ digits even if a carry occurs. Taking the first p digits for h and the last p digits for l makes $e(h) - e(l) \ge p$. Note that l may have to be normalized, making $e(h) - e(l) > p$, or l may be 0. On the other hand, if $e(x) - e(y) \ge p$, then $h = x$ and $l = y$ (or $l = -y$ for subtraction) obviously satisfies the specification.

Multiplication:

The exact product of two floating point numbers can be represented with $2p$ digits. From the leading non zero digit choose the leading p digits as the high order part h and the balance of at most p digits as the low order part l.

Division:

For division of x by y, we term a partial quotient q to be the result of rounding the exact result x/y towards positive or negative infinity to a certain length. The corresponding remainder is defined by $r = x - qy$. The traditional long-hand division of x by y can be used to determine at each step a partial quotient q and the corresponding remainder r.

In the trivial case, the exact quotient x/y is exactly representable as a floating point number, so choose q to be the exact quotient and $r = 0$. For the general case, let us assume that $q > 0$ without loss of

generality. Choose q to be the round-to-zero (truncated) result, that is $q < x/y < q + ulp(q)$. Then $0 < x/y - q < ulp(q)$ and thus $0 < |r| < |y| \cdot ulp(q)$.

Any floating point number $z \ne 0$ is an integral multiple of $ulp(z)$, that is $z = n \cdot ulp(z)$ for some integer n with $|n| < b^p$. Thus qy is an integral multiple of $ulp(q) \cdot ulp(y)$. Since $|x| > q \cdot |y|$ by assumption, x is also an integral multiple of $ulp(q) \cdot ulp(y)$, and so is $r = x - qy$. Since $|r| < |y| \cdot ulp(q) = i \cdot ulp(y) \cdot ulp(q)$ where $|y| = i \cdot ulp(y)$, r can be represented by a floating-point number.

Square Root:

For the square root of $x \ge 0$, we term a partial root q of x to be the result of rounding the exact result $\sqrt{x}$ towards positive or negative infinity to a certain length. The corresponding remainder is defined by $r = x - q^2$. The traditional long-hand square root process can be employed to visualize the selection of the partial root q and the corresponding remainder r.

Let q be the round-to-nearest value of $\sqrt{x}$, and assume that $r = x - q^2$ does not underflow. It is readily noted that $e(q)$ cannot be of opposite sign as $e(x)$ and can be of no greater magnitude unless $e(x) = 0$, $e(q) = 1$. q can never overflow or underflow and r can never overflow, but can underflow.

By the same argument as for division, x as well as q^2 must be integral multiples of $ulp^2(q)$. Hence $r = x - q^2$ is also an integral multiple of $ulp^2(q)$, say $r = i * ulp^2(q)$. We can also write $q = j * ulp(q)$ for some integer j. Our proof will be completed by showing $|i| \le |j|$.

Since q is the round-to-nearest value of $\sqrt{x}$,

$$q - \frac{\text{ulp}(q)}{2} \le \sqrt{x} \le q + \frac{\text{ulp}(q)}{2}.$$

So now

$$\begin{aligned} r &= x - q^2 \\ &= (\sqrt{x} + q) * (\sqrt{x} - q), \end{aligned}$$

and then

$$\begin{aligned} |r| &\le (2q + \frac{\text{ulp}(q)}{2}) * \frac{\text{ulp}(q)}{2} \\ &= q * \text{ulp}(q) + \frac{\text{ulp}^2(q)}{4}. \end{aligned}$$

Since $r = i * \text{ulp}^2(q)$,

$$|i| * \text{ulp}^2(q) \le |j| * \text{ulp}^2(q) + \frac{\text{ulp}^2(q)}{4},$$

so then $|i| \le |j|$ since i and j are integers, and thus r must be representable in the same format as q.

Given that a result pair q, r for the exact square root of x exists, we wish to specify that r must correspond to less than $ulp(q)$. Note that if the remainder r is positive, x must fall between q^2 and $(q+ulp(q))^2 = q^2 + 2q \cdot ulp(q) + ulp^2(q)$. Therefore, $r < 2q \cdot ulp(q) + ulp^2(q)$, and since r is an integral multiple of $ulp^2(q)$, we obtain $r \leq 2q \cdot ulp(q)$. Similarly, a negative remainder must satisfy $r > -2q \cdot ulp(q) + ulp^2(q)$.

3 High Order Part Rounding

For all five exact floating point operations $+, -, *, /,$ and $\sqrt{\ }$, the preceding specifications require that the exact answer lie in the interval $[h - ulp(h), h + ulp(h)]$ or $[q - ulp(q), q + ulp(q)]$. A consistant way of implementing these operations is to always choose the round to-nearest result as the first floating point number of the returned result, that is, the second component (l or r) can then be shown to always be exactly representable as a floating point number.

Note that if $|e(x) - e(y)| > p + 1$ for addition, the only way to exactly represent the sum is to return the original operands x and y as result, and this is obtained only by round to-nearest. The situation is analogous for subtraction. For multiplication, on the other hand, any one ulp rounding will work since the product can always be represented with $2p$ contiguous digits.

In $+, -, *$, note that the condition $e(h) - e(l) \geq p$ is equivalent to the requirement $|l| < ulp(h)$. This further guarantees that for $l \neq 0$, h will be one of the two floating-point numbers bracketing the exact result of the add, subtract or multiply operation unless h is a power of the base. In a two digit decimal floating-point system, we may obtain for $x = 0.98$ and $y = 0.005$ the result $x + y = 0.985 = h + l$ with $h = 1.0$ and $l = -0.015$. Rather than disallow such results, we have chosen the criterion corresponding to $|l| < ulp(h)$ as sufficient in view of its simplicity and reasonable extension to the divide and square root operations.

In $+, -, *$, if the round to-nearest result is chosen for h, the stricter requirement $|l| \leq \frac{1}{2} ulp(h)$ is always satisfied. Similarly, in division, if the round to-nearest result is chosen for q, the stricter requirement $|r| \leq \frac{1}{2} |y| ulp(q)$ is satisfied. In this and the following examples, we will use a two digit decimal floating point system. Consider the following problem. The division $99/120$ may produce $q = 0.82$ and $r = 0.60$ or $q = 0.83$ and $r = -0.60$, where the rounding is of $1/2$ ulp accuracy in both cases, and where the limit for r is attained in both cases.

Note that in division, the difference $e(x) - e(r)$ must sometimes be allowed to be $p - 1$. At other times it is required to be at least p. Essentially, this means that the required relationship must be $|r| < |y| \cdot ulp(q)$ and not $|r| < ulp(x)$, as seen below.

For 99 divided by 6.0 we should obtain either $q = 16$ or 17 with $r = 3.0$ or $r = -3.0$, respectively. Thus the exponent difference $p - 1 = 1$ must be allowed between x and r in this case. For 110 divided by 6.0, we should obtain either $q = 18$ or 19 with $r = 2.0$ or -4.0, respectively. In this case an allowed exponent difference between x and r of $p - 1 = 1$ would not rule out the result $q = 16$, $r = 14$ and many other such undesirable results.

For the square root, simply choosing q as a partial root of the floating-point number x does not always lead to a corresponding remainder r representable in the same floating point format. Consider that $\sqrt{9200} = 95.91...$, and the partial root $q = 95$ has a corresponding remainder $r = 9200 - 95^2 = 175$. Here though, the partial root $q = 96$ would yield a corresponding remainder $r = 9200 - 96^2 = -16$.

4 Exception Handling

The following table gives a brief overview of the exceptions that can occur in each of the five exact floating point operations. The appearance of an argument name in the table indicates that this output argument can overflow or underflow, respectively. 'X' indicates an invalid operation can occur.

operation	overflow		underflow			invalid operation
$+, -$	h		l		h & l	
$*$	h	h & l	l		h & l	
$/$		q	r	q	q & r	X
$\sqrt{\ }$			r			X

There are different ways of treating an exception. Without requiring any specific actions, we propose that overflowed and underflowed arguments are scaled appropriately. Additionally, we propose that for $+, -, *$, both output arguments are scaled simultaneously in such a case so that their sum also appears scaled. The exception handling defined in the IEEE standard 854 may be followed [2].

5 Applications and Conclusion

The described exact floating point operations can be used as elementary tools to control and increase the accuracy of numerical computations. For example, they can be used to implement multiple-precision arithmetic.

Exact addition and subtraction operations are extremely useful in applications employing repeated add/subtract operations. In particular, the ill effects of leading digit cancellation can thus be avoided. This may be of great advantage in iterative refinement methods. Together with the exact floating point multiplication as presented, this allows the implementation of an exact dot product which will not be rounded until the final exact result has been computed, thus incurring only a single rounding error. Such a dot product is an invaluable tool in matrix and vector computations.

Certain exact floating point operations were previously available in hardware in the late 60's, e.g. on UNIVAC 1100 series computers for single precision. The operations described here are (at least partially) supported by hardware on many existing architectures including IEEE 754 [1] and IBM /370 arithmetic. The multiply and add instruction of the IBM RISC 6000 allows recovery of the low order term for multiply and the remainder for divide and square root. On processors which do not support these operations in hardware, they can be simulated in software using methods from [3, 4, 5, 6].

The inclusion of these operations in new programming languages such as Fortran 90 would serve to encourage manufacturers of floating point processors to fully incorporate these operations in future hardware designs. By placing such exact floating point arithmetic in the floating point unit, hardware designers could relegate integer multiply and divide to the floating point unit, and then for the rest of the processor adopt a simplistic RISC oriented design.

References

[1] ANSI/IEEE Standard 754-1985: IEEE Standard for Binary Floating- Point Arithmetic, ANSI/IEEE, New York (1985)

[2] ANSI/IEEE Standard 854-1987: IEEE Standard for Radix-Independent Floating-Point Arithmetic, ANSI/IEEE, New York (1987)

[3] Dekker, T. J.: A floating-point technique for extending the available precision, Numerical Mathematics 18, 224-242 (1971).

[4] Kahan, W.: Further Remarks on Reducing Truncation Errors, Commun. ACM 8, 40 (1965).

[5] Linnainmaa, S.: Analysis of some known methods of improving the accuracy of floating-point sums, BIT 14, 167-202 (1974).

[6] Møller, O.: Quasi double-precision in floating-point addition, BIT 5, 37-50 (1965).

Session 2:

Multiplication

Chair:
Simon Knowles
INMOS, GB

Shallow Multiplication Circuits *

Michael S. Paterson

Department of Computer Science
University of Warwick
Coventry, CV4 7AL, England

Uri Zwick

Mathematics Institute
University of Warwick
Coventry, CV4 7AL, England

Abstract

Ofman, Wallace and others used carry save adders to design multiplication circuits whose total delay is proportional to the logarithm of the length of the two numbers multiplied. An extension of their work is presented here.

The first part presents a general theory describing the optimal way in which given carry save adders can be combined into carry save networks.

In the second part, two new designs of basic carry save adders are described. Using these building blocks and the above general theory, the shallowest known theoretical circuits for multiplication are obtained.

1 Introduction

We examine theoretical ways to speed up multiplication circuits. The general approach used is the one suggested by Ofman [10] and Wallace [13]. The present paper extends previous results reported in [11],[12].

The model we use is that of *dyadic Boolean circuits*. A dyadic Boolean circuit is an acyclic circuit composed of dyadic (i.e., two-input) Boolean gates.

There are ten non-trivial types of dyadic gates: eight of them of the form $(x^a \wedge y^b)^c$ (where $x^0 = x$ and $x^1 = \overline{x}$ and $\wedge$ denotes the AND operation), and the other two of the form $(x \oplus y)^c$, where $\oplus$ denotes the XOR operation. The first eight types include the usual AND, OR, NAND and NOR gates.

Two different cases may be considered. In the first we assume that all gate types can be used and in the second that only the eight AND-like types can be used. Our method can also be used to construct fast multiplication circuits in cases where fewer gate types, e.g., only NAND gates, are allowed.

We allow the gates to be connected in an arbitrary acyclic manner and assume that each has unit delay, i.e., the output of a gate is stabilised one unit of time after its two inputs are stabilised. The total delay of a circuit is the time from the moment at which all the inputs are stable until all the outputs are stable. In this model the total delay corresponds to the length of the longest directed path from an input to an output.

This model ignores many practical considerations. No attention is paid for example to possible VLSI layouts of these circuits. Simplifying assumptions are made: that no delays are introduced on connecting wires and that the delay of a gate is not influenced by its surroundings. The model enables however a theoretical investigation of the inherent delay needed to perform multiplication. Subsequent work may reveal ways of making the constructions described in this work more practical. The basic ideas of Ofman and Wallace, on which this work is based, are already of practical use (see e.g. [1]).

The above Boolean circuit model is one of the principal models used in the theory of computational complexity. For a summary of the extensive literature available on this subject the reader is referred to [4],[6],[14].

The first step in the Ofman-Wallace approach is to design a *carry save adder (CSA)*. The simplest CSA receives three input numbers and avoids carry propagation by outputting the sum of them as the sum of two numbers. Such a device will be called a $CSA_{3 \to 2}$. The striking discovery of Ofman, Wallace and others (see also [2],[5]) was that CSA's can have constant delay, independent of the size of the input numbers. The second step is to construction a network of CSA's that reduces the sum of n input numbers to the sum of only two. Such a network requires only a logarithmic number of layers and its total delay is therefore logarithmic. The two remaining numbers may be added using a *carry look ahead* adder (see [3],[8]) which also has logarithmic delay.

Since multiplication of two n-bit numbers involves little more than adding n numbers, the above approach yields logarithmic depth multiplication circuits.

In sections 4 and 5 of this work we present improved designs of CSA's. In section 3 we describe a general method of combining CSA's into shallow networks.

2 Bit Adders and Carry Save Adders

The simplest $CSA_{3 \to 2}$ is obtained by using an array of FA_3's (3-bit full adders) as shown in Fig. 2.1. In fact any *bit adder (BA)* could be used to construct a CSA.

A bit adder is a unit with k input bits and ℓ output bits, where $\ell < k$. Each input and output bit has an as-

---

*This work was partially supported by the ESPRIT II BRA Programme of the EC under contract # 3075 (ALCOM).

28

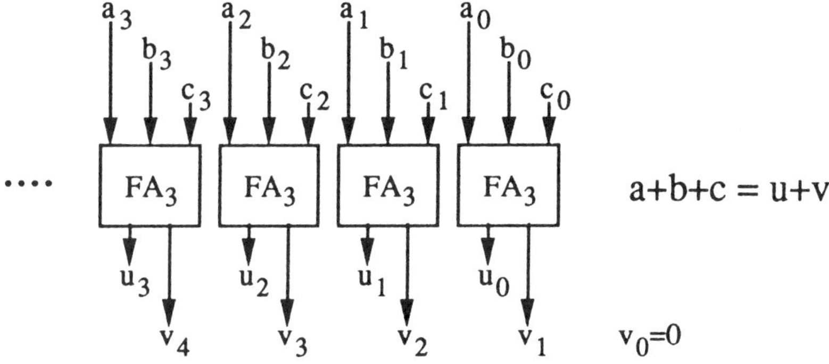

Figure 2.1: Constructing a $CSA_{3\to2}$ using FA_3's.

sociated significance. If the k input bits are denoted by $x_1, \ldots, x_k$ and their significances are $a_1, \ldots, a_k$, and if the ℓ output bits are denoted by $y_1, \ldots, y_\ell$ and their significances are $b_1, \ldots, b_\ell$ then the relation $\sum_{i=1}^{\ell} y_i 2^{b_i} = \sum_{j=1}^{k} x_j 2^{a_j}$ holds.

The simplest bit adder is the 3-bit full adder FA_3. The significance of the three input bits is 0 and the significances of the two outputs bits are 0 and 1 respectively. More generally, if $k = 2^m - 1$ then an FA_k which receives k input bits, and outputs m bits containing the binary representation of their sum could be constructed. The significance of all the inputs to an FA_k is again 0, and the significances of the m outputs are $0, 1, \ldots, m - 1$.

The fastest CSA networks which can be constructed using $CSA_{3\to2}$'s have depths asymptotic to $3.71 \log_2 n$ or $5.42 \log_2 n$, depending on whether all dyadic gates or only the AND-like gates are used (see [11],[12]). In order to get our best-performing CSA's we need to consider slightly more general BA's. If $\sum_{i=0}^{r} c_i 2^i = 2^m - 1$ we denote by $FA_{c_0,\ldots,c_r}$ the bit adder with $k = \sum_{i=0}^{r} c_i$ inputs, where c_0 of them have significance 0, c_1 of them have significance 1, and so on. The unit $FA_{c_0,\ldots,c_r}$ will have m outputs with significances $0, 1, \ldots, m - 1$. Since every number $0 \leq x < 2^m$ has a unique binary representation of length m, the output of this BA for any input vector is uniquely defined.

In section 4 we describe an efficient implementation of an $FA_{5,1}$. A $CSA_{6\to3}$ could be built using this $FA_{5,1}$ as illustrated in Fig. 2.2. The CSA's constructed in this way give rise to the shallowest known multiplication circuits. These constructions use both AND-like and exclusive-or (XOR) gates. Their asymptotic delay is about $3.57 \log_2 n$ time units (excluding the time needed for the final addition).

In section 5 we describe an efficient implementation of an $FA_{7,4}$ using only AND-like gates. A $CSA_{11\to4}$ could be built using this $FA_{7,4}$ in a similar way to that shown in Fig. 2.2. The CSA's constructed in this way yield the shallowest known multiplication circuits that use only AND-like gates. Their asymptotic delay is about $4.95 \log_2 n$ time units (excluding again the time needed for the final addition).

3 Constructing CSA networks

We are given a $CSA_{k\to\ell}$ unit G that accepts its k inputs at times $x_1, \ldots, x_k$ and delivers its ℓ outputs at times $y_1, \ldots, y_\ell$, where $k > \ell$. Our task is to compose copies of G into a network that reduces the sum of n numbers to the sum of only ℓ. We would like the delay of this network to be as small as possible.

Without loss of generality we may assume that $0 = x_1 \leq \ldots \leq x_k$ and $y_1 \leq \ldots \leq y_\ell$. In a real 'causal' device no useful output can be given until after the first input is received, and no input can be relevant unless it precedes the final output. Hence we may assume that $y_1 > x_1$ and $y_\ell > x_k$.

The *characteristic polynomial of* G is defined to be $g(z) = \sum_{j=1}^{l} z^{y_j} - \sum_{i=1}^{k} z^{x_i}$. It is easy to see that $g(1) = \ell - k < 0$ and $g(\infty) = \infty$. Hence the equation $g(z) = 0$ has at least one real root greater than 1. We call the smallest such root the *principal root of* G and denote it by λ_G. The asymptotic depth of the networks we construct depends on the principal root: the larger this root, the shallower the circuit.

As a consequence of the causality, if all the inputs up to time t are zero, then all the outputs up to time t are zero as well. If the number of inputs still to be given after time t is larger than the number of outputs still to be produced, then we can get a CSA unit $G^{[t]}$ by fixing all the inputs to G up to time t to zero and ignoring the outputs up to that time. If for some $t > 0$ a unit $G^{[t]}$ can be obtained for which $\lambda_{G^{[t]}} \geq \lambda_G$ we say that G could be *improved*, since we could use the smaller unit $G^{[t]}$ instead of G to construct circuits which are asymptotically at least as shallow. If G cannot be improved in this way then we say that it is *reduced*.

For a polynomial $f(z) = \sum_{i=0}^{m} f_i z^i$ we define $f \succ 0$ (respectively $f \succeq 0$) if $f_i > 0$ (respectively $f_i \geq 0$) for $0 \leq i \leq m$ and also write, for example, $f \preceq g$ if $(g - f) \succeq 0$.

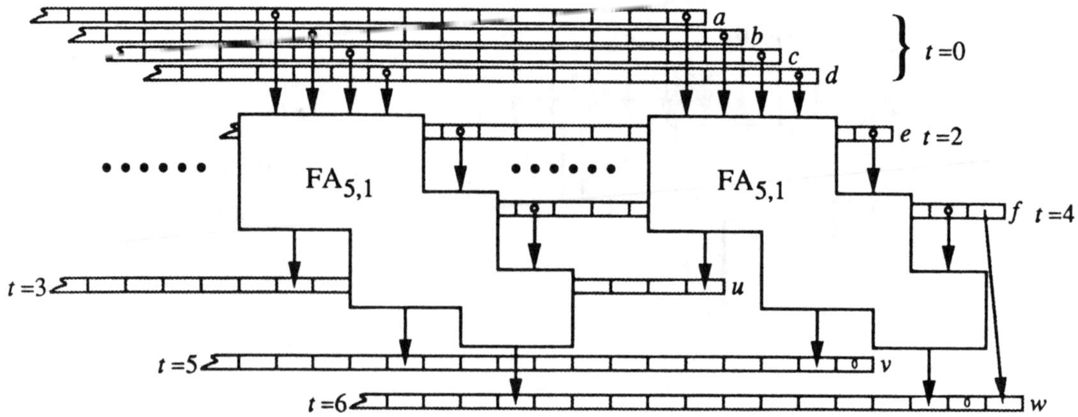

Figure 2.2: Constructing a $CSA_{6\to 3}$ using $FA_{5,1}$'s.

Lemma 3.1 *If G is reduced and $g(z) = (z - \lambda_G)h(z)$ then $h \succ 0$.*

Proof : For a polynomial $f(z) = \sum_{i=0}^{m} f_i z^i$ we define $f^{[t]}(z) = \sum_{i=t}^{m} f_i z^i$. Note that if $G^{[t]}$ is a functional CSA then its characteristic polynomial is $f^{[t]}(z)$.

We have $h_{m-1} = g_m > 0$ where $m = \deg(g)$. Suppose that for some t, $0 \le t < m - 1$, we have $h^{[t+1]} \succ 0$ but $h_t \le 0$. Then

$$g^{[t+1]}(z) = h_t z^{t+1} + (z - \lambda_G)h^{[t+1]}(z) < 0$$

for $1 \le z < \lambda_G$. Hence $\lambda_{G^{[t+1]}} \ge \lambda_G$, which contradicts the assumption that G is reduced. $\qquad\square$

We can interpret the equation $\lambda_G h(z) + g(z) = zh(z)$ as an assertion that if $\lambda_G h_i$ data items are available at time i, for $0 \le i < m$, and (some of them) are input to a copy of the unit G, then the result is that h_i items are available at time $i + 1$ for $0 \le i < m$. This suggests that we may be able to reduce the number of items by a factor of λ_G at every time unit. The delay of a network for the carry save addition of n numbers in this case would be about $\log_{\lambda_G} n$ time units.

There is however a small obstacle to be overcome. The numbers $\lambda_G h_i$ and h_i are in general non-integral. We will use integers to approximate the real numbers that we encounter and show that this does not affect the validity of our results.

A copy of G that receives its inputs at times $t+x_1,\ldots,t+x_k$ (and yields its outputs at times $t+y_1,\ldots,t+y_\ell$) is said to be *based* at time t. The essentials of a network could be described by specifying the number of CSA units that are based at each given time. The outputs produced at time t can be supplied as inputs at time t in an arbitrary manner. If more items are consumed at time t than produced then the network may receive some external inputs at time t. If more items are produced than consumed, then the network produces some external outputs at time t.

For every $N > 0$ we construct a CSA-network in which $c_t = \lceil A/\lambda^t + b\rceil$ copies of G are based at time t for $0 \le t \le T$ where $\lambda = \lambda_G$, $A = N/(\lambda h_0)$, $b = 1/(\lambda - 1)$ and $T = \lceil \log_\lambda A\rceil \le \lceil \log_\lambda N\rceil$. The coefficient of z^t in the *characteristic function*, $\left(\sum_{t=0}^{T} c_t z^t\right) g(z)$, of the network, gives the number of inputs or outputs required or supplied at time t. A negative number denotes inputs while a positive number denotes outputs.

We claim that this network accepts at least N inputs at or after time 0, and produces only a constant number of outputs, no later than time $T + m \le \lceil \log_\lambda N\rceil + m$. This follows from the next lemma.

Lemma 3.2

$$N + \left(\sum_{t=0}^{T} c_t z^t\right) g(z) \preceq z^{T+1}(b + 2)h(z).$$

Proof :

$$N + \sum_{t=0}^{T} \lceil A/\lambda^t + b\rceil z^t g(z) =$$

$$N + \sum_{t=0}^{T} \lceil A/\lambda^t + b\rceil z^{t+1} h(z) - \sum_{t=0}^{T} \lceil A/\lambda^t + b\rceil \lambda z^t h(z)$$

$$= N + \lceil A/\lambda^T + b\rceil z^{T+1} h(z) - \lceil A + b\rceil \lambda h(z) +$$

$$\sum_{t=1}^{T} \left(\lceil A/\lambda^{t-1} + b\rceil - \lceil A/\lambda^t + b\rceil \lambda\right) z^t h(z)$$

$$\preceq \lceil A/\lambda^T + b\rceil z^{T+1} h(z)$$

$$\preceq z^{T+1}(b + 2)h(z).$$

We used the fact that $\lceil B + b\rceil < \lceil B/\lambda + b\rceil \lambda$ for any $B > 0$. The number b was chosen to ensure this. $\qquad\square$

The constant number of outputs produced by these networks could be reduced to two using an additional fixed delay, independent of N.

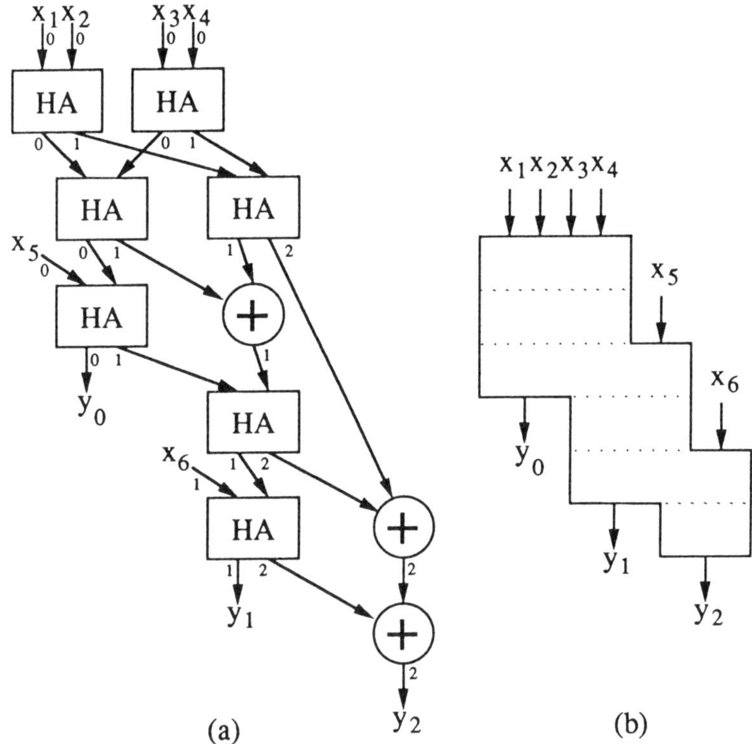

Figure 4.1: An implementation of an $FA_{5,1}$.

4 A $6 \rightarrow 3$ Carry Save Adder

An implementation of an $FA_{5,1}$ is given in Fig. 4.1(a). The implementation uses seven *half adders (HA)* and three XOR gates. A *HA* is composed of an XOR gate and an AND gate. The left output of an *HA* with inputs a, b is $a \oplus b$ (the sum) and the right output is $a \wedge b$ (the carry). The three separate XOR gates used in Fig. 4.1(a) could in fact be replaced by OR gates.

In order to verify the validity of this implementation we imagine at first that the three XOR gates are replaced by *HA*'s. Notice that the connections between the *HA*'s respect the significances of the inputs and outputs. The significance associated with each wire in the circuit is written next to that wire in Fig. 4.1(a). The inputs $x_1, \ldots, x_5$ have significance 0 while x_6 has significance 1. It is easy to check that the carry output of the three *HA*'s that we used to replace the XOR gates are always zero so the *HA*'s could be replaced by XOR gates or by OR gates.

Note that the input x_5 is supplied to this unit two units of time after $x_1, \ldots, x_4$ are supplied and that y_0 is then obtained one unit of time later, even before x_6 needs to be supplied. The outputs y_1 and y_2 are obtained one and two units of time after x_6 is supplied. This behaviour is depicted in Fig. 4.1(b). The $CSA_{6 \rightarrow 3}$ constructed using this $FA_{5,1}$ will have the same delay characteristics.

The results of the previous section give us the optimal way of combining these $CSA_{6 \rightarrow 3}$'s into networks. The delay of these networks for the (carry save) addition of n numbers will be approximately $\log_\lambda n \simeq 3.57 \log_2 n$ time units where $\lambda \simeq 1.21486$ is the root of the polynomial equation $\lambda^6 + \lambda^5 - \lambda^4 + \lambda^3 - \lambda^2 - 4 = 0$.

5 An $11 \rightarrow 4$ Carry Save Adder

The $CSA_{6 \rightarrow 3}$ described in the previous section relied heavily on the use of XOR gates. An XOR gate could always be replaced by three AND-like gates with a total delay of two time units. Better results are obtained however by using a completely different design.

Khrapchenko [9] gave the design of an FA_7 with which a $CSA_{7 \rightarrow 3}$ with the characteristics given in Fig. 5.2 could be constructed. He also described networks based on this CSA with asymptotic delay of $5.12 \log_2 n$. The networks that he described were not optimal however. Using the designs of section 3, or even the less general designs described in [11],[12], better networks of delay $5.07 \log_2 n$ can be obtained.

In this section we give an implementation of an $FA_{7,4}$ using which the preceding results can be further improved. Since the design of this unit is based on Khrapchenko's design, we give a concise summary of his construction in Fig. 5.1.

In Figs. 5.1 and 5.3 we use the following notation. We denote by $\mathbf{S}_k^A$ the symmetric function of k variables which takes the value 1 for inputs $x_1, \ldots, x_k$ if and only if $\sum x_i \in A$. For example, $\mathbf{S}_7^{4567}$ stands for the majority function on seven variables. For conciseness we write U_A for $S_3^A(u)$ where $u = (x_1, x_2, x_3)$, and V_A for $S_4^A(v)$ where $v = (x_4, x_5, x_6, x_7)$, and so on.

Notation: $x = \underbrace{x_1 x_2 x_3}_{u} \underbrace{x_4 x_5 x_6 x_7}_{v}$

$$
\begin{array}{lll}
4666666 & y_0 & = & \mathbf{S}_7^{1357} = U_{02}V_{13} \vee U_{13}V_{024} \\
5667777 & y_1 & = & \mathbf{S}_7^{2367} = U_{23}V_{04} \vee U_{12}V_1 \vee U_{01}V_2 \vee U_{03}V_3 \\
5666666 & y_2 & = & \mathbf{S}_7^{4567} = V_4 \vee U_{123}V_{34} \vee U_{23}V_{234} \vee U_3 V_{1234}
\end{array}
$$

$$
\begin{array}{lll}
233 & U_{01} & = & \overline{U}_{23} \\
244 & U_{02} & = & \overline{U}_{13} \\
233 & U_{12} & = & \overline{U}_{03}
\end{array}
$$

$$
\begin{array}{lll}
122 & U_3 & = & x_1(x_2 x_3) \\
233 & U_{03} & = & \overline{x}_1(\overline{x}_2\overline{x}_3) \vee x_1(x_2 x_3) \\
244 & U_{13} & = & \overline{x}_1(\overline{x}_2 x_3 \vee x_2\overline{x}_3) \vee x_1(\overline{x}_2\overline{x}_3 \vee x_2 x_3) \\
233 & U_{23} & = & x_1(x_2 \vee x_3) \vee x_2 x_3 \\
122 & U_{123} & = & x_1 \vee (x_2 \vee x_3)
\end{array}
$$

$$
\begin{array}{lll}
4444 & V_1 & = & \overline{x}_4\overline{x}_5(\overline{x}_6 x_7 \vee x_6\overline{x}_7) \vee (\overline{x}_4 x_5 \vee x_4\overline{x}_5)\overline{x}_6\overline{x}_7 \\
4444 & V_2 & = & V_{234}\overline{V}_{34} \\
4444 & V_3 & = & (\overline{x}_4 x_5 \vee x_4\overline{x}_5)x_6 x_7 \vee x_4 x_5)(\overline{x}_6 x_7 \vee x_6\overline{x}_7) \\
4444 & V_{13} & = & \overline{V}_{024}
\end{array}
$$

$$
\begin{array}{lll}
2222 & V_4 & = & x_4 x_5 x_6 x_7 \\
3333 & V_{04} & = & \overline{x}_4\overline{x}_5\overline{x}_6\overline{x}_7 \vee x_4 x_5 x_6 x_7 \\
3333 & V_{34} & = & (x_4 x_5 \vee x_6 x_7)(x_4 x_6 \vee x_5 x_7) \\
4444 & V_{024} & = & (\overline{x}_4\overline{x}_5 \vee x_4 x_5)(\overline{x}_6\overline{x}_7 \vee x_6 x_7) \vee (\overline{x}_4 x_5 \vee x_4\overline{x}_5)(\overline{x}_6 x_7 \vee x_6\overline{x}_7) \\
3333 & V_{234} & = & (x_4 \vee x_5)(x_6 \vee x_7) \vee (x_4 \vee x_6)(x_5 \vee x_7) \\
2222 & V_{1234} & = & x_4 \vee x_5 \vee x_6 \vee x_7
\end{array}
$$

Figure 5.1: Khrapchenko's construction of an FA_7.

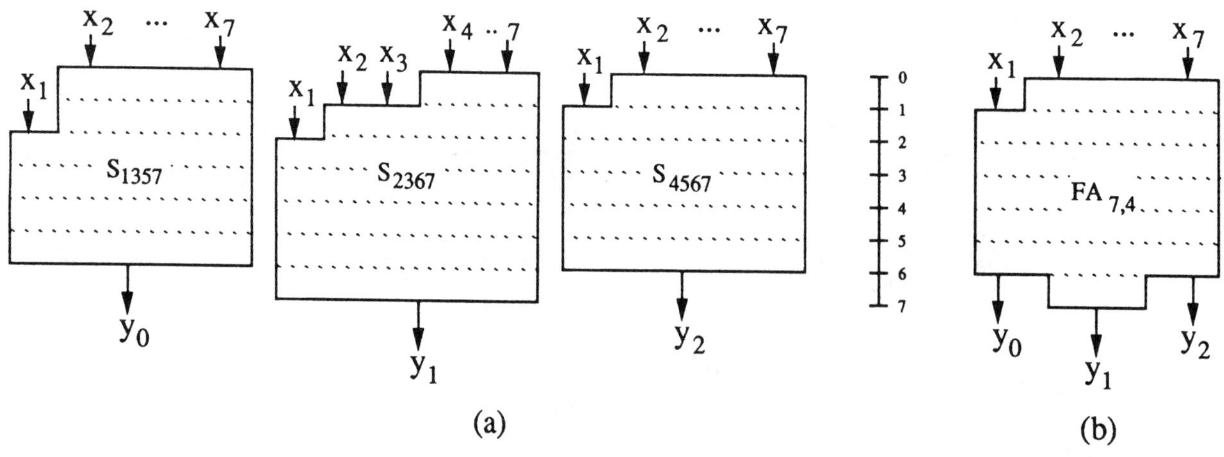

Figure 5.2: The delay characteristics of Khrapchenko's construction.

Notation: $x = \underbrace{x_1 x_2 x_3 \overbrace{x_4 x_5 x_6 x_7}^{v}}_{s} \, x_8 x_9 x_{10} x_{11}$

where $u = x_1 x_2 x_3$ and $v = x_4 x_5 x_6 x_7$ and $t = x_8 x_9 x_{10} x_{11}$.

4666666	y_0	$=$	S_{1357}
78899996666	y_1	$=$	$S_{0145} T_{13} \vee S_{2367} T_{024}$
89999997777	y_2	$=$	$(S_{4567} T_{04} \vee S_{2345} T_1) \vee (S_{0123} T_2 \vee S_{0167} T_3)$
78888886666	y_3	$=$	$(S_{234567} T_{34} \vee S_{67} T_{1234}) \vee T_{234}(S_{4567} \vee T_4)$
56666664444	$S_{4567} \vee T_4$	$=$	$(U_{23} V_{234} \vee U_{123} V_{34}) \vee ((U_3 V_{1234} \vee V_4) \vee T_4)$
5667777	S_{0145}	$=$	$\overline{S}_{2367}$
5666666	S_{0167}	$=$	$\overline{S}_{2345}$
5666666	S_{0123}	$=$	$\overline{S}_{4567}$
5666666	S_{2345}	$=$	$S_{234567} \overline{S}_{67}$
4555555	S_{234567}	$=$	$U_{123} V_{1234} \vee (U_{23} \vee V_{234})$
4555555	S_{67}	$=$	$U_{23} V_4 \vee U_3 V_{34}$

Figure 5.3: The new $FA_{7,4}$ construction.

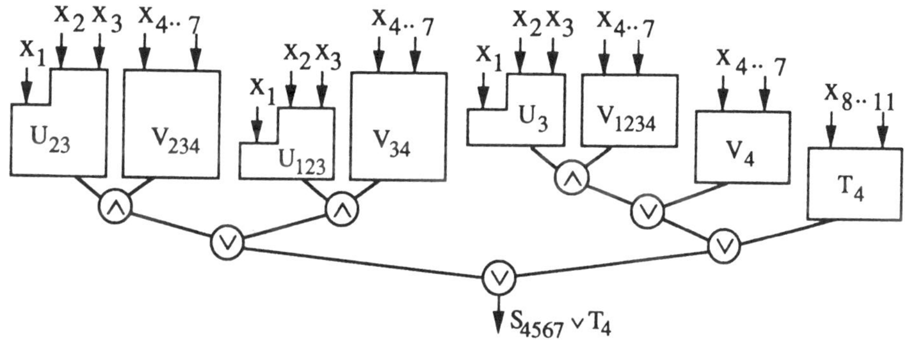

Figure 5.4: The final stages in the computation of $S_{4567} \vee T_4$.

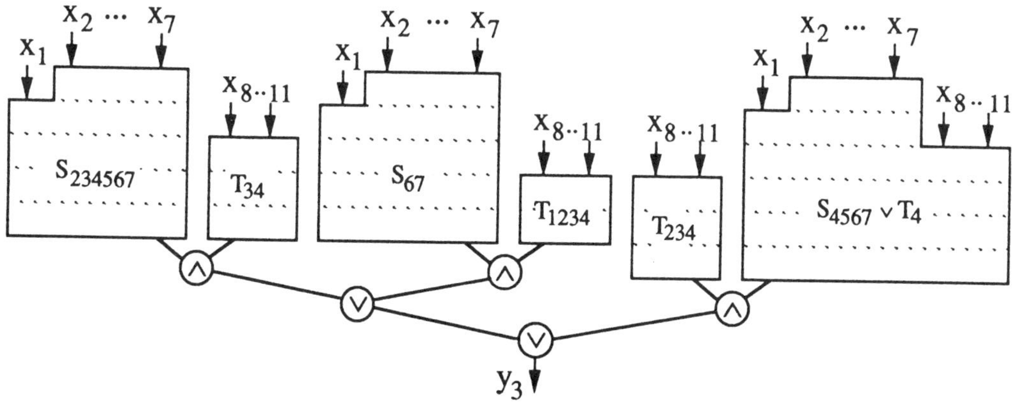

Figure 5.5: The final stages in the computation of y_3.

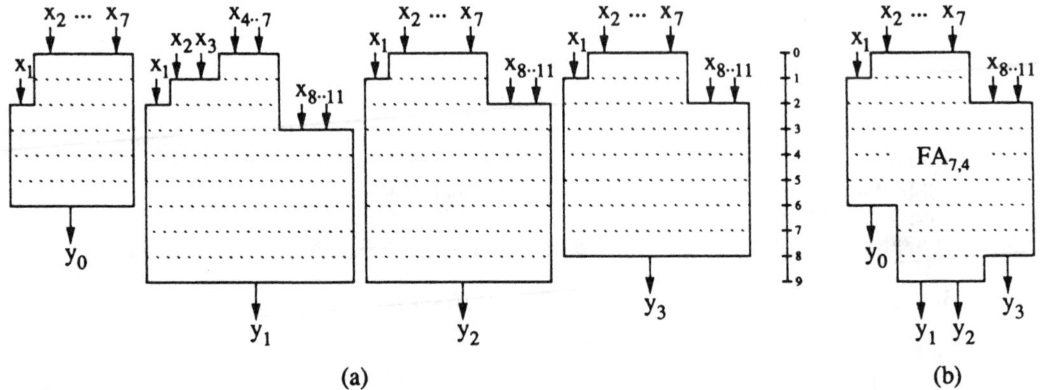

Figure 5.6: The delay characteristics of the new $FA_{7,4}$.

The numbers given to the left of each formula in Fig. 5.1 are the depths of the variables in that formula. They will be called *delay vectors*. The delay characteristics of the three output bits y_0, y_1, y_2 that compose Khrapchenko's FA_7 are described in Fig. 5.2(a). In [12] it is shown that the optimal way of combining these three units into a single unit is as presented in Fig. 5.2(b).

The construction of our new $FA_{7,4}$ is given in Fig. 5.3. Figures 5.4 and 5.5 depict the final stages in the construction of y_3. The delay characteristics of y_0, y_1, y_2, y_3 are shown in Fig. 5.6(a) and we see that we can fit them all into the unit outlined in Fig. 5.6(b).

By the results of section 3, we can combine the new $CSA_{11\to 4}$'s into networks of asymptotic depth $\log_\lambda n \simeq 4.95 \log_2 n$ where $\lambda \simeq 1.15041$ is the principal root of the equation $2\lambda^9 + \lambda^8 + \lambda^6 - 4\lambda^2 - \lambda - 6 = 0$.

6 Concluding remarks

We have presented a general construction and some specific designs which yield circuits for carry save addition which are faster than those previously published. Although we have only given asymptotic results here, the same methods provide efficient networks for small numbers of inputs. There is a polynomial time algorithm which, for any CSA G and any n, gives an optimal-depth network of G's for the carry save addition of n inputs.

Our constants will no doubt be improved before long, but the techniques provide a simple construction method which may be of more durable value.

References

[1] El Gamal A., Gluss D., Ang P-H., Greene J., Reyneri J., "A CMOS 32 bit Wallace tree multiplier-accumulator," *1986 ISSCC Digest of Technical Papers*, pp. 194-195.

[2] Avizienis A., "Signed-digit number representation for fast parallel arithmetic," *IEEE Trans. Elec. Comp.* Vol. EC10 (1961), pp. 389-400.

[3] Brent R., "On the addition of binary numbers," *IEEE Trans. on Comp.*, C-19 (1970), pp. 758-759.

[4] Boppana R., Sipser M., "The complexity of finite functions," in *Handbook of Theoretical Computer Science Vol. A: Algorithms and Complexity*, ed. van Leeuwen, Elsevier/MIT Press, 1990, pp. 757-804.

[5] Dadda L., "Some schemes for parallel multipliers," *Alta Frequenza*, Vol. 34 (1965), pp. 343-356.

[6] Dunne P.E., *The complexity of Boolean networks*, Academic Press, 1988.

[7] Karatsuba A., Ofman Y., "Multiplication of multi-digit numbers on automata," *Soviet Physics Dokl.*, Vol. 7 (1963), pp. 595-596.

[8] Khrapchenko V.M., "Asymptotic estimation of addition time of a parallel adder," *Problemy Kibernet.*, Vol. 19 (1967), pp. 107-122 (in Russian). English translation in *Syst. Theory Res.*, Vol. 19 (1970), pp. 105-122.

[9] Khrapchenko V.M., "Some bounds for the time of multiplication," *Problemy Kibernet.*, Vol. 33 (1978), pp. 221-227 (in Russian).

[10] Ofman Y., "On the algorithmic complexity of discrete functions," *Doklady Akademii Nauk SSSR*, 145 pp. 48-51 (in Russian). English translation in *Sov. Phys. Doklady*, Vol. 7 (1963) pp. 589-591.

[11] Paterson M.S., Pippenger N., Zwick U., "Faster circuits and shorter formulae for multiple addition, multiplication and symmetric Boolean functions," *Proceedings of the 31st Ann. IEEE Symp. on Found. of Comp. Sci.*, St. Louis 1990.

[12] Paterson M.S., Pippenger N., Zwick U., "Optimal carry save networks," *Boolean function complexity: Selected papers from the LMS symposium, Durham 1990*. To appear, Cambridge Univ. Press, 1991.

[13] Wallace C.S., "A suggestion for a fast multiplier," *IEEE Trans. Electronic Comp.* EC-13 (1964) pp. 14-17.

[14] Wegener I., *The complexity of Boolean functions*, Wiley-Teubner Series in Computer Science, 1987.

A Radix-4 Modular Multiplication Hardware Algorithm Efficient for Iterative Modular Multiplications

Naofumi Takagi
Department of Information Science
Kyoto University
Kyoto 606, Japan

Abstract

A fast radix-4 modular multiplication hardware algorithm is proposed. It is efficient especially in applications, such as encryption/decryption in RSA cryptosystem, where modular multiplications are carried out iteratively. Each subtraction for the division for residue calculation is embedded in the repeated multiply-addition. Numbers are represented in a redundant representation and addition/subtractions are performed without carry propagation. A serial-parallel modular multiplier based on the algorithm has a regular cellular array structure with a bit slice feature suitable for VLSI implementation.

1 Introduction

In encryption/decryption in RSA cryptosystem [1], modular multiplications with a large modulus (longer than 500-bit) are carried out iteratively. Design of a fast algorithm for modular multiplication with a large modulus is the key to developing a high performance encryption/decryption circuit for such a cryptosystem. In this paper, we propose a fast modular multiplication hardware algorithm which is efficient especially in applications where modular multiplications are carried out iteratively.

Various algorithms for modular multiplication have been proposed and some of them have been realized [2]. Most of them are classified into two methods, i.e., "division-after-multiplication" and "division-during-multiplication". In an n-bit modular multiplication by the former method, an ordinary n-bit multiplication is carried out first and then a 2n-bit by n-bit division for residue calculation is performed. In the latter method, each subtraction step for the division for residue calculation is embedded in the repeated multiply-addition [3]. The latter requires a less amount of hardware than the former does [4]. However, in general, the latter requires more addition/subtractions than the former does [5]. In either method, the key point to increasing the computation speed is to perform addition/subtractions of long numbers fast.

We can perform addition/subtractions of long numbers fast without carry propagation by the use of a redundant representation, such as the carry save form and the signed-digit representations. Vandemeulebroecke et al proposed a division-after-multiplication algorithm with the redundant binary representation, i.e, the radix-2 signed-digit representation where addition/subtractions for the division as well as for the multiplication are performed without carry propagation [5]. Preparata and Vuillemin showed an efficient division-during-multiplication algorithm with redundant representations as an application of their cellular division method [6]. Morita proposed a similar algorithm with the carry save form based on a higher radix, independently [4]. Recently, we developed a radix-2 and a radix-4 division-during-multiplication algorithm with the redundant binary representation [7].

Preparata et al's algorithm, as well as Morita's, overcame the drawback of the division-during-multiplication method. Namely, the number of addition/subtractions required by them is almost the same as that required by the division-after-multiplication method. In particular, the radix-4 version of Morita's algorithm is very efficient, because we can generate the multiples of the modulus required for residue calculation, as well as the multiples of the multiplicand, by negating (complementing) and/or shifting them. However, in these algorithms, the operands have to be in the ordinary binary representation, while the intermediate results are in redundant representations. This fact decreases the computation speed of iterative multiplications where the product of the former multiplication is used as the operands of the next multiplication, because a time-consuming conversion which includes a carry-propagate addition of long numbers is required at each multiplication. Our former algorithms do not have this drawback but require more addition/subtractions.

In this paper, we propose a new radix-4 modular multiplication algorithm with a redundant binary representation. It is a kind of division-during-multiplication method. The number of required addition/subtractions is as the same as that required by Morita's radix-4 algorithm and is about the half of that required by our former radix-4 algorithm. The multiplicand, as well as the multiplier, can be in a redundant representation. The product is also in the same redundant representation as the operands. Hence, the product can be used as either of the operands of the next multiplication in iterative multiplications. Namely, we can keep the intermediate results in the redundant representation, and perform the conversions only at the beginning and the end of the whole iterative multiplications. We do not need the time-consuming conversion of long numbers at each multiplication. Therefore, our algorithm is more efficient than Preparata and Vuillemin's and Morita's in iterative multiplications for, e.g., RSA encryption/decryption.

A serial-parallel modular multiplier based on the proposed algorithm has a regular cellular array structure with a bit slice feature suitable for VLSI implementation. The depth of its combinational circuit part is a constant independent of n, the length of the modulus, and therefore, it can operate with a fast clock. Its amount of hardware is proportional to n. It seems easy to implement a high performance RSA encryption/decryption circuit based on the multiplier on a VLSI chip using today's technology.

In the next section, we describe redundant representations for a residue class based on the redundant binary representation. We propose a radix-4 modular multiplication hardware algorithm, and show its correctness in Section 3. In Section 4, we consider a serial-parallel modular multiplier based on the algorithm. In Section 5, we apply the multiplier to RSA encryption/decryption. Section 6 is a conclusion. We explain how the proposed algorithm has been derived, in Appendix.

2 Redundant Representations for a Residue Class

We consider multiplication in a residue class $Z_Q = \{0, 1, ..., Q-1\}$ where $2^{n-1} \leq Q < 2^n$. We assume that the multiplicand and the multiplier are represented in a redundant representation, and calculate the product represented in the same redundant representation. We use a redundant representation based on the redundant binary representation, i.e., the radix-2 signed-digit representation. We also represent all partial products (intermediate results) in another redundant representation which is also based on the redundant binary representation. We perform all calculations for modular multiplication in the redundant binary representation. For the partial products, we use a representation which is more redundant than that for the operands and the product. This is one of the key points to getting our very efficient algorithm.

The redundant binary representation has a fixed radix 2 and a digit set $\{\bar{1}, 0, 1\}$, where $\bar{1}$ denotes -1 [8]. An n-digit redundant binary number $A = [a_{n-1}a_{n-2}...a_0]$ ($a_i \in \{\bar{1}, 0, 1\}$) has the value $\sum_{i=0}^{n-1} a_i \cdot 2^i$. Hereafter, we use A to denote both a representation and its value. Note that there may be several redundant binary numbers which have a certain value. Using this redundancy, we can add two redundant binary numbers without carry propagation. For the details of carry-propagation-free addition, see, e.g., [9]. We can get a negation of a redundant binary number by changing the signs of all nonzero digits in it.

For the multiplicand, the multiplier and the product,

we represent $\alpha \in Z_Q$ by an n-digit redundant binary number, A, which satisfies $-d_1 \cdot Q < A < d_1 \cdot Q$ and $A \equiv \alpha \pmod{Q}$. For the partial products, we represent $\beta \in Z_Q$ by an $(n+2)$-digit redundant binary number, B, which satisfies $-d_2 \cdot Q < B < d_2 \cdot Q$ and $B \equiv \beta \pmod{Q}$. d_1 and d_2 can be any numbers which satisfy the following conditions.

$$\frac{9}{16} \leq d_1 \leq \frac{55}{96}$$

$$\frac{9}{4} \leq d_2 \leq \frac{55}{24}$$

$$2 \cdot d_1 + 3 \cdot d_2 \leq 8$$

$$4 \cdot d_1 \geq d_2$$

We will show how the above conditions have been derived, in Appendix.

We can convert (the ordinary unsigned binary representation of) α ($\in Z_Q$) to the former redundant representation by comparing α with $Q/2$ down to the $(n-5)$th position and subtracting Q from α if the former is larger. (The i-th position of a number means the i-th position from the least significant position, which has the weight 2^i when the number is in radix-2.) We can perform the subtraction in the redundant binary representation without carry propagation. On the other hand, we can convert A to α by an ordinary binary subtraction and an addition. Namely, we calculate $A^+ - A^-$ or $A^+ - A^- + Q$ and chose the former as α if it is non-negative, where A^+ and A^- are n-bit binary numbers formed from positive and negative digits of A respectively. We need subtraction and addition with carry propagation in this conversion. Note that these conversions are required only at the beginning and the end of the whole iterative multiplications.

3 A Radix-4 Modular Multiplication Algorithm

We consider a modular multiplication in the former redundant representation shown in the previous section. Namely, the multiplicand X and the multiplier Y, as well as the product P are n-digit redundant binary numbers whose absolute values are less than $d_1 \cdot Q$, and $P \equiv X \times Y \pmod{Q}$ holds.

The algorithm is based on the following recursion equation.

$$P_j := 4 \cdot P_{j+1} + \hat{y}_j \cdot X - 4 \cdot c_j \cdot Q$$

Initially, we set $P_{\lfloor n/2 \rfloor + 1}$ to 0. $P_{-1}/4$ is the product.

$\hat{y}_j$ is the j-th digit of the recoded multiplier $\hat{Y}$. We recode the multiplier Y to a $(\lfloor n/2 \rfloor + 1)$-digit radix-4 signed-digit number $\hat{Y} = [\hat{y}_{\lfloor n/2 \rfloor}...\hat{y}_0]$ ($\hat{y}_j \in \{\bar{2}, \bar{1}, 0, 1, 2\}$) which has the same value as Y [7,10]. ($\bar{2}$ denotes -2.) Table 1 shows a recoding rule of the multiplier. Each $\hat{y}_j$ depends on only five digits of Y, i.e., $y_{2j+1}, y_{2j}, y_{2j-1}, y_{2j-2},$ and y_{2j-3}. We let $\hat{y}_{-1}$ be 0, by regarding y_j as 0 for $j < 0$. We can obtain $\hat{y}_j \cdot X$ by negating and/or shifting X.

In the calculation, we represent each partial product P_j in the latter redundant representation shown in the

Table 1: A recoding rule of the multiplier

(a) Stage 1

$y_{2j+1}\backslash y_{2j}$	$\bar{1}$	0	1
$\bar{1}$	$\bar{1},1$	$*0,\bar{2}/\bar{1},2$	$0,\bar{1}$
0	$0,\bar{1}$	$0,0$	$0,1$
1	$0,1$	$*1,\bar{2}/0,2$	$1,\bar{1}$

$*$: y_{2j-1} is non-negative. / Otherwise.

(b) Stage 2

$yt_j\backslash yu_j$	$\bar{1}$	0	1
$\bar{2}$	$\times$	$\bar{2}$	$\bar{1}$
$\bar{1}$	$\bar{2}$	$\bar{1}$	0
0	$\bar{1}$	0	1
1	0	1	2
2	1	2	$\times$

$\times$: Never occurs.

previous section. Namely, we represent P_j by an $(n+2)$-digit redundant binary number which satisfies $-d_2 \cdot Q < P_j < d_2 \cdot Q$. We select c_j from $\{\bar{2},\bar{1},0,1,2\}$, by comparing $4 \cdot P_{j+1} + \hat{y}_j \cdot X$ with $\pm 2 \cdot Q$ and $\pm 6 \cdot Q$ down to the $(n-4)$th position. Figure 1 shows the Robertson's diagram for the algorithm. We can obtain $-4 \cdot c_j \cdot Q$ by complementing and/or shifting Q. We perform the additions for the recursion equation in the redundant binary representation without carry propagation.

The algorithm is as follows.

Algorithm [MODMUL]

(Inputs)

Modulus Q

(an n-bit binary number, $2^{n-1} \leq Q < 2^n$)

Multiplicand X

(an n-digit redundant binary number,

$-d_1 \cdot Q < X < d_1 \cdot Q$)

Multiplier Y

(an n-digit redundant binary number,

$-d_1 \cdot Q < Y < d_1 \cdot Q$)

(Output)

Product P

(an n-digit redundant binary number,

$-d_1 \cdot Q < P < d_1 \cdot Q, P \equiv X \times Y \pmod{Q}$)

(Algorithm)

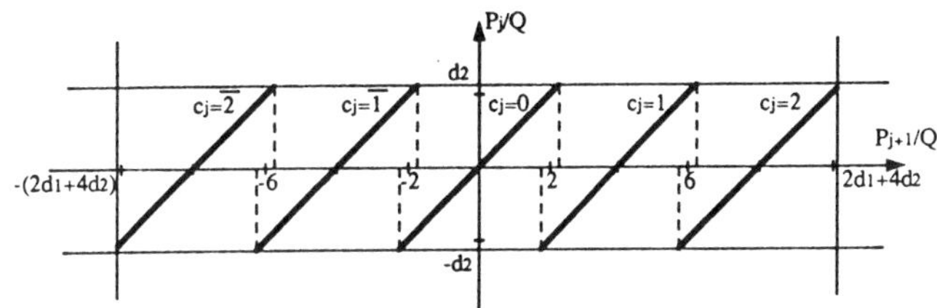

Figure 1: Robertson's diagram for [MODMUL]

Step 1: $P_{\lfloor n/2 \rfloor + 1} := 0$

Step 2: for $j := \lfloor n/2 \rfloor$ down to -1 do

begin

Calculate $\hat{y}_j$

$R_j := 4 \cdot P_{j+1} + \hat{y}_j \cdot X$

(redundant binary addition)

$c_j :=$

$$\begin{cases} \bar{2} & if \quad T(R_j) \leq -T(6 \cdot Q) \\ \bar{1} & if \quad -T(6 \cdot Q) < T(R_j) \leq -T(2 \cdot Q) \\ 0 & if \quad -T(2 \cdot Q) < T(R_j) < T(2 \cdot Q) \\ 1 & if \quad T(2 \cdot Q) \leq T(R_j) < T(6 \cdot Q) \\ 2 & if \quad T(R_j) \geq T(6 \cdot Q) \end{cases}$$

$P_j := R_j - 4 \cdot c_j \cdot Q$

(redundant binary addition)

end

Step 3: $P := P_{-1}/4$ $\qquad\square$

R_j is an $(n+4)$-digit redundant binary number, because of its value. Table 2 shows a computation rule for this addition. (sx_i is the i-th digit of $\hat{y}_j \cdot X$ and is $\bar{x}_{i-1}$ or $\bar{x}_i$ or 0 or x_i or x_{i-1} accordingly as $\hat{y}_j$ is $\bar{2}$ or $\bar{1}$ or 0 or 1 or 2. ($\bar{x}_i$ is 1 or 0 or $\bar{1}$ accordingly as x_i is $\bar{1}$ or 0 or 1.) We need a special computation rule at the most significant two positions, in order to let R_j be an $(n+4)$-digit number.)

$T(R_j)$ is the most significant 8 digits of R_j. $T(2 \cdot Q)$ is the most significant 5 bits of Q. We can calculate $T(6 \cdot Q)$ from the most significant 6 digits of Q so that $|6 \cdot Q - T(6 \cdot Q)| < 2^{n-4}$. (We can either calculate it beforehand and store it or calculate it at each iteration step.) In the determination of c_j, each boundary can be included in either of the corresponding regions.

Since Q is a binary number, the second redundant binary addition for obtaining P_j is easier than the first one. Table 3 shows a computation rule for this addition. (sq_i is the i-th digit of $-4 \cdot c_j \cdot Q$ and is q_{i-3} or q_{i-2} or 0 or q'_{i-2} or q'_{i-3} accordingly as c_j is $\bar{2}$ or $\bar{1}$ or 0 or 1 or 2. (q'_i is 1 or 0 accordingly as q_i is 0 or 1.) $sq_{n+3}sq_{n+2}$ is 01 or 00 or 00 or $\bar{1}1$ or $\bar{1}0$ according to c_j. We need a bit complicated rule at the most significant four positions, in order to make P_j an $(n+2)$-digit number.)

Table 2: A computation rule for the first addition

(a) Stage 1

r_{n+3}, rt_{n+2}

$p_{n+1} \backslash p_n$	$\bar{1}$	0	1
$\bar{1}$	$*\bar{1},\bar{1}/\times$	$\bar{1},0$	$*0,\bar{1}/\bar{1},1$
0	$*0,\bar{1}/\bar{1},1$	$0,0$	$*1,\bar{1}/0,1$
1	$*1,\bar{1}/0,1$	$1,0$	$*\times/1,1$

$*$: p_{n-1} is non-negative. / Otherwise.

$\times$: Never occurs.

o for $0 \leq i \leq n+1$

ru_{i+1}, rt_i

$p_{i-2} \backslash sx_i$	$\bar{1}$	0	1
$\bar{1}$	$\bar{1},0$	$*0,\bar{1}/\bar{1},1$	$0,0$
0	$*0,\bar{1}/\bar{1},1$	$0,0$	$*1,\bar{1}/0,1$
1	$0,0$	$*1,\bar{1}/0,1$	$1,0$

$*$: Both p_{i-3} and sx_{i-1} are non-negative. / Otherwise.

(b) Stage 2

o for $0 \leq i \leq n+2$

r_i

$rt_i \backslash ru_i$	$\bar{1}$	0	1
$\bar{1}$	$\times$	$\bar{1}$	0
0	$\bar{1}$	0	1
1	0	1	$\times$

$(ru_0 = 0)$

$\times$: Never occurs.

The most significant n digits of P_{-1} is the product. (Note that P_{-1} is an $(n+2)$-digit redundant binary number and its least significant two digits are 0.)

Figure 2 shows an example of a modular multiplication according to the algorithm [MODMUL]. $\lfloor n/2 \rfloor + 2$ clock cycles are required to carry out a modular multiplication excluding the I/O.

In the rest of this section, we show the correctness of the algorithm [MODMUL].

First, we show that $P_j \equiv X \times \hat{Y}_j \pmod{Q}$ holds for all j's ($\lfloor n/2 \rfloor + 1 \geq j \geq -1$), where $\hat{Y}_j$ is the most significant $\lfloor n/2 \rfloor - j + 1$ digits of $\hat{Y}$, i.e., $\sum_{i=j}^{\lfloor n/2 \rfloor} \hat{y}_i \cdot 4^{i-j}$. Note that $\hat{y}_{-1}$ is 0 and that $\hat{Y}_{-1} = 4 \cdot \hat{Y}$. We can prove this fact by induction on j.

When $j = \lfloor n/2 \rfloor + 1$, $P_{\lfloor n/2 \rfloor + 1} = 0$ and $\hat{Y}_{\lfloor n/2 \rfloor + 1} = 0$. Hence, the equation holds.

Assume that $P_{j+1} \equiv X \times \hat{Y}_{j+1} \pmod{Q}$ holds. Since

Table 3: A computation rule for the second addition

(a) Stage 1

o for $0 \leq i \leq n+1$

pu_{i+1}, pt_i

$r_i \backslash sq_i$	0	1
$\bar{1}$	$0,\bar{1}$	$0,0$
0	$0,0$	$1,\bar{1}$
1	$1,\bar{1}$	$1,0$

(b) Stage 2

o $pv := 2(r_{n+3} + sq_{n+3}) + (r_{n+2} + sq_{n+2}) + pu_{n+2}$
(pv must be $\bar{1}$ or 0 or 1.)

o for $0 \leq i \leq n+1$

$\dot{p}_i$

$pt_i \backslash pu_i$	0	1
$\bar{1}$	$\bar{1}$	0
0	0	1

($pu_0 = 0$, if c_j is non-negative. $pu_0 = 1$, otherwise.)

(c) Stage 3

p_{n+1}, p_n

$pv \backslash \dot{p}_{n+1}\dot{p}_n$	$\bar{1}$-	$0\bar{1}$	00	01	1-
$\bar{1}$	$\times$	$\times$	$\times$	$\bar{1},\bar{1}$	$\bar{1},\dot{p}_n$
0	$\dot{p}_{n+1},\dot{p}_n$				
1	$1,\dot{p}_n$	$1,1$	$\times$	$\times$	$\times$

$\times$: Never occurs.

o for $0 \leq i \leq n-1$ $p_i := \dot{p}_i$

$P_j = 4 \cdot P_{j+1} + \hat{y}_j \cdot X - 4 \cdot c_j \cdot Q$, $P_j \equiv 4 \cdot X \times \hat{Y}_{j+1} + \hat{y}_j \cdot X \pmod{Q}$. Hence, $P_j \equiv X \times \hat{Y}_j \pmod{Q}$ holds, and the fact has been proved.

Next, we show that $-d_2 \cdot Q < P_j < d_2 \cdot Q$ holds for all j's. Again, we can prove this fact by induction on j. Recall that $\frac{9}{4} \leq d_2 \leq \frac{55}{24}$, that $2 \cdot d_1 + 3 \cdot d_2 \leq 8$, and that $Q \geq 2^{n-1}$.

When $j = \lfloor n/2 \rfloor + 1$, $P_{\lfloor n/2 \rfloor + 1} = 0$. Hence, the inequality holds.

Assume that $-d_2 \cdot Q < P_{j+1} < d_2 \cdot Q$ holds. Since $R_j = 4 \cdot P_{j+1} + \hat{y}_j \cdot X$, $-(4 \cdot d_2 + 2 \cdot d_1) \cdot Q < R_j < (4 \cdot d_2 + 2 \cdot d_1) \cdot Q$ holds. (Note that $-d_1 \cdot Q < X < d_1 \cdot Q$ and that $-2 \leq \hat{y}_j \leq 2$.) Now, we have to consider the following five cases.

(1) $T(R_j) \leq -T(6 \cdot Q)$

$-(4 \cdot d_2 + 2 \cdot d_1) \cdot Q < R_j < -6 \cdot Q + 2 \cdot 2^{n-4}$,

$Q = 100101011$ (299)
$X = 10\bar{1}\bar{1}00101$ (165)
$Y = \bar{1}01100001$ ($-159 \equiv 140 \pmod{299}$) $\Rightarrow \hat{Y} = \bar{1}2\bar{2}01$

```
4P₅            0  0  0  0  0  0  0  0  0  0  0  0  0
(ŷ₄=1̄)    +             0  1̄  0  1  1  0  0  1̄  0  1̄
           ─────────────────────────────────────────
R₄             0  0  0  0  1̄  1  0  1̄  0  0  1̄  0  1̄
(c₄=0)     +   0  0  0  0  0  0  0  0  0  0  0  0  0(0)
           ─────────────────────────────────────────
P₄                0  0  0  1̄  0  1̄  0  0  1̄  0  1̄
4P₄            0  0  0  1̄  0  1̄  0  0  1̄  0  1̄  0  0
(ŷ₃=2)     +             1  0  1̄  1̄  0  0  1  0  1  0
           ─────────────────────────────────────────
R₃             0  0  0  0  1̄  0  1̄  0  1̄  1  0  1̄  0
(c₃=0)     +   0  0  0  0  0  0  0  0  0  0  0  0  0(0)
           ─────────────────────────────────────────
P₃                0  0  1̄  0  1̄  0  0  1̄  0  1̄  0
4P₄            0  0  1̄  0  1̄  0  0  1̄  0  1̄  0  0  0
(ŷ₂=2̄)    +             1̄  0  1  1  0  0  1̄  0  1̄  0
           ─────────────────────────────────────────
R₂             0  1̄  0  1  0  1̄  1  1̄  1̄  0  0  1̄  0
(c₂=1̄)    +   0  0  1  0  0  1  0  1  0  1  1  0  0(0)
           ─────────────────────────────────────────
P₂                0  1̄  0  1  1̄  0  0  0  1̄  1̄  0
4P₂            0  1̄  0  1  1̄  0  0  0  1̄  1̄  0  0  0
(ŷ₁=0)     +             0  0  0  0  0  0  0  0  0  0
           ─────────────────────────────────────────
R₁             0  1̄  0  1  1̄  0  0  1̄  1  1̄  0  0  0
(c₁=2̄)    +   0  1  0  0  1  0  1  0  1  1  0  0  0(0)
           ─────────────────────────────────────────
P₁                1  1̄  0  1  1̄  0  0  0  0  0  0
4P₁            1  1̄  0  1  1̄  0  0  0  0  0  0  0  0
(ŷ₀=1)     +             0  1  0  1̄  1̄  0  0  1  0  1
           ─────────────────────────────────────────
R₀             1  1̄  0  1  0  1̄  1  1̄  0  1  1̄  1  1̄
(c₀=2)     +   1̄  0  1  1  0  1  0  1  0  0  1  1  1(1)
           ─────────────────────────────────────────
P₀                0  0  0  1  1̄  0  1  1̄  1  0  1
4P₀            0  0  0  1  1̄  0  1  1̄  1  0  1  0  0
(ŷ₋₁=0)    +             0  0  0  0  0  0  0  0  0  0
           ─────────────────────────────────────────
R₋₁            0  0  0  1  1̄  0  1  0  1̄  1  1̄  0  0
(c₋₁=0)    +   0  0  0  0  0  0  0  0  0  0  0  0  0(0)
           ─────────────────────────────────────────
P₋₁               1  1̄  1̄  1  1̄  0  0  1̄  1̄  0  0
```

$P = 1\bar{1}\bar{1}1\bar{1}00\bar{1}\bar{1}$ (77)

Figure 2: An example of modular multiplication

and hence, $-(8 + d_2) \cdot Q < R_j < -(8 - d_2) \cdot Q$.
Since $c_j = \bar{2}$, $P_j = R_j + 8 \cdot Q$.
Therefore, $-d_2 \cdot Q < P_j < d_2 \cdot Q$ holds.

(2) $-T(6 \cdot Q) < T(R_j) \leq -T(2 \cdot Q)$

$-6 \cdot Q - 2 \cdot 2^{n-4} < R_j < -2 \cdot Q + 2 \cdot 2^{n-4}$,

and hence, $-(d_2 + 4) \cdot Q < R_j < -(4 - d_2) \cdot Q$.
Since $c_j = \bar{1}$, $P_j = R_j + 4 \cdot Q$.
Therefore, $-d_2 \cdot Q < P_j < d_2 \cdot Q$ holds.

(3) $-T(2 \cdot Q) < T(R_j) < T(2 \cdot Q)$

$-2 \cdot Q - 2^{n-4} < R_j < 2 \cdot Q + 2^{n-4}$,

and hence, $-d_2 \cdot Q < R_j < d_2 \cdot Q$.
Since $c_j = 0$, $P_j = R_j$.

Therefore, $-d_2 \cdot Q < P_j < d_2 \cdot Q$ holds.

(4) $T(2 \cdot Q) \leq T(R_j) < T(6 \cdot Q)$

$-d_2 \cdot Q < P_j < d_2 \cdot Q$ holds,

from a similar discussion to (2).

(5) $T(R_j) \geq T(6 \cdot Q)$

$-d_2 \cdot Q < P_j < d_2 \cdot Q$ holds,

from a similar discussion to (1).

Thus, $-d_2 \cdot Q < P_j < d_2 \cdot Q$ holds in any case, and the fact has been proved.

From the above facts, we get $P_{-1} \equiv 4 \cdot (X \times Y) \pmod{Q}$ and $-d_2 \cdot Q < P_{-1} < d_2 \cdot Q$. Since P_{-1} is obtained by the calculation of $4 \cdot P_0 + 0 - 4 \cdot c_{-1} \cdot Q$, its least significant two digits are 0. Hence, we can calculate $P := P_{-1}/4$ by just taking the most significant n digits of P_{-1}. Then, P satisfies $P \equiv X \times Y \pmod{Q}$ and $-\frac{1}{4} \cdot d_2 \cdot Q < P < \frac{1}{4} \cdot d_2 \cdot Q$. Since $4 \cdot d_1 \geq d_2$, $-d_1 \cdot Q < P < d_1 \cdot Q$ holds.

Thus, the obtained P satisfies $P \equiv X \times Y \pmod{Q}$ and $-d_1 \cdot Q < P < d_1 \cdot Q$. Therefore, the algorithm [MODMUL] is correct.

4 A Serial-Parallel Modular Multiplier

A serial-parallel modular multiplier based on the algorithm [MODMUL] has a regular cellular array structure with a bit slice feature suitable for VLSI implementation. Figure 3 shows a block diagram of the multiplier. Here, we assume that the multiplier performs one iteration step in each clock cycle. The multiplier consists of four registers and a combinational circuit part. The registers are for storing redundant binary numbers X, Y, and P_j, and a binary number Q. The register for Y is a shift register, and Y is shifted with two positions to the left in each clock cycle. When n is odd, initially, we have to attach 0 to Y from the left (to the most significant position). The combinational circuit part consists of a $\hat{y}_j$ calculating circuit, a c_j selecting circuit, and circuits for the slices. The circuit for the slices are composed of a negate-shift-and-select circuit for generating $\hat{y}_j \cdot X$, a redundant binary adder for calculating R_j, a complement-shift-and-select circuit for generating $-4 \cdot c_j \cdot Q$, and a simpler redundant binary adder for calculating P_j. The depth of the combinational circuit part is a constant independent of n. The amount of hardware of the whole multiplier is proportional to n.

Figure 4 illustrates a block diagram of the combinational circuit part of the slice for a middle position (the region enclosed with the dashed line in Figure 3). It consists of four basic cells, i.e., an XSEL for the generation of $\hat{y}_j \cdot X$, an RBA1 for the redundant binary addition for R_j, a QSEL for the generation of $-4 \cdot c_j \cdot Q$, and an RBA2 for the simpler redundant binary addition for P_j. According to our CMOS logic designs, the depth and the gate count of these cells are 3 and 5 (28 transistors), 4 and 7 (40 transistors), 2 and 2 (20 transistors), and 3 and 5 (24 transistors), respectively. We can shorten the clock period by performing the calculation of $\hat{y}_j$ in the previous cycle concurrently with the addition for

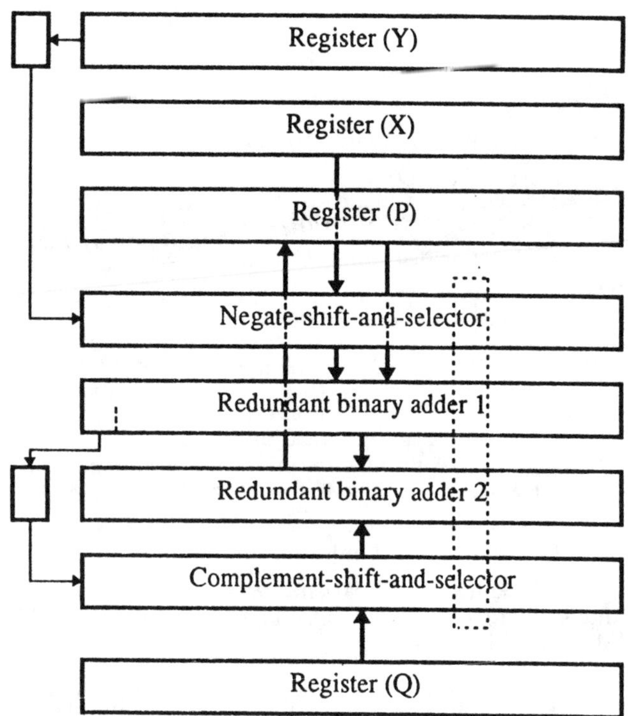

Figure 3: A block diagram of a modular multiplier

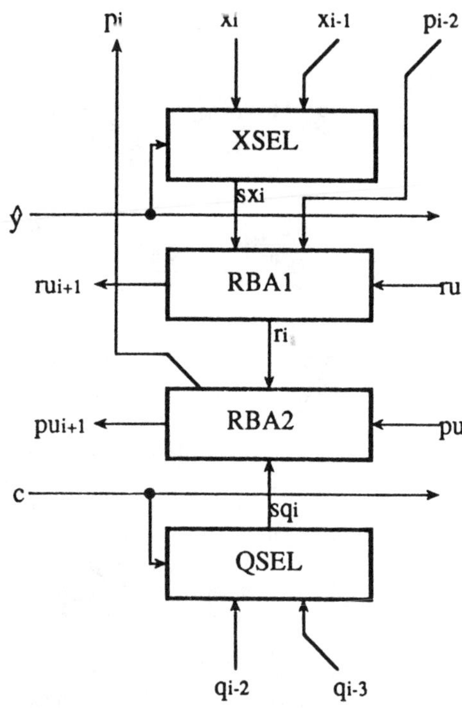

Figure 4: A block diagram of a slice of the multiplier

calculating P_{j+1}. When we prepare $T(6 \cdot Q)$ beforehand, the depth of the combinational circuit part becomes 19. (However, in practice, we need buffers for driving long lines for $\hat{y}_j$ and c_j.) The gate count of a slice is 19 (112 transistors). The gate count of the combinational circuit part of the whole multiplier is about 19n (112n transistors) for a large n. The total number of bits for the registers is about 7n. A 512-digit modular multiplier will consist of about 100,000 transistors including the buffers. It is expected to operate with about 33MHz clock and to carry out 512-digit multiplication in about $7.8\mu sec$ (excluding I/O), when it is fabricated with $2\mu m$ CMOS technology. It may operate with faster clock, if it is fabricated with today's advanced technologies.

Once a multiplier is fabricated, the length of its registers, adders and etc. is fixed. Assume that the length of the register for the modulus is n'-bit. The multiplier can perform any modular multiplication with a modulus which is shorter than or equal to n'-bit. When the modulus is shorter than n'-bit, we put the operands in the registers from the left side and fill the rest with 0's. The number of required clock cycles for a multiplication depends on the length of the operands but does not depend on the length of the registers.

As the advance of VLSI technologies, we may realize a faster multiplier which performs more than one iteration steps in each clock cycle.

5 Application to RSA Encryption/ Decryption

The proposed multiplier is efficient especially in applications where modular multiplications are performed iteratively. In such applications, we keep intermediate results in the redundant representation and convert only the final result to the ordinary representation.

As an example of such an application, we consider RSA encryption/decryption. In RSA encryption, we calculate $cipher := (message)^e \bmod Q$, where e is an encryption key. Decryption is carried out in the same way. A simple way to perform modular exponentiation is to repeat squaring and multiplication as shown below [1]. (We assume e is represented by a k-bit binary number $[1e_{k-2}...e_0]$.)

$$C := M$$
$$\text{for } i := k - 2 \text{ down to } 0 \text{ do}$$
$$\quad \text{begin}$$
$$\quad\quad C := C^2 \bmod Q$$
$$\quad\quad \text{if } e_i = 1 \text{ then } C := C \cdot M \bmod Q$$
$$\quad \text{end}$$

Here, M is the message and the calculated C is the cipher.

$2k - 2$ multiplications are required in the worst case. (About $1.5k$ multiplications are required in the average.) We keep C in the redundant representation (the former one shown in Section 2) during the calculation, and convert only the final C to the ordinary binary representation. We can convert the input into the redundant representation in the same way as the calculation of P from P_0 in the algorithm [MODMUL]. Therefore, we do not need additional hardware for this conversion. On the other hand, we need a carry propagate adder for the conversion of the result from the redundant representation to the ordinary binary.

In encryption of a message block sequence, we can perform the exponentiation concurrently with the input of the next message block, and the conversion and the output of the calculated cipher block. Namely, pipeline processing for continuous blocks is possible. The processing speed is dominated by the exponentiation speed. When the size of the message block (the length of the modulus) is 512-bit and the length of the encryption key is also 512-bit, the throughput for encryption with 33MHz clock will be at least 65kbps, which is more than 6 times as large as that of the fastest actual RSA chip listed in [2]. Note that we have used a very simple exponentiation algorithm and have assumed the worst case. (Recently, Shand et al reported a very high performance RSA encryption/decryption system based on an efficient modular multiplication algorithm cooperated with the Chinese remaindering [11]. Our multiplier may have comparable performance when it is cooperated with the Chinese remaindering.)

6   Concluding Remarks

We have proposed a radix-4 modular multiplication hardware algorithm, which is efficient especially in applications where modular multiplications are performed iteratively. In the algorithm, we represent numbers in redundant representations, and perform modular additions without carry propagation. We use a more redundant representation for representing intermediate results than for the operands. This enables us to reduce the number of addition/subtractions required in the "division-during-multiplication" method. This technique might be useful for design of new efficient arithmetic algorithms with a redundant representation.

A serial-parallel modular multiplier based on the proposed algorithm has a regular cellular array structure with a bit slice feature suitable for VLSI implementation. It seems easy to fabricate an RSA encryption/decryption circuit based on the proposed multiplier on a VLSI chip using today's technology, which is expected to have a throughput of several times as large as that of the fastest actual RSA chip.

Acknowledgements

This work was done while the author was staying at Computer Systems Laboratory, Stanford University. The author would like to thank Professor Shuzo Yajima of Kyoto University who gave the author the chance to stay at Stanford. The author also would like to appreciate the hospitality of Professor Michael J. Flynn of Stanford University.

References

[1] R.L.Rivest, A.Shamir and L.Adleman: "A method for obtaining digital signatures and public-key cryptosystems", Commun. ACM, vol.21, no.2, pp.120-126, Feb. 1978.

[2] E.F.Brickell: "A survey of hardware implementations of RSA", Lecture Notes in Computer Science, vol.435, G.Brassard Ed., 'Advances in Cryptology - CRYPTO'89 Proceedings', pp.368-370, Springer-Verlag, 1990.

[3] E.F.Brickell: "A fast modular multiplication algorithm with application to two key cryptography", D.Chaum et al Eds., 'Advances in Cryptology, Proceedings of CRYPTO 82', pp.51-60, Plenum Press, New York, 1983.

[4] H.Morita: "A fast modular-multiplication algorithm based on a higher radix", Lecture Notes in Computer Science, vol.435, G.Brassard Ed., 'Advances in Cryptology - CRYPTO'89 Proceedings', pp.387-399, Springer-Verlag, 1990.

[5] A.Vandemeulebroecke, E.Vanzieleghem,T.Denayer and P.G.A.Jespers: "A new carry-free division algorithm and its application to a single-chip 1024-b RSA processor", IEEE J. Solid-State Circuits, vol.25, no.3, June 1990.

[6] F.P.Preparata and J.E.Vuillemin: "Practical Cellular Dividers", IEEE Trans. Comput., vol.C-39, no.5, pp.605-614, May 1990.

[7] N.Takagi and S.Yajima: "A modular multiplication hardware algorithm with a redundant representation", Report of Technical Group on Computation, Institute of the Electronics, Information and Communication Engineers of Japan, COMP89-103, Jan. 1990. (also to appear in IEEE Trans. Comput.)

[8] A.Avizienis: "Signed-digit number representations for fast parallel arithmetic", IRE Trans. Electron. Comput., vol.EC-10, no.3, pp.389-400, Sep. 1961.

[9] N.Takagi, H.Yasuura and S.Yajima: "High-speed VLSI multiplication algorithm with a redundant binary addition tree", IEEE Trans. Comput., vol.C-34, no.9, pp.789-796, Sep. 1985.

[10] N.Takagi: 'Studies on hardware algorithms for arithmetic operations with a redundant binary representation', Doctoral dissertation, Dept. Info. Sci., Kyoto Univ., Aug. 1987.

[11] M.Shand, P.Bertin and J.E.Vuillemin: "Hardware speedups in long integer multiplication", Proc. 2nd Annual ACM Symp. on Parallel Algorithms and Architectures - SPAA '90, pp.138-145, July 1990.

Appendix: Derivation of the Algorithm

Assume that the multiplicand, the multiplier and the product are represented in n_1-digit redundant binary numbers with the value larger than $-d_1 \cdot Q$ and smaller than $d_1 \cdot Q$. $2 \cdot d_1$ must be larger than or equal to 1. Hence, $d_1 \geq \frac{1}{2}$. Since $2^{n-1} \leq Q < 2^n$, $n_1 = n + \lceil \log_2 d_1 \rceil$.

In general, a radix-4 modular multiplication algorithm is based on the following recursion equation.

$$P_j := 4 \cdot P_{j+1} + \hat{y}_j \cdot X - c1_j \cdot Q$$

Initially, we set $P_{\lfloor n_1/2 \rfloor + 1}$ to 0. P_0 is the product. $\hat{y}_j$ ($\in \{\bar{2}, \bar{1}, 0, 1, 2\}$) is the j-th digit of the recoded multiplier $\hat{Y}$. $-c1_j \cdot Q$ is for the residue calculation.

We let $c1_j$ be $2^{r1} \cdot c_j$ ($r1 \geq 0$) and select c_j from $\{\bar{2}, \bar{1}, 0, 1, 2\}$, so that we can obtain $c1_j \cdot Q$ by complementing and/or shifting Q. Namely, we rewrite the recursion equation as follows.

$$P_j := 4 \cdot P_{j+1} + \hat{y}_j \cdot X - 2^{r1} \cdot c_j \cdot Q$$

In the calculation, we represent each partial product P_j by an n_2-digit redundant binary number which satisfies $-d_2 \cdot Q < P_j < d_2 \cdot Q$. $n_2 = n + \lceil \log_2 d_2 \rceil$. In order that P_j stays in this range, the following inequality must hold.

$$4 \cdot d_2 + 2 \cdot d_1 - 2^{r1} \cdot 2 \leq d_2 \qquad (1)$$

Furthermore, in order that we can determine c_j by evaluating only several digits of $4 \cdot P_{j+1} + \hat{y}_j \cdot X$ and Q, the following inequality must hold. (There must exist an overlap between each contiguous regions in the Robertson's diagram shown in Figure 1.)

$$2^{r1} - d_2 < d_2 \qquad (2)$$

According to the above calculation, we obtain P_0 which satisfies $-d_2 \cdot Q < P_0 < d_2 \cdot Q$. We have to convert it to P which satisfies $-d_1 \cdot Q < P < d_1 \cdot Q$ and $P \equiv P_0 \pmod{Q}$. We can perform this conversion by the following calculation.

$$P := P_0 - c2 \cdot Q$$

As in the case of $c1_j$, we let $c2$ be $2^{r2} \cdot c$ ($r2 \geq 0$) and select c from $\{\bar{2}, \bar{1}, 0, 1, 2\}$, so that we can obtain $c2 \cdot Q$ by complementing and/or shifting Q. The following two inequalities must hold.

$$d_2 - 2^{r2} \cdot 2 \leq d_1 \qquad (3)$$

$$2^{r2} - d_1 < d_1 \qquad (4)$$

Inequality (4) is the condition that we can determine c by evaluating only several digits of P_0 and Q.

From (2), $d_2 > 2^{r1-1}$. From (4), $d_1 > \frac{1}{2}$. Substituting these to (1), we get $r1 > 1$. Since the increment of $r1$ causes the increment of n_2, we should select $r1$ as small as possible. Here, we select 2 as $r1$. Then, from (1) and (2), we get $3 \cdot d_2 + 2 \cdot d_1 \leq 8$ and $d_2 > 2$, respectively. From these, $d_1 < 1$. From this and (4), we get $r2 = 0$.

Substituting $r1 = 2$ and $r2 = 0$ to the recursion equation and the equation for the conversion, we get the following equations.

$$P_j := 4 \cdot P_{j+1} + \hat{y}_j \cdot X - 4 \cdot c_j \cdot Q$$
$$P := P_0 - c \cdot Q$$

Here, c_j and c are selected from $\{\bar{2}, \bar{1}, 0, 1, 2\}$.

We can rewrite the second equation as follows.

$$4 \cdot P := 4 \cdot P_0 - 4 \cdot c \cdot Q$$

When $d_2 \leq 4 \cdot d_1$, we can perform the final conversion using the circuit for the iteration step. Then, we can combine the above equations and get the following equations for our algorithm.

$$P_j := 4 \cdot P_{j+1} + \hat{y}_j \cdot X - 4 \cdot c_j \cdot Q$$
$$P := P_{-1}/4$$

Substituting $r1 = 2$ and $r2 = 0$ to the inequalities, we get $\frac{1}{2} < d_1 < \frac{7}{12}$, $2 < d_2 < \frac{7}{3}$, $2 \cdot d_1 + 3 \cdot d_2 \leq 8$, and $4 \cdot d_1 \geq d_2$. We can select d_1 and d_2 so that they satisfy these conditions. In any case, n_1 and n_2 become n and $n+2$, respectively. The larger d_2 is, the fewer the number of digits to be looked into in the determination of c_j is.

Now, we consider how many digits we should look into for determining c_j. Assume that we compare $4 \cdot P_{j+1} + \hat{y}_j \cdot X$ with $\pm 2 \cdot Q$ and $\pm 6 \cdot Q$ down to the k-th position. We should select k as large as possible to reduce the hardware. k must satisfy the following condition.

$$2 \cdot 2^k \leq (d_2 - 2) \cdot Q$$

Since Q can be 2^{n-1}, $k \leq n - 2 + \log_2(d_2 - 2)$ must hold. Since $d_2 < \frac{7}{3}$, k is at most $n - 4$. In reverse, k is $n - 4$ when $d_2 \geq \frac{9}{4}$. Hence, we get the condition $d_2 \geq \frac{9}{4}$.

Putting it all together, we get $\frac{9}{16} \leq d_1 \leq \frac{55}{96}$, $\frac{9}{4} \leq d_2 \leq \frac{55}{24}$, $2 \cdot d_1 + 3 \cdot d_2 \leq 8$, and $4 \cdot d_1 \geq d_2$.

High-Speed Multiplier Design Using Multi-Input Counter and Compressor Circuits

Mayur Mehta* and Vijay Parmar
Advanced Micro Devices
5204 E. Ben White Blvd.
Austin, TX 78741

Earl Swartzlander, Jr.
Dept. of Electrical and Computer Engineering
University of Texas at Austin
Austin, TX 78712

Abstract

Multiplication represents one of the major bottlenecks in most digital processing systems. Depending on the word-size, several partial products are added to evaluate the product. The well-known shift-and-add algorithm uses minimal hardware but has unacceptable performance for most applications. Several parallel fast multiplication schemes have been suggested using several levels of blocks containing full adders.

This paper presents the design of a fast multiplier implemented using either (7,3) parallel counter or (7:3) compressor circuits for implementation in CMOS technology. The resulting 16 by 16-bit multiplier has less delay than conventional fast multipliers, although the gate count is about 10% higher.

1 Introduction

Multiplication is inherently a slow operation as a large number of partial products are added to produce the product. For example, in a 16 by 16-bit multiplication, 16 partial products are added. Modified Booth encoding [1] can be used to reduce the number of partial products. A modified Booth encoding algorithm looking at groups of three bits at a time will reduce the number of partial products in our example from 16 to eight. That is still a large number and will involve a substantial delay in comparison with other functional units in the system such as adders. In applications like digital signal processing, this delay is unacceptable, particularly in the context of ever increasing throughput requirements. Recent Reduced Instruction Set Computing processors have also started to include multiplier units, since many applications make extensive use of multiplication. Researchers have developed several fast multiplication approaches.

Wallace [2] suggested the idea of pseudo-adders, which are essentially arrays of full adders without rippling carries, that take three inputs and reduce them to two equivalent outputs. Wallace used pseudo-adders at several levels in the summation of the partial products. Thus, for 16 by 16-bit multiplication without Booth encoding, where there are 16 partial products, the Wallace multiplier uses five pseudo-adders to reduce the 16 partial products to eleven at the first level. It uses three pseudo-adders at the second level to reduce to eight partial products, two at the next level to get six partial products, two more pseudo-adders to get four partial products, and one each at the following two levels finally producing two partial products. At this point, a fast carry propagate adder (CPA) such as a carry lookahead adder, carry skip adder, carry select adder, etc. is used to determine the final product. Thus, a Wallace multiplier has a total delay equivalent to six full adder delays plus one CPA delay. A shift-and-add algorithm has 15 CPA delays for the same case, although it uses less hardware. The array multiplier scheme used by other researchers [3], makes use of only one pseudo-adder at each level. It reduces the number of partial products by one at each level. Therefore, for the 16 by 16-bit multiplication case cited above, it takes 14 full adder delays and one CPA delay.

Dadda [4], [5] generalized the idea of using full adders to reduce the partial product matrix by introducing the concept of (n,m) parallel counters. An (n,m) parallel counter is a combinational network with n inputs and m outputs where the outputs express the count of the number of inputs that are ONEs. Thus, a full adder is a (3,2) parallel counter. Dadda aimed at reducing the height of the partial product matrix by application of suitable parallel counters. He further theorized that depending on the available parallel counters, a sequence of numbers exists which could be used to determine the appropriate height of the partial product matrix at each level. At every level, he used just enough parallel counters to reduce the height of the partial product matrix to the next lower number in the sequence. When using (3,2) parallel

---

* Currently with Ross Technology, Inc. Austin, TX 78736

43

counters, the sequence is 2, 3, 4, 6, 9, 13, 19, etc. A 16 by 16-bit multiplier using Dadda's scheme has the same delay as the Wallace multiplier and requires fewer gates, but has a less regular structure and might be more difficult to lay out in VLSI.. Dadda's scheme can be implemented with parallel counters other than (3,2) counters.

In the Dadda (n,m) parallel counter, n is the number of inputs with equal weight. The n bits must come from the same column of the partial product matrix. Stenzel, *et al.* [6] extended this idea to include parallel counters with input bits coming from multiple columns, which are described as $(c_{k-1}, c_{k-2}, \ldots, c_0, d)$ parallel counters, where k is the number of input columns, c_i is the number of input bits in the column of weight 2^i, and d is the number of output bits. Counters like (5, 5, 4), (2, 2, 2, 3, 5) and (3, 3, 3, 3, 6) were suggested and the Wallace or Dadda scheme was used to reduce the partial product matrix height to two rows. The design included a 4 by 4 array implemented with a ROM for the generation of the partial product matrix instead of the AND gate arrays used by others. This reduces the initial number of partial products. Appropriate multi-input parallel counters were used to reduce the height of the partial product matrix. The major drawback of this design was the use of ROMs for implementing the parallel counters. While a ROM based approach is practical for LSI technology, it is impractical with current VLSI technology since ROMs are slow and occupy substantial area. Multi-column parallel counters implemented using combinational logic have large delay due to the need to propagate the carry across several columns.

In implementing multipliers the word sizes are generally multiples of two. Researchers have explored this idea by using a (4:2) compressor to convert four bits to two. This block is really a (5,3) counter that takes four input bits, one intermediate carry input bit from the previous column, and generates an intermediate carry output with weight two and two output bits with weights one and two. Recently Santoro and Horowitz [7] implemented a 64 by 64 array multiplier with (4:2) compressors realized using pairs of (3,2) counters. Earlier Shen and Weinberger designed a (4:2) compressor without using (3,2) counters [8]. They used exclusive-OR gates to derive the sum bit from the five inputs. The intermediate carry is generated using the four inputs, while the carry output is based on the four inputs and the intermediate carry from the previous block. More recently, Nagamatsu, *et al.* [9] used this approach with minor modifications to implement a 32 by 32 bit multiplier using a (4:2) compressor circuit realized with 0.8-μm CMOS.

Swartzlander [10] introduced a methodology to design a counter with 2k+1 inputs, using two k input counters and a $(1+\lfloor \log_2(k) \rfloor)$ stage ripple carry adder. Using this philosophy a (7,3) counter can be designed using two (3,2) counters and a two stage ripple carry adder (which is equivalent to two (3,2) counters).

This paper, describes a multiplication scheme using a (7,3) counter circuit that offers substantial improvement over the one implemented with (3,2) counters. To illustrate its utility, the logic design of a 16 by 16-bit multiplier using (7,3) parallel counters is described. It has fewer gate delays than the Wallace and Dadda schemes for a 10% increase in gate count. It has slightly more gate delays than the (4:2) compressor based 16 by 16-bit multiplier with a slightly smaller gate count. Since the counter based design has fewer inter cell data lines, it is expected to be easier to lay out and may be faster due to the reduced capacitive loading on the interconnections. A design for a multiplier based on (7:3) compressor circuits exhibits almost identical delay and area characteristics to the based on (7,3) counters.

The next Section presents the design of the (7,3) counter. This methodology is useful for designing similar blocks with a higher number of inputs which would be suitable for large word-size multipliers. The (7,3) counter is also compared with a (7,3) counter designed using (3,2) counters. After that, a (7:3) compressor which is designed by modifying the (7,3) counter is described. Then, practical application of these circuits is demonstrated by considering a 16 by 16-bit multiplier design. Some of the decisions in the design are scalable to larger word-sizes. The design is compared with similar implementations using the Wallace, Dadda, and Nagamatsu schemes.

2   (7,3) Parallel Counter

Figure 1 shows the logic diagram of the proposed (7,3) counter. The seven inputs are divided into two groups (X_0, X_1, X_2, X_3) and (X_4, X_5, X_6). Internal signal A is the carry for two, three, or four ONEs in the first group, while internal signal B is the carry for two or three ONEs in the second group. Finally, internal signal C is generated by ORing the carry for 4 ONEs in the first group (along with A) and 1-1, 1-3, 3-1, and 3-3 interactions between the two groups. A, B, and C are combined using a conventional full adder to get outputs C_1 and C_2, with weights of two and four respectively. This circuit has been tested exhaustively through logic simulations to confirm that it correctly implements the described functionality.

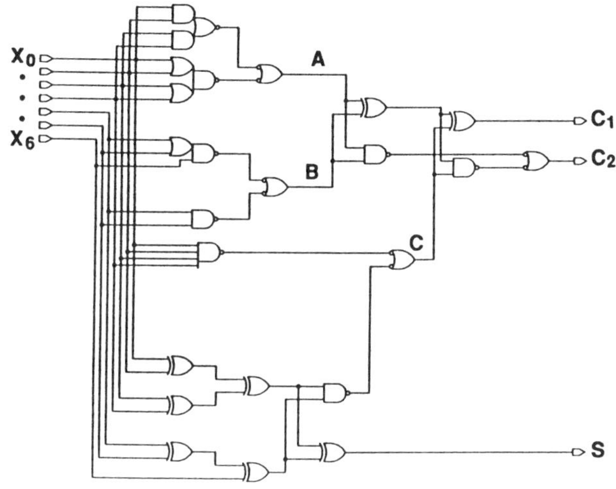

Figure 1. (7,3) Parallel Counter.

Figure 2 shows the block diagram of a (7,3) parallel counter implemented using (3,2) counters. First two (3,2) counters take six of the seven inputs and generate two sum and two carry outputs. The sum outputs are combined with the seventh input in another (3,2) counter to generate the S output of the (7,3) counter. The carry output of this (3,2) counter is combined with the carry outputs from the two first level counters using a fourth (3,2) counter to yield C_1 (which is the sum output) and C_2 (which is the carry output), with weights of two and four respectively. This circuit has six exclusive-OR delays. Assuming that an exclusive-OR gate has twice the delay of a normal gate, and assuming that complex AND-AND-NOR and OR-OR-NAND gates have 1.5 times the delay of a normal gate, the proposed (7,3) counter has four exclusive-OR delays which is a 33 percent improvement over the one using (3,2) counters while using approximately the same number of gates. The next Section presents a (7:3) compressor circuit designed by modifying the (7,3) counter.

3 (7:3) Compressor

In designing a (4:2) compressor, a problem arises since a sum output with a weight of one and a carry output with a weight of two are not enough to convey the maximum possible count of four. This problem is circumvented by generating an intermediate carry output which is fed into the next block in the adder array. Thus a (4:2) compressor is effectively a (5,3) counter. A block diagram of a (4:2) compressor is shown in Figure 3. The first box takes the four inputs and generates one sum output S' and two carry outputs C_{out} and C'. C' is generated only when S' is zero. The next box takes S',

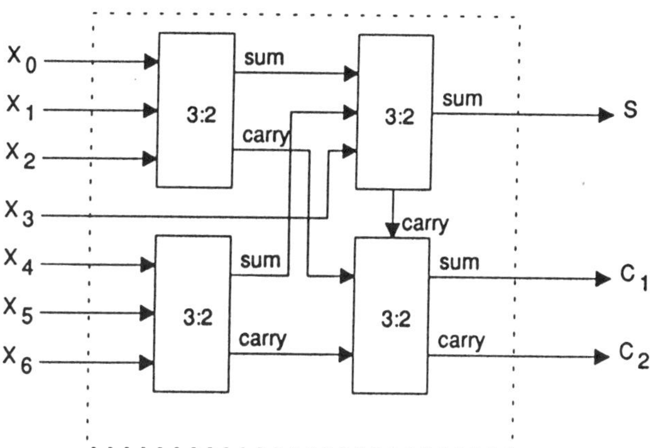

Figure 2. (7,3) Counter Implementation with (3,2) Counters.

C_{in}, and C' as inputs and generates S and C. Since S and C cannot represent the case when C', S', and C_{in} are all ONEs, C' is forced to be ZERO when S' is ONE. Carries in such cases are accounted for by C_{out} which is essentially the C_{in} to the next (4:2) compressor. One advantage of this scheme is that the intermediate carry output, C_{out}, is generated only by the four inputs and is used in the following block to generate C and S. This avoids carry propagation across more than two blocks.

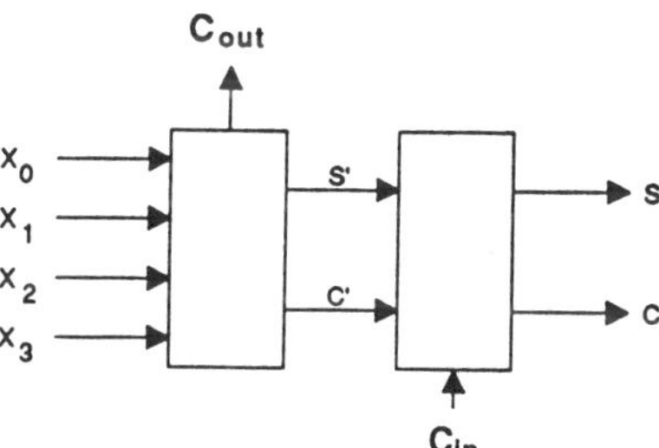

Figure 3. (4:2) Compressor.

A similar idea was used in designing the (7:3) compressor. Figure 4 shows a block diagram of this compressor. Of course, three outputs are enough to count seven bits, and the intermediate carry outputs could have been avoided. They were retained to explore the

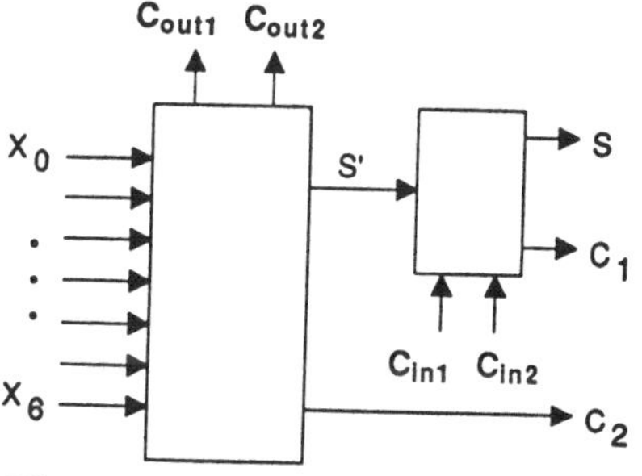

Figure 4. (7:3) Compressor.

possibilities of reducing the overall delay and gate count of the multiplier. The left box takes inputs X_0-X_6 and generates the sum output S' and carry outputs C_{out1}, C_{out2}, and C_2. C_{out1} and C_{out2} are intermediate carries which are fed into the next (7:3) compressor block in the adder array as C_{in1} and C_{in2}. In the next box, S', C_{in1}, and C_{in2} are added using a conventional (3:2) counter to generate S and C_1. S, C_1, and C_2 are the three output bits. Note that this circuit differs from the (7,3) parallel counter in that C_1 and C_2 both have a weight of two.

Figure 5 shows the logic diagram of the (7:3) compressor. This circuit is very similar to the (7,3) counter circuit. C_{out1} is generated only from X_0, X_1, X_2, and X_3 for cases when two, three, or four of them are ONEs. Similarly, C_{out2} is generated from X_4, X_5, and X_6. This is useful when this block is used with five inputs as C_{out2} is not generated in that case. Usually, the next block in such cases is a Nagamatsu (4:2) compressor which takes only one intermediate carry input. C_2 takes care of the carries not accounted by C_{out1} and C_{out2}.

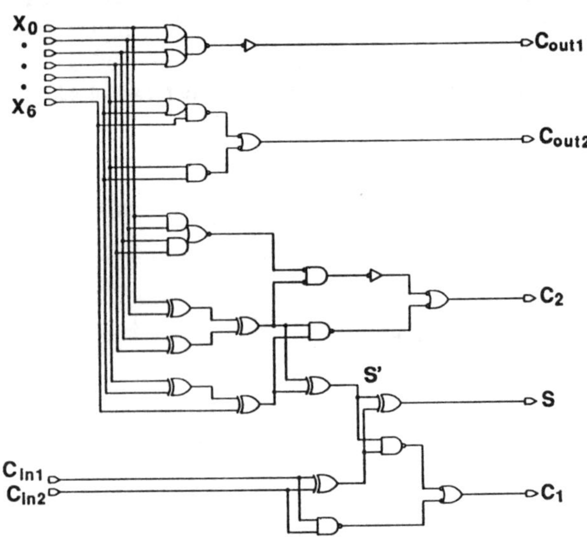

Figure 5.   (7:3) Parallel Compressor.

4   Implementation

A CMOS 16 by 16-bit multiplier has been designed using the (7,3) counter and (7:3) compressor circuits described above. The multiplier uses 16-bit magnitudes, but could be modified to accept twos complement numbers by the Baugh and Wooley approach [11]. Booth encoding is not used here, although the approach is applicable to multipliers using modified Booth encoding [1]. The parallel structure closely resembles the Wallace scheme with some modifications. Figures 6 and 7 show

the partial product matrices for multipliers using (7,3) counters and (7:3) compressors respectively.

The partial product matrix is generated using an array of AND gates. In the Wallace multiplication scheme, each row of the partial product matrix is input to an array of adders (compressors). As shown in Figures 6 and 7, this was modified slightly to reduce the overall delay. Thus, partial products included in the first block are the ones in the top seven rows plus some of the bit products from the two rows below. This is shown by the solid line which steps down on the left. The same procedure is repeated for the second block at the top level. As a result of this modification, four bit products are left at the lower right corner (columns 14, 15 and 16) of the partial product matrix. These are reduced to one partial product using two half adders. This results in seven partial products at second level which are reduced to three and from then to two using arrays of full adders (i.e., (3:2) counters). If this modified approach were not used, there would have been eight partial products at the second level and four at the next level. The (4:2) compression would have been required at the third level which would have added to the delay and the gate count.

Half-adders (2,2), full adders (3,2), (4,3) counters or (4:2) compressors, (5,3) counters or (5:3) compressors, and (6,3) counters or (6:3) compressors were used at the right and left end of the top two levels. (4,3), (5,3) and (6,3) counters are modified (7,3) counters with slightly reduced gate count. Similarly, (5:3) and (6:3) compressors are modified (7:3) compressors.

5   Discussion

The complexity of a variety of different implementations of a 16 by 16-bit multiplier is shown on Table 1. The Table shows the number of components (i.e., adders, compressors, and counters) and the total equivalent gate count for several multipliers. The equivalent gate count is the sum of number of "simple" gates (i.e., inverters, 2-input and 3-input NAND and NOR gates), 1.5 times the number of complex gates, and two times the number of exclusive-OR gates. It is expected that an implementation using (7,3) counters will be easier to lay out due to the absence of intermediate carries.

The multiplier delay can be compared by evaluating the number of gate delays to reduce the number of partial products from 16 to two. Note however that the total time to perform the multiplication must also include the time to generate the bit products (one gate delay) and the time to sum the two words in a carry propagate adder (ten

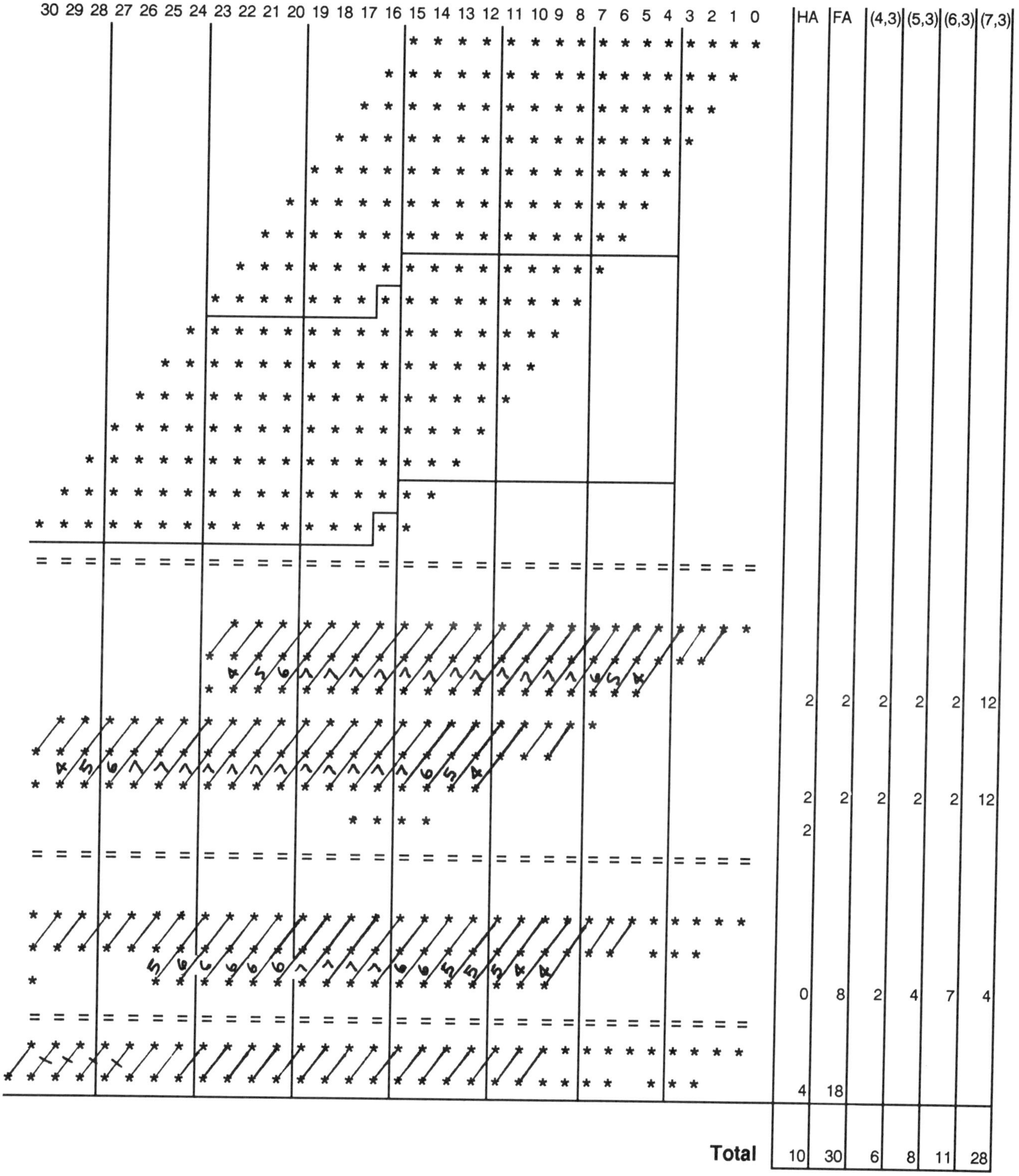

Figure 6. 16 by 16-Bit Multiplier Implemented with (7,3) Counters.

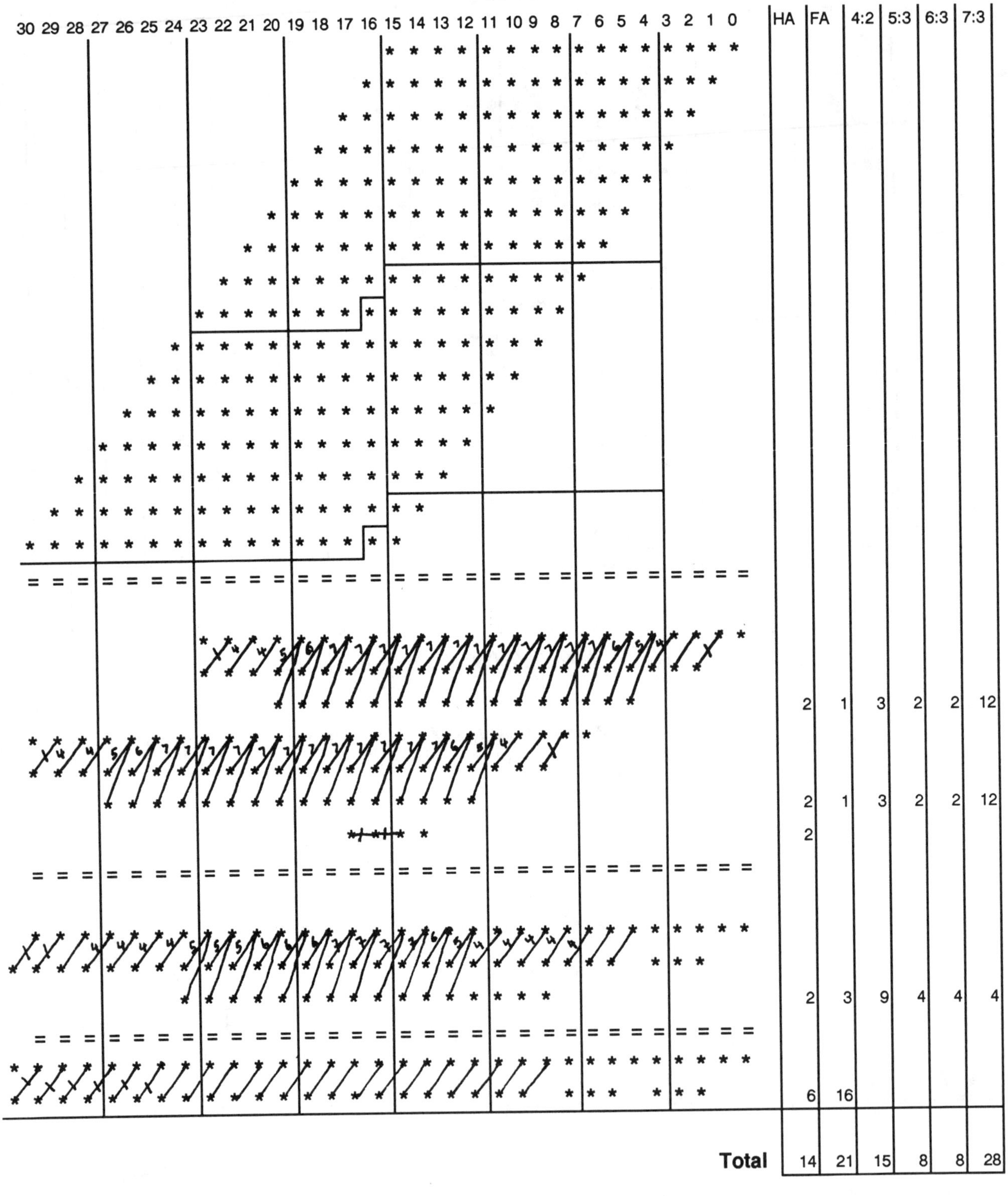

Figure 7. 16 by 16-Bit Multiplier Implemented with (7:3) Compressors.

Table 1. Comparison of the Complexity of Various 16 by 16-Bit Multipliers.

	Wallace	Dadda	(4:2) Compressor	(7:3) Compressor	(7,3) Counter
Half Adder	35	16	18	14	10
Full Adder	200	196	18	21	30
(4:2) Compressor			95	15	
(5:3) Compressor				8	
(6:3) Compressor				8	
(7:3) Compressor				28	
(4,3) Counter					6
(5,3) Counter					8
(6,3) Counter					11
(7,3) Counter					28
Total Gate Count	1575	1452	1641	1650	1623

delays for a carry lookahead adder). The Wallace and Dadda multipliers both use six levels of full adders (four gate delays each) to reduce the number of partial products from 16 to two. As each full adder has two exclusive-OR gate delays (two "simple" gate delays each), their delay is 24. The (4:2) compressor approach uses three exclusive-OR gates per level, using three levels to reduce the partial products to two giving a total of 18 gate delays. Both of the new schemes (i.e., using (7:3) compressors and using (7,3) counters) have a delay of eight exclusive-OR gate delays plus one full adder delay for a total of 20 "simple" gate delays. Implementations based on (4:2) compressors, (7,3) counters, and (7:3) compressors improve upon the Wallace and Dadda schemes in terms of overall delay.

One of the advantages of the multiplier implementation using (7,3) counters over the implementation with (4:2) compressors is the reduced number of interconnections. This has been established in IBM RS/6000 Floating Point Unit design [12] which uses (7,3) counters to implement a 56 by 56 multiplier. The (4:2) compressor reduces four bits to two at each level, while the (7,3) counter reduces seven bits to three. Clearly a multiplier implemented with (4:2) compressors will require more blocks at the early stages of the bit product reduction process than one using (7,3) counters. At lower levels, as the number of rows of partial products goes down, implementations using (4:2) compressors become more

efficient. For example, a 56 by 56 bit multiplier requires eight (7,3) counter arrays at the top level to reduce the number of partial products to 24 or 14 (4:2) compressor arrays to reduce the number of partial products to 28. At the next level, it requires three (7,3) counter arrays to reduce to twelve partial products or seven (4:2) compressor arrays to reduce to 14 partial products. In addition to the increased number of interconnections because of the greater number of blocks, (4:2) compressors also require connections between adjacent compressors for intermediate carries. The major difference between implementations using (7,3) counters and (7:3) compressors is that the compressor based implementations require interconnections between adjacent blocks for the intermediate carries.

The (4:2) compressor based approach has a slightly lower delay with a gate count that is slightly higher than that of the (7,3) parallel counter implementation. Saving two gate delays out of a total of 31 (this includes one delay for the AND gates that generate the bit product matrix, 20 delays for the reduction to a two row matrix, and ten delays for a carry lookahead adder) may be less significant than the speed variation due to gate loading which is expected to be less for the (7,3) parallel counter implementation.. Multipliers implemented with (4:2) compressors are expected to be harder to lay out than those implemented with (7,3) counters as there are more interconnections.

6 Conclusion

A novel scheme for parallel multiplication using (7,3) counter circuits has been designed and discussed. A (7:3) compressor circuit was investigated as an extension of the (4:2) compressor concept. This study indicates that parallel multipliers implemented using (7,3) counters have better performance than those implemented using (7:3) compressors. They exhibit identical delay characteristics while the counter implementation requires fewer gates and lays out better. Although the (7,3) counter implementation uses more gates than the Wallace and Dadda schemes, it achieves a lower delay. The (7,3) counter based implementation compares favorably with the (4:2) compressor implementation in terms of gate count, although it has slightly higher delay for the 16 by 16-bit multiplier example.

References

[1] H. Sam and A. Gupta, "A Generalized Multibit Recoding of Two's Complement Binary Numbers and Its Proof with Application in Multiplier Implementations," *IEEE Transactions on Computers*, Vol. 39, pp. 1006-1015, 1990.

[2] C. S. Wallace, "A suggestion for a Fast Multiplier," *IEEE Transactions on Electronic Computers*, Vol. EC-13, pp. 14-17, 1964.

[3] K. Hwang, *Computer Arithmetic: Principles, Architecture, and Design*, New York: Wiley, 1979.

[4] L. Dadda, "Some Schemes for Parallel Multipliers," *Alta Frequenza*, Vol. 34, pp. 349-356, 1965.

[5] L. Dadda, "On Parallel Digital Multipliers," *Alta Frequenza*, Vol. 45, pp. 574-580, 1976.

[6] W. J. Stenzel, W. J. Kubitz, and G. H. Garcia, "A Compact High-Speed Parallel Multiplication Scheme," *IEEE Transactions on Computers*, Vol. C-26, pp. 948-957, 1977.

[7] M. R. Santoro and M. A. Horowitz, "SPIM: A Pipelined 64 x 64-bit Iterative Multiplier," *IEEE Journal of Solid-State Circuits*, Vol. 24, pp. 487-493, 1989.

[8] D. T. Shen and A. Weinberger, "4-2 Carry-Save Adder Implementation Using Send Circuits," *IBM Technical Disclosure Bulletin*, Vol 20, pp. 3594-3597, 1978.

[9] M. Nagamatsu, *et. al.*, "A 15-ns 32 x 32-b CMOS Multiplier with an Improved Parallel Structure," *IEEE Journal of Solid-State Circuits*, Vol. 25, pp. 494-497, 1990.

[10] E. E. Swartzlander, Jr., "Parallel Counters," *IEEE Transactions on Computers*, Vol. C-22, pp. 1021-1024, 1973.

[11] C. R. Baugh and B. A. Wooley, "A Two's Complement Parallel Array Multiplication Algorithm," *IEEE Transactions on Computers*, Vol. C-22, pp. 1045-1047, 1973.

[12] R. K. Montoye, E Hokenek, and S. L. Runyon, "Design of the IBM RISC System/6000 Floating-Point Execution Unit," *IBM Journal of Research and Development*, Vol. 34, pp. 59-70, 1990.

A High-Radix Hardware Algorithm
for Calculating the Exponential M^E Modulo N

Holger Orup

Computer Science Department
Aarhus University
DK-8000 Aarhus C, DENMARK
e-mail: orup@daimi.aau.dk

Peter Kornerup

Dept. of Math. and Computer Science
Odense University
DK-5230 Odense M, DENMARK
e-mail: kornerup@imada.ou.dk

Abstract

In a class of crypto systems fast computation of modulo exponentials is essential. The popular RSA protocol uses operands of more than 500 bits to achieve a sufficient security. We present a parallel version of a well known exponentiation algorithm that halves the worst case computing time. It is described how a high radix modulo multiplication can be implemented by interleaving a serial-parallel multiplication scheme with an SRT division scheme. The problems associated with high radices are efficiently solved by the use of a redundant representation of intermediate operands. We show how the algorithms can be realized as a highly regular VLSI circuit. Simulations indicate that a radix 32 implementation of the algorithms is able of computing 512 bit operand exponentials in 3.2 msec. This is more than 5 times faster compared to other known implementations.

1 Introduction

The concept of two-key crypto systems was introduced by Diffie and Hellman [5] in 1976. In 1978 Riverst, Shamir and Adleman [13] published an encryption scheme based on computing exponentials. The RSA scheme realizes encryption as put forth by Diffie and Hellman. Both encryption of a message and decryption of an encrypted message are done by computing an exponential $M^E \bmod N$, $M \in [0, N[$. The message is denoted M and the key (E, N). Modulus N is a product of two very large primes and the security of the system depends on the length of the keys. To achieve a sufficient security key lengths of 500-600 bits are necessary.

For an implementation of the RSA protocol it is crucial to calculate modulo exponentials in a rate corresponding to the transmission rate between transmitter and receiver to avoid the encryption to be a bottle neck in the communication system. Brickell [4] has made a survey of hardware implementations of RSA. In his paper chips from Cryptech [7] are the fastest, with a computing time of 512 bit messages corresponding to a rate of $17\frac{\text{Kbit}}{\text{sec}}$. Thorn EMI [15] has made chips that encrypt 512 bit messages at $29\frac{\text{Kbit}}{\text{sec}}$.

In this paper we will present a method for obtaining rates of more than $150\frac{\text{Kbit}}{\text{sec}}$. Section 2 describes how we construct a new and faster algorithm for performing exponentials by splitting the computation into parallel multiplications. The implementation of a high radix modulo multiplication is elaborated in sections 3, 4 and 5, where it is shown how to interleave a multiplication with a SRT division scheme and how the algorithm can be realized in VLSI design. Section 6 discusses the performance of a VLSI implementation. The summary indicates how the methods can be generalized to even higher radices.

2 Exponentiation

A commonly known algorithm for performing an exponentiation is named *Russian Peasant* [8]. It performs the computation as a number of multiplications and squarings proportional to the bit length of the exponent. The algorithm can be modified to perform a modulo exponentiation by substituting the multiplications and squarings for modulo multiplications and modulo squarings [13]. Below is shown a variant in which the exponent E is read from the least significant bit. The i'th bit of E is denoted e_i. In the curly brackets is an invariant for the loop.

If we denote the bit length of M, E and N by n the worst case time is

$$T[\text{Exp}, n] = 2nT[\text{Mult}, n]$$

and the average time is

$$T[\text{Exp}, n] = \frac{3}{2}nT[\text{Mult}, n],$$

where it is assumed that the computing time for squaring and multiplication is identical.

Observing that the three statements in the loop are independent of each other it is possible to speed up the algorithm by performing *two* multiplications in *parallel*. In this way the computing time is reduced to

$$T[\text{Exp}, n] = nT[\text{Mult}, n], \qquad (1)$$

51

```
Algorithm:
    Modulo exponentiation.
Stimulation:
    E, M, N, where E ≥ 0 and 0 ≤ M < N.
Response:
    X = M^E mod N
Method:
    i := 0;
    X := 1; Y := M;
    WHILE i < n DO
        {* X · Y^(E div 2^i) ≡_N M^E *}
        IF e_i = 1 THEN
            X := (X · Y) mod N
        END;
        Y := (Y · Y) mod N;
        i := i + 1;
    END;
```

Algorithm 1: *Variant of Russian Peasant for modulo exponentiation.*

which is independent on the number of 1–bits in E. It is hard to imagine how an exponentiation can be performed with less than n squarings.

Note that by distributing the modulo reduction to squaring and multiplication the bit-length of intermediate operands is bounded. This makes the exponentiation algorithm feasible to implement for large values of n.

3 Multiplication

To implement the exponentiation algorithm mentioned above we need an efficient way to perform modulo multiplication. Several algorithms have been presented [3] [2] [1] [10] [7]. All of them use radix 2. We will follow the approach in e.g. [3], where the modulo reduction is further distributed in the algorithm for multiplication. A similar approach is followed in [9] where a radix 4 algorithm is elaborated.

The usual way of multiplying is by scanning the multiplier serially from the *least* significant digit and parallelly adding a multiple of the multiplicand followed by a *right* shift of the partial product [14]. This gives a maximal carry ripple length of the parallel additions corresponding to the length of the multiplicand.

In the algorithm shown below we scan the multiplier from the *most* significant digit and the partial product is *left* shifted. In an ordinary multiplication this would result in a maximal carry ripple length equal to the sum of the multiplier length and the multiplicand length. But since we perform a modulo reduction on the partial product in every iteration, the length of the intermediate operands will be limited to n plus a few digits. Assume we scan the multiplier k bits at a time, corresponding to processing a digit in radix 2^k, we can

express the serial-parallel multiplication scheme as in Algorithm 2.

```
S := 0; i := n' − 1;
WHILE i ≥ 0 DO
        S := (2^k S + a_i B) mod N;
        i := i − 1;
END;
```

Algorithm 2: *Serial-parallel multiplication with integrated modulo reduction.*

In this algorithm S is the accumulator, n' the number of radix 2^k digits, a_i is digit number i of the multiplier, B the multiplicand and N the modulus.

The modulo reduction can be carried out by *interleaving the multiplication with a division*. Division is usually performed by inspecting the partial remainder and subtracting a multiple of the divisor, followed by a *left* shift of the resulting partial remainder. In Algorithm 3 the variable S has the dual role of a partial remainder in a SRT division scheme [14] and of a partial product in a serial-parallel multiplication. This is the reason why we want to scan the multiplier from the most significant digit and left shift the partial product. Note that this

```
S := 0; i := n' − 1;
WHILE i ≥ 0 DO
        q := Estimate(S div N);
        S := 2^k S + a_i B − 2^k q N;
        i := i − 1;
END;
Correction of S;
```

Algorithm 3: *Modulo multiplication with quotient estimation.*

method is only feasible if we are able to generate the multiples $a_i B$ and qN rapidly.

Instead of calculating the *exact* value of the quotient digit, which is a tedious task for long operands, we *estimate* a value of q. This, of course, implies that the final result is not necessarily completely modulo reduced, and a correction must be performed after the loop by subtracting N until S belongs to the correct interval. It is then required that the precision of the estimate is chosen such that the range of S does not diverge during a computation. Assuming that

$$
\begin{aligned}
a &\in \{0, 1, \ldots, 2^k - 1\} \\
B &\in [0; 2N[\\
q &\in \{0, 1, \ldots, q_{max}\}
\end{aligned}
$$

we can derive a range restriction, which must be satisfied by the estimated q, in the following way

$$2^k S + aB - 2^k qN \leq q_{max} N$$
$$S - qN \leq \frac{q_{max} N - aB}{2^k} \qquad (2)$$
$$S - qN \leq \frac{q_{max} - 2(2^k - 1)}{2^k} N.$$

As we shall see later we can construct hardware that generates multiples qN efficiently if the range of q belongs to $[0; 42]$. Since q is non negative it is required that S is non negative. We achieve this by restricting $S - qN$ to be non negative. The restriction then implies that the maximal radix 2^k is 16, i.e. the scan factor is limited to 4.

However, we are able to increase the scan factor. The idea is to reduce the contribution of the multiple aB in (2) by choosing a larger divisor. Recall that we just want to modulo reduce the partial product, so we can choose an easily generated multiple of N, e.g. $2^r N$. This gives the following range restriction

$$2^k S + aB - 2^k q 2^r N \leq q_{max} 2^r N$$
$$S - q 2^r N \leq \frac{q_{max} 2^r N - aB}{2^k} \qquad (3)$$
$$S - q 2^r N \leq \frac{q_{max} 2^r - 2(2^k - 1)}{2^k} N.$$

Since we want to generate multiples aB with the same hardware as qN, a is limited to 42 and the maximal radix is therefore 32, corresponding to a scan factor of 5. To achieve this, restriction (3) implies that r must be greater than or equal to 1.

We are now ready to present the final algorithm for modulo multiplication. Note that the final corrections can be made by iterating two extra cycles, while setting $a_i = 0$ and furthermore assuming $r = k$.

The final result is read from S discarding the $2k$ least significant bits, and belongs to the interval $[0; 2N[$. A further reduction is not necessary since, according to the stimulation conditions, we can directly start up a new multiplication in the exponentiation algorithm with inputs in the interval $[0; 2N[$. When the exponentiation algorithm terminates the result will also belong to $[0; 2N[$, here a reduction is necessary. This reduction is easily carried out while outputting the result serially. The correctness of Algorithm 4 is proven in [11]. The time complexity is:

$$T[\text{Mult}, n] = (\left\lceil \frac{n+1}{k} \right\rceil + 2) T[\text{iteration}] \qquad (4)$$

In the rest of this paper we will describe how to perform the central operations of the loop, and take a closer look at the hardware architecture of the multiplication unit.

Algorithm:
  Modulo multiplication.

Stimulation:
  $A = a_{n'-1} a_{n'-2} \cdots a_0 a_{-1} a_{-2}$,
  where $a_i \in [0; 2^k - 1]$,
    $a_{-1} = a_{-2} = 0$,
    $n' = \lceil \frac{n+1}{k} \rceil$;

  B, where $B \in [0; 2N[$;
  N, where $N \in]2^{n-1}; 2^n[$;
  k, where $k \geq 3$;
  r, where $r = k$.

Response:
  S div $2^{2k} \equiv_N AB$ and
  S div $2^{2k} \in [0; 2N[$.

Method:
  $S := 0; i := n' - 1$;
  WHILE $i \geq -2$ DO
    $\{*S \equiv_N (A \text{ div } 2^{k(i+1)}) \cdot B*\}$
    $q := \text{Estimate}(S \text{ div } 2^r N)$;
    $S := 2^k S + a_i B - 2^{k+r} qN$;
    $i := i - 1$;
  END;

Algorithm 4: *Modulo multiplication.*

4 Calculation of $2^k S + aB - q 2^{k+r} N$

Because we are dealing with very long operands we use redundant carry save adders. This implies that the result of a multiplication is represented in *two* words. To avoid an area or time consuming carry-completing adder in the circuit we represent the multipliers and multiplicands in carry save form during the *complete* computation of an exponential. The modulus is represented in 2'complement form and in one word. We get the expression

$$2^k (S_s + S_c) + a(B_s + B_c) - q 2^{k+r} N \qquad (5)$$

The multiplier digit a in radix 32 is recoded from the carry-save representation of A through two levels. The 5 digit positions of A is interpreted as three digits in radix 4: d_2, d_1, d_0, where $d_2 \in [0; 2]$ and $d_1, d_0 \in [0; 6]$. These are first recoded into digit sets $[-1; 1]$ repectively $[-1; 3]$, possibly generating and absorbing carries. Secondly these digits are recoded into $\alpha_2, \alpha_1, \alpha_0$ with $\alpha_i \in [-1; 2]$, where α_2 may absorb a carry without generating a carry out. Hence the radix 32 digit a is represented as:

$$a = 4^2 \alpha_2 + 4\alpha_1 + \alpha_0, \quad \alpha_i \in \{-1, 0, 1, 2\}, \qquad (6)$$

thus aB can be computed as the sum of three shifted versions of B in a multiplexing network as shown in

Figure 5. The result is again represented in carry save form $(aB)_s$ and $(aB)_c$. The computation of $-qN$ is performed the same way, noting that $q \in [0; 42]$ can be represented in radix 4 as in (6).

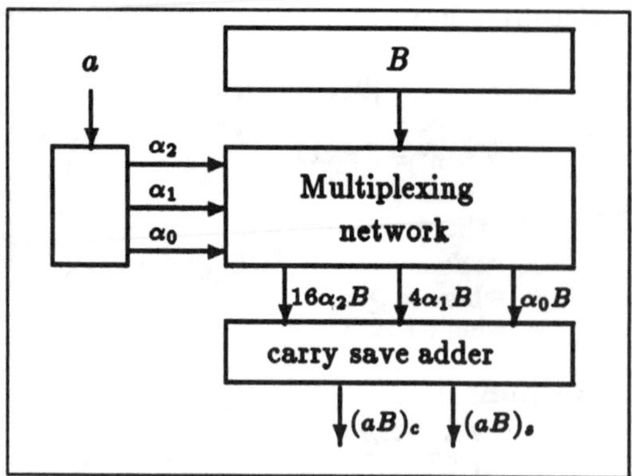

Figure 5: *Unit for generating a multiple.*

Expression (5) is now expanded to the form

$$
\begin{aligned}
&2^k(S_s + S_c) \\
&+((aB_s)_s + (aB_s)_c) + ((aB_c)_s + (aB_c)_c) \\
&+2^{k+r}((-qN)_s + (-qN)_c)
\end{aligned}
\quad (7)
$$

The cost of keeping the operands in carry save form is that we get an expression of eight terms instead of three terms. To reduce the hardware we perform the computation in a pipelined fashion, and share a single unit for generating multiples and a single 4-2 adder for summing terms. The adder is constructed of two carry save adders. In Figure 6 the timing of the computation of expression (7) is illustrated. Each row shows the activity of a hardware component by enclosing the activity in a box. In the left column the components are described. A single iteration of the loop in Algorithm 4 is computed by six cycles of the hardware. The iterations are overlapped, as illustrated by the dashed lines, resulting in a throughput of one iteration per 3 hardware cycles. (U_s, U_c), (V_s, V_c) and (S_s, S_c) denote registers in the pipeline for saving results in carry save form. The computation is performed from left to right.

5 Estimation of S div $2^r N$

Several implementations or suggestions of how to implement the quotient estimation have been presented in the past. According to [1] the estimate can be found by multiplying a few of the most significant digits of the partial remainder S and the reciprocal of the divisor $2^r N$. This assumes that the necessary amount of digits of $\frac{1}{2^r N}$ is part of input to the chip or that it is computed on the chip. The standard approach for SRT division, or as extended in [6], is to use table lookup, implemented as a PLA circuit. This method seems to be infeasible for radices as high as in this paper. We will follow the approach in [3] which is based on what we identify as a "parallel exhaustive search" for the quotient digit. In parallel we compute the sign of resulting partial remainders when the quotient digit assumes all possible values, i.e. we perform in parallel

$$S - q2^r N, \quad q \in \{0, 1, \ldots, q_{max}\},$$

where S is in carry save form. The sign is detected as the carry out of the computation. A high carry out indicates a non negative partial remainder. Then we determine the quotient digit as the smallest value of q which results in a positive remainder. Since the sign computation involves a carry ripple we only use a few of the most significant digits of S and $-q2^r N$, and consequently the quotient digit will just be an *estimate*. The necessary number of digits is determined by the range restriction (3). In Figure 7 is shown a single cell in the quotient estimation unit for calculating the sign of $S - q2^r N$. There is one cell for each possible value of q. In the figure we denote by p the position of the most

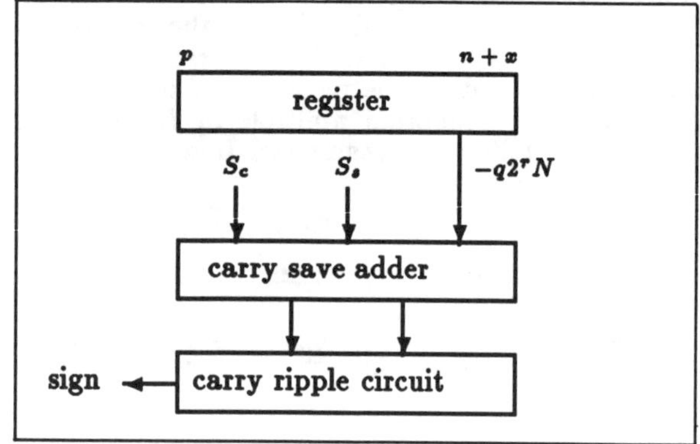

Figure 7: *A cell in the quotient estimation unit.*

significant bit, the sign bit, in S_s, S_c and in $-q2^r N$. According to restriction (3) $p = n + r + \lceil \log_2 q_{max} \rceil$ and for $q_{max} = 42$ this gives $p = n + r + k + 1$. n is the bit length of N, and $n + x$ denotes the position of the least significant bit in the quotient estimation. Since the weights of the discarded parts of S_s, S_c and $-q2^r N$ are all positive and belongs to $[0; 2^{n+x}[$, this computation gives the following range for the estimated value of q:

$$
\begin{aligned}
S - q2^r N &= S' + (-q2^r N)' + \varepsilon \\
&\in [0; 2^r N + 3 \cdot 2^{n+x}[,
\end{aligned}
$$

where S' and $(-q2^r N)'$ denotes S and $-q2^r N$ with the bits from 0 to $n + x - 1$ set to zero. Now x can be determined from the range restriction (3), where the redundant digit set $\{0, 1, \ldots, q_{max}\}$ is assumed for the multiplier:

$$
\begin{aligned}
2^r N + 3 \cdot 2^{n+x} &< \frac{q_{max} 2^r - 2q_{max}}{2^k} N \\
2^x &< \frac{1}{6}(\frac{2^r - 2}{2^k} q_{max} - 2^r),
\end{aligned}
$$

where the smallest value 2^{n-1} of N has been inserted.

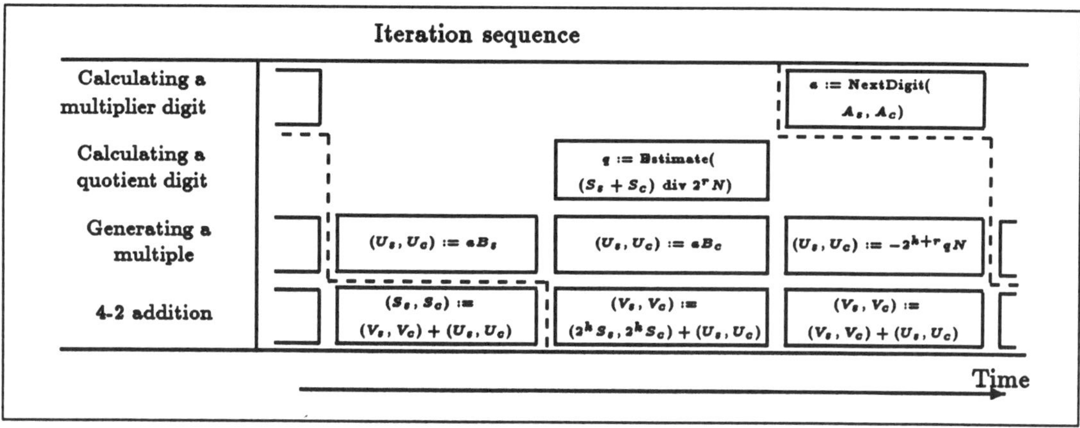

Figure 6: *Timing diagram for an iteration in the multiplication algorithm. The iterations are overlapped.*

Substituting 5 for r and k, and 42 for q_{max} we get $x \leq 0$. Thus the necessary number of bits in the quotient estimation is $p - (n + x) + 1 = 12$ in the case of radix 2^5.

This method for quotient estimation seems very area consuming but compared to the size of the multiple generating unit and the 4-2 adder for operands longer than 500 bits this area is reasonable. As usual in VLSI design a high degree of concurrency results in faster circuits at the cost of area.

6 Performance

Equation (1) and (4) gives an expression for the computing time of an exponentiation

$$T[\text{Exp}, n] = n\left(\left\lceil \frac{n+1}{k} \right\rceil + 2\right)T[\text{iteration}].$$

As explained in Figure 6 an iteration in the multiplication loop is performed in three cycles through the hardware. Anticipating that the quotient estimation unit is the critical path in the circuit we get

$$T[\text{Exp}, n] = n\left(\left\lceil \frac{n+1}{k} \right\rceil + 2\right)3T[\text{quotient estimation}].$$

A cell of the quotient estimate unit has been designed in a 2μ CMOS process and simulations shows a delay less than 20 ns. For $n = 512$ and $k = 5$ we achieve a computing time of 3.2 ms, corresponding to a bit rate of

$$\frac{n}{T[\text{Exp}, n]} = 159\frac{\text{Kbit}}{\text{sec}}.$$

Compared to the hardware implementation from Thorn EMI [15] with a bit rate of 29 $\frac{\text{Kbit}}{\text{sec}}$ this design improves the speed by a factor of more than 5.

The calculation on computing time assumes that we have implemented the parallel version of the exponentiation algorithm. This can be done in two ways: By replicating the multiplication unit or by pipelining a single multiplication unit. With respect to area the last approach is preferable. Observing that the two parallel multiplications have the modulus N and the multiplicand B in common we only need to add extra registers for a multiplier in carry save form and latches to implement the pipeline. An iteration in the multiplication loop now consist of six clock cycles at approximately 10ns. Clock frequencies as high as this can be hard to achieve. A way to avoid the use of a clock to synchronize the circuit is to use self-timed circuit schemes [16].

The VLSI design of the hardware components shows high regularity and the area for wiring is minimized through the use of carry save adders. At the expense of regularity and area, the speed of quotient estimation can be increased by replacing the carry ripple circuit in Figure 7 by carry look-ahead circuit.

We can obtain a rough estimate of the area by comparing the proposed architecture to the one described in [12], which has been laid out using a silicon compiler: The area is approximately 200 mm^2 in a 1.2μ CMOS process technology for a chip capable of modulo exponentiating 561 bit operands. The architecture presented here include *one* unit for generating multiples and *one* 4-2 adder where [12] include two of each and additionally a 561 bit ripple adder. The adder is used to convert the result of a multiplication from carry save form to a non redundant representation. Taking into account the extra latches for pipelining a multiplication unit we believe that the area will be less than 200 mm^2 in a 1.2μ process if we use the silicon compiler and its library cells. We can reduce the area significantly by making a full custom layout of the design, since the library cells are designed in a conservative manner using static registers and static logic gates. Another way to reduce the area (and the computing time) is by choosing a smaller process technology, e.g. 0.8μ which approximately halves the area.

All of the fastest implementations in Brickells survey [4] include more than one chip. Cryptechs 712 bit solution [7] comprise 6 chips, where each chip contains a datapath for 120 bit. Thorn EMIs 768 bit solution [15] comprise a controller chip and 3 datapath chips for 256 bit each.

7 Summary

We have presented a way to speed up a well known exponentiation algorithm by performing two multiplications in parallel, and we have shown how these multiplications can be performed efficiently using high radices. Further more we have developed a highly regular hardware architecture, based on the redundant carry save addition technique, implementing the multiplication algorithm with a radix of 32. Simulations indicates a resulting speed improvement of more than 500% compared to other known implementations.

Currently we are working on generalizing the multiplication to even higher radices. By expressing the quotient and multiplier in a symmetric redundant digit set it seems simple to modify the hardware architecture to radix 64.

References

[1] Paul Barrett. Implementing the Riverst Shamir and Adleman public key encryption system on a standard digital signal processor. In *Advances in Cryptology - CRYPTO '86*, pages 311–323, 1986.

[2] G.R. Blakely. A Computer Algorithm for Calculating the Product AB Modulo M. *IEEE Trans. Computers*, C-32:497–500, 1983.

[3] Ernest F. Brickell. A fast modular multiplication algorithm with applications to two key cryptography. In D. Schaum, R.L. Riverst, and A.T. Sherman, editors, *Advances in Cryptology, Proceedings of Crypto '82*, pages 51–60, New York, 1982. Plenum Press.

[4] Ernest F. Brickell. A Survey of Hardware Implementations of RSA. In Gilles Brassard, editor, *Advances in Cryptology - CRYPTO '89*, pages 368–370. Springer-Verlag, 1990.

[5] W. Diffie and M.E. Hellman. New directions in cryptography. In *IEEE Trans. on Info. Theory*, volume IT-22(6), pages 644–654, Nov. 1976.

[6] Jan Fandrianto. Algorithm for high speed shared radix 8 division and radix 8 square root. In *Proceedings of the 9th Symposium on Computer Arithmetic*, pages 68–75. IEEE, 1989.

[7] Frank Hoornaert, Marc Decroos, Joos Vandewalle, and René Govaerts. Fast RSA-Hardware: Dream or Reality ? In *Advances in Cryptology - EUROCRYPT '88*, pages 257–264, 1988.

[8] Donald E. Knuth. *The Art of Computer Programming - Seminumerical Algorithms*, volume 2. Addison-Wesley, 2. edition, 1981.

[9] Hikaru Morita. A Fast Modular-multiplication Algorithm based on a Higher Radix. In Gilles Brassard, editor, *Advances in Cryptology - CRYPTO '89*, pages 387–399. Springer-Verlag, 1990.

[10] G.A. Orton, M.P. Roy, P.A. Scott, and L.E. Peppard. VLSI implementation of public-key encryption algorithms. In *Advances in Cryptology - CRYPTO '86*, pages 277–301, 1986.

[11] Holger Orup and Erik Svendsen. VICTOR. Forbedringer og videreudviklinger af VICTOR - en integreret kreds til understøttelse af RSA-kryptosystemer. Computer Science Department of Aarhus University - Internal report, 1990.

[12] Holger Orup, Erik Svendsen, and Erik Andreasen. VICTOR an Efficient RSA Hardware Implementation. In I.B. Damgård, editor, *Advances in Cryptology - EUROCRYPT '90*, pages 245–252. Springer-Verlag, 1991.

[13] Ronald L. Riverst, A. Shamir, and L. Adleman. A method for obtaining digital signatures and public-key cryptosystems. In *Communications of the ACM*, volume 21, pages 120–126, Feb. 1978.

[14] Norman R. Scott. *Computer Number Systems & Arithmetic*. Prentice-Hall, 1985.

[15] THORN EMI. RSA Evaluation Board. Technical Report 10, Thorn EMI Central Reasearch Laboratories, 1988.

[16] T.E. Williams, M. Horowitz, R.L. Alverson, and T.S Yang. A self-timed chip for division. In Paul Losleben, editor, *Advanced Research in VLSI. Proceedings of the 1987 Stanford Conference*, pages 75–95. The MIT Press, 1987.

Session 3:

Inner Products

Chair:
Svetoslav Markov
Bulgarian Academy of Science

Arithmetic for Digital Neural Networks

D. Zhang, G. A. Jullien, and W. C. Miller
VLSI Research Group
Department of Electrical Engineering
University of Windsor
Windsor, Ontario, CANADA N9B 3P4

Earl Swartzlander, Jr.
Dept. of Electrical and Computer Engineering
University of Texas
Austin, TX 78712
USA

Abstract

This paper describes the implementation of large input digital neurons using designs based on parallel counters. The implementation of the design uses a two-cell library, in which each cell is implemented using "Switching Trees" which are pipelined binary trees of n-channel transistors. Results obtained from initial switching trees realized with a 3-μm CMOS process indicate that the design is capable of being pipelined at 40 MHz sample rates with better performance expected for more advanced technologies. It appears feasible to develop a wafer scale implementation with 2000 neurons (each with 1000 inputs) that would perform over 3×10^{12} additions/second.

1 Introduction

Over the past several years, neural networks have become a subject of great interest due to their massive parallelism, modularity, and robustness. Current research in the area of neural networks embraces modeling, algorithms, architectures, and implementations [1], [2].

There have been a variety of implementation approaches for neural networks, including software and hardware [3], electronic and optical implementations [4], VLSI [5], and discrete component realizations. The real promise for applications of neural networks currently lies in specialized hardware, in particular VLSI or WSI implementations that achieve high levels of computational performance with modest development effort [6].

The high interconnectivity and relatively low signal precision requirements of neural networks appear well tailored to analog realizations. With the increasing size of neural networks, however, it is becoming increasingly difficult to achieve acceptable matching of gains and thresholds on the analog integrated circuits. On the other hand digital realizations, which superficially appear to be inferior in terms of computational density, have advantages of flexibility in terms of programming for a variety of architectures and learning strategies, as well as simplifying the memory function. Currently, there is a strong development effort, to implement neural networks in VLSI using digital technology [7]-[9].

In digital implementations the requirements for parallel digital arithmetic computation are severe. As an example, for a fully connected layered network with two layers and n neurons per layer, each neuron is required to form an inner product of n elements using 1-bit binary input lines and m-bit weights (e.g., m=16). Thus, a key problem for neuron design is to build an efficient multi-input gated adder structure suitable for m-bit inputs. An interesting approach is to use a parallel counter [10]-[12], and this paper presents such an implementation, using a new pipelined dynamic CMOS logic family based on *Switching Trees.*

2 Neural Networks Implemented with Parallel Counters

A three input neural network is shown in Figure 1. The k-th digital neuron on the i-th layer of the n neuron/layer neural network evaluates the following inner product:

$$Y_{i+1,k} = F_{nl}\left(W_{i,n,k} + \sum_{j=0}^{n-1} Y_{i,j}\, W_{i,j,k}\right)$$

Where: $Y_{i,j}$ (for j=0,1,...,n-1) is the input vector and $W_{i,j,k}$ is the weight vector, and F_{nl} is a non-linear function. When the input vector consists of one bit binary components, the sum of products reduces to a multioperand addition. A quasi serial [11], [12] realization of a 1000 input neuron is shown on Figure 2. The weights, $W_{i,j,k}$, are stored in recirculating shift registers that are accessed beginning with the LSB. A 1010 input parallel counter sums the 1000 data inputs, $W_{i,j,k}$ (for j=0, 1,...,999), the weight that sets the neuron's threshold $W_{i,1000,k}$, and the nine carry bits that arise in the counting process. A process was given in [10] for the implementation of 2n + 1 input parallel counters with

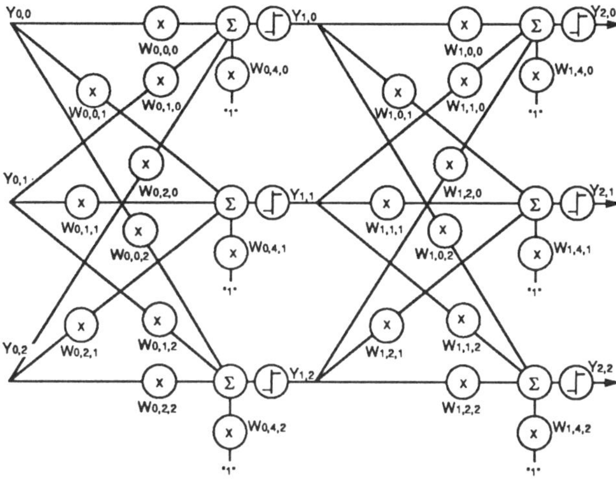

Figure 1. Example of a Two Layer Fully Connected Neural Network.

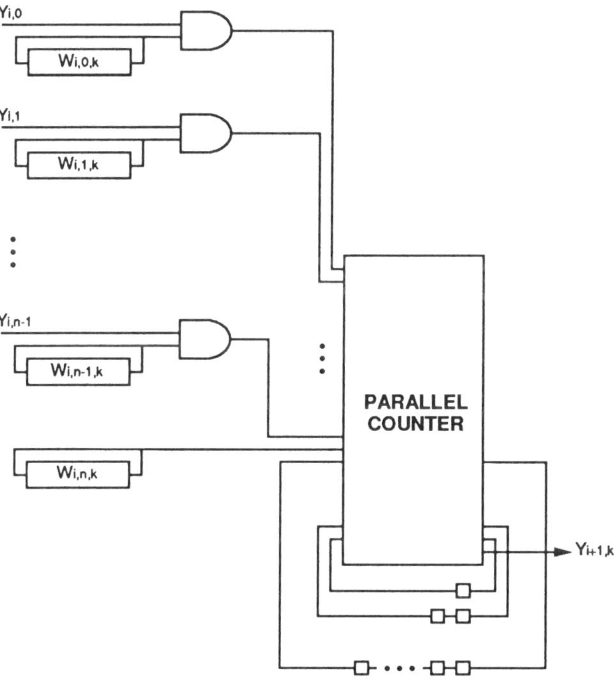

Figure 2. Artificial Neuron Implementation.

two n input parallel counters. This process can be applied recursively resulting in an implementation that uses only full adders (which are in fact three input parallel counters). A 1010 input parallel counter constructed with the recursive procedure uses 1000 full adders and requires nine adder delays to generate the LSB of the count and 17 adder delays to generate the MSB of the count. In succeeding sections of this paper, switching trees are used to construct a significantly faster counter that is comparable in complexity.

3 Switching Tree Implementation

Implementation of the parallel counter is based on the application of switching trees [13]. An unminimized switching tree is a binary decision tree, directly implemented as an n-channel MOSFET logic block in a complex CMOS pipelined dynamic gate circuit. We denote the tree as a graph, $G=\{X, V, L\}$, where X is the edge set (NMOS transistors), V is the vertex set (interconnecting nodes), and L is the link set (shorting links) of G. Let $\{x_i, x'_i \mid i=0, 1, ..., n-1\} \in X$, where x_i and x'_i are defined as two different directions, $\diagdown$ (n-channel transistor driven by the logical complement) and $\diagup$ (n-channel transistor driven by the logical true), respectively. Let $\{v_i \mid i=0, 1, ..., n\} \in V$, where v_0 is a root and v_n is a leaf. Let $\{v_i \cdots v_j \mid i,j =0, 1, ..., n\} \in L$, where a link, $v_i \cdots v_j$, is denoted as a dotted line from v_i to v_j.

A path, P, is a connection from the root, v_0, to a leaf, v_n, constructed by the edges and the links. A 1-path corresponds to programmed bits in the truth table of the logic function. In a switching tree, all true paths are 1-path and complement paths do not terminate at the bottom edge. A height of the tree is the number of edges of the longest path (only including the maximum edges). A link, $v_i \cdots v_j$, is an connection between v_i and v_j if these vertices are equivalent. Note that all leaves can be linked.

Given the truth table of a binary function, a switching tree can be constructed, and by applying simple graph theory rules, a minimized tree can be obtained. It is interesting that the same concept was introduced by Shannon [14] for implementing relay logic over 50 years ago. The current minimization procedures and optimization constraints are, however, somewhat different. The switching tree can be directly implemented by a form of pipelined dynamic logic. By using such a direct mapping procedure, the design is optimized for specific silicon cost functions such as area and time. A direct constraint is the height of the switching tree. This height is optimized by balancing the conflicting requirements of area minimization, charge sharing, noise immunity, pipeline cycle time and pipeline latency against the efficiencies inherent in the decomposition of the problem at hand. The designs presented in this paper are based on a switching tree height of seven transistors; this number yields immediate efficiencies in the parallel counter decomposition, with a respectable 25 ns pipeline period for a 3-μm CMOS process.

The pipeline structure used here is based on work by Yuen and Svennson [15], in true single phase (TSP) clocking strategies. The implementation shown here only uses n-channel transistor logic networks (the switching tree). The p-channel output stage is simply an inverter function instead of a p-channel tree. Although the TSP clock strategy was originally employed to obtain clocking periods of only a few ns, the present implementation trades such high clock rates for increased complexity of switching logic at each evaluation node, and hence a much smaller number of series nodes to complete a given logic function. The latency of these systems will typically be lower than that of applying the TSP clocking to minimum complexity evaluating nodes (e.g., 2-input gates), even though the clock rate is slower. From simulation experiments, it also has been determined that the pipelining of complex single evaluation nodes can be performed at higher throughput rates than by using Domino logic evaluation stages between pipeline latches. These design decisions are based on simulation and a large amount of silicon verification performed over the past several years with the 3-μm CMOS process. The simulations are based on mask extracted data with semi-empirical (level-3) SPICE models tuned from fabrication tests of individual transistors. A more aggressive (i.e., sub-micron CMOS) technology may be expected to improve the performance significantly.

4 Counter Cell Design

A single stage dynamic logic cell is defined as a building element for switching tree implementations. Each node of the logic is pipelined; a switching tree structure is used as the complex path to ground for that node. Given the 7-transistor height for the switching tree, two cells serve as the basis for the large parallel counter design.

4.1 Seven Input Parallel Counter Cell

A seven input parallel counter with three outputs is the primary building block for large counter implementation. Its switching tree structure and the corresponding domino logic cell are shown in Figure 3. The inputs are x_0, x_1, x_2, x_3, x_4, x_5, and x_6, and the outputs are s_2 (MSB), s_1, and s_0 (LSB). This cell is used to perform most of the reduction in the large parallel counter. It is attractive because each use of it reduces the number of bits in the bit matrix by four (there are seven inputs and three outputs) whereas each use of a full adder in [10] only reduces the number of bits by one.

4.2 Four Bit Word Adder Cell

The switching tree for a two word by four bit adder and its domino logic cell are shown in Figure 4. This cell is used to perform horizontal reduction in the counter. The

cell receives two four bit words X (= x_3, x_2, x_1, x_0) and Y (= y_3, y_2, y_1, y_0) and a carry c_0 as inputs and produces a five bit output word, c_4, (MSB) s_3, s_2, s_1, s_0 (LSB).

5 Parallel Counter Structure

Based on the seven input counter and the four bit word adder cells, a large parallel counter design using switching trees is developed in this section. A two stage reduction process is employed: first the seven input parallel counter cells are used to reduce the initial 1010 row matrix to a matrix with no more than three elements in each column (i.e., no more than three rows where the second and third rows may be sparse). Then the four bit word adder cells are used to complete the reduction to a single row.

5.1 Initial Reduction

The design process for the large parallel counter is similar to that used by Dadda to develop high speed parallel multipliers [16], [17]. Specifically since the seven input counter has three outputs, the following sequence is used to set the maximum height of the columns in the bit matrices: 3, 7, 15, 35, 79, 183, 427, etc. Each entry is determined by multiplying seven times the integer quotient of the previous entry divided by three and adding any remainder from the division. Thus the third entry was determined as $7 \lfloor 7/3 \rfloor + 1 = 15$. Instead of strictly following Dadda's method which involves reducing the height of each column to no more than the appropriate value from the sequence, greater levels of reduction were performed in the early stages of the reduction. This minimizes the number of latches required to pipeline the counter.

The initial reduction is illustrated by the spread sheet of Figure 5. Each **Bold** entry indicates the number of one bit data located in that column of the matrix. For example, the input matrix (MATRIX 0) has 1010 rows in column 1. Immediately below that entry is the indication that 144 seven input parallel counter cells (indicated as 7:3 counters on Figure 5) are used producing 146 outputs in column 1, and 144 outputs each in columns 2 and 3.

The three sets of outputs from each group of parallel counters are connected with a diagonal line on the diagram. Thus MATRIX 1 has 144 rows in columns 2 and 3 and 146 rows in column 1, etc. Strict application of Dadda's method would have left column 1 with 182 entries and columns 2 and three with 138 entries each. MATRIX 1 for this approach has 434 data whereas Dadda's approach produces a MATRIX 1 with 458 points. Six stages of reduction employing 252 seven input

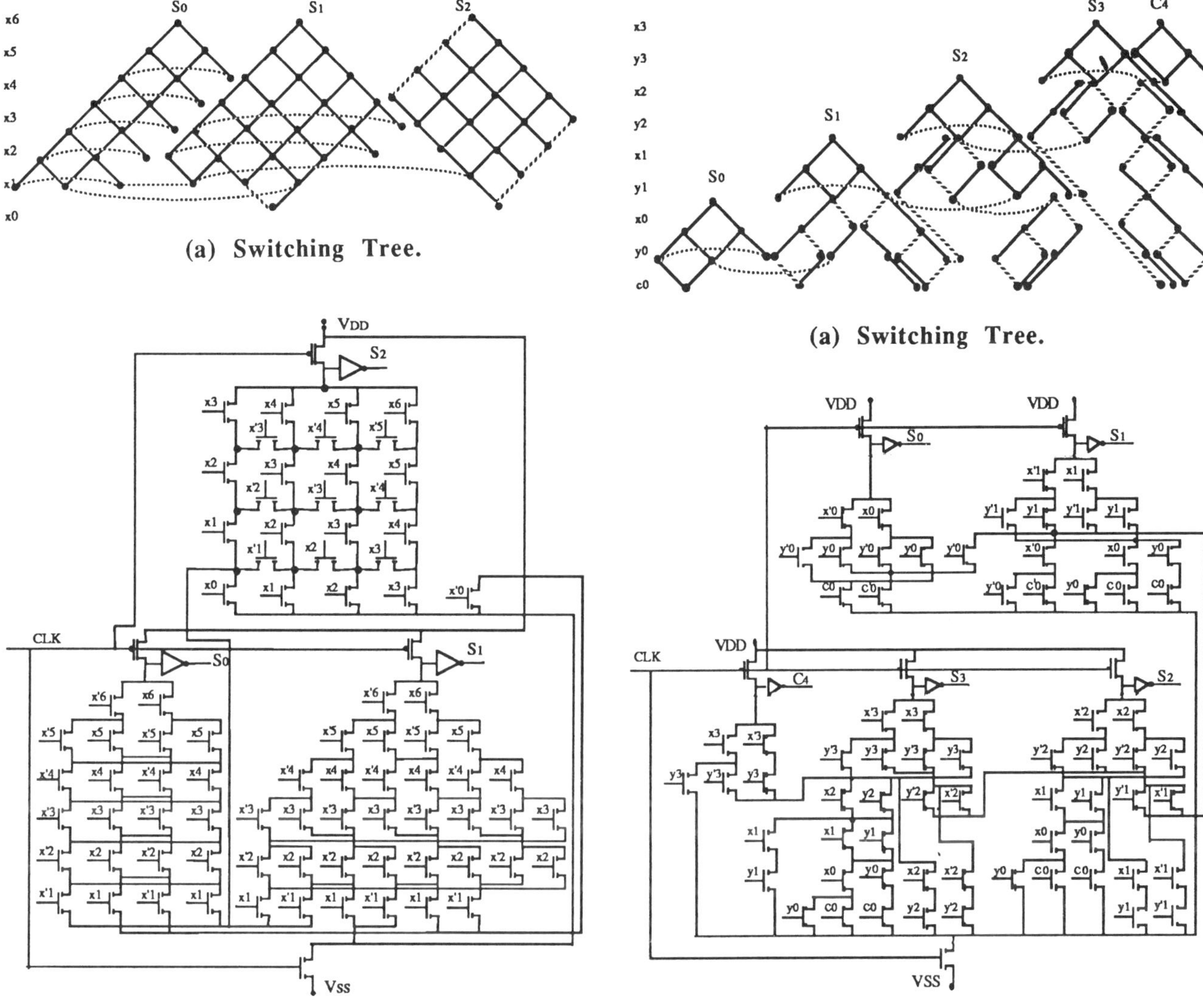

(a) Switching Tree.

(a) Switching Tree.

(b) Domino Logic.

(b) Domino Logic.

Figure 3. Seven Input Parallel Counter.

Figure 4. Four Bit Word Adder.

parallel counter cells produce MATRIX 6 which is ten columns wide and no more than three rows deep. This output is sufficient to express a count of 2003 which indicates that not all bits can be one simultaneously. This error results from the use of seven input counters in places where there are fewer than seven inputs.

5.2 Final Reduction

The final reduction shown in Figure 6 uses four four bit word adders in three steps to reduce MATRIX 6 from Figure 5 to a single eleven bit word (MATRIX 10). At the present time, the design process for this final reduction is a cut and try exercise in geometric reduction.

5.3 Counter Performance

This counter is implemented with 252 seven input parallel counters and four four bit word adders. Since each of these cells is three to four times the complexity of a full adder, the switching tree implementation and the full adder implementation (which requires 1000 full adders) are comparable in total complexity. The switching tree cells are comparable in delay to a full adder, so that a non-pipelined implementation of the switching tree counter (which requires ten cell delays) should be nearly twice as fast as the full adder implementation (which requires 17 full adder delays). Pipelined implementations should be able to operate at comparable rates

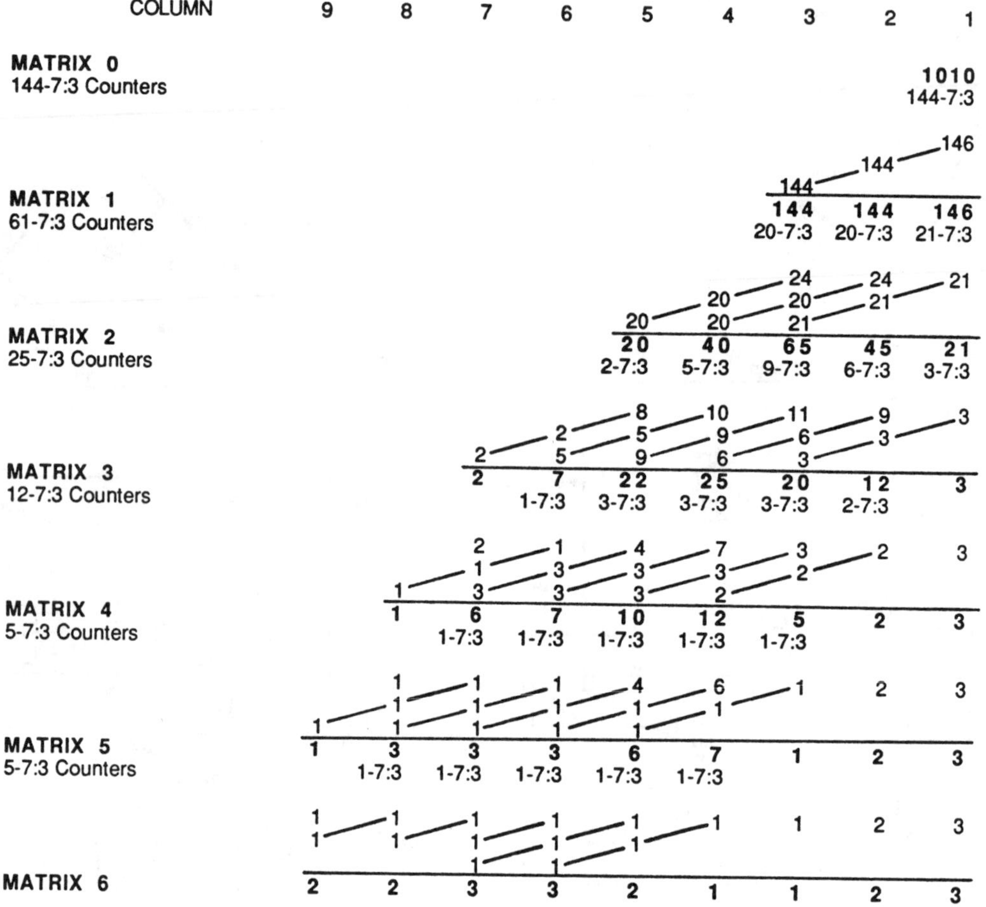

Figure 5. First Stage in the Bit Matrix Reduction Process.

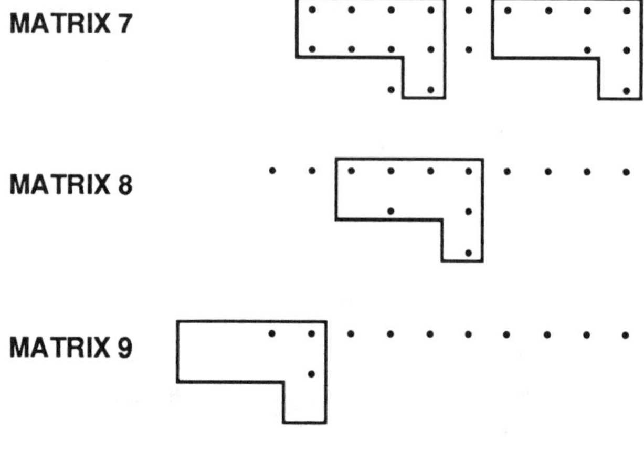

Figure 6. Second Stage in the Bit Matrix Reduction Process.

5.4 Neural Network Performance

Each cycle of the neural network requires 25 cycles of the parallel counter (16 for the 16 bit weights and nine to assimilate the carries). At a 25 ns clock rate, the time for a neuron cycle is 625 ns. If a wafer scale integration neural network is constructed with 2000 neurons (1000 on each of two layers) as appears feasible [14], the wafer should be able to attain 3×10^9 neuron cycles/second. Since each neuron cycle involves summing 1000 weighted data, the total computation rate is in excess of 3×10^{12} additions/second.

Conclusions

In this paper, a novel switching tree structure for parallel counter arithmetic of digital neural networks is introduced. Only two types of cells are used in the counters, with a total of 256 cells required (252 seven input parallel counters and four four bit word adders). Simulation results suggest that the structure can achieve a total computation rate in excess of 3×10^{12} additions/second.

Acknowledgements

The first three authors acknowledge support from the Natural Sciences and Engineering Research Council of Canada for funding their portion of this work.

References

[1] R. Eckmiller and C. V. D. Malsburg, *Neural Computers*, Springer Verlag, 1988.

[2] C. Lau and B. Widrow, eds., "Special Issues on Neural Networks, I and II," *Proceedings of the IEEE* Vol. 78, September and October, 1988.

[3] J. Ghosh and K. Hwang, "Mapping Neural Networks onto Message-Passing Multicomputers," *Journal of Parallel and Distributed Computing*, Vol. 5, 1988.

[4] R. H. Nielsen, "Performance Limits of Optical, Electro-Optical, and Electronic Neurocomputers," *Proceedings of the SPIE*, Vol. 634, pp. 277-306, 1986.

[5] H. P. Graf and P. deVegvar, "A CMOS Implementation of a Neural Network Model," *Proceedings of the Stanford Conference on Advanced Research on VLSI*, MIT Press, pp. 351-367, 1987.

[6] M. A. Sivilotti, M. R. Emerling and C. Mead, "VLSI Architectures for Implementation of Neural Networks," *Proceedings of the AIP Conference*, Vol. 151, P. 408, 1986.

[7] H. P. Graf, L. D. Jackel and W. E. Hubbard, "VLSI Implementation of a Neural Network Model," *Computer*, pp. 41-49, March, 1988.

[8] D. Zhang, G. A. Jullien and W. C. Miller, "Mapping Neural Networks onto Systolic Arrays," *IEEE Transactions on Circuit and Systems*, (in print), 1990.

[9] M. Griffin, *et al.*, "An 11-Million Transistor Neural Network Execution Engine," *IEEE International Solid-State Circuits Conference Digest of Technical Papers*, pp. 180, 181, 313, 1991.

[10] E. E. Swartzlander, Jr., "Parallel Counters," *IEEE Transactions on Computers*, Vol. C-22, pp. 1021-1024, 1973.

[11] E. E. Swartzlander, Jr., "The Quasi-Serial Multiplier," *IEEE Transactions on Computers*, Vol. C-22, pp. 317-321, 1973.

[12] E. E. Swartzlander, Jr., B. K. Gilbert, and I. S. Reed "Inner Product Computers," *IEEE Transactions on Computers*, Vol. C-27, pp. 21-31, 1978.

[13] D. Zhang, G. A. Jullien and W. C. Miller, "Switching Tree Structures for VLSI Implementations," submitted to *IEEE Transactions on Computers*, 1990.

[14] C. E. Shannon, "A Symbolic Analysis of Relay and Switching Circuits," *AIEE Transactions*, Vol. 57, pp. 713-723, 1938.

[15] J. Yuan and C. Svennson, "High-Speed CMOS Circuit Technique," *IEEE Journal of Solid-State Circuits*, Vol. 24, pp. 62-71, 1989.

[16] L. Dadda, "Some Schemes for Parallel Multipliers," *Alta Frequenza*, Vol. 34, pp. 349-356, 1965.

[17] L. Dadda, "On Parallel Digital Multipliers," *Alta Frequenza*, Vol. 45, pp. 574-580, 1976.

Exact Accumulation of Floating-Point Numbers

Michael Müller

Christine Rüb

Wolfgang Rülling

Max-Planck-Institut
für Informatik
D-6600 Saarbrücken

Max-Planck-Institut
für Informatik
D-6600 Saarbrücken

Fachhochschule
Furtwangen
D-7743 Furtwangen

Abstract

We present a new idea for designing a chip which computes the exact sum of arbitrary many floating-point numbers, i.e. it can accumulate the floating-point numbers without cancellation. Such a chip is needed to provide a fast implementation of Kulisch-arithmetic. This is a new theory of floating-point arithmetic which makes it possible to compute least significant bit accurate solutions to even ill conditioned numerical problems. Our approach avoids the disadvantages of previously suggested designs which are too large, too slow or consume too much power. The crucial point is a technique for a fast carry resolution in a long accumulator. It can also be implemented in software.

1 Introduction

Everybody who has implemented an algorithm using arithmetic on "real" numbers knows the problems arising from approximating reals by floating-point numbers. Even in very simple geometric algorithms, inaccurate calculations can lead to a crash of the program or to completely wrong results.

The major reasons for inexact calculations in computers are cancellation occurring in subtractions and truncation after multiplications. Also, during the huge computations performed on modern vector processors very small errors have enough time to accumulate until the result has nothing in common with the desired solution of the given problem. Hence, calculating exactly is much more important in those computers, but they still supply no remedy against this problem.

To avoid these problems, Kulisch and Miranker (cf. [14], [15]) derived a new theory of floating-point arithmetic. They propose to provide, besides the four standard basic operations $+$, $-$, $\cdot$, $/$, the inner product of two vectors (denoted as $*$) as a fifth basic operation. This operation should, as well as the standard basic operations, be available in the three rounding modes rounding to nearest ($\square$), rounding towards $+\infty$ ($\triangle$) and rounding towards $-\infty$ ($\triangledown$). These three rounding modes, implemented in a way such that $x \circledast y = \bigcirc(x \star y)$ for $\bigcirc \in \{\square, \triangledown, \triangle\}$ and $\star \in \{+, -, \cdot, /, *\}$, are necessary to derive reliable error estimations (cf. [13]). Here, $\circledast$ denotes the operation actually performed by the machine.

We will refer to the operation $\circledast$ as the **rounded exact inner product** or, for short, **exact inner product**.

This theory allowed the development of new algorithms for the numerical standard problems. These algorithms compute bounds for the (real) solution of the given problem. In general, the bounds are adjacent floating-point numbers, and in this case the result is called **least significant bit accurate**. Even for very ill conditioned problems where traditional numerical methods fail to give a reasonable answer, the new algorithms mostly compute least significant bit accurate results. Also the computed intervals are proven to contain the actual result, i.e. one can rely on them.

To implement these algorithms, one needs the exact inner product and there exist extensions of Pascal and Fortran (PASCAL-SC [4,9] and FORTRAN-SC [6]) and subroutine packages (ACRITH [10], ARITHMOS [20]) providing this operation in software. In [3] one can find a good overview over the state of the art in this field. But as far as we know there still exists no chip for this purpose and thus the acceptance of the new methods is quite small since there is no really fast implementation of the exact inner product. Especially for high performance vector computers a software realization seems not to be competitive.

There have already been several proposals how to design such a chip ([24], [16], [7], [12], [25]), but most of them seem to have no chance of being realized in the near future because they are either too large, consume too much power since in most cycles the whole circuit is active, or too slow since one has to wait a hundred or more cycles to obtain the result after the last input has been accepted. Thus, we present a new strategy to compute the exact inner product of two vectors which promises to avoid the essential disadvantages of the other concepts.

This paper is structured as follows. In section 2 we describe the algorithms which form the basis of our design. In section 2.1 we outline the basic ideas of our algorithms, and in section 2.2 we describe the algorithms in more detail. Section 2.2.1 is dedicated to accepting a new summand and section 2.2.2 treats the problem of rounding the sum to get a floating-point result. Section 3 briefly describes how we intend to realize the algorithms of section 2 in hardware and in section 4 we show how the algorithm can be implemented in software.

64

2 The Algorithm

Suppose that we want to compute the exact inner product of two vectors $v = (v_1, \ldots, v_n)$ and $w = (w_1, \ldots, w_n)$, and that the components of the vectors are represented with $\overline{m}$ bits of mantissa and an exponent range of $[\bar{e}_1 .. \bar{e}_2]$, $\bar{e}_1 \leq \bar{e}_2$. To compute the inner product, we first have to compute the n exact products $v_i \cdot w_i =: z_i$, $1 \leq i \leq n$, where the n products z_i are each represented with $m = 2 \cdot \overline{m}$ bits of mantissa and an exponent range of $[e_1 .. e_2] = [2 \cdot \bar{e}_1 .. 2 \cdot \bar{e}_2]$. Then we have to add the n products z_i, $1 \leq i \leq n$, and round the exact sum according to the desired rounding mode.

The exact product of two floating-point numbers can be computed by standard techniques and we thus concentrate on the exact, i.e. cancellation-free, summation of floating-point numbers. We use the following technique which is also used by [24], [16], [7], [12] and [25]:

> Each summand that arrives at the chip is expanded to a fixed-point number with $m + e_2 - e_1 + 1$ digits. The expanded summands are then added using pipelining. After the last summand has been accepted, remaining carries are removed if necessary, and the computed sum is rounded to the required floating-point format.

The problems that arise when this technique is implemented in hardware originate from the length of the expanded summands. E.g. if the components of the input vectors are single-precision floating-point numbers (we refer to the IEEE standard 754, cf. [11]), the length of their mantissas is 24 and the length of their exponents is 8. Thus, the exact products of the components have mantissas of length 48 and exponents of length 9, and the expanded summands have a length of $48 + 2^9 = 560$. If the components of the input vectors are double-precision floating-point numbers, then their mantissas have a length of 53 and their exponents have a length of 11. Thus, the expanded summands have a length of $106 + 2^{12} = 4202$.

In our design we want to avoid the three essential disadvantages of the previous layouts mentioned in the introduction, i.e. the design should be small, only a small part of the circuit should be active in each cycle to reduce power consumption and the result should be available only a short time after the last summand has been accepted. How this can be achieved is shown in the rest of this section.

2.1 The Basic Ideas

As mentioned above, we reduce the computation of the inner product to the summation of products computed elsewhere. This is done by adding the summands into a long accumulator.

Intuition tells us that it is not necessary to consider all accumulator digits when we add a number covering only a small part of it. This is clearly true if the number is positive and can be added without producing a carry at the left end of the range covered. If this locality would hold for all additions, we could maintain the contents of the accumulator in an ordinary storage, read the part affected by the addition, use a moderately sized adder to do the main work and write the modified section back to the storage. To make this idea work in all cases, we have to solve two major problems:

1. What do we do with the carries?

2. How do we treat negative summands?

Let us first ignore negative summands and concentrate on carry handling. A first approach could be to accumulate the carries as suggested by Kirchner and Kulisch (cf. [12]). But this means that the accumulated carries have to be resolved at the end of the addition which takes a lot of time. Thus, we looked for a way to get immediately rid of the carry resulting from an addition.

Let us inspect carry handling during a conventional addition w.l.o.g. in the binary system. Suppose we add two numbers a and b where b has many leading zeros. Consider what happens to a carry when we reach the leading zeros of b, cf. Figure 1. As long as the corresponding digits in a are 1, the bits in the sum become 0 and the carry propagates. When we reach a 0 in the binary representation of a this digit of the sum becomes 1 and the carry disappears. The following bits of the sum are the same as the leading bits of a. Hence, in order to resolve a carry we only need to find the next zero in a and toggle all bits between this position and the position from which the carry originated.

$$
\begin{array}{l}
XXX\ldots XX0111\ldots 11XXXX\ldots XXX \\
+ 1 \\
\hline
=XXX\ldots XX1000\ldots 00XXXX\ldots XXX
\end{array}
$$

Figure 1: Carry resolution.

To make use of this observation, we use the following strategy. We divide the accumulator into several blocks. To add a new summand, we perform ordinary additions for the blocks intersected by the mantissa of the summand. The less significant blocks remain unchanged. If the most significant addition yields a carry, this carry has to be added to the more significant blocks as sketched above. To avoid stepping through all bits of the more significant blocks, we maintain for every block some information about its contents which describes whether the block consists only of ones or whether there is a zero in it. Thus we know whether the carry passes through a block or whether it is accepted in it. The first block that contains a zero has to be incremented by 1. The blocks consisting only of ones that we have skipped should be inverted, but this is too time and power consuming. Hence, we invert only the contents information and register in a second information bit that the contents of this section of the accumulator are not the contents of the storage but only zeros. This information always has to be considered when we want to read something from the storage and has to be updated when we write a block to the storage. The storage is organized such that a row of the storage corresponds to a block of the accu.

To avoid running sequentially through all information bits, we use the adder for carry propagation as follows. We define a binary number which represents the states of the blocks of the accu. The least significant bit of the number corresponds to the least significant block of the accu. A bit is one iff the corresponding block of the accu contains only ones. To this number we add the carry, i.e. we add a number having a 1 only at the position corresponding to the block first entered by the carry. Figure 2 illustrates this procedure. In the result of this addition the bits different from the bits in the input number are exactly the bits corresponding to the rows passed by the carry (marked with a bar in Fig. 2) and the row which absorbs the carry (marked with a cross). Hence, it is easy to update the information bits and to address the row which absorbs the carry.

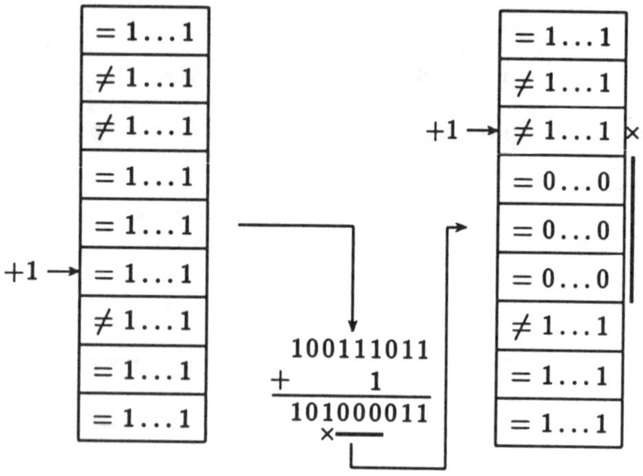

Figure 2: The use of the adder to propagate the carry over the blocks.

Now we have shown how carries can be treated if all summands are positive. Next we show how we will handle negative summands. When expanded to the fixed-point format, the negative numbers are represented in two's complement and thus require all the leading digits being filled up with ones. At first sight this seems to be incompatible with our effort of working only locally, but this is not true. Again we consider the way we perform arithmetic manually. Suppose we want to subtract b from a. If we use binary representation, we compute $a + (-b)$ representing $-b$ in two's complement which we get by inverting all digits of b and adding 1. If a corresponds to a section of the accumulator and $a \geq b$, then this method of subtracting two numbers can also be applied locally to our accumulator since only local changes are necessary. If $a < b$, then the result of the subtraction is a negative number and we cannot replace some section of the accumulator by a negative number.

But consider the way a negative number is represented in two's complement. The sign bit has value -2^n where

n is the number of digits used to represent the number, not counting the sign bit. Hence, if we manipulate a section of the accumulator in the usual way, a negative sign bit of the result can be interpreted as a negative carry, i.e. from the more significant bits we have to subtract this carry. This negative carry can be handled nearly the same way as a positive carry. The only difference is that it is accepted by a 1 and propagated by a 0. Hence, the meaning of the two information bits associated with a row has to be extended in the following way. One bit indicates whether all bits in the row are equal and if they are, the other bit indicates whether the row has to be interpreted as constant 1 or constant 0. The propagation of a negative carry over the blocks is similar to the propagation of a positive carry: a block containing only zeros is represented as a zero, other blocks are represented as one and we subtract the carry from the number constructed in this way. In the following section we give a more detailed description of the addition algorithm and the rounding process.

2.2 Some Details of the Algorithms

In describing the algorithms for adding and rounding, we will widely abstract from a concrete hardware realization, but some considerations are influenced by the floating-point system we use. To keep things simple, we describe the algorithm for the binary system but the principles can also be applied to any other number system, e.g. the decimal system. Since we want to construct a chip that fits together with other hardware, we have to consider existing standards. A commonly accepted standard for the representation of floating-point numbers is the IEEE floating-point standard 754 (cf. [11]). This standard uses sign magnitude representation for the mantissas. Hence, we choose the following **floating-point format** for our abstract description: The summands are represented as floating-point numbers consisting of m mantissa digits for the absolute value of the mantissa, a sign bit and an integer exponent in the range $[e_1 .. e_2]$. We do not require the summands to be normalized.

Let the accumulator be realized by L digits (including the sign bit) representing the accumulator contents in two's complement. L must be at least $e_2 - e_1 + m + 1$. To be able to add k-times the largest representable summand, there must be $\lceil \log k \rceil$ additional overflow bits. Since it is not hard to make enough overflow bits available, we will always assume that L is chosen large enough such that a new summand can be added without producing an overflow. As we already stated in section 2.1, the accumulator is subdivided into several blocks. Here we assume that the accumulator is divided into blocks of size s and that we are able to add s-bit numbers. We also assume that $L = n \cdot s$ for some $n \in \mathbb{N}$.

2.2.1 The Algorithm for Adding a New Summand:
The algorithm for adding a new summand works as follows. First we "convert" the summand into fixed-point representation. This can be done by adjusting the mantissa to the corresponding position at the accumulator. Since we can address each block of the accumulator individually, we can perform the positioning in two steps. One step is a coarse positioning done

by addressing the blocks of the accu that are intersected by the mantissa digits (cf. Figure 3). The other step, the fine positioning, is done by a shifting circuit. The output of this circuit is stored in a register and we know (from the exponent of the summand) which part of the register corresponds to which block of the accu. For each of these blocks of the accu we perform a usual addition stepping through them from right to left as indicated in Figure 3. Each rectangle represents a block of s bits and each addition is an s-bit addition. If in the last addition a carry occurs, we have to pass all blocks that cannot accept the carry and add it to the first accepting block.

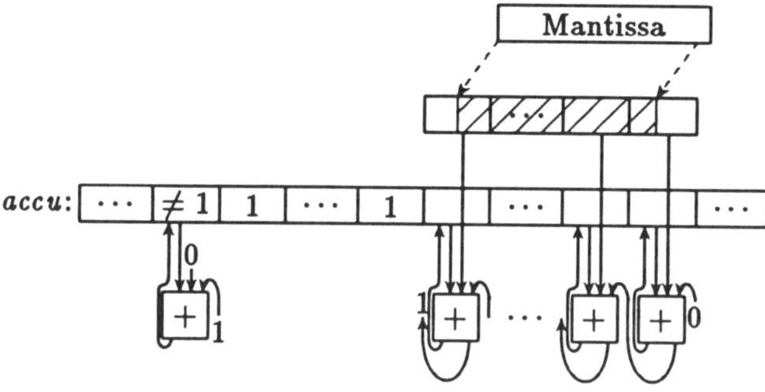

Figure 3: Illustration of the addition algorithm if we add a positive summand and when after the local additions a carry remains to be resolved.

The address of the accu block containing the least significant mantissa digit is given by the more significant bits of the exponent, provided that s is a power of 2. The shift distance is $s - 1$ at most. Then the shifted mantissa digits cover at most $r = \lceil (m + s - 1)/s \rceil$ blocks of the accu.

If the summand is negative, then the mantissa has to be subtracted. This is done by adding the two's complement, i.e. by adding the inverted output of the shifter plus one. Here it is not necessary to represent the sign bit explicitly since subtracting a positive $r \cdot s$-bit number from a positive $r \cdot s$-bit number always yields a number that can be represented in two's complement using $r \cdot s$ bits plus a sign bit. Hence, no overflow or underflow occurs. Furthermore, the sign bit of the result is 1 iff there is no carry entering the column of the sign bit. This means that we have to resolve a negative carry iff the summand is negative and no carry resulted from the last one of the r addition. The procedure for carry resolution is already described in section 2.1.

2.2.2 Rounding and Normalization: Given the contents of the accumulator, we have to perform the specified rounding at the end of the summation. Rounding towards $-\infty$ can clearly be done by truncation; this corresponds to omitting a nonnegative part of the sum. Rounding towards $+\infty$ can be performed by truncating and computing the next larger floating-point number if the truncated part was $\neq 0$. Rounding towards zero is now trivial. Rounding to nearest is a little bit harder. Again we truncate the part of the contents of the accu that won't fit into the mantissa. If what we truncated

was less than $100\ldots00$, then we are done. If it was more than $100\ldots00$, we compute the next larger floating-point number. If it was equal to $100\ldots00$ and the mantissa is even, we are done, too, but if the mantissa is not even, we also have to compute its successor. Computing the successor of a floating-point number can usually be achieved by adding 1 to its mantissa. If the resulting number is not representable in the given floating-point format a, suitable exception has to be signaled.

In order to determine the significant part of the accu we need to know the position of the most significant bit (m.s.b.), i.e. the first bit different from the sign bit. To find out whether the part of the accu that will be truncated is greater, equal or less than $100\ldots00$, it is also helpful to know the position of the least significant bit (l.s.b.), i.e. the rightmost bit that is 1.

To determine the m.s.b. and the l.s.b. we use the same principle that we used to find the position where a carry is accepted. A negative carry is accepted at the rightmost 1 which is to the left of the position from which the carry originated. Hence, the position of the l.s.b. of a binary number is the position where a negative carry, which enters from the right, is accepted. It can be found by subtracting 1 from the number. If the number is not zero this subtraction causes exactly one digit, namely the l.s.b., to change from 1 to 0. Hence, we can determine the l.s.b. in two steps. To determine the block, we define a binary number which represents the states of the blocks of the accu as we already did for the carry resolution. A bit in this number is 0 iff the corresponding block contains only zeros. To this number we add $1\ldots1$ which is equivalent to subtracting 1. To determine the position within the block we apply the same procedure to the contents of the block. Determining the m.s.b. is similar; we have to find the left-most bit which is different from the sign bit.

Let Σ denote the computed exact sum, i.e. the value of the contents of the accumulator. Having computed the m.s.b. and the l.s.b., we extract the significant part of the accu and compute the appropriate exponent to get $\bigtriangledown(\Sigma)$ with the mantissa represented in two's complement. If the sign is negative, we mainly have to invert the mantissa and to add 1 to obtain the sign magnitude representation of $\bigtriangledown(\Sigma)$.

Now we have treated rounding towards $-\infty$ and next we will explain what modifications are necessary to support the other rounding modes. For any floating-point number x let $succ(x)$ denote the next larger floating-point number. Since $\bigtriangledown(\Sigma) \leq \Sigma \leq succ(\bigtriangledown(\Sigma))$, the rounded contents $\bigcirc(\Sigma)$ of the accu are contained in $\{\bigtriangledown(\Sigma), succ(\bigtriangledown(\Sigma))\}$ for any rounding mode $\bigcirc$. Hence, it is sufficient to compute $\bigtriangledown(\Sigma)$ and $succ(\bigtriangledown(\Sigma))$ and decide which of both values is the right one. Considering the remarks at the beginning of this section this decision can easily be made.

3 Realization in Hardware

In this section we turn to the hardware implementation of the algorithms presented in section 2. As one can see from the previous discussion, the circuit consists only of well known components: an adder for s-bit nubmers, a shifter, a storage and some decoders.

The sizes of the modules (i.e. the number of bits they must be designed for) are fixed by the choice of the floating-point system and the choice of the size s of a block of the accumulator. The choice of s influences the performance and space requirement of the circuit significantly. If we choose s small, the adder and the shifter are small, but on the other hand we have to use the adder and the shifter many times to process a summand. If we choose s large, things are the other way round: the adder and the shifter are large, but we need only few additions and shifts to process a summand. It is also important to choose s as a power of 2 since then the computation of the shift distance and the blocks affected by the addition of a new summand are trivial; they merely require selecting the more, less resp. significant bits of the exponent.

It seems that s should be chosen as large as possible to achieve a high performance. This is because the reduction of the number of additions seems to gain much more than the loss of speed resulting from the longer delay time of the larger modules makes up. To get a better feeling for a good choice of s we consider the following example.

Example: The input numbers shall be (cf. the beginning of section 2) exact products of double precision floating-point numbers corresponding to the IEEE standard 754, cf. [11]. This requires an accumulator of about 4200 bits plus a sufficient number (e.g. 60) of additional bits to avoid overflow (cf. section 2.1). The mantissas of the products have a length of $m = 106$.

To calculate the number of additions needed to process a summand (depending on s) we recall from section 2.2.1 that the number of blocks intersected by a mantissa of length m is $r = \lceil (m + s - 1)/s \rceil$ at most. We need one addition for each of those blocks plus $\lceil (n - r)/s \rceil$ to find which block accepts a carry plus one for removing the carry, i.e. we need $r + \lceil (n - r)/s \rceil + 1$ additions. The expression $\lceil (n - r)/s \rceil$ results from the fact that a carry can run over $n - r$ blocks at most, where n is the number of blocks of the accumulator. Table 1 shows the number of additions required and the number $n = \lceil 4260/s \rceil$ of blocks of the accumulator if we choose $s = 2^k$ for some k.

k	$s = 2^k$	r	# additions	n
4	16	8	26	267
5	32	5	11	134
6	64	3	5	67
7	128	2	4	34

Table 1: The effect of choosing $s = 2^k$ for some k.

For $s = 128$ we need the least possible number of additions and hence this seems to be the best value to achieve a high speed. Choosing s larger than 128 is not useful since we need at least 4 additions in the worst case: it can always happen that the mantissa intersects two blocks of the accu and that we need two additions to remove a carry. But $s = 64$ might also be a good choice if there is not enough space to realize a 128-bit adder and shifter. ∎

4 Implementation in Software

The algorithm presented above can also be used to obtain a fast software implementation of the long accumulator. Almost all operations we use are available in an ordinary computer. The only problem is to find in a binary number x the w.l.o.g. left-most bit which equals 1. This corresponds to compute $\lfloor \lg x \rfloor$, and not all processors provide this operation. But as shown in [8] $\lfloor \lg x \rfloor$ can be computed in constant time using ordinary arithmetic and bitwise Boolean operations as follows. Let b be the wordlength and consider a partitioning of the word x into blocks of $s = \lceil \sqrt{b+1} \rceil$ bits. Define C to have 1's precisely in the left-most bit position of each block. Evaluating the expression $(C - [(C - x \wedge \overline{C}) \wedge C]) \vee (x \wedge C)$, where $\overline{C}$ is the bitwise complement of C, yields a number in which the left-most bit of a block is 1 iff the corresponding block of x contains a 1, and all other bits in the number are 0. Multiplying this number with a suitable constant compresses all the leading bits of the blocks into the rightmost $\lceil b/s \rceil$ bits. Now, by table look-up we can find the number of the block of x, which contains the left-most 1 and by a second look-up we can find the position within this block. Hence $\lfloor \lg x \rfloor$ can be computed using about 4 bitwise Boolean operations, 2 subtractions, 1 multiplication and 2 table look-ups.

References

[1] B. BECKER, "Efficient Testing of Optimal Time Adders", TR 04/1985, SFB 124, Universität des Saarlandes

[2] B. BECKER, R. KOLLA, "On the Construction of Optimal Time Adders", TR 07/1987, SFB 124, Universität des Saarlandes

[3] G. BOHLENDER, "What Do We Need Beyond IEEE Arithmetic?", in [22]

[4] G. BOHLENDER, L. RALL, CH. ULLRICH, J. WOLFF VON GULDENBERG, "PASCAL-SC: A Computer Language for Scientific Computation", Academic Press, New York, 1987

[5] R. P. BRENT, H. T. KUNG, "A Regular Layout for Parallel Adders", *IEEE Trans. on Comp.*, C-31, pp. 260–264, March 1982

[6] J. H. BEHLER, S. M. RUMP, U. W. KULISCH, M. METZGER, CH. ULLRICH, W. WALTER, "FORTRAN-SC: A Study of a FORTRAN Extension for Engineering/Scientific Computation with Access to ACRITH", *Computing 39*, pp. 93–110, 1987

[7] P. R. Capello, W. L. Miranker, "Systolic Super Summation", *IEEE Transactions on Computers*, Vol. 37, No. 6, pp. 675–676, June 1988

[8] M.L. Fredman, D.E. Willard, "BLASTING Through The Information Theoretic Barrier With FUSION TREES", *Proc. of the 22nd Annual ACM Symp. on Theory of Computing*, pp. 1–7, 1990

[9] R. Hammer, M. Neaga, D. Ratz, "PASCAL-SC: New Concepts for Scientific Computation and Numerical Data Processing", Institute for Applied Mathematics, University of Karlsruhe, Federal Republic of Germany

[10] IBM, "High-Accuracy Arithmetic Subroutine Library (ACRITH)", General Information Manual, 3rd ed., GC 33-6163-02, IBM Corporation, 1986

[11] American National Standards Institute / Institute of Electrical & Electronic Engineers, "A Standard for Binary Floating-Point Arithmetic", ANSI/IEEE Std. 754-1985, New York, August 1985

[12] R. Kirchner, U. W. Kulisch, "Accurate Arithmetic for Vector Processors", *Journal of Parallel and Distributed Computing*, Vol. 5, pp. 250–270, 1988

[13] U. W. Kulisch, "Grundlagen des Numerischen Rechnens — Mathematische Begründung der Rechnerarithmetik", Mannheim, Bibliographisches Institut, 1976

[14] U. W. Kulisch, W. L. Miranker, "Computer Arithmetic in Theory and Practice", New York, Academic Press, 1981

[15] U. W. Kulisch, W. L. Miranker (eds.), "A New Approach to Scientific Computation", New York, Academic Press, 1983

[16] P. Lichter, "Realisierung eines VLSI-Chips für das Gleitkomma-Skalarprodukt der Kulisch-Arithmetik", Diplomarbeit, Fachbereich 10, Angewandte Mathematik und Informatik, Universität des Saarlandes, March 1988

[17] M. Müller, Ch. Rüb, W. Rülling, "Exact Addition of Floating-Point Numbers", TR 05/1990, SFB 124, Universität des Saarlandes

[18] Th. Ottmann, G. Thiemt, Ch. Ullrich, "Numerical Stability of Geometric Algorithms (Extended Abstract)", *Proceedings of the 3rd Annual ACM Symp. on Computational Geometry*, pp. 119–125, 1987

[19] S. M. Rump, H. Böhm, "Least Significant Bit Evaluation of Arithmetic Expressions in Single-Precision", *Computing 30*, pp. 189–199, 1983

[20] SIEMENS, "ARITHMOS, (BS2000) Benutzerhandbuch", SIEMENS AG, U2900-J-Z87-1, 1986

[21] J. Sklansky, "Conditional Sum Addition Logic", *IRE-EC 9*, pp. 226–231, June 1960

[22] Ch. Ullrich (ed.), "Computer Arithmetic and Self-Validating Numerical Methods", Academic Press, June 1990

[23] J. Vuillemin, L. Guibas, "On Fast Binary Addition in MOS Technologies", *ICCC 82*, pp. 147–150

[24] Th. Winter, "Ein VLSI-Chip für Gleitkomma-Skalarprodukt mit maximaler Genauigkeit", Diplomarbeit, Fachbereich 10, Angewandte Mathematik und Informatik, Universität des Saarlandes, April 1985

[25] T. Yilmaz, J.F.M. Theeuwen, R.J.W.T. Tangelder, J.A.G. Jess, "The Design of a Chip for Scientific Computation", Eindhoven University of Technology, 1989 and Euro Asic, Grenoble, Jan. 25–27, 1989

Fast Hardware Units for the Computation of Accurate Dot Products

Andreas Knöfel

Institute for Applied Mathematics
University of Karlsruhe
W-7500 Karlsruhe

1. Motivation

Matrix and vector operations based on dot product expressions occur in almost all scientific and engineering applications. The lack of the popular programming languages and computer architectures to provide operators for these data types and corresponding accurate hardware instructions forced users to emulate the vector operations by constructing loops with scalar floating point instructions. Cancellation and immediate rounding in these loops cause uncertain and inaccurate numerical results and aggrevate an error analysis.

The investigation of these effects at the Institute for Applied Mathematics lead to the definition of scalar and vector operations by semimorphism [4], being the basis for a properly defined arithmetic. Operations defined by a semimorphism deliver a result that differs from the correct answer by at most one rounding. The programming languages FORTRAN [10] and PASCAL [11] have been extended for an easy access to numerical data types as well as the corresponding operators with a scalar and vector arithmetic based on semimorphisms. Problem solving routines delivering a verified enclosure of the result have been developed by means of these new programming languages showing the usefulness of the arithmetic. The entire runtime system for all arithmetic operators was written in software due to the uncertain floating point operations delivered by the computer systems. In 1985, the IEEE society defined a standard for scalar floating point operations [9]. Floating point processors based on this standard have been developed for many computers and became faster and faster. The accurate vector operations, however, remained on the software emulation level and their execution time decreased more and more compared with the algorithms using the given scalar arithmetic.

Several accurate hardware vector arithmetic units [2], [3], [7] were proposed in the recent years. For an overview see [1]. They particularly address the vector computer area. For personal computers up to main frames in the following sections two hardware units with high performance, short pipelines and additional hardware less than one multiplier are described.

2. The principle of operation

Several principles for the computation of accurate dot products have been developed in the past, but the principle with the most flexibility is the Long Accumulator (LA) [6]. In the LA algorithm the intermediate operations are executed exactly. At first the products are computed to double length. Then they are accumulated into a fixed-point register, called LA. To ensure an exact accumulation at any time, the LA must have

$$S = K + 2E_{max} + 2|E_{min}| + 2l + 1$$

digits of the base b of the input floating point format with l mantissa digits and an exponent between E_{min} and E_{max}. K additional digits are for intermediate overflows and one digit is interpreted as the sign of the LA. After the final accumulation the exact dot product value is rounded back into the input floating point format or completely stored for further computations. Figure 1 shows a Long Accumulator. Several products are aligned according their exponent due to accumulation.

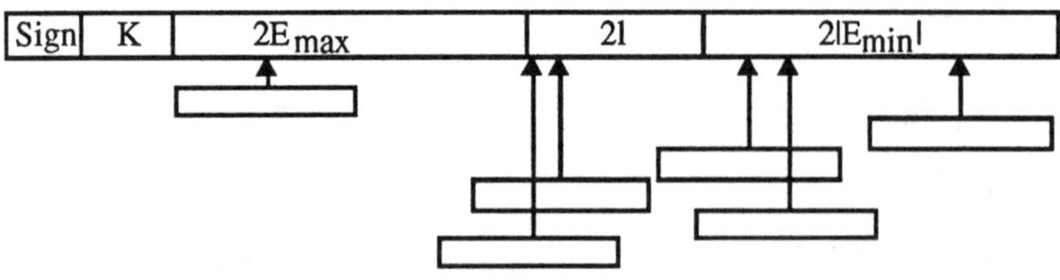

<u>Figure 1:</u> Long Accumulator

In the accumulation step the double length products are shifted to the corresponding position in the LA where they are accumulated. For customary floating point formats like IBM /370 [8], IEEE [9] the products intersect only with 3-5% of the LA. Thus, it is useful to keep the accumulator in a memory and perform only a local accumulation of the product by loading the intersecting accumulator words. After the local accumulation a carry or borrow may propagate over many digits as shown in figure 2.

In software implementations of the LA handling the carry is accumulated in a loop into succeeding words until it is resolved. M. Müller, Ch. Rüb and W. Rülling at the University of Saarbrücken developed a fast solution to accelerate the carry handling [5]. The entire carry processing can be performed by only one addition, since a carry skips an accumulator word only if all digits are equal to (b-1), for b being the base of the number system. After the carry skip, the value of this word is zero. Thus, no addition has to be executed, because the result is predictable. The same effect occurs for borrows. They skip a word only if all digits are zero and after the skip all digits are (b-1). A two bit wide register, keeping two flags, is attached to each accumulator word. One flag indicates, whether all digits of the corresponding LA word are zero and the other one, whether all digits are (b-1). The carry skip over word boundaries is handled by a simple logic that toggles the flags and delivers the address of the accumulator word where the carry is resolved, since one digit is less than (b-1) there. This carry resolve word is incremented in one final addition. The same principle works for the borrows and only the final accumulation has to be performed, as shown in figure 3. The flags have to be actualized after each addition when the word is stored back into memory. Therefore, a word is read from the memory only if both flags are reset. In the other cases a constant is read into the adder.

After the accumulation of the last product the result can be read from the accumulator. Here the flags are also important to find the leading accumulator word being the first word whose flag is not equal to the sign of the LA. This leading accumulator word delivers the leading digits and the exponent of the result. As the rounding is concerned it is important if there are trailing digits not equal to zero. This could also be done by a logical OR of all trailing ZERO-flags.

3. Hardware description

3.1 A sequential circuit

Using a LA stored as a sequence of words with the mentioned flags, a fixed number of accumulation steps is necessary to perform the local accumulation of the product, one step for the carry skip handling and one final carry resolve step. All these steps could be pipelined if the LA is kept in a dual-port-RAM. After the first word, intersecting with the least significant part of the mantissa has been read from the RAM and accumulated, the next word where the next mantissa part must be accumulated can be read from the RAM. The results from the adder and the actualized flags can be written back via the other port of the RAM at the same time. There are no address conflicts because the memory addresses increase and the execution time for the addition lays between the read and the write access to the same address.

It is known in advance, how many accu words intersect with the mantissa. Therefore, the carry skip process and the prediction of the carry resolve address can be performed in parallel to the local accumulation, starting with a fixed offset to the address where the least significant mantissa part is accumulated. At that time it is not known whether a

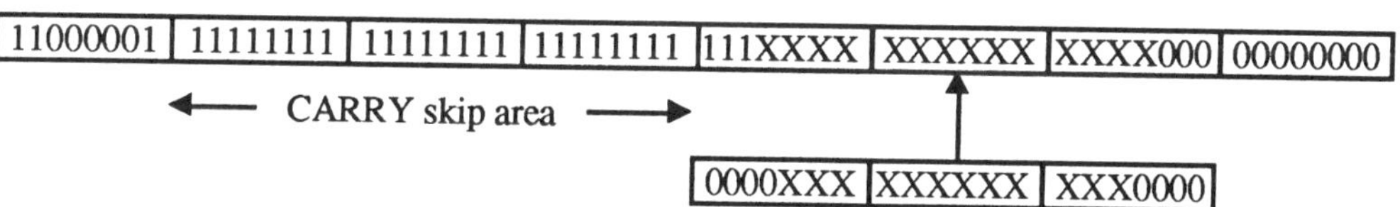

Figure 2: Local accumulation with carry skip

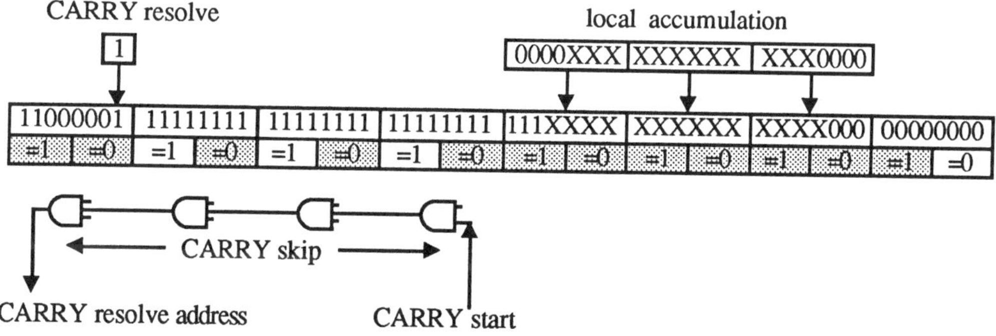

Figure 3: Carry handling with flag logic

final carry or borrow has to be resolved. Therefore, only a copy of the flags is toggled on the way to determine the carry resolve address. To ensure a proper flag setting at any time, the flag values after an addition must be actualized in both the copy and the original flag registers. The final addition can be integrated without any wait-cycles into the read-add-write process, since the carry output of the adder can be added to the carry resolve word in any way. Like in a carry-select-adder the copy of the flags becomes the actual flag setting only if there was a carry after the local mantissa accumulation. This pipelining is shown in figure 4 for a computer with 32 bit data busses, IEEE double input numbers and a 64 bit wide RAM to keep the LA. The accumulation is perfomed in seven cycles within an eight cycle pipeline with the main steps

- operand read,
- multiply & shift and
- accumulate.

In the step "read" the memory accesses and the loop counter and address pointer administration must be handled which could not be performed in eight cycles for many computers. Therefore, eight cycles are a lower bound for the entire pipelining.

Figure 5 shows a block diagram of the circuit. The additional components for the exact accumulation are the RAM with the flags and an address incrementer and a wider shifter and adder. The multiplier must compute the exact product. In almost all computers, however, the lower product part is cut off and ignored. Therefore, the additional amount for the exact multiplication is only a wider output bus. The unit can be designed as a coprocessor or part of a main processor and fits onto one chip being applicable for main frames as well as work stations or PC´s.

3.2 A parallel accumulation circuit

The progress in hardware design and technology allowed the integration of concepts into computers that seemed to be applicable only for vector processors. Cache memories,wide data busses, RISC instruction sets, parallel execution units and also pipelined floating point operations are the highlights of this new generation of computers. Eight cycles for one pipeline step is too slow for them. But with a few changes in the design the accumulation step can be performed in two cycles and thus may be adapted to the high I/O-bandwidth and fast multiplication times of these computers.

The addresses of all LA-words where an addition must be performed can be predicted. If the accu address is known where the least significant part of the product is accumulated the next part of the product is accumulated to the subsequent accu word and so on. The start address for the carry skip and resolve address prediction has a fixed offset from this start address and therefore the carry resolve address can be delivered in advance. All words influenced may be read in parallel into a sufficiently wide adder and accumulated there in one step as shown in figure 6.

This could only be performed with a multi-port-RAM, but up to six-port-RAMs are state-of-the-art. The width of the LA-words give the designer a parameter to decrease the number of ports down to three. The advantage is that one address decoder manages all ports. The lowest intersecting word is passed always via port 1, it´s successor via port 2 and so on. The drivers of each port must be designed to deliver a value according to the flags. All port drivers and the start signal for the carry skip are controlled by one

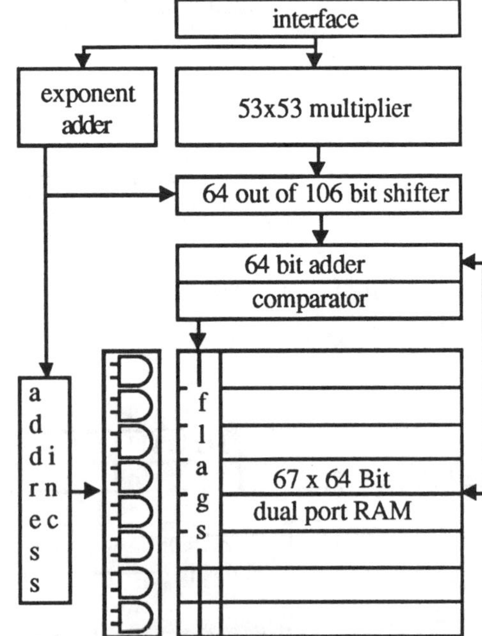

Figure 5: Block diagram

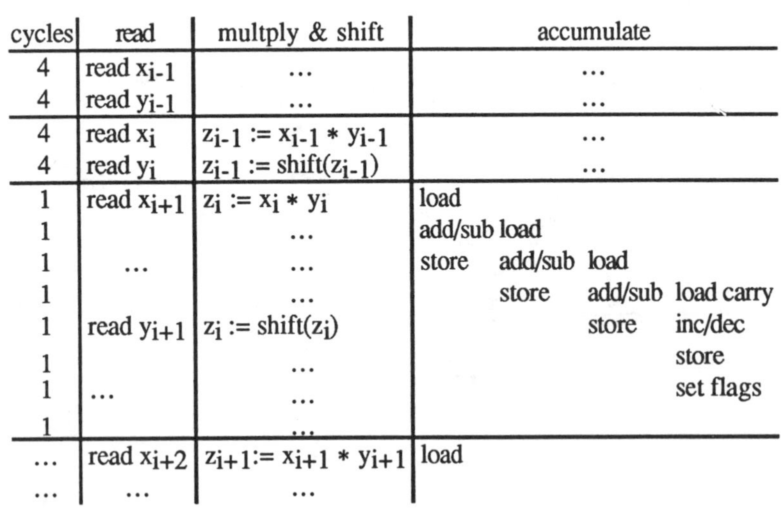

cycles	read	multply & shift	accumulate
4	read x_{i-1}	…	…
4	read y_{i-1}	…	…
4	read x_i	$z_{i-1} := x_{i-1} * y_{i-1}$	…
4	read y_i	$z_{i-1} := shift(z_{i-1})$	…
1	read x_{i+1}	$z_i := x_i * y_i$	load
1		…	add/sub load
1	…	…	store add/sub load
1		…	store add/sub load carry
1	read y_{i+1}	$z_i := shift(z_i)$	store inc/dec
1		…	store
1	…	…	set flags
1		…	
…	read x_{i+2}	$z_{i+1} := x_{i+1} * y_{i+1}$	load
…	…	…	

Figure 4: Pipelined accumulation and the overall dot product pipeline
… indicate that the execution is still in process

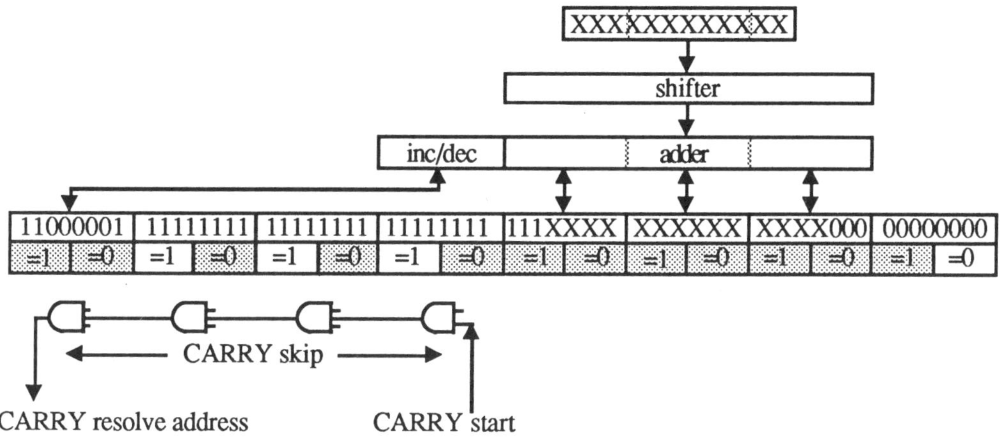

Figure 6: Parallel accumulation

decoder. The carry skip logic determines the carry resolve address, toggles a copy of the flags and then drives the last port where the carry-resolve-word is passed through. This hardwired multi-port driver can be designed very fast so that all words influenced are written to the adder in one cycle. Figure 7 shows the modified block diagram of the RAM.

The entire process with the steps
- decoding of the address
- predicting the carry resolve address and toggle a copy of the flags
- loading the words to the adder
- accumulation
- restore the result and actualize flags

could be performed in two cycles, as shown in figure 8. The accumulation and flag actualization must therefore be performed very fast using sophisticated adders and tree structures in the comparators and decoders. Figure 9 shows a tree structured path to determine the carry resolve word.

4. Comparison with other circuits

The presented circuits are compared with the proposed solutions of P.R. Cappello andW.L. Miranker [2], R. Kirchner and U. Kulisch [3] and T. Yilmaz, J.F.M. Theeuwen, R.J.W.T. Tangelder and J.A.G. Jess [7]. For all circuits the total amount of transistors has been estimated for the IEEE-DOUBLE format with design data of the Institute for Microelectronics Stuttgart. The result of this estimation and other relevant properties are listed in table 1.

circuit	# transistors	pipeline length	# cycles accumulation
[2]	700000 K	≈ 4200	1
[3]	750 K	≈ 70	1
[7]	300 K	≈ 70	2
sequential	200 K	24	8
parallel	200 K	6	2
multiplier	80 K		

Table 1: Comparison between several circuits.

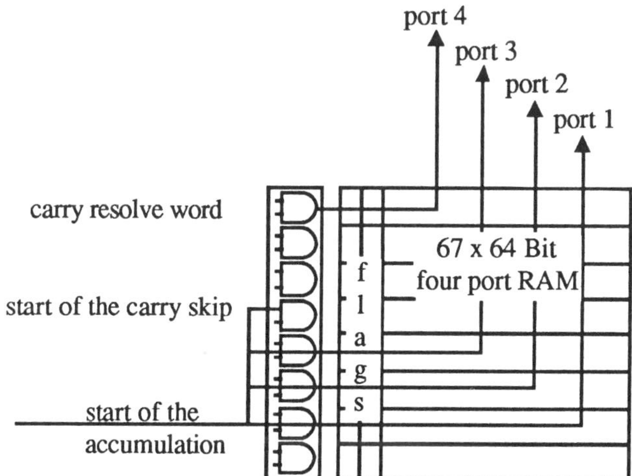

Figure 7: Multi port RAM unit for the parallel accumulation

read x_{i-1}		
read y_{i-1}		
read x_i	$z_{i-1} := x_{i-1} * y_{i-1}$	
read y_i	$z_{i-1} := shift(z_{i-1})$	
read x_{i+1}	$z_i := x_i * y_i$	determine addresses
read y_{i+1}	$z_i := shift(z_i)$	load / accumulate / store
read x_{i+2}	$z_{i+1} := x_{i+1} * y_{i+1}$	determine addresses
...	...	load / accumulate / store

Figure 8: The dot product pipeline with the parallel accumulation

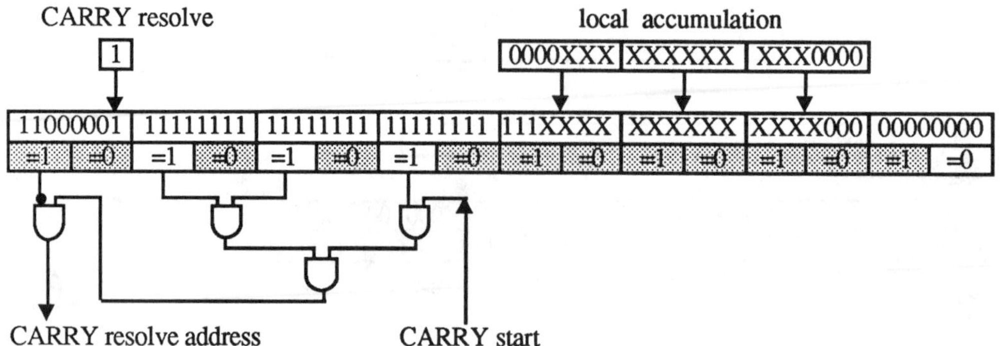

Figure 9: Tree structured carry resolve logic

5. Conclusion

The circuits presented in this article are easy to implement
with available techniques as one chip and deliver a high
performance solution for dot product computations. Beside
the RAM for the LA all units can be used for scalar
operations to avoid a hardware overhead for scalar and
vector units. The additional hardware amount for a
combined scalar and vector computation unit is about
120 K transistors and therefore also applicabable for PC´s.
The entire data flow from the input interface down to the
accumulation has been discussed in contrast to former
proposals, where only the accumulation was handled. The
pipeline of the first unit has 24 stages and the pipeline of
the second unit only 6 stages for the entire process. Thus,
these units are also applicable for short vectors and for
computations with complex numbers.

References

[1]   G. Bohlender: "What Do We Need Bejond IEEE
      Arithmetic? ". In: Ullrich (ed): Computer Arithmetic
      and Self-Validing Numerical Methods, Academic
      Press, New York, 1990

[2]   P.R. Cappello, W.L. Miranker: "Systolic Super
      Summation". IEEE Transactions on Computers,
      EC 37-(6), 1988

[3]   R. Kirchner, U. Kulisch: "Arithmetic for Vector
      Processors". Proc. of the 8th Symp. of Computer
      Arithmetic (ARITH-8), IEEE Computer Society,
      1987

[4]   U. Kulisch, W.L. Miranker: "Computer Arithmetic
      in Theory and Practice". Academic Press, New York,
      1981

[5]   M. Müller, Ch. Rüb, W. Rülling: "Exact Addition
      of Floating Point Numbers". Proc. of the 10th
      Symp. of Computer Arithmetic (ARITH-10), IEEE
      Computer Society, 1991

[6]   S.M. Rump: "Kleine Fehlerschranken bei Matrix-
      problemen". Ph.D.thesis, Universität Karsruhe, 1980

[7]   T. Yilmaz, J.F.M. Theeuwen, R.J.W.T. Tangelder,
      J.A.G. Jess: "The Design of a Chip for Scientific
      Computation". Euro ASIC, Grenoble, 1989

[8]   International Business Machines Corporation: "IBM
      System /370: Principles of Operations", IBM
      GA22-7000

[9]   American National Standard Institute; Institute of
      Electrical and Electronics Engineers: "IEEE Standard
      for Binary Floating Point Arithmetic". ANSI/IEEE
      Std 754-1985, New York, 1985

[10] International Business Machines Corporation: "IBM
      High Accuracy Arithmetic - Extended Scientific
      Computation, Reference", IBM SC33-6462-00

[11] R. Klatte, U. Kulisch, M. Neaga, Ch. Ullrich,
      D. Ratz: "PASCAL-XSC Sprachbeschreibung und
      Beispiele". Springer, Heidelberg, 1991

Session 4:

Residue Arithmetic

Chair:
Magdy Bayoumi
University of Southwestern Louisiana

A General Division Algorithm for Residue Number Systems[*]

Jen-Shiun Chiang

Mi Lu

Electrical Engineering Department
Texas A&M University
College Station, Texas   77843-3128
U.S.A.

Abstract

We present in this paper a novel general algorithm for signed number division in Residue Number Systems (RNS). A parity checking technique is used to accomplish the sign and overflow detection in this algorithm. Compared with conventional methods of sign and overflow detection, the parity checking method is more efficient and practical. Sign magnitude arithmetic division is implemented using binary search. There is no restriction to the dividend and the divisor (except zero divisor), and no quotient estimation is necessary before the division is started. In hardware implementations, the storage of one table is required for parity checking, and all the other arithmetic operations are completed by calculations. Only simple operations are needed to accomplish this RNS division. All these characteristics have made our algorithm simple, efficient, and practical to be implemented on a real RNS divider.

1   Introduction

Residue Number Systems (abbreviated as RNS) are attractive to many people. A RNS is composed of moduli that are independent of each other. A number in the RNS is represented by the residue of each modulus, and arithmetic operations are accomplished based on each modulus. Since the moduli are independent of each other, there is no carry propagation among them, and it is easy to implement RNS computations on a multi-ALU system. The operation based on each modulus can be performed by a separate ALU, and all the ALU's can work concurrently. These characteristics allow RNS computations to be completed more quickly — an attractive feature for people who need high speed arithmetic operations [1, 2, 3].

Overflow detection, sign detection, number comparison, and division in RNS are very difficult and time consuming [4, 5]. These shortcomings limited most of the previous RNS applications to additions, subtractions, and multiplications.

The general division algorithms can be classified into two groups [6]: multiplicative algorithms and subtractive algorithms. The multiplicative algorithms compute the reciprocal of the divisor; the quotient is obtained by the multiplication of this reciprocal and the dividend. Subtractive algorithms recursively subtract the multiple of the denominator from the numerator until the difference becomes less than the denominator. The multiple is then the quotient.

There are several RNS division algorithms that are classified as multiplicative algorithms [5, 7, 8]. These multiplicative algorithms use the mixed radix number conversion to find the reciprocal of the divisor and to compare numbers. Iteratively, the approximate quotient is made closer to the accurate one. Due to the involvement of the mixed radix number conversion, the arithmetic calculation is very complicated and needs a lot of stored tables. Among these multiplicative algorithms, Kinoshita's algorithm [8] uses mixed radix numbers to approximate the quotient, and requires either a decimal divider or the storage of a very large table. Banerji's algorithm [7] also uses the mixed radix number approach and requires a lot of storage. Chren [5] criticizes that the standard deviation of the mean of the execution time needed in this algorithm is high. Chren's algorithm [5] is modified from Banerji's. Chren made some effort to reduce the storage and to improve the standard deviation of the mean execution time, but the storage and the computation time needed by the mixed radix number conversion are still expensive.

On the other hand, there are several algorithms classified as subtractive algorithms [4, 9, 10]. These subtractive algorithms use the conventional division approach, and no mixed radix number conversion is necessary. Therefore, the arithmetic calculations are not complicated. However, its number comparison and sign detection consume a lot of time and hardware. Szabo's algorithm [4] is not a general division algorithm but a scaling algorithm. Keir et al. [9] present two algorithms, both of which involve the binary expansion of the quotient. The speed of Keir's first algorithm is not desirable. His second algorithm uses look-up tables so that the hardware requirements are huge. Lin's algorithm [10] is a modification of the

---

[*]This research is partially supported by the National Science Foundation under Grant No. MIP-8809328.

well known CORDIC division algorithm, but it needs a lot of comparators, which is not practical for general computing applications.

In this paper we use parity checking for sign and overflow detection. Compared to conventional methods, the parity checking method is more efficient and practical. Based on the extension of overflow and sign detection techniques, a new signed RNS division algorithm is presented. Basically, this is a subtractive division algorithm using an efficient method to detect overflows and compare numbers. We use sign magnitude arithmetic for RNS division. In this division algorithm, binary search is used. There are no restrictions to dividends and divisors (except zero divisor), and no quotient estimation is necessary before the division is executed. In a hardware implementation, only the storage of one table is required to perform the parity checking, and almost all other correlated arithmetic operations are completed by calculations. All these characteristics have made our algorithm simple, efficient, and practical.

2 Residue Codes

2.1 Residue Numbers and the Arithmetic of Residue Numbers

The RNS representation of an integer is defined as follows. Let $\{m_1, m_2, ..., m_n\}$ be a set of positive numbers all greater than 1. The m_i's are called moduli and the n-tuple set $\{m_1, m_2, ..., m_n\}$ is called the modulus set. Consider an integer number X. For each modulus in set $\{m_1, m_2, ..., m_n\}$ we have $x_i = X \bmod m_i$ (denoted as $|X|_{m_i}$). Thus a number X in RNS can be represented as

$$X = (x_1, x_2, ..., x_n),$$

given a specific modulus set $\{m_1, m_2, ..., m_n\}$. In order to avoid redundancy, the moduli of a residue number system must be pair-wise relatively prime.

Let $M = \prod_{i=1}^{n} m_i$. It has been proved, in [4], that if $0 \le X < M$, the number X corresponds one to one with the RNS representation. If the result of one calculation exceeds M, we say that overflow occurs. All the numbers should be within the dynamic range M (i.e., $0 \le X < M$). Then, the RNS arithmetic can be performed.

If there are two numbers X and Y, the representations in RNS are as follows,

$$X = (x_1, x_2, ..., x_n)$$

and

$$Y = (y_1, y_2, ..., y_n).$$

We use $\otimes$ to represent the operator of additions, subtractions, and multiplications. The arithmetic in RNS can be represented as follows,

$$X \otimes Y = (z_1, z_2, ..., z_n)$$

where

$$z_i = |x_i \otimes y_i|_{m_i}.$$

From the definition of the mod operation, all moduli are positive. x_i may be less than y_i, which yields $x_i - y_i < 0$. In mod operation, if $x_i - y_i < 0$, then z_i is defined as

$$z_i = m_i + (x_i - y_i). \tag{1}$$

2.2 Number Comparison for Unsigned Numbers

As we know, number comparison and overflow detection in RNS are very difficult. It is necessary to find methods that are efficient, practical, and easily implemented.

Let parity indicate whether an integer number is even or odd. We say two numbers are of the same parity if they are both even or both odd. Otherwise the two numbers are said to be of different parities. We will use the properties of the parities of numbers to accomplish the number comparison.

In a survey of Soviet research on residue number systems [11], Miller et al. defined a function called the *core* function and explored its properties as follows:

Let $m_1, m_2, ..., m_n$ be the relatively prime moduli of a residue number system with product M. For fixed integers $w_1, w_2, ..., w_n$, the core R_N of an integer is defined as follows:

$$R_N = \sum_{i=1}^{m} w_i \left\lfloor \frac{N}{m_i} \right\rfloor,$$

where $\lfloor X \rfloor$ denotes the greatest integer function which is not greater than X. The coefficients w_i are fixed for the moduli set and do not depend on the integer N.

Theorem 1 *Let the moduli m_i and the core R_M be odd. Let $(a_1, a_2, ..., a_n)$ and $(b_1, b_2, ..., b_n)$ be the residue representations of integers $A, B \in [0, M)$. Then $A + B$ causes an overflow if*

(i) $(a_1 + b_1, ..., a_n + b_n)$ is odd, and A and B have the same parity; or

(ii) $(a_1 + b_1, ..., a_n + b_n)$ is even and A and B have different parities.

Let the interval $[0, M/2]$ represent positive numbers and the interval $(M/2, M)$ represent negative numbers.

Theorem 2 *If the moduli m_i and the core R_M are odd, and $(a_1, a_2, ..., a_n)$ is the residue representation of a non-zero integer $A \in [0, M)$, then A is positive if and only if*

$$(|2a_1|_{m_1}, ..., |2a_n|_{m_n})$$

is even.

According to the theory of core functions, if the core function of an RNS number is known, it is easy to detect overflows and the sign of the number. However, it is very difficult to find the core function in RNS by the method given in [11]. Discarding the core function and revising the theorems mentioned by

Miller et al. [11], the following theorems express the properties we need for comparison of unsigned numbers. Consider the whole dynamic range, $[0, M)$ of positive numbers from 0 to $(M - 1)$. Let all m_i's in the modulus set $(m_1, m_2, ..., m_n)$ be odd numbers, and let $X = (x_1, x_2, ..., x_n)$ and $Y = (y_1, y_2, ..., y_n)$ be two RNS numbers. Suppose $Z = X - Y = (z_1, z_2, ..., z_n)$, then we have the following theorem.

Theorem 3 *Let X and Y have the same parity and $Z = X - Y$. $X \geq Y$, iff Z is an even number. $X < Y$, iff Z is an odd number.*

Proof:

If $X \geq Y$, then $X - Y \geq 0$ and Z equals $X - Y$. We know from the mathematical axioms that the two numbers are with the same parity, and the result of the subtraction should be an even number. Therefore, $X \geq Y$ implies that Z is an even number.

On the other hand, suppose that Z is an even number and X and Y are with the same parity. If $X < Y$, then $X - Y < 0$. From equation (1) we have $Z = X - Y + M$. Since M is an odd number and $X - Y$ is even, Z must be an odd number. This contradicts the assumption that Z is even. Therefore, if Z is an even number and X and Y are with the same parity, then $X \geq Y$.

If $X < Y$, then $X - Y < 0$, from (1) we have $Z = X - Y + M$. Since m_i's are all odd numbers, M should be an odd number. In addition, $(X - Y)$ is an even number and this implies that Z is an odd number. Therefore, it is obvious that if $X < Y$, Z is an odd number.

On the other hand, suppose that Z is an odd number and X and Y are with the same parity. If $X \geq Y$, then $X - Y \geq 0$. Since X and Y are with the same parity, Z must be an even number. This contradicts the assumption that Z is an odd number. Therefore, if Z is an odd number and X and Y are with the same parity, then $X < Y$.

Theorem 3 shows us a method to compare two numbers if the parities of these two numbers are the same. Similarly, if the parities of two numbers are different, then the following theorem can tell us which one is bigger.

Theorem 4 *Let X and Y have different parities and $Z = X - Y$. $X \geq Y$, iff Z is an odd number. $X < Y$, iff Z is an even number.*

Proof:

If $X \geq Y$, then $X - Y \geq 0$ and Z equals $X - Y$. We know that the two numbers have different parities, and the result of the subtraction should be an odd number. Therefore, $X \geq Y$ implies that Z is an odd number.

On the other hand, suppose that Z is an odd number and X and Y have different parities. If $X < Y$, then $X - Y < 0$. From equation (1) we have $Z = X - Y + M$. Since M is an odd number and $X - Y$ is odd, Z must be an even number. This contradicts the assumption that Z is odd. Therefore, if Z is an odd number and X and Y have different parities, then $X \geq Y$.

If $X < Y$, then $X - Y < 0$, from (1) we have $Z = X - Y + M$. Since m_i's are all odd numbers, M should be an odd number. In addition, $(X - Y)$ is an odd number and this implies that Z is an even number. Therefore, $X < Y$ implies that Z is an even number.

On the other hand, suppose that Z is an even number and X and Y have different parities. If $X \geq Y$, then $X - Y \geq 0$. Since X and Y have different parities, Z must be an odd number. This contradicts the assumption that Z is an even number. Therefore, if Z is an even number and X and Y are with different parities, then $X < Y$.

Table 1 is referred to when performing parity checking for number comparisons. The decimal numbers under the entry "#" are corresponding to the residue numbers for modulus set $(3, 5, 7)$, and the parities of them are given under the entry P.

The following is an example to illustrate the above theorems.

Example 1

Let the moduli be $m_1 = 3$, $m_2 = 5$, $m_3 = 7$, and hence $M = 3 \cdot 5 \cdot 7 = 105$. Consider $X_1 = (0, 3, 5)$ and $Y_1 = (1, 3, 0)$. From calculation we have $Z_1 = X_1 - Y_1 = (2, 0, 5)$. Look up Table 1, the parities of X_1, Y_1, and Z_1 are odd, even, and odd respectively. From Theorem 4 we know $X_1 > Y_1$.

In the decimal number system, $X_1 = 33$, $Y_1 = 28$, and $Z_1 = 5$, and the result is obvious.

Note that if the number M is big, the parity table may be huge.

2.3 Signed Numbers and the Properties

The method used to represent negative numbers in RNS is similar to that used in conventional radix number systems. Letting the dynamic range be M, we can define the positive and negative numbers as follows [4].

Definition 1 *If the dynamic range $M = \prod_{i=1}^{n} m_i$, then the range of a positive number X is defined as $0 \leq X \leq \lfloor \frac{M}{2} \rfloor$, and the range of a negative number Y is defined as $\lfloor \frac{M}{2} \rfloor < Y < M$. Like the radix number system, the negative numbers, $-1, -2, ..., -(\lfloor \frac{M}{2} \rfloor - 1), -\lfloor \frac{M}{2} \rfloor$, are represented by the numbers, $(M - 1), (M - 2), ..., (\lfloor \frac{M}{2} \rfloor + 2), (\lfloor \frac{M}{2} \rfloor + 1)$, respectively.*

Notice here, 0 is considered as a positive number.

From Definition 1 we can find that the complement of X is $M - X$. In a similar way, the representation of the complement of a number in RNS can be found in the following lemma.

Lemma 1 *Let the modulus set be $\{m_1, m_2, ..., m_n\}$, and the corresponding modulus set of a positive number X in RNS be $\{x_1, x_2, ..., x_n\}$. $-X$ in RNS can be represented by the complement of X which is equal to $\{(m_1 - x_1)_{m_1}, (m_2 - x_2)_{m_2}, ..., (m_n - x_n)_{m_n}\}$.*

Proof:

From Definition 1, $-X$ in RNS corresponds to $M - X$. Applying equation (1), the corresponding modulus set of $-X$ is $\{(m_1 - x_1)_{m_1}, (m_2 - x_2)_{m_2}, ..., (m_n - x_n)_{m_n}\}$. The dynamic range of RNS can be divided into two halves, one for positive numbers and the other for negative numbers, as described in Definition 1. If the

#	(3	5	7)	P	#	(3	5	7)	P	#	(3	5	7)	P	#	(3	5	7)	P	#	(3	5	7)	P
0	0	0	0	0	21	0	1	0	1	42	0	2	0	0	63	0	3	0	1	84	0	4	0	0
1	1	1	1	1	22	1	2	1	0	43	1	3	1	1	64	1	4	1	0	85	1	0	1	1
2	2	2	2	0	23	2	3	2	1	44	2	4	2	0	65	2	0	2	1	86	2	1	2	0
3	0	3	3	1	24	0	4	3	0	45	0	0	3	1	66	0	1	3	0	87	0	2	3	1
4	1	4	4	0	25	1	0	4	1	46	1	1	4	0	67	1	2	4	1	88	1	3	4	0
5	2	0	5	1	26	2	1	5	0	47	2	2	5	1	68	2	3	5	0	89	2	4	5	1
6	0	1	6	0	27	0	2	6	1	48	0	3	6	0	69	0	4	6	1	90	0	0	6	0
7	1	2	0	1	28	1	3	0	0	49	1	4	0	1	70	1	0	0	0	91	1	1	0	1
8	2	3	1	0	29	2	4	1	1	50	2	0	1	0	71	2	1	1	1	92	2	2	1	0
9	0	4	2	1	30	0	0	2	0	51	0	1	2	1	72	0	2	2	0	93	0	3	2	1
10	1	0	3	0	31	1	1	3	1	52	1	2	3	0	73	1	3	3	1	94	1	4	3	0
11	2	1	4	1	32	2	2	4	0	53	2	3	4	1	74	2	4	4	0	95	2	0	4	1
12	0	2	5	0	33	0	3	5	1	54	0	4	5	0	75	0	0	5	1	96	0	1	5	0
13	1	3	6	1	34	1	4	6	0	55	1	0	6	1	76	1	1	6	0	97	1	2	6	1
14	2	4	0	0	35	2	0	0	1	56	2	1	0	0	77	2	2	0	1	98	2	3	0	0
15	0	0	1	1	36	0	1	1	0	57	0	2	1	1	78	0	3	1	0	99	0	4	1	1
16	1	1	2	0	37	1	2	2	1	58	1	3	2	0	79	1	4	2	1	100	1	0	2	0
17	2	2	3	1	38	2	3	3	0	59	2	4	3	1	80	2	0	3	0	101	2	1	3	1
18	0	3	4	0	39	0	4	4	1	60	0	0	4	0	81	0	1	4	1	102	0	2	4	0
19	1	4	5	1	40	1	0	5	0	61	1	1	5	1	82	1	2	5	0	103	1	3	5	1
20	2	0	6	0	41	2	1	6	1	62	2	2	6	0	83	2	3	6	1	104	2	4	6	0

Table 1: Parity Table for Modulus Set (3,5,7)

moduli are all pair-wise prime and are all odd numbers, then the maximum positive number is $\frac{M-1}{2}$. A negative number's magnitude can be found by applying Definition 1 and Lemma 1; it must fall in the positive range. In this case, the unsigned number comparisons described in Theorem 3 and 4 are applicable. The following definition is to define overflows in the positive range of the RNS.

Definition 2 *Suppose that there are two positive numbers in RNS, $X = (x_1, x_2, ..., x_n)$ and $Y = (y_1, y_2., , , .y_n)$. Overflow exists if $X + Y > \frac{M-1}{2}$.*

Notice that Definition 2 considers the case that both X and Y are positive numbers. This definition is referred to when we discuss overflow detection in the addition of two numbers.

Corollary 1 *The overflow detection theory in Definition 2 applies to the addition of only two numbers.*
Proof:
The maximal number in Definition 2 is $(\frac{M-1}{2})$, and the maximal sum of two numbers is $(M - 1)$ which is within the dynamic range. If there are 3 numbers or more, the maximal sum of those numbers is greater than $(M - 1)$ which is out of the dynamic range, and by the definition of RNS the sum is not correct. Therefore, the overflow detection theory described in Definition 2 is correct only for two-number additions.

2.4 Multiplicative Inverse

Consider the number $|b|_m$, Szabo and Tanaka [4] define the multiplicative inverse as follows.

Definition 3 *If $0 \leq a < m$ and $|ab|_m = 1$, a is called the multiplicative inverse of b mod m, and is denoted as $|b^{-1}|_m$.*

Notice that the multiplicative inverse of a number does not always exist. The following theorem from [4] describes the condition of its existence and the proof is omitted.

Theorem 5 *The quantity $|b^{-1}|_m$ exists if and only if the greatest common divisor of b and m, $gcd(b, m)$, is equal to 1, and $|b|_m \neq 0$. In this case $|b^{-1}|_m$ is unique.*

We have already developed several efficient methods (given in Theorem 3, Theorem 4, and Definition 2) for number comparison and overflow detection in order to perform the addition of two positive numbers. We use these theorems to derive the division algorithm for signed RNS numbers in the following section.

3 Division Algorithm

We now present a division algorithm in RNS using sign magnitude arithmetic and binary search.

3.1 Descriptions of the Algorithm

Given two numbers, dividend X and divisor Y, the division in RNS is to find the quotient $Z = \lfloor \frac{X}{Y} \rfloor$, where $\lfloor \frac{X}{Y} \rfloor$ denotes the greatest integer which is not greater than $\frac{X}{Y}$. As mentioned before, this algorithm is classified as a subtractive algorithm. Therefore, it is necessary to detect the sign in the subtraction and the overflow in the addition. Theorem 3 and 4 provide an efficient way to perform the number comparison. The absolute value of the dividend and the divisor are used when performing the division calculation, and the the overflow in the addition of two numbers is detected by applying Definition 2. In addition, the signs of the dividend and the divisor need to be detected, and the negative numbers need to be complemented. After finishing the division on two absolute values, it is necessary to transfer the quotient to the proper representation(positive or negative) in RNS. Given modulus set $(m_1, m_2, ..., m_n)$ with dividend $X = (x_1, x_2, ..., x_n)$ and divisor $Y = (y_1, y_2, ..., y_n)$, we are to find the quotient Z, where $Z = \lfloor \frac{X}{Y} \rfloor$. The dynamic range, M, of the RNS is $M = \prod_{i=1}^{n} m_i$. Corollary 1 tells us that the overflow detection can be applied only to the addition of two numbers, a special case of which is the addition of two equal numbers. In other words, multiplying a number by 2 is allowed, and our algorithm is developed on this basis (see Part II below).

This algorithm can be divided into four parts. Part I detects the signs of the dividend and the divisor and transfers them to positive numbers. Part II finds 2^j, such that $Y \cdot 2^j \leq X < Y \cdot 2^{j+1}$, and finds the difference between 2^j and the quotient. Part III deals with the case $Y \cdot 2^j \leq X \leq \frac{M-1}{2} < Y \cdot 2^{j+1}$ which is from Part II, and finds out the difference between 2^j and the quotient. Part IV transfers the quotient to the proper representation in RNS (positive or negative).

Part I.

The largest number in the positive range of the RNS is $\frac{M-1}{2}$, and for convenience we set a variable $M_p = \frac{M-1}{2}$. Using Theorem 3 and 4, we compare the dividend and the divisor with M_p. If the dividend or the divisor is less than or equal to M_p, then the dividend or the divisor is positive, and nothing needs to be changed. On the other hand, if the dividend or the divisor is greater than M_p, then the dividend or the divisor is negative, and it should be complemented. If either the dividend or the divisor is negative, we have to set the sign variable, SIGN, to 1. SIGN will be used to convert the quotient to a proper form in Part IV.

Part II.

We find the proper 2^j such that $Y \cdot 2^j \leq X < 2^{j+1}$ in the following way. Two variables, LowerBound and UpperBound, are set to record the range in which the value of the quotient is to be found. The LowerBound and the UpperBound will dynamically change, as the algorithm is executed. In iteration j, LowerBound= 2^j and UpperBound= 2^{j+1}, we repeatedly compare $(2^j \cdot Y)$ with X and detect whether $(2^{j+1} \cdot Y)$ is greater than $\frac{M-1}{2}$, until we find some j, denoted as $\hat{j}$, such that $Y \cdot 2^j \leq X < Y \cdot 2^{\hat{j}+1}$. QuotientBase records the LowerBound when the procedure halts, and it is unchanged until the end of the division operation. QuotientExt records the difference between the exact quotient value and the QuotientBase, and the initial value of QuotientExt is set to 0. The final value of the quotient is equal to QuotientBase+QuotientExt.

In each iteration, the UpperBound is updated by doubling its value. Two cases may occur when the above procedure halts. In one case, (UpperBound·Y) is smaller than $\frac{M-1}{2}$. Then a binary search starts to find the difference between QuotientBase and the quotient, and we need $\hat{j} - 1$ steps to finish this part, since 2^j integers exist in the range $\left[2^j, 2^{j+1}\right)$. In each step of the binary search, we have to compare X with $\left(Y \cdot \frac{\text{UpperBound}+\text{LowerBound}}{2}\right)$.

Here, $\left(Y \cdot \frac{\text{UpperBound}+\text{LowerBound}}{2}\right)$ for each modulus, $\left[|2^{-1}|_{m_i} \cdot |Y|_{m_i} \cdot (|\text{UpperBound}|_{m_i} + |\text{LowerBound}|_{m_i})\right]_{m_i}$, is to be found. Hence, the multiplicative inverse of 2, $|2^{-1}|_{m_i}$, needs to be pre-

pared. If $\left(X - Y \cdot \frac{\text{UpperBound}+\text{LowerBound}}{2}\right) < 0$, then set UpperBound= $\frac{\text{UpperBound}+\text{LowerBound}}{2}$ and QuotientExt= 2·QuotientExt. Otherwise set LowerBound= $\frac{\text{UpperBound}+\text{LowerBound}}{2}$ and QuotientExt= (2·QuotientExt+1). When this procedure is finished, go to Part IV.

In the other case, UpperBound·Y may be greater than $\frac{M-1}{2}$ and overflow thus occurs. Then we go to Part III.

Part III.

If there is an overflow, it means that $(Z \cdot Y)$ lies between $Y \cdot 2^j$ and $\frac{M-1}{2}$. We therefore update UpperBound as $\frac{\text{UpperBound}+\text{LowerBound}}{2} = 2^j + 2^{j+1}$, and LowerBound as 2^j. QuotientExt is updated as QuotientExt=(QuotientExt·2). Then we examine whether $\left(\frac{\text{UpperBound}+\text{LowerBound}}{2} \cdot Y\right)$ overflows again. Continue this procedure until $\left(\frac{\text{UpperBound}+\text{LowerBound}}{2} \cdot Y\right)$ does not overflow. If $\left(\frac{\text{UpperBound}+\text{LowerBound}}{2} \cdot Y\right)$ does not overflow and $\left(X - \frac{\text{UpperBound}+\text{LowerBound}}{2} \cdot Y\right) \geq 0$, set LowerBound= $\frac{\text{UpperBound}+\text{LowerBound}}{2}$ and QuotientExt=(QuotientExt·2 + 1), and detect overflow again. If $\left(\frac{\text{UpperBound}+\text{LowerBound}}{2} \cdot Y\right)$ does not overflow and $\left(X - \frac{\text{UpperBound}+\text{LowerBound}}{2} \cdot Y\right) < 0$, set UpperBound= $\frac{\text{UpperBound}+\text{LowerBound}}{2}$ and QuotientExt=(QuotientExt·2), and perform the similar operations as defined in the binary search in Part II. As in Part II, after $\hat{j} - 1$ steps go to Part IV. In the above procedure, if (UpperBound − LowerBound) = 1, then the job is finished. Let the quotient equal LowerBound and go to Part IV to get the proper quotient expression (as a positive number or a negative number).

Part IV.

The absolute value of Quotient equals the sum of QuotientExt and QuotientBase. From Part I, the exact quotient may be negative. Therefore, if SIGN=1, the absolute value of the found quotient should be complemented.

3.2 The Correctness of the Algorithm

The absolute value of a quotient is equal to the quotient of the absolute value of the dividend and the absolute value of the divisor. The sign of the quotient depends on the signs of the dividend and the divisor. If the signs of the dividend and the divisor are different, then the quotient is negative. Otherwise, the quotient is positive.

The dynamic range of the positive numbers in RNS, $\left[0, \frac{M-1}{2} = M_p\right]$, can be divided into several subintervals. The boundaries of the subintervals are as follows:

$$0, (2^0 \cdot Y), (2^1 \cdot Y), (2^2 \cdot Y), ..., (2^j \cdot Y), ..., (2^n \cdot Y), M_p.$$

We are to find the proper j such that $2^j \cdot Y < X <$

$2^{j+1} \cdot Y$, i.e., to find the subinterval $\left[2^j, 2^{j+1}\right)$ in which the quotient lies. Usually M_p is not a power of 2, therefore, if $2^n \cdot Y < X < M_p$, i.e., when $j = n$, the search for j has to be stopped, and a binary search is to start. In this case two variables, UpperBound and LowerBound, are set as UpperBound$=2^{(n+1)}$, and LowerBound$=2^n$. Then we continuously calculate $\frac{\text{UpperBound}+\text{LowerBound}}{2}$ for the quotient estimation, and compare $\left(\frac{\text{UpperBound}+\text{LowerBound}}{2} \cdot Y\right)$ with X. Recursively substituting the UpperBound or LowerBound with this estimated value, we can reduce the range in which the quotient lies, and finally find the quotient value. Notice that initially, UpperBound$= 2\cdot$LowerBound, hence

$$\frac{\text{UpperBound} + \text{LowerBound}}{2} = \text{LowerBound} + \frac{\text{LowerBound}}{2}. \tag{2}$$

If UpperBound is updated with this value, then the new estimation is equal to

$$\frac{\text{LowerBound} + \text{LowerBound} + \frac{\text{LowerBound}}{2}}{2} = \text{LowerBound} + \frac{\text{LowerBound}}{4}.$$

If UpperBound remains the same, as big as $2\cdot$LowerBound, and LowerBound is updated with the value in (2), then the new estimation will be

$$\text{LowerBound} + \frac{\text{LowerBound} + \frac{\text{LowerBound}}{2}}{2} = \text{LowerBound} + \frac{\text{LowerBound}}{2} + \frac{\text{LowerBound}}{4},$$

and so on. It is easy to find that the quotient can be represented by

$$\text{LowerBound} + \sum_i \lambda_i \frac{\text{LowerBound}}{2^i}, \quad \text{with } \lambda_i = 0 \text{ or } 1.$$

The first term is then recorded as QuotientBase in the algorithm and the second term QuotientExt. The quotient in the division is found as QuotientBase+QuotientExt which is hence correct.

3.3 Division Algorithm

The flow chart of the algorithm is shown in Figure 1, and the detailed division algorithm is as following.

```
/* Suppose that m1, m2, ..., and mN are
/* N moduli which are pair-wise prime and
/* all odd numbers. Let M=m1*m2*m3*...*mN,
/* and Mp=(M-1)/2 be the largest positive
/* number. We use the equation
/* Dividend/Divisor=Quotient+
/* Remainder/Divisor, and the Quotient is
```

The symbols used in the flowchart are listed below each followed by the corresponding variable used in the algorithm.

X: Dividend,
Y: Divisor,
Z: Quotient,
UB: UpperBound,
LB: LowerBound,
QE: QuotientExt,
QB: QuotientBase,
B: Bounding,
$\overline{Z}$: COMPLEMENT(Z).

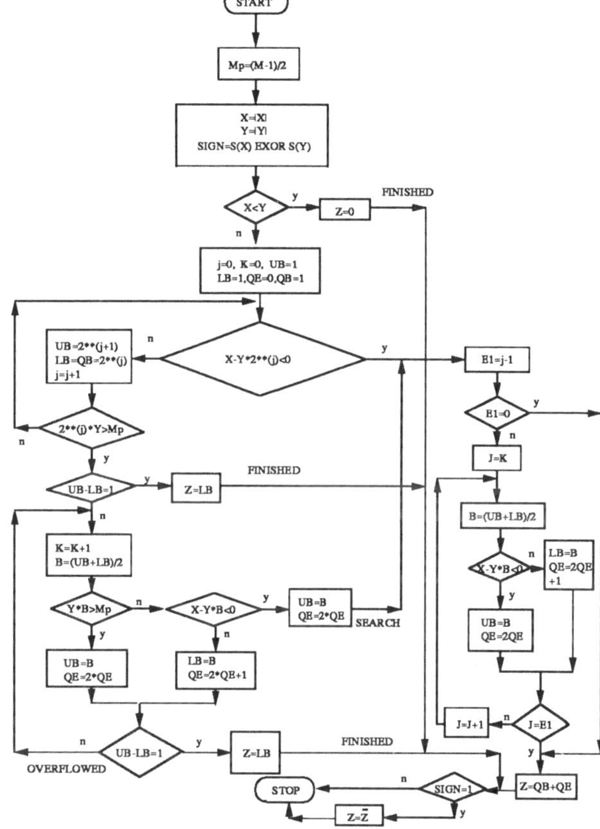

Figure 1: Flowchart of the Division Algorithm

```
/* is to be found.

START PROCEDURE;
/* Check the signs of Dividend and Divisor
/* and set a register SIGN as
/* SIGN=[S(Dividend) EXOR S(Divisor)] to
/* save the sign of the Quotient, where
/* S(..) means the sign of .. . Define S(..)
/* =1 for a negative number, and S(..)=0 for
/* a positive number.
   Mp=(M-1)/2
   S(Dividend)=0, S(Divisor)=0
   IF Dividend < 0
```

```
  THEN Dividend=COMPLEMENT(Dividend)
        /* COMPLEMENT(Dividend)=M-Dividend
    S(Dividend)=1
  END IF;
  IF Divisor < 0
    THEN Divisor=COMPLEMENT(Divisor)
      S(Divisor)=1
  END IF;
  SIGN=S(Dividend) EXOR S(Divisor)
  IF Dividend<Divisor
      THEN Quotient=0, GO TO FINISHED;
/* Find out which interval the quotient
/* falls in, i.e., find j, such that
/* 2**j<Quotient<2**(j+1)
  j=0, K=0, UpperBound=1, LowerBound=1,
  QuotientExt=0, QuotientBase=1
    WHILE (Dividend-Divisor*2**(j))>=0
      THEN DO
      UpperBound=2**(j+1)
      LowerBound=QuotientBase=2**(j)
      j=j+1
      IF 2**(j)*Divisor>Mp
        THEN
          IF (UpperBound-LowerBound)=1
          THEN Quotient=LowerBound,
              GO TO FINISHED;
          END IF;
OVERFLOW:K=K+1
          Bounding=(UpperBound+LowerBound)/2
          IF Divisor*Bounding>Mp
          THEN
            UpperBound=Bounding
            QuotientExt=2*QuotientExt
          ELSE
            IF (Dividend-Divisor*Bounding)<0
            THEN
              UpperBound=Bounding
              QuotientExt=2*QuotientExt
              GO TO SEARCH
            ELSE
              LowerBound=Bounding
              QuotientExt=2*QuotientExt+1
            END IF;
          END IF;
          IF (UpperBound-LowerBound)=1
          THEN
            Quotient=LowerBound
            GO TO FINISHED;
          ELSE
            GO TO OVERFLOW;
          END IF;
      END IF;
    END WHILE;
/* Binary search for QuotientExt
SEARCH:E1=j-1
    IF E1<>0
      THEN
        FOR J=K TO E1
        DO
          Bounding=(UpperBound+LowerBound)/2
          IF (Dividend-Divisor*Bounding)<0
            THEN
              UpperBound=Bounding
```

```
              QuotientExt=2*QuotientExt
            ELSE
              LowerBound=Bounding
              QuotientExt=2*QuotientExt+1
            END IF;
        END FOR;
      END IF;
/* The found quotient equals the sum of
/* QuotientBase and QuotientExt. If SIGN=1
/* it means that Quotient is a negative
/* number, then the final Quotient is the
/* complement of the found Quotient.
Quotient=QuotientBase+QuotientExt
FINISHED:IF SIGN=1
        THEN Quotient=COMPLEMENT(Quotient)
        END IF;
END PROCEDURE;
```

3.4 Example of the Division Algorithm

Suppose that moduli are $m_1 = 3$, $m_2 = 5$, and $m_3 = 7$. Given $X = (2, 1, 1) = -34$ and $Y = (2, 0, 5) = 5$, find quotient $Z = \frac{X}{Y}$.

Since the moduli are $m_1 = 3$, $m_2 = 5$, and $m_3 = 7$, we can find $M = m_1 \cdot m_2 \cdot m_3 = 105$, $M_p = \frac{M-1}{2} = 52 = (1, 2, 3)$. The multiplicative inverses of 2, which are used in the calculation of $\frac{\text{UpperBound}+\text{LowerBound}}{2}$, corresponding to m_1, m_2, and m_3 are $|2^{-1}|_{m_1} = 2$, $|2^{-1}|_{m_2} = 3$, and $|2^{-1}|_{m_3} = 4$ respectively. Parity checking uses Table 1. The quotient can be calculated in the following steps with the required variables. The short notations of these variables are listed in Figure 1.

1. $S(-34) = 1,$ $S[(2, 1, 1)] = 1,$
   $S(5) = 0,$ $S[(2, 0, 5)] = 0,$
   SIGN=1,   SIGN=1,
   COMP$(-34) = 34.$ COMP$[(2, 1, 1)] = (1, 4, 6).$

2. $34 > 5 \cdot 2^0,$ $(1, 4, 6) > (2, 0, 5),$
   $j = 0.$ $j = 0.$

3. $34 > 5 \cdot 2^1,$ $(1, 4, 6) > (2, 0, 5) \cdot (2, 2, 2)$
       $= (1, 0, 3),$
   $j = 1.$ $j = 1.$

4. $34 > 5 \cdot 2^2,$ $(1, 4, 6) > (1, 0, 3) \cdot (2, 2, 2)$
       $= (2, 0, 6),$
   $j = 2.$ $j = 2.$

5. $5 \cdot 2^3 > 34 > 5 \cdot 2^2,$ $(2, 0, 5) \cdot (2, 2, 2) = (1, 0, 5)$
       $> (1, 4, 6) > (2, 0, 6),$
   QB=$2^2,$ QB=$(1, 4, 4),$
   $J = 0,$ $J = 0,$
   UB= $2^3,$ UB= $(2, 2, 2) \cdot (2, 2, 2) \cdot$
       $(2, 2, 2) = (2, 3, 1),$
   LB= $2^2,$ LB= $(2, 2, 2) \cdot (2, 2, 2)$
       $= (1, 4, 4),$
   B= $\frac{2^3+2^2}{2} = 6,$ B= $\frac{(2,3,1)+(1,4,4)}{2} = (0, 1, 6),$
   QE= $2 \cdot 0 + 1 = 1,$ QE= $(0, 0, 0) + (1, 1, 1)$
       $= (1, 1, 1),$
   Set LB=B.   Set LB=B= $(0, 1, 6).$

6. $5 \cdot 2^3 > 34 > 5 \cdot 6,$ $(1, 0, 5) > (1, 4, 6) >$
       $(2, 0, 5) \cdot (0, 1, 6) = (0, 0, 2),$
   $J = 1,$ $J = 1,$

$$\begin{aligned}
&\text{UB}= 2^3, && \text{UB}= (2,3,1),\\
&\text{LB}= 6, && \text{LB}= (0,1,6),\\
&\text{B}= \tfrac{2^3+6}{2}=7, && \text{B}= \tfrac{(2,3,1)+(0,1,6)}{2}=(1,2,0),\\
&\text{QE}= 2\cdot 1+0=2, && \text{QE}= (2,2,2)\cdot(1,1,1)\\
& && +(0,0,0)=(2,2,2),\\
&\text{Set UB}=\text{B}. && \text{Set UB}=\text{B}=(1,2,0).\\
&7.\ |Z|=\text{QB}+\text{QE} && |Z|=\text{QB}+\text{QE}=(1,4,4)\\
&\quad = 2^2+2=6. && +(2,2,2)=(0,1,6).\\
&8.\ \text{SIGN}=1, && \text{SIGN}=1,\\
&\quad Z=\text{COMP}(6), && Z=\text{COMP}(0,1,6),\\
&\quad = -6. && = (0,4,1).
\end{aligned}$$

3.5 Discussions

This algorithm requires four parts of calculation. Constant time is needed in Part I to find the absolute values of the dividend and the divisor, and in Part IV to transfer the absolute value of the quotient, $|Z|$, to the proper form. In Part II and III, our algorithm needs $(2\cdot\log_2 Z)$ steps to finish the division operation. The first $\log_2 Z$ steps find the range which the quotient falls in, and the second $\log_2 Z$ steps find the difference between QuotientBase and the quotient. Each step needs several RNS addition and subtraction operations, one RNS multiplication, and a table look-up for the parities. The RNS arithmetic operations do not need quotient estimation, base extension, or mixed radix number conversion, which makes this algorithm very fast and easy to implement compared to previous proposals.

4 Conclusions

We have presented a division algorithm which needs only simple RNS arithmetic operations, and can be easily implemented. This is a general division algorithm, with no restrictions to either dividend or divisor. No estimation of the quotient is required before the division is executed. These characteristics make the calculation less complicated, more efficient, and speedier.

We also presented a very good and easy technique for overflow detection and number comparison. In the traditional way of detecting overflow and comparing numbers in RNS, mixed radix numbers have to be used. This is time consuming and requires complex hardware. Our method is more efficient and less complicated than the existing algorithms.

A parity-checking technique is presented in this paper for number comparisons and overflow detection. With today's advanced VLSI technology, we will have no difficulty building a parity table that lists all the moduli with parities. Some small tables may also be needed to store data such as the values of the multiplicative inverse of 2, $|2^{-1}|_{m_i}$. Except the tables mentioned above, no other table are required, and all we need is simple arithmetic calculations. This algorithm can be easily implemented on hardware and can achieve good time performance which is logarithmic to the size of the quotient.

Acknowledgements

The authors would like to thank the anonymous reviewer for his helpful comments.

References

[1] W. K. Jenkins and B. J. Leon, "The use of residue number systems in the design of finite impulse response digital filters," *IEEE Transactions on Circuits Systems*, vol. CAS-24, no. 4, pp. 199–201, 1973.

[2] F. J. Taylor, "A VLSI residue arithmetic multiplier," *IEEE Transactions on Computers*, vol. C-31, pp. 540–546, June 1982.

[3] D. D. Miller and J. N. Polky, "An implementation of the LMS algorithm in the residue number system," *IEEE Transactions on Circuits System*, vol. CAS-31, pp. 452–461, May 1984.

[4] N. S. Szabo and R. I. Tanaka, *Residue Arithmetic and Its Application to Computer Technology*. New York: McGraw-Hill, 1967.

[5] W. A. Chren Jr., "A new residue number system division algorithm," *Computers Math. Applic.*, vol. 19, no. 7, pp. 13–29, 1990.

[6] S. Waser and M. J. Flynn, *Introduction to Arithmetic for Digital System Designers*, ch. 5, p. 172. New York: Holt, Rinehart & Winston, 1982.

[7] D. K. Banerji, T. Y. Cheung, and V. Ganesan, "A high-speed division method in residue arithmetic," in *5th IEEE Symp. on Computer Arithmetic*, pp. 158–164, 1981.

[8] E. Kinoshita, H. Kosako, and Y. Kojima, "General division in the symmetric residue number system," *IEEE Transactions on Computers*, vol. C-22, pp. 134–142, February 1973.

[9] Y. A. Keir, P. W. Cheney, and M. Tannenbaum, "Division and overflow detection in residue number systems," *IRE Transactions on Electronic Computers*, vol. EC-11, pp. 501–507, August 1962.

[10] M.-L. Lin, E. Leiss, and B. McInnis, "Division and sign detection algorithm for residue number systems," *Computers Math. Applic.*, vol. 10, no. 4/5, pp. 331–342, 1984.

[11] D. D. Miller, J. N. Polky, and J. R. King, "A survey of Soviet developments in residue number theory applied to digital filtering," in *26th Midwest Symp. Circuits Systems*, August 1983.

New Approach to Integer Division in Residue Number Systems

Dragan Gamberger

Rudjer Bošković Institute

41000 Zagreb, Yugoslavia

Abstract

A new division algorithm substantially different from the known ones, and especially appropriate for the Residue Number Systems (RNS), is presented. It makes use of the fact that multiplicative inverse element of a divisor, that is relatively prime to system moduli, can be easily determined in the RNS. The number of its iterations depends only on the magnitude of the divisor and the moduli of the system. The problems in the algorithm realization are analyzed in detail and a complete solution using the incompletely specified RNS is described.

1 Introduction

It is known that complexity of integer division in the Residue Number System (RNS) represents one of the main reasons for relatively seldom usage of this number system. Practically, the RNS is used only in the applications where it is possible to avoid integer division at all, like digital filtering or number theoretic transforms. Although two different algorithms for integer division have been suggested already 23 years ago [1], and although there have been some efforts to realize them [2], not a single successful implementation of integer division has been published yet.

One of the suggested integer division algorithms makes use of the method similar to the conventional binary division. The application of this algorithm and its modifications have the main disadvantage that each iteration requires magnitude comparison. This operation is very fast in weighted number systems but relatively complicated and slow in the RNS [1,3]. Apart from that, this algorithm requires a table of the residue representations of the integer powers of 2. This table is iteratively used at the beginning of the algorithm in the computation of an approximate quotient, and later in the iterative process of successive approximation in the computation of the correct quotient [1,2].

The second integer division method suggested (division by approximate divisor) does not have this disadvantages. It makes use of the fact that in the RNS the division by a product of some system moduli can be realized relatively simply [3]. The idea is to substitute the real divisor by an approximate one, i.e. by a product of some system moduli and to use this division iteratively until a correct quotient is found. Algorithm correctness is assured only if for any divisor Y there is an approximate divisor $\tilde{Y}$ ($Y \leq \tilde{Y} < 2Y$), what

requires the RNS with specially chosen moduli. The main drawback of this algorithm is its need for special logic and look-up tables in the determination of the best approximate divisor and that the rate of convergence depends not only on the dividend and divisor magnitude but on the quality of divisor approximation, as well [1].

The objective of this paper is to present a novel division algorithm substantially different from the mentioned ones. In the section 2 of this paper the algorithm is described and the proof of its correctness is given. In the section 3 the problems of its realization are analyzed and a complete solution is presented. The unique algorithm characteristic, that the number of iterations depends only on the magnitude of divisor, is analyzed in section 4.

In the presentation the realization by read only memories (ROM's) as in [3] is supposed. Hardware and time complexity of the presented algorithms is measured by the number of necessary ROM's for its realization and the worst case number of successive memory accesses, called modular steps, respectively.

2 General Algorithm Principles

The goal of the algorithm is to compute the quotient Z

$$Z = \left\lfloor \frac{X}{Y} \right\rfloor$$

where both X and Y are positive integers and where $\lfloor a \rfloor$ denotes the greatest integer equal or less than a . The trivial case is for $Y = 1$ when it holds $Z = X$.

For the non-trivial case $Y > 1$, the idea of the algorithm is to compute a new dividend X' and a new divisor Y', both positive integers so that conditions (1) are satisfied

$$X' < X$$
$$1 \leq Y' < Y \tag{1}$$
$$Z = \left\lfloor \frac{X}{Y} \right\rfloor = \left\lfloor \frac{X'}{Y'} \right\rfloor .$$

It means that we are looking for new dividend and divisor that are smaller than the original ones but have the same quotient. Now if Y' equals 1 , then Z equals X', and if $Y' > 1$ then X' and Y' can be used as the new starting dividend and divisor, respectively. The process can be iterated. After a finite number of steps, some Y' must be equal to 1 and the corresponding X' is the result of division.

The presented idea can be applied in any number system but the problem is how to efficiently compute X' and Y' which satisfy conditions (1). A general solution to this problem is not known. A solution applicable only for the RNS defined with N moduli $m_1, m_2, \ldots, m_N$, that are different prime numbers is presented in this paper. In such RNS a number X is represented by N-tuple $X = (x_1, x_2, \ldots, x_N)$, where $x_i = X \bmod m_i$ and there is a unique representation for each positive integer in the range $X < M$, where M is the product of all system moduli. We suppose that the moduli are selected so that this condition is satisfied both for the dividend and the divisor.

In the RNS thus defined the new dividend and the new divisor can be computed in each iteration as follows:

<u>A iteration</u>: If $\gcd(Y, M) = G > 1$, where gcd denotes greatest common divisor function

$$X' = \left\lfloor \frac{X}{G} \right\rfloor \tag{2}$$

$$Y' = \frac{Y}{G}. \tag{3}$$

It is obvious that relations (2) and (3) satisfy conditions (1) and that they can be relatively easily realized in the RNS because they both are divisions by some product of system moduli.

<u>B iteration</u>: If $\gcd(Y, M) = 1$, what means that the divisor is relatively prime with system moduli, then X' and Y' can be computed by

$$X' = \left\lfloor \frac{XD}{M} \right\rfloor \tag{4}$$

$$Y' = \left\lfloor \frac{YD}{M} \right\rfloor = \frac{YD - 1}{M} \tag{5}$$

where D represents multiplicative inverse element of $Y \bmod M$

$$YD = 1 \bmod M \tag{6}$$

or in other words

$$YD = kM + 1 \quad (k \text{ is an integer}).$$

D satisfying inequality $1 < D < M$ always exists, and it is unique according to the theorem 2-2 in [1] and the fact that conditions $\gcd(Y, M) = 1$ and $1 < Y < M$ are both true in this case. The proof that the relations (4) and (5) thus defined satisfy conditions (1) is given in the appendix.

Computation of the multiplicative inverse element in the RNS is not a problem because its each modulus defines a separate finite field and the operation can be performed independently and completely in parallel for different moduli. Because weighted number systems are indefinite fields the necessity to compute the multiplicative inverse element in this algorithm is the

main reason that it can not be implemented in them. Even if weighted number systems would be used for representing finite fields, the presented algorithm could not be implemented in them because multiplicative inverse element computation could be executed only by iterative repetition of integer division itself. In order to demonstrate only the ideas of the new algorithm, in Example 1 we suppose for a moment that decade number system presents a finite field with M elements and that we know how to compute multiplicative inverse elements in it.

Example 1 Divide 502 by 15 in a decade number system with $M = 1024$;

1. iteration $X = 502$, $Y = 15$, $\gcd(15, 1024) = 1$
   $D = 751$ because of $751 * 15 = 11 * 1024 + 1$
   $X' = \lfloor 502 * 751/1024 \rfloor$, $Y' = (15 * 751 - 1)/1024$

2. iteration $X = 368$, $Y = 11$, $\gcd(11, 1024) = 1$
   $D = 931$ because of $931 * 11 = 10 * 1024 + 1$
   $X' = \lfloor 368 * 931/1024 \rfloor$, $Y' = (11 * 931 - 1)/1024$

3. iteration $X = 334$, $Y = 10$, $\gcd(10, 1024) = 2$
   $X' = \lfloor 334/2 \rfloor$, $Y' = 10/2$

4. iteration $X = 167$, $Y = 5$, $\gcd(5, 1024) = 1$ $D = 205$ because of $205 * 5 = 1024 + 1$ $X' = \lfloor 167 * 205/1024 \rfloor$, $Y' = (5 * 205 - 1)/1024$

5. iteration $X = 33$, $Y = 1$, $Z = 33$.

3 The Algorithm Implementation Analysis

Although at first glance it might seem that both A and B iterations can be easily realized in the RNS, practically it is not so. In the A iteration, for example, both relations (2) and (3) include division by a variable product of moduli. Realization of division by any defined product of moduli is not a great problem [3] but the complete solution should include hardware for all possible combinations of product of moduli and this number grows very fast with the number of moduli. For example, in the RNS with 5 moduli 30 different products of moduli are possible and according to realization suggested in [3], 395 look-up tables stored in ROM's are necessary only for this operation. The same problem exists in the integer division algorithm by an approximate divisor. In this paper the problem is solved so that the relations (2) and (3) are realized by the same hardware as relations (4) and (5) of the B iteration.

In the realization of the B iteration, as we have already noticed, the computation of the multiplicative inverse element D and the multiplication by it are not problems in the RNS. The division by M can be also easily realized because it is a constant product of moduli. The main problem is that the results of multiplication XD and YD should be presented in the RNS in which variables of the magnitude $(M - 1)(M - 2)$ can be uniquely represented. The problem might be

85

solved so that we introduce the extended RNS that besides N moduli $m_1, m_2, \ldots, m_N$ includes additional R moduli $p_1, p_2, \ldots, p_R$, where P, product of all the additional moduli p_i, is at least $M - 1$. In this case all the computations should be done in the so defined extended RNS except that the multiplicative inverse element D is computed only for the moduli m_i and then transformed by the standard base extension algorithm to the representation in the extended RNS. The drawback of this solution is its relative great hardware complexity and execution time. The execution of the B iteration in the extended RNS requires at least $2N + R + 1$ modular steps: N for the base extension algorithm, $N + R$ for division and 1 for multiplication. Only division by M in an extended RNS system with 5 main moduli and 5 additional moduli would require 65 look-up tables in ROM's.

To make the problem easier we suggest the following modifications. The relations (4), (5), and (6) can be rewritten as

$$X' = \frac{XD - (XD) \bmod P}{P} \tag{7}$$

$$Y' = \frac{YD - 1}{P} \tag{8}$$

$$YD = 1 \bmod P. \tag{9}$$

The first modification is that instead of the extended RNS we have introduced a new RNS, called auxiliary RNS defined with R moduli $p_1, p_2, \ldots, p_R$. Each modulus p_i of the auxiliary RNS should be a prime number different from all the moduli of the main system and selected so that their product P is greater than the product of all main moduli M. Multiplicative inverse element of the divisor is now computed in the auxiliary RNS and division is by the constant P. It can be easily verified in the Appendix that because of $P > M$, relations (7) and (8) satisfy conditions (1), as well.

The second modification refers to the changes in the relation (7) where we have introduced subtraction of the quantity $(XD) \bmod P$. This modification ensures that the difference is divisible by P without remainder. This fact is important because now we do not have to use division algorithm presented in [3] that requires N modular steps but we can use algorithm for division with zero remainder presented in [1,4] executable in only 1 modular step. The second condition for the application of this algorithm is also satisfied in this case because of P is always relatively prime to all the moduli of the main RNS.

We must also notice that relations (7) and (8), in contrast to the relations (4) and (5), do not have to be executed in the extended RNS although their numerators can be out of the range of uniquely defined numbers for the main RNS [4]. The reasons are that we know in advance that the final results are integers uniquely representable in the main RNS and that direct division as defined in [5] is used.

The main advantage of the suggested modification is that B iteration can be executed now in $N + R + 2$

block	number of look-up tables
a	N(N-1)/2 + R(N-1)
b	N(N-1)/2 + R(N-1)
d	R
e	R(R-1)/2 + N(R-1)
f	R(R-1)/2 + N(R-1)
g	N
h	N
i	N

Table 1: Number of look-up tables in the blocks

modular steps. The disadvantage is that magnitude $(XD) \bmod P = (X/Y) \bmod P = Z'$ must be computed and subtracted in relation (7). The complete B iteration based on relations (7), (8), and (9) is presented in Figure 1.

The iteration execution starts with the conversion of magnitudes X and Y to the auxiliary RNS. A distinct hardware is used for each conversion (blocks 'a' and 'b') so that after N modular steps both variables are presented in the auxiliary system. Thereafter the multiplicative inverse element D in this system is computed and immediately transformed back to the main RNS in R modular steps by the block 'e'. Multiplication Y by D in the main system is executed by the block 'h'. In the same block subtraction of 1 and division by P is incorporated and it is possible according to [3] because they both are constants. In parallel to the transformation of D to the main system, X is directly divided by Y in the auxiliary RNS and the result Z' is transformed to the main system by the blocks 'd' and 'f', respectively. The block 'g' waits the number D to be computed in the main RNS and then executes multiplication X by D. The final value of X' is computed by the 'i' block.

Execution time in the form of the modular steps number is given in Figure 1 on the right-hand side of each block. One can easily verify that the total is $N + R + 2$ for the computation of X' and $N + R + 1$ for Y'. Necessary number of look-up tables for the block realizations as functions of the moduli number in the main and the auxiliary systems is given in Table 1.

The total is $N(N + 2R) + R(2N + R - 2)$ look-up tables in ROM's for the whole B iteration. In the system with 5 main and 5 auxiliary moduli, for example, it makes 140 look-up tables.

Example 2 Let us show how iteration B would look like in a concrete example. The task is to compute X' and Y' from $X = 502$ and $Y = 16$ in the RNS with moduli $7, 11, 13$ and $M = 1001$ using auxiliary RNS with moduli $5, 17, 19$ and $P = 1615$.

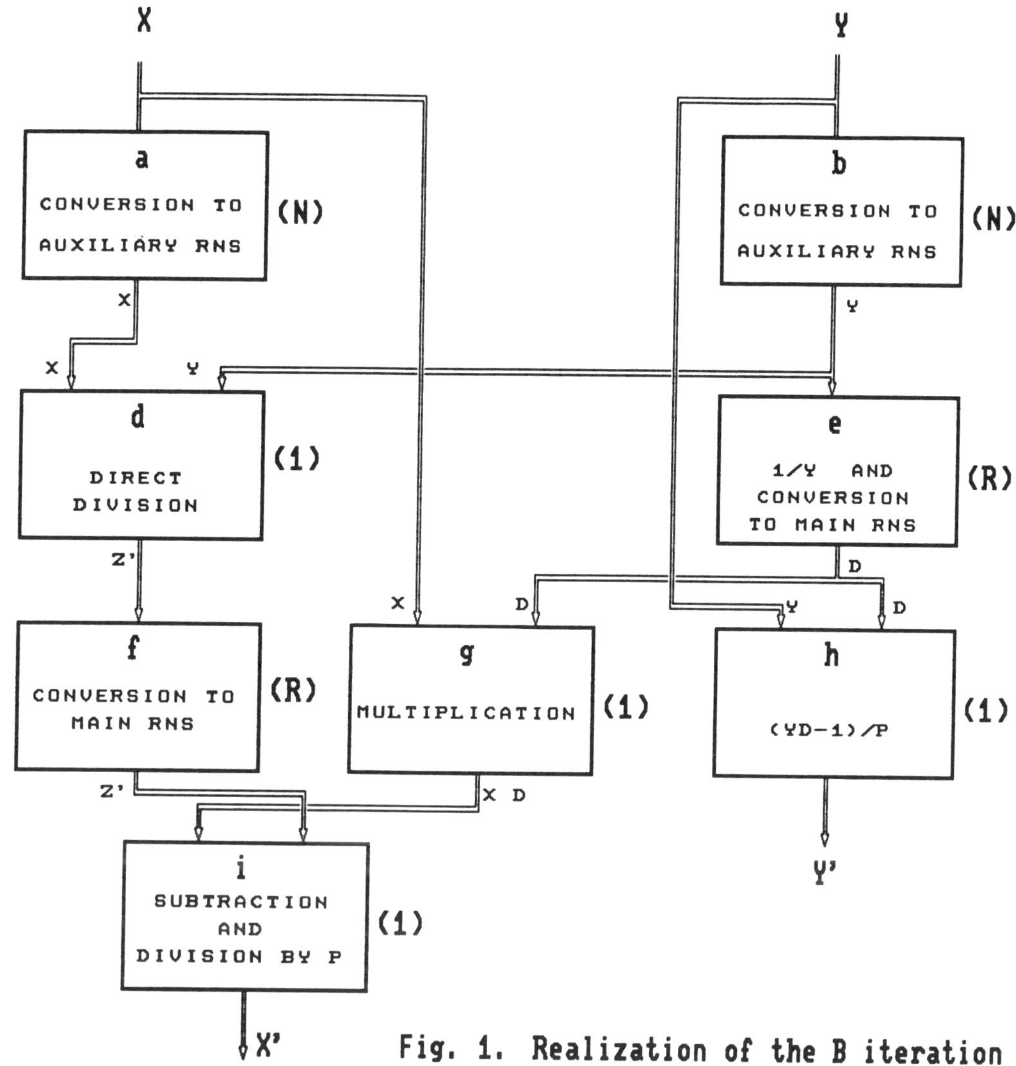

Fig. 1. Realization of the B iteration

main RNS: $m_1 = 7, m_2 = 11, m_3 = 13$ aux. RNS: $p_1 = 5, p_2 = 17, p_3 = 19$

input

$$X = (5, 7, 8) = 502$$
$$Y = (2, 5, 3) = 16$$

output of block

a: $X = (2, 9, 8) = 502$

b: $Y = (1, 16, 16) = 16$

d: $Z' = (2, 8, 10) = 637$

e: $D = (3, 2, 10) = 101$

f: $Z' = (0, 10, 0) = 637$

g: $XD = (1, 3, 2) = 50702$

h: $Y' = \frac{(2,5,3)(3,2,10) - (1,1,1)}{(5,9,3)} = (1, 1, 1) = 1$

i: $X' = \frac{(1,3,2) - (0,10,0)}{(5,9,3)} = (3, 9, 5) = 31$.

Now, we would like to show how the same hardware with slight modifications can be used for the A iteration, as well. We should firstly notice that, because of the fact that the multiplicative inverse element is by the suggested modification computed in the auxiliary RNS, the test whether the divisor is relative prime with the moduli, must be also executed in this system. If $\gcd(Y, P) = G > 1$ then the A iteration follows. Relations (2) and (3) can be rewritten as:

$$X' = \frac{X * P/G - (X \bmod G) * P/G}{P} \qquad (10)$$

$$Y' = \frac{Y * P/G}{P}. \qquad (11)$$

The differences are that the relation (2) is transformed so that division without remainder can be used and that division by a variable G is substituted by the multiplication by P/G and division by the constant P .

The computation problem of the variable P/G can be solved by application of the base extension algorithm for incompletely specified numbers presented in

[5]. This algorithm is qualified by the same time and hardware complexity as the normal base extension algorithm presented in [3] when the prime moduli greater than 2 are used. The operations of both algorithms are equal for completely specified numbers and the completely specified version in blocks 'e' and 'f' can be substituted by the more general incompletely specified version without any change in the B iteration execution. An additional characteristic of the algorithm presented in [5] is as follows: if the starting value is an incompletely specified number X in an RNS system, the result is a completely specified number XM_u presented in an other system, where M_u is the product of moduli of the starting system in which number X had unspecified value.

Figure 2 shows the realization of the complete algorithm including both A and B iterations. Each iteration starts with both values X and Y being transformed by blocks 'a' and 'b' to the auxiliary RNS. Thereafter, by a simple AND/OR logic, it is tested whether Y is relatively prime to P. The variable F is set if this is true. The value of variable F influences the operation of block 'h' and the new introduced block 'c'. If $F = 1$ then the block 'c' output is $\tilde{Y} = Y$ and the same operations, as previously described, are performed by blocks 'd'-'h'.

If Y is not relatively prime to the moduli of the auxiliary system, then $F = 0$ and Y is substituted by $\tilde{Y}$ in the 'c' block. Number $\tilde{Y}$ has the value 1 for the moduli in which number Y had value equal to 0 and the unspecified value $'u'$ for the moduli in which number Y had a value different from 0. Thus the number $\tilde{Y}$ represents an incompletely specified value 1 with the product of unspecified moduli equal P/G.

In Figure 2 it is supposed that the 'c' block is realized by a set of R look-up tables in ROM's and that its execution requires one modular step although it could be also realized by simple random logic circuits. The total time for the execution of one iteration has increased to $N + R + 3$ modular steps and the total hardware requirement is now $N(N + 2R) + R(2N + R - 1)$ look-up tables in ROM's.

The multiplicative inverse element of $\tilde{Y}$ is it itself. After conversion to the main system in the block 'e' the result is $D = P/G$. The value Y' can be computed in the block 'h' in a single modular step by the multiplication YD and division by the constant P. This function is slightly different from the one in the block 'h' for the B iteration where subtraction of constant 1 was also included. Because of that in the block 'h' both expressions should be realized and the appropriate one selected by the variable F.

Block 'd' executes in both iterations the operation of direct division of X by $\tilde{Y}$. In the case when $\tilde{Y}$ has incompletely specified value 1 with the product of unspecified moduli equal P/G then the result of division is $Z' = (X) \bmod G$ with the same unspecified moduli. After conversion to the main system in block 'f' the result is $Z' = (X \bmod G) * P/G$. The result of multiplication in block 'g' is $XD = X * P/G$. This way

we have prepared all the values according to (10) that the same operation in block 'i' as in B iteration can calculate the value of X'.

Example 3 Let us show the iteration A appearing in a concrete example. The task is to compute X' and Y' from $X = 502$ and $Y = 15$ in the RNS with moduli $7, 11, 13$ and $M = 1001$ using auxiliary RNS with moduli $5, 17, 19$ and $P = 1615$.

$$
\begin{array}{ll}
\text{main RNS} & \text{aux. RNS} \\
m_1 = 7, m_2 = 11, m_3 = 13 & p_1 = 5, p_2 = 17, p_3 = 19
\end{array}
$$

input
$$X = (5, 7, 8) = 502$$
$$Y = (1, 4, 2) = 15$$
output of block

a:
$$X = (2, 9, 8) = 502$$
b:
$$Y = (0, 15, 15) = 15$$
$$F = 0$$
c:
$$\tilde{Y} = (1, u, u) = 1$$
d:
$$Z' = (2, u, u) = 2$$
e:
$$D = (1, 4, 11) = 323$$
f:
$$Z' = (2, 8, 9) = 646$$
g:
$$XD = (5, 6, 10) = 16214$$
h:
$$Y' = \frac{(1,4,2)(1,4,11)}{(5,9,3)} = (3, 3, 3) = 3$$
i:
$$X' = \frac{(5,6,10)-(2,8,9)}{(5,9,3)} = (2, 1, 9) = 100 \ .$$

Figure 2 presents other necessary elements of the division hardware, as well: input registers X and Y, output register Z, control logic, and test logic for $Y = 1$. The division begins with the START signal which loads starting values into the input registers and triggers control logic. Afterwards this logic generates a series of L signals with a period of $N + R + 3$ modular steps. Their function is to load values X' and Y' into the input registers. At the end, when condition $Y = 1$ is detected, the S signal stops this series of pulses and generates STOP signal that loads the output register.

It follows a complete division example in the RNS.

Example 4 Divide 502 by 15 in the RNS with moduli $7, 11, 13$ and $M = 1001$ using auxiliary RNS with moduli $5, 17, 19$ and $P = 1615$.

1. iteration $X = (5, 7, 8) = 502$, $Y = (1, 4, 2) = 15$, $S = 0$

   in aux. RNS $X = (2, 9, 8)$, $Y = (0, 15, 15)$,
   $F = 0$, $\tilde{Y} = (1, u, u)$, $Z' = (2, u, u)$,
   $D = (1, u, u)$

   in main RNS $Z' = (2, 8, 9) = 646$,
   $D = (1, 4, 11) = 323$,
   $X' = \frac{(5,7,8)(1,4,11)-(2,8,9)}{(5,9,3)} = (2, 1, 9) = 100$,
   $Y' = \frac{(1,4,2)(1,4,11)}{(5,9,3)} = (3, 3, 3) = 3$

2. iteration $X = (2, 1, 9) = 100$, $Y = (3, 3, 3) = 3$, $S = 0$

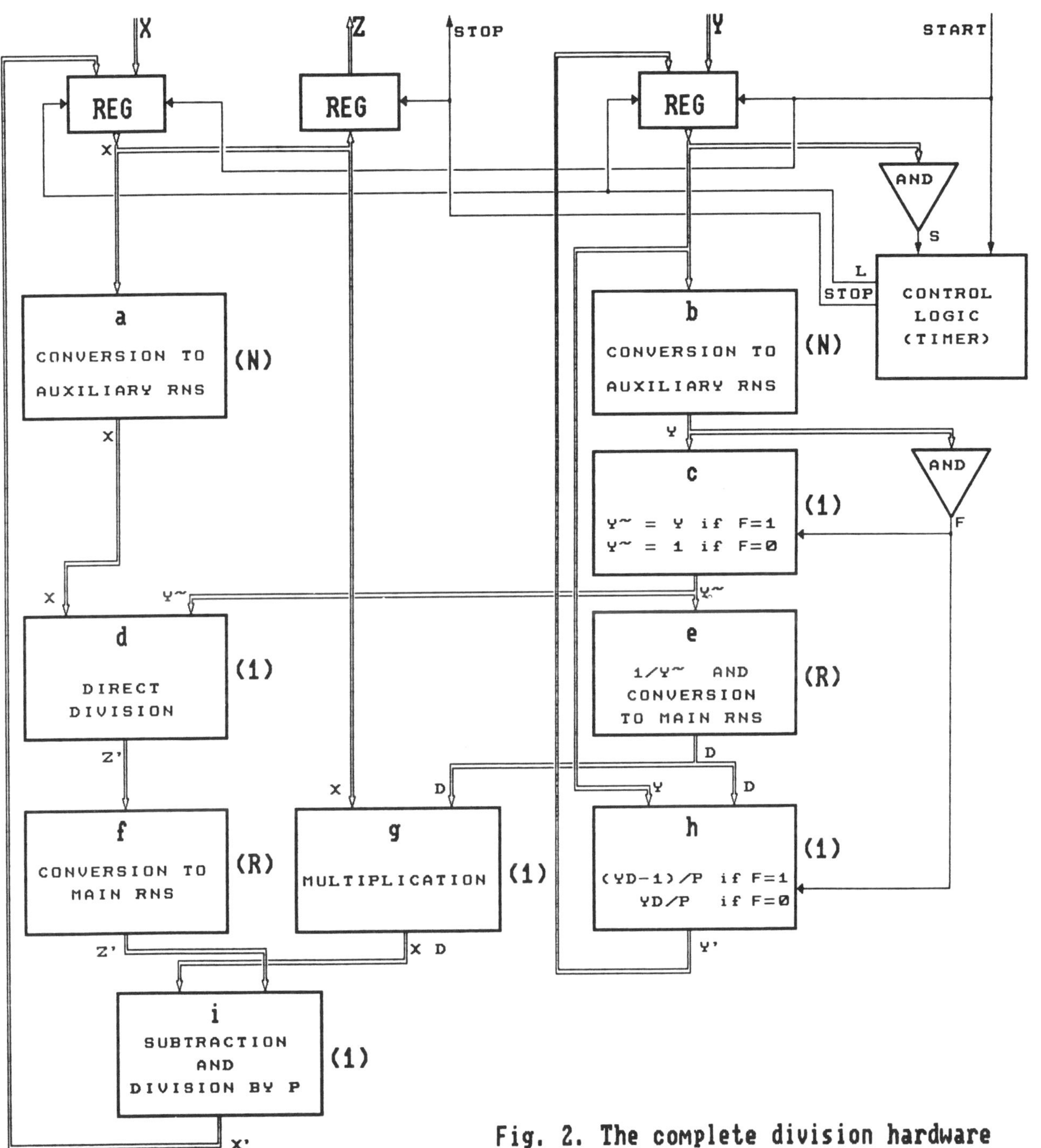

Fig. 2. The complete division hardware

in aux. RNS $X = (0, 15, 5)$, $Y = (3, 3, 3)$,
$F = 1$, $\tilde{Y} = (3, 3, 3)$, $Z' = (0, 5, 8)$,
$D = (2, 6, 13)$

in main RNS $Z' = (4, 10, 5) = 1110$,
$D = (6, 10, 11) = 1077$,
$X' = \frac{(2,1,9)(6,10,11) - (4,10,5)}{(5,9,3)} = (3, 0, 1) = 66$,
$Y' = \frac{(3,3,3)(6,10,11) - (1,1,1)}{(5,9,3)} = (2, 2, 2) = 2$

3. iteration $X = (3, 0, 1) = 66$, $Y = (2, 2, 2) = 2$,
$S = 0$

in aux. RNS $X = (1, 15, 9)$, $Y = (2, 2, 2)$,
$F = 1$, $\tilde{Y} = (2, 2, 2)$, $Z' = (3, 16, 14)$,
$D = (3, 9, 10)$

in main RNS $Z' = (5, 0, 7) = 33$,
$D = (3, 5, 2) = 808$,
$X' = \frac{(3,0,1)(3,5,2) - (5,0,7)}{(5,9,3)} = (5, 0, 7) = 33$,
$Y' = \frac{(2,2,2)(3,5,2) - (1,1,1)}{(5,9,3)} = (1, 1, 1) = 1$

4. iteration $X = (5, 0, 7) = 33$, $Y = (1, 1, 1) = 1$,
$S = 1$, $Z = (5, 0, 7) = 33$.

4 The Unique Characteristic of the Algorithm

In all known division algorithms the number of iterations depends on the magnitude of the quotient Z . On the contrary, the number of iterations in the presented algorithm depends only on the divisor magnitude. It follows the fact that in relations (3) and (5) the value of the new divisor depends only on the value of the previous divisor.

Theoretically, the maximum number of iterations needed for the division by the divisor Y might be $Y - 1$. The real number of iterations for the divisor Y depends on the magnitude of the number system used. Generally, the same divisor needs different numbers of iteration in various number systems. Although a great divisor might possibly need less iterations than a small one, it is normal that the number of iterations grows with the magnitude of the divisor. An exact method to determine the real number of iterations for a given system is not known. Table 2 presents results obtained by simulation in an auxiliary RNS with moduli $43, 47, 53, 59, 61$ and $P = 385499687$.

The described algorithm characteristic can lead to time and hardware savings in the cases when all divisors are of a small value or when more numbers should be divided by the same divisor.

The theory of the presented algorithm presents a base on which fast division by constant (scaling) can be realized in both, the RNS and weighted number systems. Actually, if the multiplicative inverse element of a number can be computed in advance and memorized, then division by this number reduces to one multiplication and one division. This division is in the RNS division by a product of system moduli and for a weighted systems it is division by a product of bases, both easily realizable.

divisor range	mean number of iterations	max number of iterations
$10 - 19$	2.2	4
$100 - 109$	4.9	9
$1000 - 1009$	6.2	9
$10000 - 10009$	10.4	15
$100000 - 100009$	9.9	19
$10^6 - 10^6 + 9$	13.1	18
$10^7 - 10^7 + 9$	14.7	18

Table 2: Mean and maximum number of iterations in the RNS with $P = 385499687$ for divisors of different magnitude.

5 Conclusion

As it is true that some simple algorithms in weighted systems are hard to realize in RNS, it seems that sometimes the opposite is true, as well. Such an example is the computation of the multiplicative inverse element in the RNS of a number relatively prime with all the system moduli. It can be realized in only one step by small independent look-up tables, as fast and simple as addition, subtraction, and multiplication. A new approach to integer division in the RNS makes use of this fact.

The presented algorithm can be realized in a few different ways and among them the described one has some significant advantages: both iterations are realized with the same hardware, each iteration can be executed in the $N + R + 3$ modular steps, and the necessary control logic is very simple. Although in the presented realization we succeeded to significantly reduce the number of necessary ROM's, a great hardware complexity is still the main problem of all the integer division algorithms in the RNS.

Regardless of the way the new algorithm is realized, its disadvantage is that, except by simulation, we do not know the way to estimate the mean and maximum number of necessary iterations. At the same time the fact that the number of iterations depends only on the divisor magnitude might be very interesting in some special purpose applications.

Appendix

Proof that the relations (4) and (5) satisfy conditions (1).

The conditions $X < X$ and $Y' < Y$ are satisfied because of the fact that $D < M$. The third condition $\lfloor X'/Y' = Z \rfloor$ is also satisfied if we can prove that in relation

$$X' = ZY' + Q'$$

condition $0 \leq Q' < Y'$ is true where X' and Y' are new dividend and divisor, respectively, and Z the correct result of division of the original dividend and divisor.

From

$$X' = \left\lfloor \frac{XD}{M} \right\rfloor$$

$$D = \frac{1 + Y'M}{Y}$$

$$X = ZY + Q \quad (0 \le Q < Y)$$

follows

$$X' = \left\lfloor \frac{X + ZYY'M + QY'M}{YM} \right\rfloor$$

$$Q' = \left\lfloor \frac{X + QY'M}{YM} \right\rfloor .$$

Obviously, $Q' \ge 0$ and because of $X \le M - 1$ and $Q \le Y - 1$

$$Q' \le \left\lfloor \frac{M - 1 + YY'M - Y'M}{YM} \right\rfloor =$$

$$= \left\lfloor Y' + \frac{M - 1 - Y'M}{YM} \right\rfloor .$$

Because of $Y' \ge 1$ it follows

$$Q' \le \left\lfloor Y' - \frac{1}{YM} \right\rfloor = Y' - 1.$$

So it is proved that condition $0 \le Q' < Y'$ is satisfied and that the presented algorithm is correct.

References

[1] N. S. Szabo, R. I. Tanaka, *Residue Arithmetic and Its Applications to Computer Technology,* McGraw-Hill, 1967.

[2] E. Kinoshita, H. Kosako, Y. Kojima, "General Division in the Symmetric Residue Number System," *IEEE Trans. on Computers,* Vol. C-22, pp. 134-142, 1973.

[3] G. A. Jullien, "Residue Number Scaling and Other Operations Using ROM Arrays," *IEEE Trans. on Computers,* Vol. C-27, pp. 325-336, 1978.

[4] R. T. Gregory, E. V. Krishnamurthy, *Methods and Applications of Error-Free Computation,* Springer-Verlag, 1984.

[5] D. Gamberger, "Incompletely Specified Numbers in the Residue Number System - Definition and Applications" *Proceedings of the 9th Symposium on Computer Arithmetic,* pp. 210-215, Santa Monica, U.S.A., 1989.

Small Moduli Replications in the MRRNS

N. Wigley, G.A. Jullien, D. Reaume, W.C. Miller

VLSI Research Group, University of Windsor
Windsor, Ontario, Canada N9B 3P4

Abstract

The use of finite polynomial rings to design algorithms for the processing of digital signals has received considerable attention in recent years. The authors have recently introduced a method which utilizes the bit patterns of the input data to directly compute the polynomial coefficients; this technique is straightforward and preserves, in part, some of the magnitude information. The new magnitude information coding allows different scaling and conversion algorithms than those required for standard RNS decoding. This paper discusses new polynomial mapping strategies involving replications of *very small* rings modulo 3, 5 and 7 including a scaling and conversion algorithm for such a mapping.

1. Introduction

Finite rings can offer considerable advantages over binary arithmetic in performing integer arithmetic. The most visible use of such finite rings is in the coding of integers as elements of a set of rings, with relatively prime moduli, allowing large dynamic range closed operations (addition, multiplication) to be carried out by a set of parallel small ring calculations. This is known as the Residue Number System (RNS) [6]. If the calculations are carried out simultaneously, the independence of the calculations (there are no dynamic range carries) allows a relaxation in synchronization requirements that can have considerable advantages in VLSI implementations [3].

If M is a positive integer which factors as $M = \prod m_i$, with the $\{m_i\}$ all pairwise relatively prime, then an isomorphism between the ring Z_M and the direct-product ring:

$$Z_{m_1} \times Z_{m_2} \times \ldots \times Z_{m_K}$$

is well-known (the inverse isomorphism is known as the Chinese Remainder Theorem (CRT)). Any computation which can be embedded into Z_M can thus be considered as a computation in the direct-product ring. In the latter, the computations are performed simultaneously and independently, which leads to a great simplification in hardware and allows the pipelining of data through each of the arithmetic units.

Some disadvantages of the RNS are:

1. The necessarily large size imposed on the modulus M in order to obtain the embedding of the calculation (M must accommodate the full dynamic range of the calculation, even though the smaller moduli m_k need not).

2. The residues of the input data must be computed (mod m_k) for each k.

3. The resulting answers must be assembled, to yield the correct answer, by means of the CRT or the equivalent Mixed Radix Method.

4. Scaling can pose considerable difficulties, partly because magnitude information has been completely lost in the finite-ring representation.

An optional method of proceeding is based on the use of finite polynomial rings, where the polynomials have coefficients which are considered in Z_M, and the indeterminates are used to indicate bit information (or, occasionally, the complex unit). This method is known as the *Modulus Replication RNS* [10], or *MRRNS*, and consists of encoding integer data as polynomials in several indeterminates[9].

2. The MRRNS technique

In the *MRRNS* technique the integer data are mapped to a polynomial ring. This idea has been explored by other authors [1][2][7], but the work normally assumes fairly complicated mapping strategies between overspanned complex numbers and polynomial coefficients. In the original *MRRNS* technique, the data are first rewritten as polynomials in some fixed radix 2^β. For theoretical purposes we use the indeterminate X in place of this radix, thus enabling us to represent the data as polynomials in X. The coefficients of these polynomials are then integers which are smaller in magnitude than 2^β. This mapping (which is trivial) is then followed by a mapping to the direct-product ring

$Z_m \times Z_m \times \ldots \times Z_m$. The computations can

now be carried out using independent linear pipelines, each computing over the same ring, at the end of which the inverse of the original mapping is applied. The forward and inverse mappings (to and from the direct product ring) are themselves linear pipelines, and so we have rendered the complete system as a set of linear pipelines operating over a finite ring, modulo m. The output of the inverse mapping stage is a redundantly coded number (here the redundancy arises because of the magnitude overlap of the polynomial coefficients) which is converted back into a non-redundant representation using a combination of scaling and binary addition. For complex data we may use the *QRNS* mapping strategy [4] prior to the *MRRNS* mapping, and invert the *QRNS* mapping after the inverse *MRRNS* mapping. In this way complex computations are mapped to the direct product ring, again yielding to linear pipeline implementation.

A disadvantage of the technique is a large redundancy in the finite-ring representation of data, which results in a need for considerable replication of the hardware used to perform the computations. This replication is ameliorated, however, by the *repeated* use of very small moduli, allowing for the simple design of computational hardware. Moreover, linear pipelining and its attendant advantages in fault tolerance, testing and skewed clocking, make fabrication using wafer scale integration look encouraging. In this paper we use a different mapping strategy in which the data are written as polynomials in several variables, each representing a different power of 2. This will have the effect of increasing the dynamic range of the calculation, though it will increase the redundancy of the computational hardware. A major advantage of this scheme is that it allows us to utilize very small moduli, namely 3, 5, and 7. Thus all the data will be mapped to polynomials whose coefficients lie in the ring Z_{105}, and can thus be treated as elements of the three rings Z_3, Z_5, and Z_7. This in turn allows a very simple design of the hardware necessary for the computations within these small rings. Clearly the method will also work for larger moduli; in fact, the number of parallel channels will drastically reduce. Our concern in this paper, however, is to show that residue calculations over very small rings can still perform large dynamic range arithmetic. By way of example, we will use the *MRRNS* technique for the computation of a 1024-point *FFT* using a multiplexed radix 4 computational element and moduli 3, 5, and 7.

3. The Mapping Strategy

We write the integers representing the real and imaginary parts of the data, together with the coefficients of the *FFT*, as polynomials in the variables W, X, Y and Z, where $W = 2$, $X = 4$, $Y = 16$, and $Z = 256$. With this notation, any positive integer $< 2^{16}$ can be written in a unique fashion as a sum:

$$\sum_{i_1, i_2, i_3, i_4 \in \{0,1\}} a_{i_1 i_2 i_3 i_4} W^{i_1} X^{i_2} Y^{i_3} Z^{i_4} \qquad (1)$$

with the coefficients equal to 0 or 1. Similarly, any negative integer $> -2^{16}$ can be written in the same form with coefficients 0 or -1 (note that the use of 0 and ± 1 implies a signed bit representation of the coefficients). To obtain representations for complex integers we should like to use the *QRNS* method, but we cannot inasmuch as the moduli 3 and 7 do not support a complex unit. To avoid this obstacle and yet still preserve a channel-independent mode of multiplication, we use an additional indeterminate, which we call T, to represent the complex unit j. The indeterminate T cannot satisfy the polynomial equation $T^2 + 1 = 0$ (because 3 and 7 do not support roots of -1). Instead, we use the polynomial $T(T^2 - 1) = 0$ to define the mapping. This polynomial always has three roots in any finite ring Z_m, provided $m > 2$; the penalty we pay for this modification is a 50% increase in the number of rings in the direct product, the advantage is that these rings are very small. Since each of the 'bit indeterminates' also form 1st order polynomials, we may use the same 3 root polynomial to form the direct product mapping. The amazing feature about this mapping is that the complex operator and the bit operators are interchangeable, allowing a variety of binary representations of complex numbers to be simply mapped to the direct product ring. This map is performed by evaluating each of the five variables W, X, Y, Z and T at each of the three roots 0, $+1$ and -1. This results in $3^5 = 243$ results for each of the moduli 3, 5, and 7. Observe that the map is very simple, consisting of nothing more difficult than sign changes and additions.

As an illustration, Figure 1 depicts the forward mapping of an 8-bit integer map to three indeterminates: W, X, Y, producing 27 elements for each bit. The mapping elements for bit-5 are shown explicitly. Each mapping layer corresponds to a separate bit; the monomials corresponding to that bit position are shown alongside the map layer. Note that the mapping will be performed for each of the three moduli: 3, 5, 7. This will result in 81 parallel computations over small ring moduli (although the mapping is shown as 3-dimensional in Fig. 1, the implementation uses independent channels). An inner product on the bit mappings results in the 81 inputs to the computational inner

product required by the algorithm.

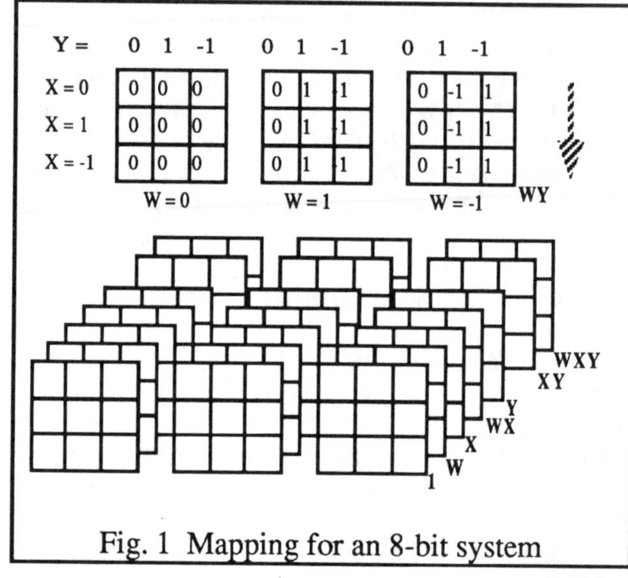

Fig. 1 Mapping for an 8-bit system

In terms of complex numbers we may map the Y indeterminate to j and treat the 8-bit number as a concatenation of a 4-bit real and a 4-bit imaginary Gaussian number. If we decide to map X to j, then the 8-bit sequence represents a 'digit division multiplex' type of decimation, between the real and imaginary parts. If we wish to code the sequence as an 8-bit real number, then each of the indeterminates will represent a power of 2. Only at the inverse map, when the indeterminates are replaced by the quantity they represent, will their relevance become clear. If the forward map is formally treated as a map on *three* coefficients, namely the coefficients of 1, W, and W^2, even though the coefficient of W^2 is always zero, we can use this formalism to invert the above map; this inverse is performed after the computation of the required algorithm (in this paper a radix-4 DFT computational element) has been performed independently in each of the direct product rings. In a hardware implementation we need only consider the two coefficient forward map; the inverse map will, in general, use three coefficients.

We now set $T = j$, so that $T^2 = -1$. (This is allowed at this time since the complex multiplications have already been performed. Moreover, the isomorphism involving T has been accomplished both forward and backward.) This blends two of the three streams into one, each stream being the coefficients of the real and imaginary polynomials which represent the final result as polynomials over the rings Z_3, Z_5, and Z_7. By using the *CRT*, and a combined scaling algorithm, we can finally combine these coefficients to give coefficients in the ring Z_{105}, the input wordlengths having been been selected in such a way that modular overflow is either not possible or has very low probability. The scaling and conversion algorithm is presented in the next section.

4. Scaling and Decoding

Each coefficient in the inverse mapped polynomial represents a weighting by a specific power of 2. Table 1 shows the monomial equivalences for the first seven powers of 2 weightings.

Weight	Equivalences			
2^1	W			
2^2	W^2	X		
2^3	W^2X	X^2	Y	
2^4	W^2Y	XY		
2^5	W^2Z	XZ	W^2Y^2	XY^2
2^6	W^2XY	X^2Y	Y^2	Z
2^7	W^2XZ	X^2Z	YZ	

Table 1

The representation is redundant since each of the coefficients is in the range [-52,52] while the weightings are in ascending powers of 2. Conversion is performed by summing coefficients that have the same power of 2; it turns out that these additions do not cause overflow (a proof of which will be presented in a later publication) and so the additions can be

carried out over the rings, simply extending the inverse mapping computational array. We now have polynomials in powers of 2 which have coefficients given by their residues modulo 3, 5, and 7, respectively. These coefficients are then decoded by means of the *CRT* (mixed radix conversion is preferred in our implementation), and will all lie in the interval [-52, 52]. Some rare exceptions to this are allowed as discussed in the error analysis below.

We now perform the scaling by using a factor 2^s. A low error method of applying this scale factor is to use the recursive relationship:

$$\tilde{C}_i = \frac{\tilde{C}_{i-1}}{2^\gamma} + C_i; \quad \gamma = \begin{cases} 1 & \text{for } 0 \le i \le s \\ \\ 0 & \text{for } i \ge s \end{cases}$$

Using this method, the coefficients corresponding to the first s powers of 2 are processed using 5-bit additions.

The error is limited to $\displaystyle\sum_{i=1}^{s} 2^{-i} = \frac{1 - 2^{-s}}{1 - 2^{-1}} - 1 < 1.$

The recovery array is more clearly evident in Fig. 2. The total number of bits of the full dynamic range conversion is B, the number of scaling bits is s. The ⬬ blocks represent least significant bit removal (divide by 2). A very important point is that we are doing most of the work in linear small ring pipelines with the final output generated by standard binary adders. The only RNS type of structure we require is the 3 ring converter to map each coefficient to a mod 105 ring. The 3 rings total only 8-bits, and the conversion can be performed with a single 8 input circuit. This circuit can also provide mapping to any weighted magnitude protocol.

For example, a redundant mapping protocol could be implemented if redundant addition is desired for the conversion process. Our new technique of applying switching trees to a dynamic CMOS implementation yields efficient implementations of such small input bit circuits.

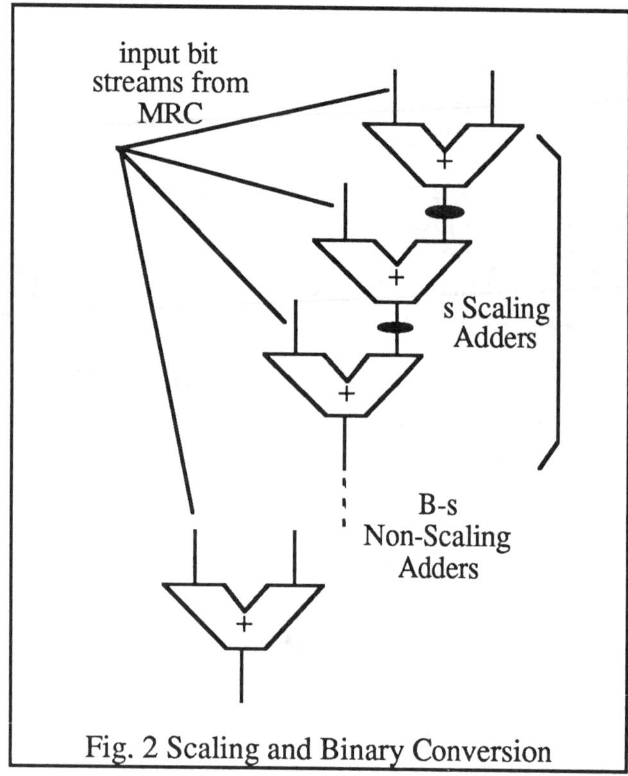

Fig. 2 Scaling and Binary Conversion

5. Experimental Error Analysis

For the purpose of measuring scaling error we have simulated the *MRRNS* technique with input data consisting of random complex integers. We assumed complex integers whose real and imaginary parts consist each of 14 bit random integers (plus an additional sign bit). For the twiddle factors we used approximations of 15 bits. The distribution of relative errors was measured in the sense of relative root mean square, and the results are given in Graph 1. The average relative root mean square error (*RRMS* error) was 1.96×10^{-4}. A comparison was made against a *QRNS* system using moduli 61, 53, 41, 37, and 29 with twiddle factors quantized to 14-bits. This comparison yielded a 1.08×10^{-3} *RRMS* error for the *QRNS* system versus a 1.96×10^{-4} *RRMS* error for the *MRRNS* technique. This demonstrates the ability of a very small ring system to offer significant dynamic range. The hardware difference is considerable. The *QRNS* system

has much smaller redundancy, but the lack of an efficient general multiplication structure [8] (compared to fixed multiplication) and the very large overhead, and awkward structure, associated with converting 6-bit modulus systems [3] make the *MRRNS* technique much more attractive.

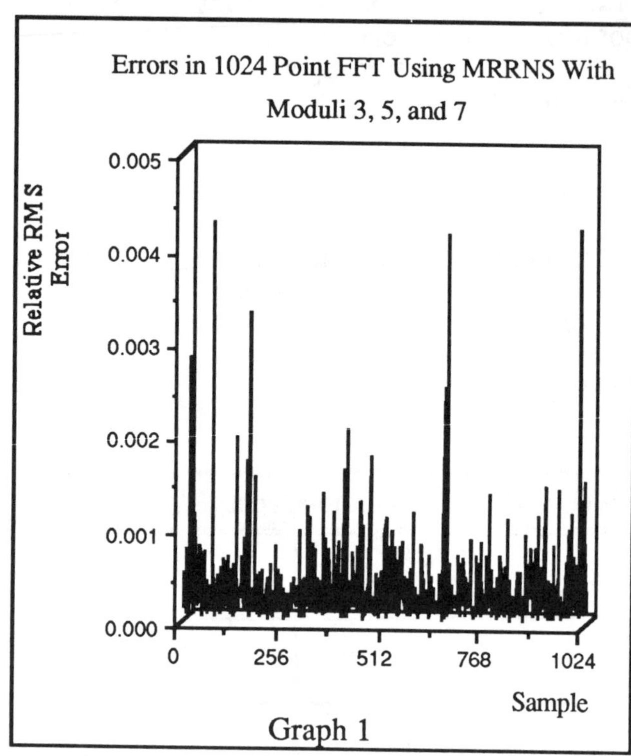

Graph 1

6. VLSI Implementation Considerations

Our preferred approach to the implementation is to use switching trees implementing dynamic pipelined logic. Because of space limitations we cannot discuss the concept in any detail. The reader is directed to [11] for an introduction to some of the basic concepts of dynamic logic. In brief, dynamic logic differs from static logic in that the parasitic capacitance of the evaluation (logic output) node is precharged to a logic '1' level. After pre-charge, a network of transistors is used to evaluate the node. For our purposes, if the network provides a path to ground for the evaluation node, then this is considered to be a logic '1' output. If the network provides an open circuit between the evaluation node and ground, this is considered to be a logic '0'. The logic function of the transistor network is therefore to discharge the evaluation node for every minterm that is included in the boolean algebra expression of the switching function. The basic concept behind switching tree cells is to implement the transistor network as a look-up table, but to construct the table as a minimized tree. The minimization is based on the electrical characteristics required of the switching circuit (i.e. the discharge and charge sharing characteristics in evaluation and pre-charge, respectively) and does not involve either the usual Boolean algebra minimization or the concept of logic gate primitives. Because we will minimize the table directly, an initial choice has to be made regarding the dimensionality of the table decomposition (it is to be noted here that a normal ROM is decomposed into 2-dimensions - rows and columns).

The dimensionality of the decomposition directly leads to the number of series transistors in the switching circuit, and hence to the speed performance. The decoders represent the first stage of a two-stage circuit implementation process (decode, look-up) and their complexity will determine whether the two stages are to be treated as a single pipelined combinational logic circuit, or as a two-stage pipeline. Clearly the former is preferable, since the latency will be doubled if a two-stage pipeline look-up has to be performed. This determination is decided purely on the complexity of the decoders. For n input lines the decoders of the n-dimensional table are simply inverters and will not require an additional pipeline stage. It is interesting to note that the 1-dimensional table is multiplexer logic, and the n-dimensional table is a binary decision-tree.

We have determined that a discharge chain with 6 transistors in series yields dynamic pipeline speeds in excess of 50MHz for a conservative 3μm $CMOS$ process. Since 6-inputs corresponds to the maximum general logic block for two 3-bit ring values, we can use a binary decision tree, and reduce the decoders to simple logic inverters. We therefore construct a binary tree of transistors in 6 variables and program the bottom of the tree according to the truth table entries (place a transistor for a '1' and remove the transistor for a '0'). The true and complement inputs to the tree are connected to the gates of the transistors such that only one path is available for a given input word. The

presence, or absence, of a bottom transistor will determine the state of the evaluation node.

6.1 A 3-bit finite ring multiplier

As an example we show the minimized switching tree for bit zero of a mod 7 finite ring multiplier. The tree is minimized using simple rules and the final minimized tree is shown in Fig. 3. The tree has been minimized from the full 6-input binary tree by applying 3 simple graphical rules [12]. This results in some of the transistors being replaced by wire connections.

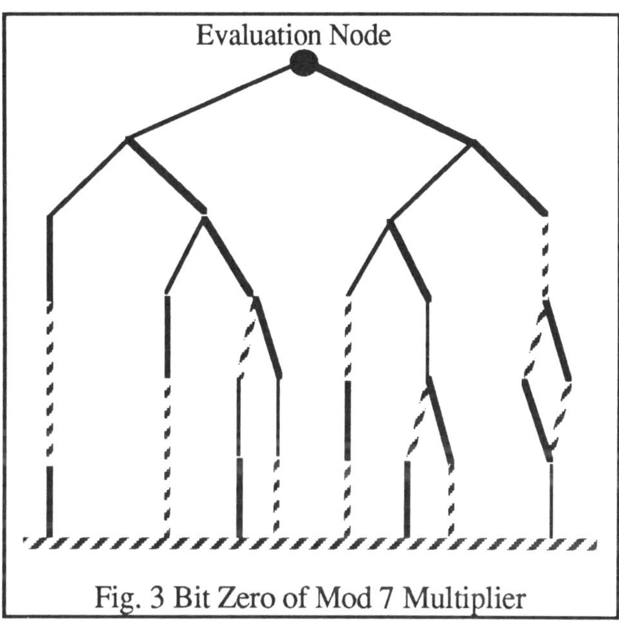

Fig. 3 Bit Zero of Mod 7 Multiplier

The thin lines indicate paths taken when the logic input at that particular level is a '0', the bold lines indicate paths taken when the logic level is a '1'. A wire is shown as a hatched line. Each level corresponds to one dimension of the table (accessed by one of the logic inputs) with the levels ordered hierarchically in correspondence with the logic input ordering on the truth table. It is important to note that our starting point for the minimization is a diagram that can be directly implemented in silicon by simply transforming paths into transistors. Minimization procedures will therefore be directly mappable to silicon. In the case of finite ring circuits there will always be don't care states providing the modulus is not a power of 2. We use these states to advantage in minimizing the tree, and have reduced the maximum tree length to 5.

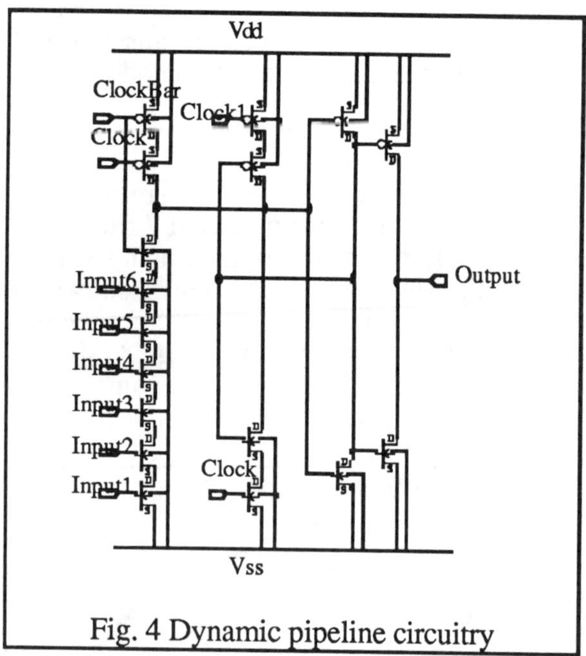

Fig. 4 Dynamic pipeline circuitry

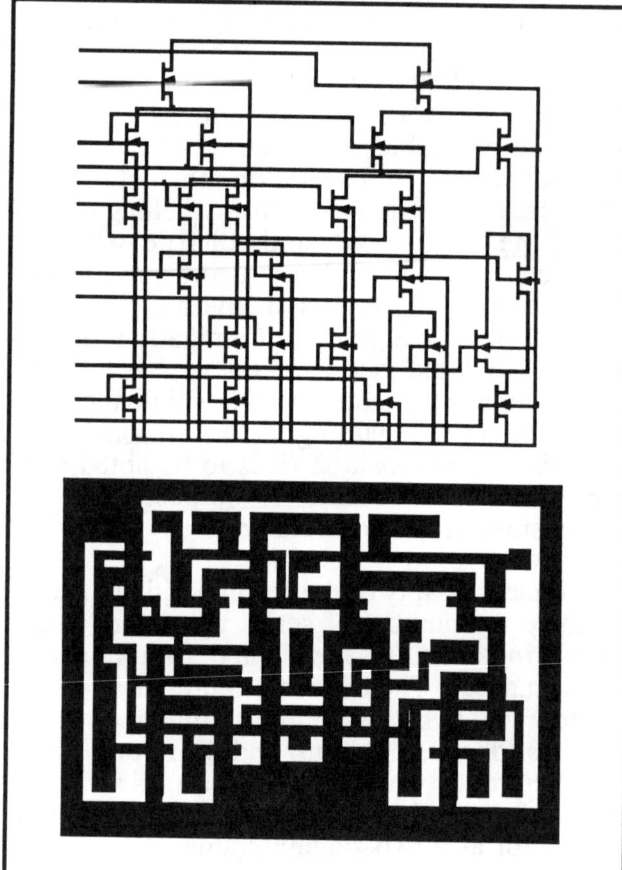

Fig. 5 Schematic and layout for bit 0 of the mod 7 multiplier

For a linear pipeline implementation, the tree can be simply embedded in a dynamic pipeline structure [5]. Our preferred technique is to use an integrated evaluate/latch structure and to connect the evaluate node to a restoring latch. This eliminates the need for any charge sharing pull-up transistors. The pipeline structure with a single 6-transistor switching path is shown in Fig. 4. In the final implementation the 6-transistor series path is replaced by the switching tree. The electrical characteristics of the tree will be virtually the same as those of the series path shown.

The transistor circuit, and a metal only layout of the tree, implemented in a $3\mu m$ $CMOS$ double metal technology are shown in Fig. 5. The tree replaces the single 6-transistor path shown in Fig. 4. A complete multiplier, including pipeline circuitry, and appropriate drivers has been sent for fabrication. The layout is shown in Fig. 6 along with SPICE simulation results for a $50MHz$ clock input. The 3 switching trees are at the bottom with pipeline circuitry and drivers placed above.

7. Conclusions

This paper has discussed mapping, scaling and conversion processes using a new mapping strategy for the modulus replication RNS ($MRRNS$).

The strategy allows direct mapping of bits of either a purely real or multiplexed bit coded complex number to a set of independent rings, defined by moduli 3, 5 and 7. Although the use of such small rings in a traditional RNS system would yield an inadequate computational dynamic range, the $MRRNS$ technique has been shown to be superior to a large $QRNS$ system operating with a computational dynamic range of over 27 bits. A classical radix-4 implementation of a 1024 FFT was used for the comparison. The scaling and conversion procedure has been shown to be a set of finite ring calculations followed by an array of ordinary binary adders. The $VLSI$ implementation of the most complex finite ring circuit required (a Mod 7 multiplier) has been shown to be easily implemented using the switching tree approach, and mask extracted simulations at $50MHz$ have been used to demonstrate the embedding of the switching trees in a dynamic pipeline/evaluate circuit with restoring latch.

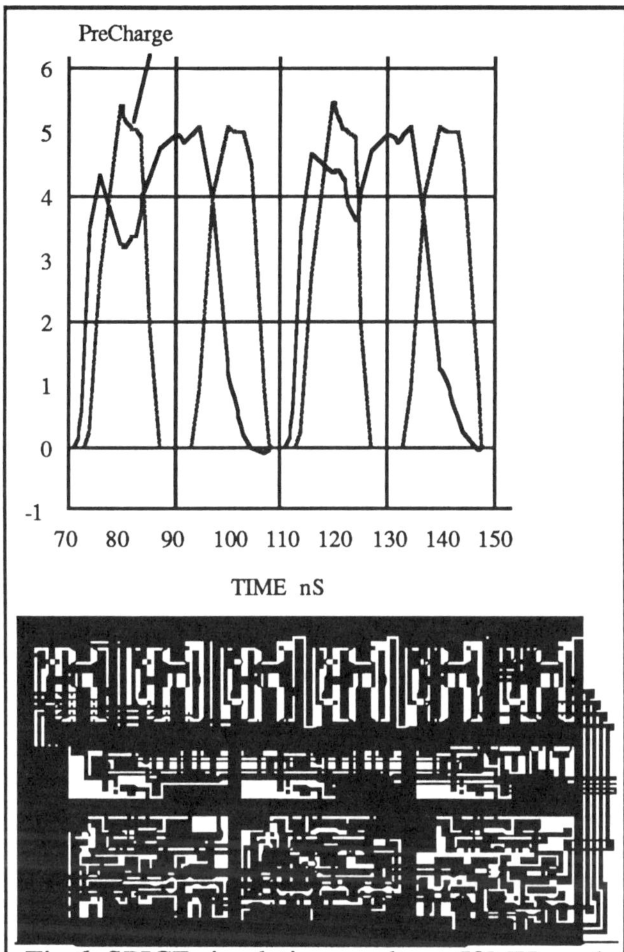

Fig.6 SPICE simulation results at 50MHz and Mod 7 multiplier layout

8. References

[1] Cozzens, J. H. and L. A. Finkelstein. "Computing the Discrete Fourier Transform Using Residue Number Systems in a Ring of Algebraic Integers." IEEE Trans. Inf. Th. **IT-31**: 580-587, 1985.

[2] Games, R. A. "An Algorithm for Complex Approximations in Z[e2πi/8]." IEEE Trans. Inform. Th. **IT-32**: 603-607, 1986.

[3] Jullien, G. A., P. D. Bird, J. T. Carr, M. Taheri and W. C. Miller. "An Efficient Bit-Level Systolic Cell Design for Finite Ring Digital Signal Processing Applications." J. VLSI Sig. Proc. **I**(3): 189-208, 1989.

[4] Jullien, G. A., R. Krishnan and W. C. Miller. "Complex Digital Signal Processing Over Finite Rings." IEEE Trans. on Circuits and Systems. **CAS-34**(No. 4 , Spe*cial Is*sue (Invited)): 365-377, 1987.

[5] Jullien, G. A. and W. C. Miller. Improved Cellular Structures for Bit-Steered ROM Finite Ring Systolic Arrays. Proceedings of the 1990 International Conference on Circuits and Systems, (Invited). , May, pp. 1414-1417., 1990.

[6] Soderstrand, M. A., W. K. Jenkins, G. A. Jullien and F. J. Taylor. Residue Number System Arithmetic: Modern Applications in Digital Signal Processing. 1986.

[7] Stouraitis, T. and A. Skavantzos. Parallel Decomposition of Complex Multipliers. 22nd. Asilomar Conf. Circ. Sys. Comp. 379-383, 1988.

[8] Taheri, M., G. A. Jullien and W. C. Miller. "High Speed Signal Processing Using Systolic Arrays Over Finite Rings." IEEE Trans. Selected Areas in Comm. **6**(3): 504-512, 1988.

[9] Wigley, N. and G. A. Jullien. Array Processing on Finite Polynomial Rings. Proceedings of the International Conference on Application Specific Array Processors. 284-295, 1990.

[10] Wigley, N. M. and G. A. Jullien. "On Moduli Replication for Residue Arithmetic Computations of Complex Inner Products." IEEE Trans. Comp. **39**(No. 8): 1065-1076, 1990.

[11] Weste, N. and Eshraghian, K. "Principles of CMOS VLSI Design: A Systems Perspective." Addison-Wesley, 1985.

[12] Jullien, G.A., Zhang, D., Miller, W.C., DelPup, L. and Grondin, R. "Designing Complex Dynamic Logic Blocks Using Switching Trees." VLSI'91, submitted.

Acknowledgements

The authors acknowledge financial support from the Natural Sciences and Engineering Research Council of Canada to carry out this research work.

Design of Residue Generators and Multi-Operand Modular Adders Using Carry-Save Adders

Stanisław J. Piestrak

Technical University of Wrocław
Institute of Engineering Cybernetics
50-370 Wrocław, POLAND
sjp@ict.pwr.pl

Abstract

In this paper, the design of residue generators and multi-operand modular adders is studied. New highly-parallel schemes using carry-save adders with end-around carry are proposed for either type of circuit. They are derived on the basis of the periodicity of the series of powers of 2 taken modulo A (A is a module). The new circuits are faster and use less hardware than similar circuits known to date.

1 Introduction

Residue arithmetic has been used in digital computing systems for various purposes for many years, see for example [1], [2], [4], and [3]. The two most important applications of residue arithmetic are:

- high-performance digital signal processing hardware which makes use of the parallel nature of the residue number systems (RNSs) arithmetic; and

- fault-tolerant digital systems protected against undetected errors using arithmetic error detecting and/or error correcting codes.

Either class of a system uses a *generator mod A*, i.e. a circuit that computes $[X]_A$, the residue modulo A, from a given binary number X. In an RNS-based digital system it is used to build a binary-to-residue converter [1], [3], while in a fault-tolerant digital system using an arithmetic error detecting code (EDC) it is used to build an encoding/decoding circuitry [5], [9], and [10]. However, since in either application the generator is the overhead which can compromise the speed of the system, it should be a fast circuit realized with the minimal amount of hardware.

Consider a generator mod A with n inputs $\{x_{n-1}, \ldots, x_1, x_0\}$ and a outputs $\{s_{a-1}, \ldots, s_1, s_0\}$, where $a = \lceil log_2 A \rceil$. The generator converts an integer $X = \sum_{j=0}^{n-1} 2^j x_j$ into $[X]_A = \sum_{j=0}^{a-1} 2^j s_j$, i.e. it computes

Research supported by the Ministry of Education of Poland, Grant DNS-T/15/070/90-2.

$$[X]_A = \Big[\sum_{j=0}^{n-1} 2^j x_j \Big]_A. \tag{1}$$

The most obvious scheme of generating $[X]_A$ is a division of X by A to find the remainder, but this technique is very slow. A more efficient algorithm given in [1], is based on the expression

$$[X]_A = \Big[\sum_{j=0}^{n-1} [2^j]_A x_j \Big]_A. \tag{2}$$

If the values of $[2^j]_A$ are directly available, $[X]_A$ may be computed by merely adding mod A those terms $[2^j]_A$ for which $x_j = 1$. The summation mod A of n residues can be performed by a $\lceil \log(n-1) \rceil$-level tree of $n-1$ two-operand binary adders mod A, a regular but costly and time-consuming scheme (by $\log n$ we always denote logarithm of n to the base 2). This approach was the basis for implementations of universal area-time efficient VLSI networks for converting an integer from binary system to residue number system [14]. The residue generator from [14] uses $\lceil n/\log n \rceil$ Brent-Kung parallel adders which perform the addition of two residues mod A in $O(\log n)$ parallel steps, each step performed in $O(\log \log A) = O(\log a)$ time. This method aimed for RNS-based systems, where n is relatively small, seems inefficient (both in terms of cost and speed) to realize encoding and decoding circuitry for arithmetic EDCs, where it is not uncommon that $n \geq 32$. An alternative method to construct the generator by using look-up table implemented with ROMs was considered in [13].

The generator mod $A = 2^a - 1$ needs a special consideration since the residue $[X]_{2^a-1}$ can be obtained by mod $2^a - 1$ addition of a-bit bytes of X, due to the identity

$$[2^{ta}]_{2^a-1} = 1. \tag{3}$$

Thus the computation of $[X]_{2^a-1}$ can be accomplished with a $\lceil \log(b-1) \rceil$-level tree of $b - 1$ adders mod $2^a - 1$, where $b = \lceil n/a \rceil$. This elegant and efficient generator was proposed by Avizienis [11], [12] to implement checking algorithms for arithmetic codes. Since the generators available for $A \neq 2^a - 1$ has been significantly more complex than for $A = 2^a - 1$, arithmetic

EDCs with the check base $A = 2^a - 1$ have been called *low − cost arithmetic codes* since then. This scheme has been recommended for implementation of encoding/decoding algorithms for low-cost arithmetic codes in all the literature known to the author, e.g. [2], [4], [5], [6], [7], [8], and [9]. Also, only low-cost arithmetic codes have been used in practical applications reported to date.

A *multi − operand modular adder* (MOMA) is another circuit frequently used in modular arithmetic applications. In particular, it can be used to build a residue generator but, as far as we know, this concept has not been considered explicitly in the literature. The design of multi-operand adders mod A was considered in [15] (only for $A = 2^a - 1$), and most recently in [19], [16], [17] (only for $A = 2^{a-1} + 1$), and [18]. The MOMA from [19] implemented by associative table look-up processing is generally more complex than any scheme using carry-save adder (CSA) network. The most efficient are the designs from [15]–for $A = 2^a - 1$, and from [18]–for any other A. Only the design from [15] uses a CSA with end-around carry (EAC).

In this paper, we show how to design new faster and less complex residue generators and MOMAs for any A. The design of either circuit is based on the extensive use of highly-parallel circuitry such as a CSA network with EAC. The design of a CSA with EAC proposed here was made possible by exploiting the periodicity of the series of powers of 2 taken mod A. Actually, the residue generators proposed here are the first ever designs which use CSAs.

2 Periodic Properties of $[2^j]_A$

The periodicity of the series of powers of 2 taken mod A is of key importance in the new design methods for the generator mod A and multi-operand adder mod A proposed in this paper. We start with the following definition adapted from [20].

Definition : The *period of the odd module* A $P(A)$ is the minimum distance between two distinct 1's in the sequence of residues of powers of 2 taken mod A. This is formally written as: $P(A) = \min\{j \mid j > 0 \quad \text{and}$

Table I. Periods $P(A)$ of odd modules $A \leq 65$.

A	3	5	7	9	11	13	15	17
$P(A)$	2	4	3	6	10	12	4	8
A	19	21	23	25	27	29	31	33
$P(A)$	18	6	11	20	18	28	5	10
A	35	37	39	41	43	45	47	49
$P(A)$	12	36	12	20	14	12	23	42
A	51	53	55	57	59	61	63	65
$P(A)$	8	52	20	18	58	60	6	12

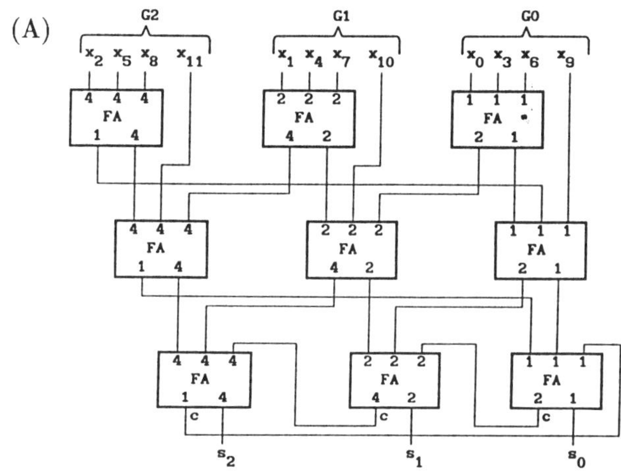

(B)

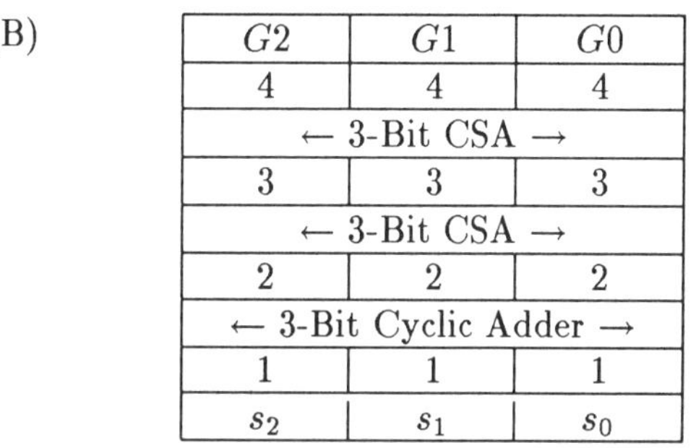

$G2$	$G1$	$G0$
4	4	4
← 3-Bit CSA →		
3	3	3
← 3-Bit CSA →		
2	2	2
← 3-Bit Cyclic Adder →		
1	1	1
s_2	s_1	s_0

Figure 1. CSA network with EAC for 4-operand adder mod 7 or 12-bit generator mod 7: (A) detailed scheme; and (B) shorthand description.

$[2^j]_A = 1\}$. Henceforth $P(A)$ will be simply called the *period of* A.

$P(A)$ can be calculated by using the following recursive equation

$$[2^i]_A = \left[\, 2[2^{i-1}]_A \,\right]_A. \tag{4}$$

However, for some A the following formulas are known to compute $P(A)$ [20]:

- $P(2^a - 1) = a$;

- $P(2^{a-1} + 1) = 2(a - 1)$;

- If $A = kB$ with k and B odd then $P(A)$ is some multiple of $P(B)$.

Now, with the concept of the period of A, we generalize the identity (3) which holds for $A = 2^a - 1$ only, to the following identity which holds for any A

$$\left[2^{tP(A)}\right]_{2^{P(A)}-1} = 1, \tag{5}$$

where t is any nonnegative integer.

3 Carry-Cave Adder Networks with End-Around Carry

The concept of *carry − save addition* has been used for many years to speed up multi-operand binary addition [15] and to implement MOMAs [15], [16], [17], and [18]. The MOMA for $A = 2^a - 1$ given in [15] (pp. 98-100) is the only example of using a *CSA with EAC* that has been reported in the literature. It is illustrated with the following example.

Example 1 : Design the 4-operand adder mod 7 using a CSA with EAC. The input 12-tuple are four residues mod 7: $X_1 = \{x_2x_1x_0\}$, $X_2 = \{x_5x_4x_3\}$, $X_3 = \{x_8x_7x_6\}$, and $X_4 = \{x_{11}x_{10}x_9\}$. The set of input bits is partitioned onto $a=3$ disjoint sets Gj with the bits of the same weight $[2^j]_7$: $G0 = \{x_0, x_3, x_6, x_9\}$, $G1 = \{x_1, x_4, x_7, x_{10}\}$, and $G2 = \{x_2, x_5, x_8, x_{11}\}$. Figure 1(A) shows a detailed scheme of the circuit, while Fig. 1(B) shows its shorthand description, where the columns Gj indicate the number of bits of weight $[2^j]_7$ in the current stage of computation. The latter scheme will be used henceforth to describe any CSA network).

The operation of the above circuit as well as any other a-bit CSA network with EAC is justified by (3) for $A = 2^a - 1$. Since $P(2^a - 1) = a$, one can expect that a similar concept can be applied to any other A, provided that a $P(A)$-bit CSA network with EAC is considered and a more general identity (5) is taken into account. Below we will show that this is indeed the case.

Suppose that we wish to add $k \geq 3$ $P(A)$-bit numbers mod $2^{P(A)} - 1$, A is an odd integer. A $P(A) - bit\ CSA\ with\ EAC$ is a natural generalization of the well known a-bit CSA with EAC used for $A = 2^a - 1$. This is because a FA with inputs of weight $[2^{P(A)-1}]_A$ generates the carry signal of weight $[2[2^{P(A)-1}]_A]_A = 1$ that can be directed to the FA with inputs of weight 1. Similarly, the a-bit binary adder with EAC used for $A = 2^a - 1$ can be generalized to a $P(A)$-bit binary adder with EAC used for any A. A remarkable feature of this circuit, which is of our particular concern, is the cyclic nature of the addition mod A with the cycle length $P(A)$. It allows any $P(A)$-bit adder with EAC to begin and end operation at *any* point of the cycle. In our applications, this adder executes at most one full cycle of length $P(A)$. Thus, in the worst case, if the $P(A)$-bit adder with EAC begins operation from Gj, its operation ends at $G(j-1)$ with the last carry signal directed to Gj (see, for example, Fig. 5). Henceforth, the $P(A)$-bit adder with EAC which goes through p ($p \leq P(A)$) stages will be called a $p - bit\ cyclic\ adder$. One limitation of the $P(A)$-bit adder is that its delay grows with $P(A)$ and it can be prohibitively large for some A. In such a case, we suggest to use the following scheme for large $P(A)$. Some other modifications of the new basic generator scheme, which allow to avoid this problem, will be given in the next section.

If $P(A)$ is large we can partition $P(A)$ cyclically ordered sets Gj onto w groups S_i, each group S_i containing *subsequent* sets Gj, $1 \leq i \leq w$. If the groups S_i are of similar sizes, the addition of two $P(A)$-bit operands can be performed in parallel by w shorter binary adders of length no more than $\lceil P(A)/w \rceil$. This implementation involves $w - 1$ extra inputs to the final converter of the generator (see Fig. 3 and Section 4), since now there are $w > 1$ sets Gj with two bits. The following example clarifies such a case.

Example 2 : A CSA network with EAC for the 32-bit generator mod 13 is given on Fig. 2. The set of 32 input bits is partitioned onto $P(13) = 12$ sets Gj. The 8-bit CSA reduces the input bits from $n = 32$ to $n' = 24$,

$G11$	$G10$	$G9$	$G8$	$G7$	$G6$	$G5$	$G4$	$G3$	$G2$	$G1$	$G0$
2	2	2	2	3	3	3	3	3	3	3	3
HA	HA	HA	HA			← 8-Bit CSA →					
2	2	2	2	2	2	2	2	2	2	2	2
← 6-Bit Adder →						← 6-Bit Adder →					
1	1	1	1	1	2	1	1	1	1	1	2

Figure 2. CSA network for 32-input generator mod 13.

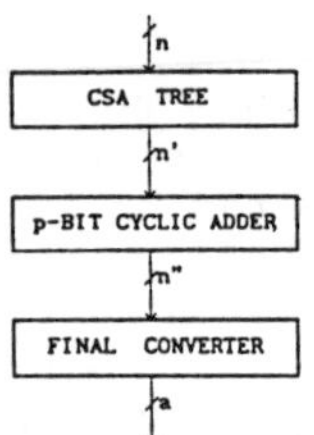

Figure 3. General structure of the new n-input generator mod A and the k-operand adder mod A.

which are uniformly distributed due to four HAs. Now a 12-bit adder with EAC could be used to reduce the bits from $n' = 24$ to $n" = 13$. However, if it is too slow a viable alternative are two 6-bit adders used as shown on Fig. 2.

4   New Generators Mod A

In this section, we propose three new schemes of the n-input generator mod A which cover the whole spectrum of A and n; two of them make use of the CSA network with EAC.

4.1   General Design Procedure

Here we will propose a new n-input generator mod A that computes (2) using a CSA network.

Let $n = uP(A) + v$, $u=0,1,2,...$ and $0 \leq v \leq P(A) - 1$. For $n > P(A)$ we have $u \geq 1$ and $v > 0$. It follows from (5) that for any j $(0 \leq j \leq P(A) - 1)$ and any nonnegative integer t

$$[2^{t(PA)+j}]_A = [2^j]_A, \tag{6}$$

which implies that $u + 1$ bits $x_j, x_{P(A)+j}, \ldots, x_{uP(A)+j}$ represent the same residue $[2^j]_A$ for any $j \leq v$. The same holds for u bits $x_j, x_{P(A)+j}, \ldots, x_{(u-1)P(A)+j}$ for any $j > v$. This leads us to the partitioning of n input bits onto $P(A)$ subsets Gj defined as $Gj = \{x_q \mid ([2^q]_A = j)$ and $(0 \leq q \leq n - 1)\}$, $j = \{0, 1, 2, \ldots, P(A) - 1\}$. The sets Gj are cyclically ordered according to increasing indices j taken mod $P(A)$; in particular, the sets $G(P(A) - 1)$ and $G0$ are adjacent and hence $G0$ follows $G(P(A) - 1)$. This leads us to a new general structure of the n-input generator mod A, shown on Fig. 3. It is designed by using the following procedure.

Procedure 1

Step 1: Partition the set of input bits $\{x_0,\ x_1, \ldots, x_{n-1}\}$ onto $P(A)$ sets $Gj = \{x_q \mid [2^q]_A = j\}$, $0 \leq j \leq P(A) - 1$.

Step 2: If $n \leq 2P(A) + 1$ assume $n' = n$ and go to Step 3. Otherwise, reduce the total number of bits from n to n' $(n' = 2P(A)$ or $2P(A) + 1)$ using a

CSA with EAC of length up to $P(A)$.

For $n' = 2P(A)$ any set Gj has two bits, whereas for $n' = 2P(A)+1$ one set, say Gj^*, has three bits while all other sets Gj have two bits.

Step 3: Reduce the number of bits from n' to $n" = P(A) + 1$ by using a p-bit cyclic adder.

  a) If $n' = n$ then $p = n' - P(A)$ and the adder starts with $G0$;

  b) If $n' = 2P(A)$ then $p = P(A)$ and the adder can start with any set Gj;

  c) If $n' = 2P(A)+1$ then $p = P(A)$ and the adder starts with Gj^*.

In any case, every set Gj but one, say Gr, contains a single bit at the end of the addition; depending on the case r equals: a) $n' - P(A)$; b) $j + 1$; or c) $j^* + 1$.

Step 4: Compute $\left[\, y_0[2^0]_A + y_1[2^1]_A + \ldots + y_r'[2^r]_A + y_r''[2^r]_A + \ldots + y_{P(A)-1}[2^{P(A)-1}]_A \,\right]_A$.

Step 4 is executed by the final converter which is nothing else but an $n"$-input generator mod A. It can be implemented using one of the schemes proposed in Subsections 4.2 and 4.3. The final converter is not used for $A = 2^a - 1$ because $n' = a$.

Basically, Proc. 1 can be used to design a generator for any A and n such that $n > P(A)$, but its efficiency grows with the ratio $n/P(A)$. The major limitation of Procedure 1 is that we cannot take the full advantage of an extensive use of the CSA network (if at all) when $n/P(A) < 3$, and in such a case most bits must be reduced by the final converter. Table I shows that there are many A for which $P(A)$ is small, e.g. for $A \in \{5, 9, 17, 21, 51\}$ we have $P(A) \le 8$, for which an efficient generator can be designed even for small n. Unfortunately, $P(A)$ can also be prohibitively large, e.g. for $A \in \{29, 37, 49, 53, 59, 61\}$ we have $P(A) \ge 28$, and for any such A, Procedure 1 can provide an efficient generator for very large n only. Thus, for any pair of n and A with a small ratio $n/P(A)$ two special schemes of the generator mod A are proposed below. They can be used to implement the whole n-input generator or only the $n"$-input final converter for the generator designed by using Proc. 1.

Example 3 : A CSA network for the 32-bit generator mod 9 is given on Fig. 4. The input bits are partitioned onto $P(9) = 6$ sets $G0 = \{x_0, x_6, x_{12}, x_{18}, x_{24}, x_{30}\}$, ..., and $G5 = \{x_5, x_{11}, x_{17}, x_{23}, x_{29}\}$. The whole CSA network is built of 25 FAs and one HA and it introduces the delay of 9Δ (Δ is the delay introduced by a single FA). The final converter which computes

$$\left[\, 1.y_0 + 2.y_1 + 4.y_2' + 4.y_2'' + 8.y_3 + 7.y_4 + 5.y_5 \,\right]_A.$$

can best be realized with a 128×4 ROM.

4.2 Generator Mod A for n≤P(A)

If $n \le P(A)$ then every coefficient $[2^i]_A$, $0 \le i \le n - 1$, represents a different residue mod A. Thus, to compute (2) we propose a special scheme, shown on Fig. 5. The set of n input bits is partitioned onto two

$G5$	$G4$	$G3$	$G2$	$G1$	$G0$
5	5	5	5	6	6
$\leftarrow$ 6-Bit CSA $\rightarrow$				$\leftarrow$ 2-Bit CSA $\rightarrow$	
4	4	4	5	4	3
$\leftarrow$ 6-Bit CSA $\rightarrow$					
3	3	3	4	3	2
$\leftarrow$ 5-Bit CSA $\rightarrow$					HA
2	2	2	3	2	2
6-Bit Cyclic Adder					
1	1	1	2	1	1
y_5	y_4	y_3	y_2'', y_2'	y_1	y_0

Figure 4. CSA network for 32-bit generator mod 9.

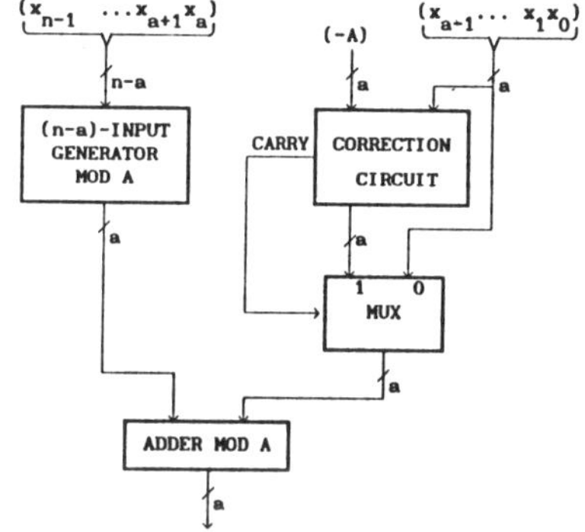

Figure 5. New generator mod A for $n \le P(A)$.

disjoint subsets $X_1 = \{x_{n-1}, \ldots, x_{a+1}, x_a\}$ and $X_2 = \{x_{a-1}, \ldots, x_1, x_0\}$. The $n - a$ bits from X_1 feed the $(n - a)$-input generator mod A that computes

$$\left[\, x_a[2^a]_A + x_{a+1}[2^{a+1}]_A + \ldots + x_{n-1}[2^{n-1}]_A \,\right]_A \quad (7)$$

and should be implemented with a PLA or ROM. The remaining a-tuple $X_2 = (x_{a-1} \ldots x_1 x_0)$ represents an integer M, $0 \le M \le 2^a - 1$, which can be treated as a residue mod A with or without overflow. To find $[M]_A$ a correction circuit with a multiplexer is employed. The correction circuit that computes $M - A$ can be implemented either as an a-bit adder with a PLA or as a ROM correction table. The carry bit generated from the correction circuit indicates whether M is greater than A. A multiplexer controlled by the carry selects the correct output from M and $M - A$.

The range of applications of the above scheme is limited, probably to $10 < n \le 10 + a$, because of the following reasons. For $n \le 10$ the whole generator can best be implemented with a single specially tailored $2^n \times a$ ROM. For $n > 10 + a$ the $(n - a)$-bit generator mod A would require more than one ROM and then a pure look-up table or any other implementation of the whole n-bit generator could be better. Nevertheless, this scheme is more efficient than any scheme based on the concept from [1], since, if implemented using modular adders, it uses a-1 modular additions less.

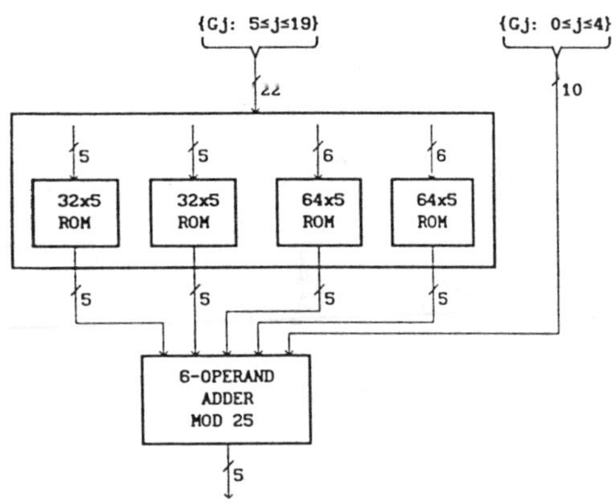

Figure 6. 32-input generator mod 25 using 6-operand adder mod 25.

4.3 Generator Mod A Using a MOMA

Here we propose the third scheme of the residue generator which handles the whole range of A and n for which two schemes proposed above are inefficient. The scheme described here first translates the groups of those bits which cannot be reduced by using a CSA network onto residues mod A. These residues are then added mod A by a MOMA. The efficiency of this scheme is a natural consequence of using a CSA-based MOMA (described in Section 5).

Example 4 : Design the 32-input generator mod 25. First consider the generator designed by using Proc. 1. The input bits are partitioned onto $P(25) = 20$ sets $Gj = \{x_j, x_{j+20} \mid 0 \leq j \leq 11\}$ and $Gj = \{x_j \mid 12 \leq j \leq 31\}$. Thus the whole generator consists of a 12-bit binary adder with inputs from the sets Gj with $0 \leq j \leq 11$ and the final converter with 21 inputs. A ROM implementation of the latter needs one $2K \times 5$ ROM and one $1K \times 5$ ROM at the first stage, and one $2K \times 5$ ROM at the second stage.

A significantly more efficient generator, shown on Fig. 6, uses the 6-operand adder mod 25. Ten bits from the sets Gj, $0 \leq j \leq 4$, are treated as two operands (the fact that they may not be residues mod 25 is immaterial here), while remaining four operands are residues mod 25 generated by four small ROMs (or PLAs) from 22 bits from the sets Gj, $5 \leq j \leq 19$. The 6-operand adder mod 25 can be designed by using Proc. 2 given in Section 5: it is built of 23 FAs, one HA, and one 256×5 ROM, and it introduces the delay of $7\Delta + d(\text{ROM})$.

4.4 Complexity Estimation

The CSA network that reduces the number of bits from n to $n"$ is built of a total of $n - n"$ FAs and a few HAs. Hardware complexity of the final converter depends on a particular scheme used. The delay $d(A, n)$ introduced by the generator designed by using Proc. 1 equals

$$d(A, n) = \theta(k).\Delta + d(p) + d(A, n"), \qquad (8)$$

where:

- $\theta(k)$, the minimum number of levels in the CSA tree with k operands (k, can be found in Table II, where: $k = u$—for $v=\{0$ or $1\}$,and $k = u + 1$—for $v > 1$.

Table II. The minimum number of levels $\theta(k)$ on a CSA tree that processes k input operands [15].

k	3	4	5÷6	7÷9	10÷13	14÷19	20÷28
$\theta(k)$	1	2	3	4	5	6	7

- $d(p)$ is the delay introduced by the p-bit binary adder; and

- $d(A, n")$ is the delay introduced by the final converter.

It is clear that the delay of the CSA tree is significantly smaller than the delay introduced by the p-bit adder and the final converter. Therefore, to improve the speed of the generator, it can be desirable to implement either the final converter or even the p-bit adder and the final generator together, with a $2^{n"} \times a$ or $2^{n'} \times a$ ROM, respectively. On the other hand, the hardware complexity of the generator can be reduced in various ways at the cost of speed. For instance, for $k > 4$ a viable alternative is to replace the CSA tree with a single bank of $P(A)$ FAs followed by a $2P(A)$-bit carry-save register (CSR) which operates in $k - 2 + p$ cycles. The trade-off is a $2P(A)$-bit CSR and extra delay of $[k - 2 - \theta(k)]\Delta$ for $n - n' - P(A)$ FAs less.

4.5 Comparison

The superiority of the designs proposed here over conventional approaches results from the following argument.

Any new generator operates as the circuit that reduces the number of input bits with weights being residues mod A from n to a. Thus, it seems appropriate to estimate the average amount of hardware required to reduce the number of bits by one. With a CSA network, one FA reduces the number of bits by one. To the author's best knowledge, there is no other scheme that does the same with less hardware and no residue generator using a CSA network has been reported in the open literature as well. For comparison, any generator based on the concept from [1] uses two a-bit adders to reduce the number of bits by one while a look-up table approach from [13] is more complex as well. These observations support our claim that any generator proposed here uses less hardware than any other existing design. The more so that the schemes used to implement the final converter, proposed in Subsections 4.2 and 4.3, are also less complex than existing designs.

The main contributors to the delay introduced by the generators proposed here are the p-bit cyclic adder and the final converter. Since the delay of the p-bit cyclic adder can be as large as $P(A)$, its partitioning onto smaller adders (suggested in Section 3) or implementing the whole generator mod A (or the final converter only) using a MOMA, can be used to improve performance of a generator. Thus, any new generator can be implemented using at most two relatively short binary adders. This clearly shows its superior speed compared to any scheme based on the concept from [1] which introduces the delay equal $2\lceil log(n - 1)\rceil.d(a).\Delta$.

Finally, we should comment a new scheme of the generator for a special case of $A = 2^a - 1$ which is very similar to the k-operand adder mod $A = 2^a - 1$ from [15]. The only difference between them is that the generator may have $n \neq ka$ inputs and needs $\lceil n/a \rceil a - n$ extra HAs for better performance. This now obvious and simple generator should be confronted with the scheme that has been commonly used to build the error checking circuitry for arithmetic codes with $A = 2^a - 1$ for about thirty years [11]. It has benn a tree of a-bit adders with EAC, despite that a significantly faster generator could have been built using CSA approach. This claim is true for all monographs on the design of fault-tolerant hardware known to the author (including [4], [5], [6], [7], and [8]) as well as for all papers on self-testing checkers for arithmetic codes with $A = 2^a - 1$, see e.g. [9]. The superiority of the CSA-based generator mod $2^a - 1$ over the conventional scheme is clear: either scheme is built of $n - a$ FAs, but the new one introduces a significantly smaller delay — $[\theta(\lceil n/a \rceil) + 2a]\Delta$ vs. $2a\lceil \log(n/a) \rceil \Delta$. (It is assumed that the delay introduced by the a-bit adder with EAC is $2a\Delta$.)

5 Multi-Operand Adder Mod A

Let $X_1 = (x_{a-1}, \ldots, x_1, x_0)$, $X_2 = (x_{2a-1}, \ldots, x_{a+1}, x_a)$, $\ldots$, and $X_k = (x_{ka-1}, \ldots, x_{(k-1)a})$ be k residues mod A. A k-operand adder mod A computes

$$[X]_A = \Big[\sum_{i=0}^{k} X_i \Big]_A \tag{9}$$

which is equivalent to

$$[X]_A = \Big[\sum_{j=0}^{a} (\sum_{i=0}^{k} x_{ia+j}) \cdot [2^j]_A \Big]_A \tag{10}$$

The new technique proposed here first performs the summation

$$\Big[\sum_{j=0}^{a} (\sum_{i=0}^{k} x_{ia+j}) \cdot [2^j]_A \Big]_{2P(A)-1} \tag{11}$$

using a CSA network with EAC, and then it performs the final evaluation of

$$\Big[\Big[\sum_{j=0}^{a} (\sum_{i=0}^{k} x_{ia+j}) \cdot [2^j]_A \Big]_{2P(A)-1} \Big]_A \tag{12}$$

using a look-up table.

5.1 Design Procedure

A new k-operand adder mod A that implements (12) employs similar ideas and the same general structure as the generator mod A, shown on Fig. 3.

The maximum of sets Gj which will eventually be used in the CSA network of the new MOMA is given by

$$q = \min\{P(A), m\}. \tag{13}$$

where $m = \lceil \log k(A-1) \rceil$ is the number of bits needed to encode the largest number resulting from the summation of k residues mod A in binary.

A new k-operand adder mod A can be designed by using the following procedure.

Procedure 2

Step 1: Find $m = \lceil \log k(A-1) \rceil$.

$G3$	$G2$	$G1$	$G0$
—	4	4	4
—	← 3-Bit CSA →		
1	3	3	2
—	←2-Bit CSA→		—
2	2	1	2
⌐—FA←—HA		←——FA←	
1	1	2	1

Figure 7. CSA network for 4-operand adder mod 5.

G7	G6	G5	G4	G3	G2	G1	G0
—	—	—	8	8	8	8	8
—	—	—	←——Two 5-Bit CSAs——→				
—	—	2	6	6	6	6	4
—	—	—	← —5-Bit CSA——→				
—	—	—	←——4-Bit CSA——→				
—	—	4	4	4	4	3	2
—	—	←——5-Bit CSA——→					HA
—	1	3	3	3	3	2	1
—	—	←——4-Bit CSA——→			HA	—
—	2	2	2	2	2	1	1
—	←——5-Bit Adder——→					—	—
1	1	1	1	1	1	1	1

Figure 8. CSA network for 8-operand adder mod 25.

Step 2: Let $\{X_i = (x_{ia-1}, \ldots x_{(i-1)a}), \ 1 \leq i \leq k\}$ be the set of input operands. Partition $n = ka$ input bits onto a subsets $Gj = \{x_{ra+j} \mid 0 \leq r \leq k-1\}$, $0 \leq j \leq a-1$, and assume $Gj = \emptyset$, $a \leq j \leq m$.

Step 3: Reduce the total number of bits from n to n' by using the CSAs with EAC of length up to q ($q = \min\{P(A), m\}$) until at most one set Gj has three bits while all other sets Gj have two or one bit.

Step 4: Reduce the number of bits from n' to n'' by using a p-bit cyclic adder (usually $n'' = q$ or $q+1$).

Step 5: Compute $\big[y_0[2^0]_A + y_1[2^1]_A + \ldots + y_r'[2^r]_A + y_r''[2^r]_A + \ldots + y_{m-1}[2^{m-1}]_A \big]_A$. by using one of the realizations of the final converter described in Section 4; $Gr = \{y_r', y_r''\}$ denotes the set with two bits, provided that such a set occurs.

5.2 Examples

The following examples illustrate the operation with and without cyclic mode of the CSA network used in the MOMA designed according to Proc. 2.

Example 5 : A CSA network for the 4-operand adder mod 5 is shown on Fig. 7. It uses seven FAs and one HA, and its delay is 5Δ. The final converter can best be implemented with a 5-input PLA. An alternative implementation of the CSA network using a single-stage 4-bit CSA and an 8-bit CSR introduces the same delay. It uses three FAs less at the cost of an 8-bit CSR.

Example 6 : A CSA network for the 8-operand adder mod 25 is given on Fig. 8. Since $P(25) = 20$ and $m = \lceil \log 192 \rceil = 8$ we have $q=\min P(25), m=8$ and the cycle does not occur. Its high-speed (HS) version uses 32 FAs, two HAs, and 256×5 ROM, and introduces the delay of $7\Delta + d(ROM)$. Its cost-effective (CE) version uses seven FAs, a 14-bit CSR, and 256×5 ROM, and introduces the delay of $11\Delta + d(ROM)$. In this case, the CE version is a viable alternative to the HS version since the former is slightly slower (by 2Δ) but allows to save a large amount of hardware.

Notes:

1. The analysis of many examples showed that if HAs are used then the p-bit cyclic adder has $p \leq a$.

2. For large k, a more viable is to implement the CSA network using a single-stage q-bit CSA with a $2q$-bit CSR ($q \leq m + 1$) which operates in $k - 2 + a$ or $k - 3 + a$ cycles.

5.3 Complexity Estimation of the New MOMA

The complexity of two basic blocks of the new MOMA: the CSA network and the final converter, will be considered separately.

The CSA network used in the new MOMA reduces $n = ka$ input bits to n", i.e. it is built of $ka - n$" FAs. Most of the FAs work in the carry-save mode, since the length of a single cyclic adder is a or a-1 only. Thus the delay of the CSA network is upper-bounded by $[\theta(k) + a]\Delta$.

Now recall that the final converter is a special case of the generator mod A, in which the CSA network can hardly be used. Since its ROM (or PLA) implementation seems the most feasible, its complexity can be appropriately estimated by the number of inputs n". Obviously, n" is upper-bounded by $P(A) + 1$ for any k. An equality holds when the CSA network operates in the cyclic mode, i.e. for any $k \geq k_c(A)$, where

$$k_c(A) = \lceil 2^{P(A)}/(A - 1)\rceil. \qquad (14)$$

For A odd we distinguish the following cases:

(a) for $A = 2^a - 1$ the cyclic mode occurs for any k;

(b) for A with $P(A) \leq 8$ the values of $k_c(A)$ are tabulated in Table III;

(c) for any other A we have $P(A) \geq 10$ and $k_c(A) \geq 32$. Thus, it seems unlikely that for any such A the cyclic mode occurs for any practical k.

Since in the case (b) we have n" ≤ 9 for any k, the implementation of the final converter with a single ROM is feasible.

Now we will consider the case (c). Suppose that $m \leq P(A)$. It implies that either n" $= m$ or n" $= m+1$. The largest number of operands (which are residues mod A) such that their sum can be encoded with m bits equals

$$k_m(A) = \lfloor (2^m - 1)/(A - 1)\rfloor. \qquad (15)$$

The values of $k_8(A)$ for some A are tabulated in Table IV. It is observed that for sufficiently large a we have $k_m(2^{a-1} + 1) = 2^{m-a+1} - 1$ and $k_m(2^a - 3) = 2^{m-a}$.

Table III. The values of $k_c(A)$ for A with $P(A) \leq 8$.

A	5	9	17	21	51
$P(A)$	4	6	8	6	8
$k_c(A)$	4	8	16	4	6

Table IV. The values of $k_8(A)$ for some A.

a	4		5				6	
A	11	13	17	19	23	29	33	61
$k_8(A)$	23	19	15	13	11	8	7	4

Table V. Characteristics of various MOMAs.

	Version	FAs	CSR	ROM	Delay
New	CE	m	$2m$	$\leq 2^{m+1}$	$[\theta(k) + a]\Delta + d(ROM)$
[18]	Basic	$2a + 7$	$4a + 10$	—	$(4k + 2a - 4)\Delta$
New	HS	$\leq (k - 1)a - m - 1$	—	$\leq 2^{m+1}$	$(k + a - 2)\Delta + d(ROM)$
[18]	Parallel	$3a + 9$	$4a + 10$	—	$(4k + a - 6)\Delta$

Note: for $k \leq 16$ we have $m=\min\{P(A), a + 4\}$ and $\theta(k) \leq 6$.

(Note that $A = 2^{a-1} + 1$ and $A = 2^a - 3$ are the smallest and the largest A that can be encoded with a bits in the case (c).) Table IV shows that for $m = 8$ these limits are attained for $a = 5$. It is clear that the final converter has no more than $a + 4$ inputs for any A and k ranging from 16 up to 31. This proves that also in the case (c) the ROM implementation of the final converter is feasible for large k and for many A, since for any $a \leq 6$ the ROM size does not exceed $1K \times a$.

5.4 Comparison

The MOMA from [15] is the most effcent scheme known for $A = 2^a - 1$; it is also the special case of Proc. 2. Therefore new MOMAs designed by Proc. 2 will be compared only for $A \neq 2^a - 1$ against the best known design obtained by using Algorithm C from [18].

Algorithm C from [18] computes (10) in the following way. During first k-2 cycles Step 2 of Alg. C is executed. Each cycle involves adding X_i and subtracting A performed by the $(a+3)$-bit CSA which are followed by the sign estimation of the partial sums, performed by the 2-bit carry-lookahead adder; the partial sums are stored in two $2(a + 3)$-bit CSRs. The final reduction involves: either $2(a + 2)$ additions performed by the $(a + 2)$-bit CSA — in a basic version; or $a + 2$ additions performed by two $(a + 2)$-bit CSAs — in a parallel version. Either version uses two $(a + 2)$-bit CSRs to store the results before the correct sum is chosen.

Table V compares the exact characteristics of the designs from [18] against the upper-bounds of the designs obtained by using Proc. 2. The comparison reveals that either version of our MOMA introduces significantly less delay than the design from [18]. Essential speed improvement can be attributed to the concept used in our design: the estimation of the sum is postponed to the last stage and, unlike in Alg. C from [18], no time is wasted for corrections and sign estimation during the carry-save stage of operation (they cause that the delay is proportional to $4k$ in Alg. C). Table V also reveals that our high-speed version becomes too complex when k grows (probably for $k > 5$). However, the complexity of our CE version and the basic version from [18] seem comparable.

6   Conclusions

The new procedures for synthesizing residue generators and multi-operand modular adders (MOMAs) using CSA are presented. First, it is observed that the periodic properties of the series of powers of 2 taken mod A, previously observed and exploited in [11] for $A = 2^a - 1$, can be extended to any A. It is shown that a CSA with EAC can be built for any A but not only for $A = 2^a - 1$ as it has been thought to date. Based on the CSA with EAC network, the first ever n-input generator mod A using CSA network is obtained. Three design schemes of the n-input generator mod A are proposed to suit various needs for A and n. Any new generator is superior to any conventional design with respect to speed and amount of hardware used. In particular, it is shown that the CSA-based generator mod $A = 2^a - 1$ should replace a scheme commonly used in the literature (built of two-operand adders mod $A = 2^a - 1$), since it offers essential speed improvement virtually at no cost. Finally, a new general procedure of a MOMA using a CSA with EAC is proposed. For any $A \neq 2^a - 1$ the new scheme shows significant speed improvement compared to the existing designs. For many A and k (k is the number of operands) the hardware reduction is also observed.

We believe that the results presented here will be beneficial for researchers and practitioners working on hardware supporting highly-reliable digital systems protected against errors by using arithmetic codes as well as high-performance RNS-based systems. In particular, they may originate interest for application of arithmetic EDCs with check bases other than $A = 2^a - 1$ since low-cost encoding and decoding circuitry can be constructed for them now. Also, these results may give a new insight into the selection criteria of moduli used to form an RNS.

References

[1] N. S. Szabo and R. I. Tanaka, *Residue Arithmetic and its Applications to Computer Technology*, McGraw-Hill, New York, 1967.

[2] F. F. Sellers, Jr., M.-Y. Hsiao, and L. W. Bearnson, *Error Detecting Logic for Digital Computers*, McGraw-Hill, New York, 1968.

[3] F. J. Taylor, "Residue arithmetic: A tutorial with examples," *Computer*, pp. 50-62, May 1984.

[4] T. R. N. Rao, *Error Coding for Arithmetic Processors*, Academic Press, New York, 1974.

[5] J. F. Wakerly, *Error Detecting Codes, Self-Checking Circuits and Applications*, North-Holland, New York, 1978.

[6] G. D. Kraft and W. N. Toy, *Microprogrammed Control and Reliable Design of Small Computers*, Prentice-Hall, Englewood Cliffs, N.J., 1981.

[7] P. K. Lala, *Fault-Tolerant and Fault-Testable Hardware Design*, Prentice-Hall, Englewood Cliffs, N.J., 1985.

[8] Y. Tohma, "Coding Techniques in Fault-Tolerant, Self-Checking, and Fail-Safe Circuits," Ch. 5 in *Fault Tolerant Computing: Theory and Techniques, Vol. I*, D. K. Pradhan, Ed., Prentice-Hall, Englewood Cliffs, N. J., 1986.

[9] D. Nikolos, A. M. Paschalis, and G. Philokyprou, "Efficient design of totally self-checking checkers for all low-cost arithmetic codes," *IEEE Trans. Comput.*, vol. C-37, pp. 807-814, July 1988.

[10] S. J. Piestrak, "Design of high-speed and cost-effective self-testing checkers for low-cost arithmetic codes," *IEEE Trans. Comput.*, vol. C-39, pp. 360-374, March 1990.

[11] A. Avizienis, "A set of algorithms for a diagnosable arithmetic unit," Jet Propulsion Lab., Calif. Inst. Technol., Pasadena, CA, *Tech. Rep.* 32-546, Mar. 1964.

[12] A. Avizienis, "Arithmetic codes: Cost and effectiveness studies for applications in digital system design," *IEEE Trans. Comput.*, vol. C-20, pp. 1322-1331, Nov. 1971.

[13] W. K. Jenkins and B. J. Leon, "The use of residue number system in the design of finite impulse response filters," *IEEE Trans. Circuits Syst.*, vol. CAS-24, pp. 191-201, Apr. 1977.

[14] R. M. Capocelli and R. Giancarlo, "Efficient VLSI networks for converting an integer from binary system to residue number system and vice versa," *IEEE Trans. Circuits Syst.*, vol. CAS-35, pp. 1425-1430, Nov. 1988.

[15] K. Hwang, *Computer Arithmetic: Principles, Architecture and Design*, Wiley, New York, 1979.

[16] C. N. Zhang, B. Shirazi, and D. Y. Y. Yun, "Parallel designs for chinese remainder conversion," in *Proc. Int. Conf. on Parallel Processing*, pp. 557-559, Aug. 17-21, 1987.

[17] L. Skavantzos, "Design of multioperand carry-save adders for arithmetic modulo $(2^n + 1)$," *Electron. Lett.*, vol. 25, No. 17, pp. 1152-1153, 17th Aug. 1989.

[18] C. K. Koc and C. Y. Hung, "Multi-operand modulo addition using carry save adders," *Electron. Lett.*, vol. 26, No. 6, pp. 361-363, 15th Mar. 1990.

[19] C. A. Papachristou, "Associative table lookup processing for multioperand residue arithmetic," *JACM*, vol. 34, pp. 376-396, Apr. 1987.

[20] V. Piuri, M. Berzieri, A. Bisaschi, and A. Fabi, "Residue arithmetic for a fault-tolerant multiplier: The choice of the best triple of bases," *Microproc. and Microprogr.*, vol. 20, pp. 15-23, 1988.

Session 5:

Floating Point Range and Precision

Chair:
Renato Stefanelli
Politecnico di Milan

Overflow/Underflow-Free Floating-Point Number Representations with Self-Delimiting Variable-Length Exponent Field

Hidetoshi Yokoo

Department of Computer Science,
Gunma University
Kiryu, Gunma 376
Japan

Abstract

A class of new floating-point representations of real numbers, based on representations of the integers is described. In the class, every representation uses a self-delimiting representation of the integers as a variable length field of exponent, and neither overflow nor underflow appears in practice. The adopted representations of the integers are defined systematically, so that representations of numbers greater than one have both exponent-significand and integer-fraction interpretations. Since representation errors are characterized by the length function of an underlying representation of the integers, superior systems in precision can be easily selected from the proposed class.

1 Introduction

For these ten years, in order to overcome the inherent problem of overflow and underflow in conventional floating-point number representation systems, several representation systems with a variable-length exponent have been developed in Japan.

An earlier work of those appears in [12], in which the exponent may be extended to the entire data word using a field that specifies the length of the exponent. Subsequently, inspired by this work, Hamada [7] found a method of realizing a variable-length exponent without any additional field to specify the length. Hamada's URR (for universal representation of real numbers) has several attractive features which include the followings.

- Neither overflow nor underflow occurs in practice.

- Specifications do not depend on data length.

- The order of numbers coincides with the lexicographic order of their representations.

- Arbitrary real values can be infinitely approximated simply by extending data length.

Despite these, the URR system has not been implemented in any real system, partly because we have found no method of performing arithmetic directly with URR data except for some trivial operations, and partly (or maybe mainly) because we have already had some commercially established number representation systems [8], [15]. The significance of URR should rather be evaluated from the following recognition that it brought. That is, since the invention of URR, we have been recognizing that there is no essential difference between the two main systems — fixed-point and floating-point — for real numbers. To understand this recognition alone, there is no need to know the details of URR, instead the following simple example will suffice.

Consider, for example, the fixed-point representation of 0.15625 in radix 2. This can be represented as 0.00101_2. This number is, in turn, represented as a floating-point form: $1.01_2 \times 2^{-3}$. If we consider numbers between 0 and 1 only, and adopt *unary encoding* to represent the exponent such as

$$
\begin{array}{cc}
-1 & 1 \\
-2 & 01 \\
-3 & 001 \\
& \vdots \\
-E & \underbrace{00\cdots0}_{E-1}1,
\end{array}
$$

then the number $1.01_{(2)} \times 2^{-3}$ can be represented as the concatenation of 001 (exponent) and 01 (economized significand[1]). The obtained floating-point representation "00101" coincides with the fractional part of the fixed-point representation. Thus, the string "00101" has both fixed-point and floating-point interpretations. The crucial point is not to make a distinction between the fixed-point and floating-point representations but to separate the use of a predetermined boundary between the exponent and significand from the use of a self-delimiting encoding of the exponent.

---

[1] For this purpose, more familiar terms *fraction* or *mantissa* are usually used. In this paper, however, we adopt the term *significand* for several reasons.

Self-delimiting encodings have been studied on the several themes of (prefix-free) representations of the integers or encodings of commas between strings [1], [4], [5], [9]. Matula and Kornerup [13] discuss an application of a certain representation of the integers to the representation of rational numbers in computers. Researchers have been — and still are — seeking efficient methods for encoding of countably infinite elements. *Unary encoding* mentioned above, which is well known also in Turing machine theory, is the simplest example of representations of the integers; however, it is less efficient. The efficiency naturally depends on the definition of itself. In the problem of floating-point representation of real numbers, the efficiency of a self-delimiting representation specifying the exponent is expected to influence the precision of the number representation system.

This paper aims to discuss a unifying approach to the use of representations of the integers as the field of exponent, and presents a class of overflow/underflow-free representations of real numbers, which includes URR as a special case. This class may serve as a good answer to an exercise in [10, p.212, Exercise 17 of Sec.4.2.1], which requires a representation whose precision decreases as the magnitude of the exponent increases.[2] As far as the abolition of overflow and underflow is concerned, the level-index system [2, 3, 11] may be preferable. However, it matters little to the present paper whether a system is fixed-point, floating-point, or level-index. Instead, our main interest is the efficient and systematic method of mapping the system to binary strings. In fact, we will define a system in our class based not only on exponent-significand decomposition of real numbers but also on integer-fraction representation. The conversion between these two bases corresponds to the generalization of the floating-point and fixed-point interpretations of a representation. In the class, representation errors can be characterized by the length function of the adopted representation of the integers. As a result, we can find improved variations of URR systematically.

2 Preliminaries

In most of conventional floating-point systems in radix 2, a real number X can be approximated by

$$X = (-1)^s F \times 2^E, \qquad (1)$$

where s is either 0 or 1, E is an integer, and F is a number which is normalized as either

$$1/2 \leq F < 1 \qquad (2)$$

or

$$1 \leq F < 2. \qquad (3)$$

A number X can be also represented as

$$X = (-1)^s(\sum_{i=0}^{n} a_i 2^i + \sum_{j=1}^{\infty} b_j 2^{-j}), \quad a_i, b_j \in \{0,1\} \quad (4)$$

---

[2]Knuth refers to [14] as an answer to the exercise.

which is reexpressed as the form

$$X = (-1)^s a_n a_{n-1} \cdots a_0 . b_1 b_2 \cdots \qquad (5)$$

in the fixed-point system. For $X \geq 1$, $\lfloor X \rfloor$, the largest integer not greater than X, is said to be the integer part of X. A binary string $b_1 b_2 \cdots$ representing $X - \lfloor X \rfloor$ is said to be the binary fraction of X. Define $|\alpha|$ to be the length of a binary string α. If we set $a_n = 1$ and $\alpha = a_{n-1} a_{n-2} \cdots a_0$ for $X \geq 1$, then the equality $n \equiv |\alpha| = E - 1$ holds for (2) and the equality $n \equiv |\alpha| = E$ holds for (3).

Although URR can be considered as a variation of floating-point representations with the normalization condition (3), no fixed-length field is used to specify the value of E. Instead, it adopts a prefix-free encoding of the integers in order to implement a variable-length exponent without any additional field to specify the length of the part. A prefix-free encoding of the integers is a special case of representations of the integers [4], which are defined in the following way.

A binary encoding or a representation R of the positive integers, $N^+ = \{1, 2, \cdots\}$, is a bijection of N^+ onto a set C of binary codewords, such that any concatenation of any members of C is uniquely decipherable. The following four conditions are important when we consider an application of representations of the integers to floating-point representation of real numbers.

(C1) *prefix condition:* A representation R is prefix-free if no codeword $R(i) \in C$ is the beginning of another codeword in C.

(C2) *lexicographic order:* A codeword $c = c_1 \ldots c_l$, $c_i \in \{0, 1\}$, is said to be lexicographically smaller than a codeword $d = d_1 d_2 \ldots d_m, d_j \in \{0, 1\}$, if $c_1 < d_1$; or if $c_i = d_i$ for $i < n$ and $c_n < d_n$ for some $n \leq l, m$; or if $c_i = d_i$ for $1 \leq i \leq l$ and $l < m$. The representation R is said to be of lexicographic order if $R(i)$ is lexicographically smaller than $R(j)$ for $i < j$.

(C3) *completeness:* Let $\{0, 1\}^*$ denote the set of all finite strings of symbols, each symbol selected from the set $\{0, 1\}$. A uniquely decipherable set C is said to be complete iff adding any new string $c \in \{0, 1\}^*$, $c \notin C$, to C gives a set $C' = C \cup \{c\}$ that is not uniquely decipherable.

(C4) *minimal property:* Let $L(i)$ denote the length of $R(i)$ in bits. The representation R is said to be minimal if $L(i) \leq L(i+1)$ for any $i \in N^+$.

The condition (C1) is a sufficient condition for unique decipherability. It is well known [6] that any uniquely decipherable representation must satisfy the Kraft inequality:

$$\sum_{i=1}^{\infty} 2^{-L(i)} \leq 1.$$

As was mentioned in the previous section, the simplest representation of the integers is *unary encoding*. It is

111

Table 1: Examples of representation of the integers.

i	$U_0(i)$	$U_{00}(i)$	$U_{011}(i)$	$\Omega(i)$
1	0	0	0	0
2	100	100	10000	100
3	101	101	10001	101
4	11000	110000	100100	110000
5	11001	110001	100101	110001
6	11010	110010	100110	110010
7	11011	110011	100111	110011
8	1110000	1101000	10100000	1101000
9	1110001	1101001	10100001	1101001
10	1110010	1101010	10100010	1101010
15	1110111	1101111	10100111	1101111
16	111100000	1110000000	101010000	11100000000
31	111101111	1110001111	101011111	11100001111

convenient for later discussions to define it as

$$U(i) = 1^{i-1}0,$$

where 1^{i-1} denotes a concatenation of $i - 1$ "1" bits, although this definition contrasts with the one used in Section 1. Both the unary encodings obviously satisfy the conditions (C1)–(C4), and the difference between them is not essential.

Now, consider the following transformations for a representation R of the integers.

Type-0 transformation: When an integer $i \in N^+$ has an ordinary binary representation 1α, where α is any string of 0's and 1's, the function R_0 is defined by

$$R_0(i) = \begin{cases} 0 & \text{for } i = 1, \\ 1R(|\alpha|)\alpha & \text{for } i \geq 2. \end{cases}$$

The function R_0 is also a representation of the integers.

Type-1 transformation: When an integer i has an ordinary binary representation 1α, the function R_1 is defined by

$$R_1(i) = R(|\alpha| + 1)\alpha.$$

The function R_1 is also a representation of the integers.

If we use U for R in the above transformations, then

$$U_0(i) = \begin{cases} 0 & \text{for } i = 1, \\ 1U(|\alpha|)\alpha = 1^{|\alpha|}0\alpha & \text{for } i \geq 2, \end{cases}$$

and

$$U_1(i) = U(|\alpha| + 1)\alpha = 1^{|\alpha|}0\alpha,$$

which mean that $U_0(i) = U_1(i)$ for any $i \in N^+$. In general, results obtained from the type-0 and type-1

transformations are not the same. However, it can be easily shown that if a representation R of the integers satisfies one of the conditions (C1)–(C4), then both the representations R_0 and R_1 also satisfy the same condition. Thus, starting from the unary encoding U, we can get infinitely many representations of the integers satisfying the conditions (C1)–(C4) after repeated applications of the type-0 and/or type-1 transformations.

If τ is a binary string and $\sigma = \tau 0$, then the representation U_σ is a representation which is obtained from U_τ by the type-0 transformation. If $\sigma = \tau 1$, U_σ is a representation which is obtained from U_τ by the type-1 transformation. In Table 1, the examples U_0, U_{00}, and U_{011} are shown in the second, third, and fourth columns, respectively.

3 New Floating-Point Representations Based on Representation of the Integers

In this section, first we limit our definition to $X \geq 1$, and then generalize it to $X < 1$. For $X \geq 1$, a floating-point representation system based on a representation of the integers will be defined by two methods: the exponent-significand pair method and the integer-fraction pair method. The relation of these two methods is summarized in Theorem 1. Since the definitions do not depend on data length, we will not note the length in this section.

3.1 Exponent-Significand Pair Method

In this method, a floating-point system based on a representation of the integers is defined as three fields. Like conventional systems, a number X of the form (1) is represented by its sign S_X, exponent, and significand, as shown in Fig.1. The last two fields specify the values of E and F in (1), respectively. The leading exponent bit is denoted by $S_{E(X)}$, which may be called as the exponent sign bit but does not exactly correspond to the sign of E. The sign bit S_X and the exponent sign bit $S_{E(X)}$ are defined in the following:

$$\begin{array}{lll} \text{if } X < -1, & \text{then } S_X = 1 \text{ and } S_{E(X)} = 0, \\ \text{if } -1 \leq X < 0, & \text{then } S_X = 1 \text{ and } S_{E(X)} = 1, \\ \text{if } 0 < X < 1, & \text{then } S_X = 0 \text{ and } S_{E(X)} = 0, \\ \text{if } 1 \leq X, & \text{then } S_X = 0 \text{ and } S_{E(X)} = 1. \end{array} \quad (6)$$

Even in the case where we restrict ourselves to $X \geq 1$, the value of E depends on the normalization condition

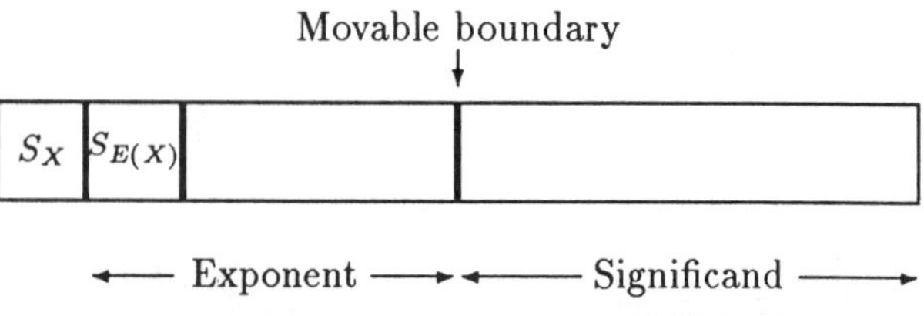

Figure 1: Format of a floating-point representation.

(2) or (3). If we adopt the condition (2), then we have $E \geq 1$ for $X \geq 1$. Under the condition (3), we have $E \geq 0$ for $X \geq 1$. Thus the following two cases are considered.

Condition (2): In this case, E is a positive integer for $X \geq 1$ and we can use a representation of the integers to specify the value of E. Specifically, we put $U_\sigma(E)$ from the 2nd bit on the exponent field. Under the condition (2), F has the form looking like

$$F = 0.1 f_2 f_3 \cdots . \tag{7}$$

Thus we can set the binary string $f_2 f_3 \ldots$ as the economized significand. Therefore, letting $\mathcal{E}_\sigma^{(2)}(X)$ denote the representation of X for the condition (2), it is encoded as

$$\mathcal{E}_\sigma^{(2)}(X) = 01 U_\sigma(E) f_2 f_3 \cdots \tag{8}$$

for $X \geq 1$, (1), and (7).

Condition (3): In order to specify the value of E, which is greater than or equal to zero, we use

$$U_\sigma^*(E) = \begin{cases} 0 & \text{for } E = 0, \\ 1 U_\sigma(E) & \text{for } E \geq 1, \end{cases} \tag{9}$$

for a representation U_σ of the integers. Under the condition (3), F has the form

$$F = 1.f_1 f_2 \cdots , \tag{10}$$

which is combined with $U_\sigma^*(E)$ to yield the representation. That is, if $\mathcal{E}_\sigma^{(3)}(X)$ denotes the system on the condition (3), then it has the bit pattern

$$\mathcal{E}_\sigma^{(3)}(X) = 01 U_\sigma^*(E) f_1 f_2 \cdots \tag{11}$$

for $X \geq 1$, (1), and (10).

3.2 Integer-Fraction Pair Method

If X has the form (5) with $s = 0$, this method defines a system in our class as the concatenation of the integer and binary fractional parts:

$$\mathcal{F}_\sigma(X) = 01 U_\sigma(\lfloor X \rfloor) b_1 b_2 \cdots \tag{12}$$

for $X \geq 1$. Then we have the following theorem.

Theorem 1 *For any representation U_σ of the integers, we have*

$$\begin{aligned} \mathcal{F}_{\sigma 0}(X) &= \mathcal{E}_\sigma^{(3)}(X), & (13) \\ \mathcal{F}_{\sigma 1}(X) &= \mathcal{E}_\sigma^{(2)}(X) & (14) \end{aligned}$$

for $X \geq 1$.

Proof: Since $X \geq 1$, X can be represented for $E \geq 0$ as

$$\begin{aligned} X &= 1.f_1 f_2 \cdots \times 2^E \\ &= 1 f_1 f_2 \cdots f_E . f_{E+1} \cdots . \end{aligned} \tag{15}$$

Then, we have from (9) and (11)

$$\mathcal{E}_\sigma^{(3)}(X) = \begin{cases} 010 f_1 f_2 \cdots & \text{for } E = 0, \\ 011 U_\sigma(E) f_1 f_2 \cdots & \text{for } E \geq 1. \end{cases}$$

From the definition of $U_{\sigma 0}$ and (15)

$$U_{\sigma 0}(\lfloor X \rfloor) = \begin{cases} 0 & \text{for } E = 0, \\ 1 U_\sigma(E) f_1 f_2 \cdots f_E & \text{for } E \geq 1. \end{cases}$$

It follows from this and (12) that

$$\mathcal{F}_{\sigma 0}(X) = \mathcal{E}_\sigma^{(3)}(X)$$

for $X \geq 1$.

Similarly, for $X \geq 1$ and $E \geq 1$, we have

$$\begin{aligned} X &= 0.1 f_2 f_3 \cdots \times 2^E \\ &= 1 f_2 f_3 \cdots f_E . f_{E+1} \cdots . \end{aligned} \tag{16}$$

Therefore,

$$U_{\sigma 1}(\lfloor X \rfloor) = U_\sigma(E) f_2 f_3 \cdots f_E .$$

This and

$$\mathcal{E}_\sigma^{(2)}(X) = 01 U_\sigma(E) f_2 f_3 \cdots$$

prove

$$\mathcal{F}_{\sigma 1}(X) = \mathcal{E}_\sigma^{(2)}(X)$$

for $X \geq 1$.   Q.E.D.

We usually interpret any system in the class as a floating-point representation consisting of sign, exponent, and significand. However, it is sometimes more convenient to interpret it as $\mathcal{F}_\sigma$ because we can uniquely identify a system only by σ without referring to the normalization condition. The above simple theorem, which not only gives the generalization of the floating-point and fixed-point interpretations but also provides a basis for the conversion of a system from/to the integer data type, is important also in this respect.

3.3 Extension

We must proceed to the case of $X < 1$. Among possible extensions, we have adopted the following one in order for the class to include URR as a special case. Since the integer-fraction pair method is not effective in $0 < X < 1$, we can extend only the exponent-significand pair method here. In the following, $\overline{S}$ denotes the 1's complement of a binary string S, i.e., if S has the form $s_1 s_2 \cdots s_N, s_i \in \{0,1\}$, then

$$\overline{S} = \overline{s_1} \overline{s_2} \cdots \overline{s_N},$$

where $\overline{s_i} = 1 - s_i$. Further, for a finite N, $\langle S \rangle$ denotes the 2's complement of S of length N. For an infinite

N, if there exists an integer j such that $s_j = 1$ and $s_{j+1} = s_{j+2} \cdots = 0$, then $\langle S \rangle$ is the concatenation of $\langle s_1 s_2 \cdots s_j \rangle$ and infinite zeros; otherwise, $\langle S \rangle = \overline{S}$.

For $0 < X < 1$, the sign bit S_X and the exponent sign bit $S_{E(X)}$ are both fixed to 0 according to (6). Under the condition (2), noting that the inequalities $0 < X < 1$ correspond to $E \leq 0$, define

$$\mathcal{E}_\sigma^{(2)}(X) = 00\overline{U_\sigma(-E+1)}f_2 f_3 \cdots \qquad (17)$$

for $0 < X < 1$ and (7). For the condition (3), define

$$\mathcal{E}_\sigma^{(3)}(X) = 00\overline{U_\sigma^*(-E-1)}f_1 f_2 \cdots \qquad (18)$$

for $0 < X < 1$ and (10), since $E \leq -1$.

To conclude the definitions, we must deal with the case $X < 0$. For either the condition (2) or (3), the extension is quite straightforward, that is, for $X < 0$

$$\mathcal{E}_\sigma^{(2)}(X) = \; <\mathcal{E}_\sigma^{(2)}(-X)>, \qquad (19)$$
$$\mathcal{E}_\sigma^{(3)}(X) = \; <\mathcal{E}_\sigma^{(3)}(-X)>, \qquad (20)$$

which are consistent with (6).

In all the definitions mentioned above, every system is defined as a semi-infinite binary string. However, we can eliminate trailing zeros to yield a finite length representation, and conversely any finite length representation is interpreted as that extended to the right with infinite extra zeros.

4   URR System

Although the URR system [7] has several desirable properties as was mentioned in Section 1, its precision decreases rapidly as a number to be represented increases. Since the original definition of URR uses a bisection method, it is not easy to consider the improvements along the line with the definition. In the proposed class, $\mathcal{E}_0^{(3)}(X)$ is identical with URR. In this section, we will first cite the original definition of URR as it is in order to emphasize the distinction between both the ways of definition. Then, we will give a brief remark on the identity of $\mathcal{E}_0^{(3)}(X)$ with URR.

In the original definition of URR, a binary string S corresponds to a semi-closed interval, say $[a, b)$, which is represented as $I(S)$. This interval is then divided into two intervals $[a, c)$ and $[c, b)$ by a third value c, which is determined by the following four steps.

(a) Rough cut

$$I(1) = [-\infty, 0) \text{ and } I(0) = [0, +\infty).$$

$$I(10) = [-\infty, -1), I(11) = [-1, 0),$$
$$I(00) = [0, 1), \text{ and } I(01) = [1, +\infty).$$

$$I(100) = [-\infty, 2), I(101) = [-2, -1),$$
$$I(110) = [-1, -0.5), I(111) = [-0.5, 0),$$
$$I(000) = [0, 0.5), I(001) = [0.5, 1),$$
$$I(010) = [1, 2), \text{ and } I(011) = [2, +\infty).$$

After the last eight subdivisions, the procedure proceeds to the step (b) for those in which the 2nd bit is equal to the 3rd bit. Otherwise, it proceeds directly to the equal-difference cut (step(d)).

(b) Double-exponential cut

Let $p^+(m)$ and $p^-(m)$ denote 2^{2^m} and 2^{-2^m}, respectively. If $m \geq 0$, the interval is partitioned recursively as:

$$I(10^{m+2}) = [-\infty, -p^+(m)) \text{ at } -p^+(m+1),$$
$$I(11^{m+2}) = [-p^-(m), 0) \text{ at } -p^-(m+1),$$
$$I(00^{m+2}) = [0, p^-(m)) \text{ at } p^-(m+1),$$
$$I(01^{m+2}) = [p^+(m), +\infty) \text{ at } p^+(m+1)$$

with incrementing m by one until one of the following intervals is obtained:

$$I(10^{m+2}1) = [-p^+(m+1), -p^+(m)),$$
$$I(11^{m+2}0) = [-p^-(m), -p^-(m+1)),$$
$$I(00^{m+2}1) = [p^-(m+1), p^-(m)),$$
$$I(01^{m+2}0) = [p^+(m), p^+(m+1)),$$

where neither 0 nor $\pm\infty$ are included at the bounds of the interval.

(c) Equal-ratio cut

The interval $I(S) = [a, b)$ is divided at $\sqrt{ab}$ as

$$I(S0) = [a, \sqrt{ab}) \text{ and } I(S1) = [\sqrt{ab}, b).$$

This is performed m times for the final value of m in step (b). After the completion of this cut, the ratio of the upper bound to the lower bound is 2.

(d) Equal-difference cut

Each interval is divided at $(a + b)/2$ as

$$I(S0) = [a, (a + b)/2) \text{ and } I(S1) = [(a + b)/2, b).$$

This is performed an arbitrary number of times until the desired binary representation is obtained.

It is obvious that $I(S0^k)$ at $k \to \infty$ converges to the lower bound of the interval. This is denoted by $V(S)$. That is, for $I(S) = [a, b)$,

$$V(S) = \lim_{k \to \infty} I(S0^k) = a.$$

For example, the representation of 7.5 in the system is as follows.

— rough cut

$$I(011) = [2, +\infty),$$

— double-exponential cut

$$I(0111) = [4, +\infty),$$

$$I(01110) = [4, 16),$$

114

– equal-ratio cut

$$I(011100) = [4, 8],$$

– equal-difference cut

$$I(0111001) = [6, 8],$$
$$I(01110011) = [7, 8],$$
$$I(011100111) = [7.5, 8].$$

Consequently, $V(011100111) = 7.5$.

Here, we note how 7.5 is represented in our class. Since $7.5 = 2^2 \times (15/8) = 7 + 0.5$, and 15/8 and 0.5 are represented by 1.111 and 0.1 respectively in the fixed-point binary system, we have

$$\mathcal{E}_0^{(3)}(7.5) = 01 \cdot 1100 \cdot 111 = 011100111$$

and

$$\mathcal{F}_{00}(7.5) = 01 \cdot 110011 \cdot 1 = 011100111$$

because $U_0^*(2) = 1100$ and $U_{00}(7) = 110011$. In these three systems, i.e., URR, $\mathcal{E}_0^{(3)}$, and $\mathcal{F}_{00}$, 7.5 is represented by 011100111.

It is not so difficult to show that the URR system falls on $\mathcal{E}_0^{(3)}(X)$ in the proposed class. As an example, we give a remark only on the case with $X \geq 2$. Let 1α denote the ordinary binary representation of the exponent E of $X \in [p^+(m), p^+(m+1))$. Under the condition (3), the length of α is m bits. Since $U_0^*(E) = 1U_0(E) = 11U(|\alpha|)\alpha = 1^{m+1}0\alpha$, we have

$$\mathcal{E}_0^{(3)}(X) = 01^{m+2}0\alpha f_1 f_2 \cdots \qquad (21)$$

for $X \geq 2$ and (10). In (21), it is quite obvious that the string $f_1 f_2 \cdots$ corresponds to the equal-difference cut. Noting that the exponent E of $X \geq 2$ can be represented as $2^m + n$ with $n = 0, 1, \ldots, 2^m - 1$, we see that URR encodes m by the unary encoding in the double-exponential cut and n as an m-bit binary number in the equal-ratio cut. Thus, it follows that the string α in (21) corresponds to the equal-ratio cut of URR.

The identity of URR and $\mathcal{E}_0^{(3)}(X)$ can be proved also in an information theoretical manner. It is known that if a representation of the integers with a given length function satisfies the conditions (C1)–(C3) then the representation is unique. Based on this fact, the present author [16] has shown that URR and $\mathcal{E}_0^{(3)}(X)$ are the same.

5 Representation Errors

When we use fixed length words to implement a system in our class, it is impossible to represent all the real numbers, and therefore representation errors may occur. Let $\Delta(X)$ denote the difference between a value X intended for the representation and the value corresponding to the bit pattern with a fixed length. We will evaluate the error characteristics by the maximum relative error $Er(X) = \Delta(X)/X$, where X corresponds to the mid-point between two discrete points.

In evaluating the error characteristics, we will use the formats $\mathcal{F}_\sigma$ instead of $\mathcal{E}_\sigma^{(2)}$ or $\mathcal{E}_\sigma^{(3)}$, because $\mathcal{F}_\sigma$s are independent of the normalization condition. If 64-bit words are used to represent $\mathcal{F}_\sigma(X)$, we have from (12)

$$\log |Er(X)| = |U_\sigma(\lfloor X \rfloor)| - 63 - \log X, \qquad (22)$$

where $\log(\cdot)$ means $\log_2(\cdot)$. We must find an efficient representation U_σ for all over the natural numbers so as to gain higher precision. Of course, however, no representation of the integers is optimal in the sense that it attains the minimum length of representation for any integer.

Let $L_\sigma(i)$ denote the length of $U_\sigma(i)$ in bits. If we set $l = \lfloor \log i \rfloor$, then we have, for example,

$$L_0(i) = 2l + 1, \qquad (23)$$
$$L_{00}(i) = \begin{cases} 1 & \text{for } i = 1 \\ l + 2\lfloor \log l \rfloor + 2 & \text{for } i \geq 2, \end{cases} \qquad (24)$$
$$L_{011}(i) = l + \lfloor \log(l+1) \rfloor \\ + 2\lfloor \log \log 2(l+1) \rfloor + 1. \qquad (25)$$

Define

$$\log^{(1)} i = \log i,$$
$$\log^{(k)} i = \log \log^{(k-1)} i \quad \text{for } k \geq 2$$

and

$$\log^m i = \sum_{k=1}^{m-1} \log^{(k)} i + 2 \log^{(m)} i. \qquad (26)$$

From the definitions of the type-0 and type-1 transformations, the difference between $L_\sigma(i)$ and $\log^m i$ for $m = |\sigma|$ is upperbounded by a constant which is independent of i. This is denoted by $L_\sigma(i) \simeq \log^m i$. Naturally, this is true of (23), (24), and (25).

In order to evaluate the efficiency of U_σ, let us consider the representations Φ and Ω of the integers which are defined by

$$\Phi(0) = 0,$$
$$\Phi(i) = 1\Phi(|\alpha|)\alpha \quad \text{for } i \geq 1 \qquad (27)$$

and

$$\Omega(i) = \Phi(|\alpha|)\alpha \quad \text{for } i \geq 1, \qquad (28)$$

where an integer $i \in N^+$ has a binary representation 1α [9]. Some examples of the representation Ω are shown in the rightmost column in Table 1. The representation Ω satisfies the conditions (C1)–(C4) and we can define new number representation systems:

$$\mathcal{E}_\Phi^{(3)}(X) = 01\Phi(E)f_1 f_2 \ldots$$

for $X \geq 1$, (1), and (10), and

$$\mathcal{F}_\Omega(X) = 01\Omega(\lfloor X \rfloor)b_1 b_2 \ldots$$

for $X \geq 1$ and (5). Following the proof of Theorem 1, it can be easily shown that $\mathcal{E}_\Phi^{(3)}(X) = \mathcal{F}_\Omega(X)$ for $X \geq 1$. Since $|\alpha| \simeq \log i$ for an integer i whose binary representation is 1α, and

$$|\Phi(i)| \simeq |\Phi(\log i)| + \log i$$

from (27), the length of Ω is represented by

$$\begin{aligned} |\Omega(i)| &\simeq \log^* i \\ &\triangleq \log i + \log^{(2)} i + \log^{(3)} i + \cdots, \end{aligned}$$

where only the positive terms are included in the sum. Comparing this with (26), we can conclude that, for any representation $\mathcal{F}_\sigma$, there exists a positive number X_m such that higher precision than $\mathcal{F}_\Omega(X)$ is never realized by $\mathcal{F}_\sigma(X)$ for $X \geq X_m$.

This observation, however, is quite unsatisfactory to compare exact relative error characteristics for a practical range of X. To make an exact comparison, we have derived the lengths of representations of the integers and plotted the relative errors as the functions of $\log X$ for $X \geq 1$. Figure 2 shows two examples of $|U_\sigma(\lfloor X \rfloor)| - 63 - \lfloor \log X \rfloor$ and $|\Omega(\lfloor X \rfloor)| - 63 - \lfloor \log X \rfloor$. In order to avoid dense and hard-to-read results, these staircase functions are plotted instead of (22) and other examples are eliminated. Figure 2 shows that the rep-

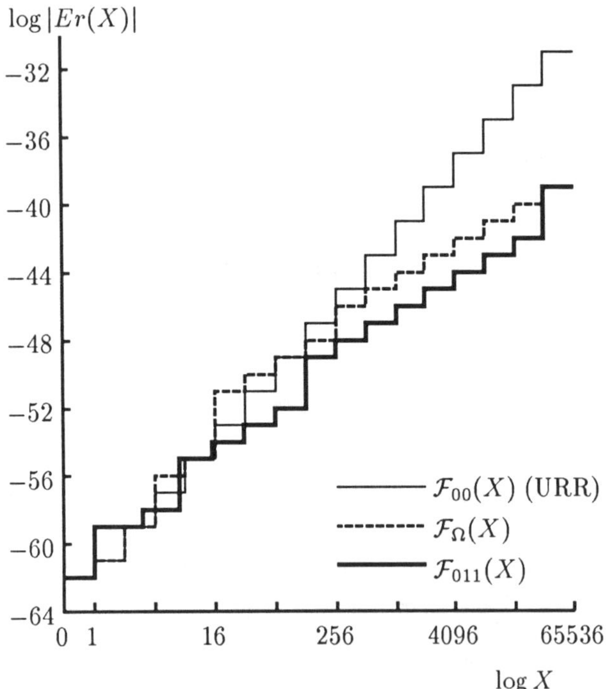

Figure 2: Comparison of representation errors.

resentation $\mathcal{F}_{011}(X) = \mathcal{E}_{01}^{(2)}(X)$ has relatively high precision for medium values of X. In comparison with other systems in the proposed class, $\mathcal{F}_{011}(X)$ has still higher precision for a wide range of X. In fact, if we imagine a representation of the integers which attains the minimum length of the three representations in Fig.2, i.e.,

$$L(i) = \min\{L_{00}(i), L_{011}(i), |\Omega(i)|\},$$

then the corresponding Kraft sum exceeds unity even for the small values of i, that is,

$$\sum_{i=1}^{2^{16}} 2^{-L(i)} > 1.$$

This suggests that it is not easy to improve the precision of the representation $\mathcal{F}_{011}(X)$.

6 Conclusion

This paper has presented a class of floating-point number representation systems with self-delimiting variable-length exponent field. Our systems are all available even for those numbers whose exponent are greater than 2^{12}, which we must give up representing in most of conventional systems.

It should be noted that, for a fixed length word, the total number of representable numbers is the same, whether the system is fixed-point, floating-pont, or any other. Thus, in addition to the development of new systems, it is important to consider the efficient and systematic use of the word. To do this and to evaluate the performance, we have incorporated the information theoretical notions such as the Kraft inequality. Consequently, we have succeeded in finding a superior system in precision, because, in the proposed class, the representation errors can be characterized by the length function of an underlying representation of the integers.

In any system of the proposed class, once determined the boundary between the exponent and significand, the arithmetic operations can be performed just as if the operands were conventional floating-point numbers. However, the determination of the boundary is slightly complicated in comparison with the conventional systems. Thus, there still remains much work for future exploration in finding more direct methods of arithmetic operations or in implementing arithmetic by hardware.

Acknowledgment

The author would like to thank Prof. M. Iri of the University of Tokyo for his continued interest and encouragement.

References

[1] Apostolico, A. and Fraenkel, A. S.: "Robust transmission of unbounded strings using Fibonacci representations," *IEEE Trans. Inform. Theory*, Vol.IT-33, No.2, pp.238-245, 1987.

[2] Clenshaw, C. W. and Olver, F. W. J.: "Beyond floating point," *J. ACM,* Vol.31, No.2, pp.319-328, 1984.

[3] Clenshaw, C. W. and Olver, F. W. J.: "Level-index arithmetic operations," *SIAM J. Numer. Anal.,* Vol.24, No.2, pp.470-485, 1987.

[4] Elias, P.: "Universal codeword sets and representations of the integers," *IEEE Trans. Inform. Theory,* Vol.IT-21, No.2, pp.194-203, 1975.

[5] Even, S and Rodeh, M.: "Economical encoding of commas between strings," *Commun. ACM,* Vol.21, No.4, pp.315-317, 1978.

[6] Gallager, R. G.: *Information theory and reliable communication,* Wiley, New York, 1986.

[7] Hamada, H.: "Data length independent real number representation based on double exponential cut," *J. Inform. Process.,* Vol.10, No.1, pp.1-6, 1986.

[8] IBM: *IBM System/370 Principles of Operation,* GA22-7000-8, 1981.

[9] Knuth, D. E.: "Supernatural Numbers," *The Mathematical Gardner* (D. A. Klarner, ed.), Prindle Weber and Schmidt, Boston, pp.310-325, 1980.

[10] Knuth, D. E.: *The Art of Computer Programming, 2: Seminumerical Algorithms,* 2nd. ed. Addison-Wesley, Reading, Mass, 1981.

[11] Lozier, D. W. and Olver, F. W. J.: "Closure and precision in level-index arithmetic," *SIAM J. Numer. Anal.,* Vol.27, No.5, pp.1295-1304, 1990.

[12] Matsui, S. and Iri, M.: "An overflow/underflow-free floating-point representation of numbers," *J. Inform. Process.,* Vol.4, No.3, pp.123-133, 1981.

[13] Matula, D. W. and Kornerup, F.: "An order preserving finite binary encoding of the rationals," *Proc. 6th IEEE Symposium on Computer Arithmetic,* pp.201-209, 1983.

[14] Morris, R.: "Tapered floating point: a new floating-point representation," *IEEE Trans. Comput.,* Vol.C-20, No.6, pp.1578-1579, 1971.

[15] Stevenson, D. et al.: "A proposed standard for binary floating-point arithmetic," *IEEE Computer,* Vol.14, No.3, pp.51-62, 1981.

[16] Yokoo, H.: "A class of number representation systems based on multiple exponential cut (in Japanese)," *Trans. of the Institute of Electronics, Information and Communication Engineers,* Vol.J72-A, No.12, pp.1998-2004, 1989.

Implementation and Analysis of Extended SLI Operations

Peter R Turner

Mathematics Department, US Naval Academy, Annapolis, MD 21402

Abstract

This paper is concerned with extended arithmetic operations, such as forming scalar products, in symmetric level index, SLI, arithmetic. Schemes for the implementation of such algorithms are described and analysed both in terms of comparative timings for these operations and their floating-point counterparts and in terms of the control of errors in the computation. It is seen that with sufficient parallelism available in the SLI processor the computation can be as fast as for floating-point operations. Also we see that the SLI operation can be modified to produce just a single rounding error from extended operations very economically.

1 Introduction

The level-index number system for computer arithmetic was first suggested in Clenshaw and Olver [1],[2]. The scheme was extended to the symmetric level index, SLI, representation in [4] and has been studied in several further papers in the last few years. Much of the earlier work is summarized in the introductory survey [3]. The primary virtue of SLI arithmetic is its freedom from overflow and underflow and the consequent ease of algorithm development available to the scientific software designer. This is not the only arithmetic system that has been proposed with this aim; for example the work of Matsui and Iri [10] and Hamada [6] suggested modified floating-point systems which share some of the properties of level-index.

Possible hardware implementations of SLI arithmetic were discussed in [13] and [14] while a software implementation incorporating some extended arithmetic was described in [16]. The error analysis of SLI arithmetic is discussed in [2] and [4] and is extended in [8], [11] and [12]. Applications and software engineering aspects of the level-index system have been discussed in [5],[9] and [15].

In this paper, we concentrate on the extended arithmetic operations, such as forming scalar products of two vectors, which are of fundamental importance in scientific computing. At the heart of any such operation is the extended addition of a finite series of terms. This operation was studied in some detail for interval arithmetic [7] and a greatly extended accumulator was designed to return the result with just a single roundoff error. In this paper we consider the implementation of extended summation algorithms for the SLI system and their analysis. We see that the time penalty associated with SLI arithmetic is reduced so sharply that, with sufficient parallelism available in the SLI processor, the summation could even be faster than conventional floating-point arithmetic. Moreover the analysis demonstrates that the objective of a single rounding error for the extended summation can be achieved at very modest cost of increasing the precision of the internal computation very slightly.

We begin with a brief review of the SLI representation and its arithmetic. In Section 2, we describe the algorithms for extended summation and their implementation. Section 3 is concerned with the timings of these algorithms and their comparison with floating-point systems. In Section 4, we present the error analysis of these operations and its implications for the control of the overall error.

The symmetric level index representation of a real number X is given by

$$X = s_X \phi(x)^{r_X} \tag{1}$$

where the two signs s_X and r_X are ± 1 and the *generalized exponential* function is defined for $x \geq 0$ by

$$\phi(x) = \begin{cases} x & 0 \leq x < 1, \\ \exp(\phi(x-1)) & x > 1. \end{cases} \tag{2}$$

It follows that for $X > 1$,

$$X = \exp(\exp(...(\exp f)...)) \tag{3}$$

where the exponentiation is performed $l = [x]$ times and $x = l + f$. The integer part, l of x is called the *level* and the fractional part, f is called the *index*. The freedom of this system from over- and underflow results from the fact that, working to a precision of no more than 5,500,000 decimal places in the index, the system is closed under the four basic arithmetic operations apart from division by zero. This is discussed briefly in [1], [4] and considered in some detail in [8].

The appropriate error measure for computation in the level index system is no longer relative error (which corresponds approximately to absolute precision in the mantissa of floating-point numbers) but *generalized precision* which corresponds to absolute precision in the index. This error measure is introduced in [1].

In order to study the extended SLI operations which are at the heart of the present work, it is desirable to remind the reader of the fundamentals of the basic arithmetic algorithms for the SLI system. The properties of the natural logarithm function reduce all such operations to addition or subtraction. The algorithms are described in some detail in [4].

The basic problem is that of finding the SLI representation $s_z\phi(z)^{r_z}$ of $Z=X\pm Y$ where X, Y are also given by their SLI representations. Without loss of generality, we may assume that $X\geq Y>0$ so that $s_z=+1$. The computation entails the calculation of the members of three short sequences which vary according to the particular circumstances. In every case, the sequence defined by

$$a_j = \frac{1}{\phi(x-j)} \qquad (j = l-1, l-2, ..., 0) \qquad (4)$$

where $l=[x]$, is computed using the recurrence relation

$$a_{j-1} = \exp(-1/a_j); \qquad a_{l-1} = e^{-f}. \qquad (5)$$

Depending on the values of r_X, r_Y and r_Z, the other sequences that may be required aregiven, for appropriate starting values, by

$$b_j = \frac{\phi(y-j)}{\phi(x-j)}, \quad \beta_j = \frac{\phi(x-j)}{\phi(y-j)}, \quad \alpha_j = \frac{1}{\phi(y-j)}$$

$$c_j = \frac{\phi(z-j)}{\phi(x-j)}, \quad h_j = \phi(z-j). \qquad (6)$$

The ranges of their values are always suitably bounded and the terms are computed according to recurrence relations similar to (5) using exponentials and logarithms of special arguments. See [4] for full details of the algorithms and [13], [14] for possible schemes for hardware implementation. The particular implementation of extended operations presented in Section 2 is based on the modified CORDIC approach of [14].

2 Extended Algorithms and their Implementation

In this section, we first describe the mathematical algorithm for extended summation and then consider how this algorithm may be implemented in an efficient manner on a machine with a high degree of parallelism available in its (hypothetical) SLI processor. The operation of summing the elements of a sequence or the components of a vector plays a central role in much scientific computing especially in the formation of scalar products and in quadrature.

The specific task then is to find the SLI representation $s_z\phi(z)^{r_z}$ of $Z = \sum_{i=0}^{N} X_i$ where each component X_i is given by its SLI representation. Thus we seek s_z, z, r_z such that

$$s_z\phi(z)^{r_z} = \sum_{i=0}^{N} s_i\phi(x_i)^{r_i} \qquad (7)$$

where we shall assume that $X_0>0$ is the largest component in the sum. (Of course, our algorithm must identify this maximum element and the time for this operation must be included in our timing comparisons in the next section.)

In all cases, the computation begins with the calculation of the terms of the sequence

$$a_j = \frac{1}{\phi(x_0-j)}$$

using the recurrence relation (5). The rest of the algorithm is divided into two cases depending on whether $X_0\geq 1$. If $X_0\geq 1$, then the b-sequence is computed for each term X_i ($i\geq 1$) for which $r_i=+1$ to yield $b_{i,0}$. The α-sequence is computed for terms with $r_i=-1$ and the product $a_0\alpha_{i,0}$ is formed. The important observation here is that, given sufficient parallelism in the processor, all of these sequences may be computed simultaneously.

The results of these calculations can be combined to yield

$$c_0 = 1 + \sum_{r_i=+1} s_i b_{i,0} + \sum_{r_i=-1} s_i a_0 \alpha_{i,0} \qquad (8)$$

from which the remaining terms of the c-sequence may be computed as usual, together with any of the h_j that may be necessary. For the case where $X_0<1$, in which case all terms in the sum are "small" similar computation with the β-sequences is used to obtain

$$c_0' = \frac{1}{c_0} = 1 + \sum_{i=1}^{N} s_i \beta_{i,0}. \qquad (9)$$

The recurrence relations for the c- and h-sequences are

included in the following description of the complete algorithm. Those for the b- and β-sequences are included in the discussion of possible hardware implementations of the algorithm. The detailed derivations of these sequences can be found in [7] and [9].

Algorithm 1 Extended SLI summation

Input: $s_i\phi(x_i)^{r_i}$ $(i=0,1,...,N)$

    with $s_0=1$ and $\phi(x_0)^{r_0} > \phi(x_i)^{r_i}$ $(i \geq 1)$

    $x_0 = [x_0] + f_0$

Compute: $a_j = 1/\phi(x_0-j)$ $(j = [x]-1,...,0)$ using (4)

  If $r_0=+1$ (*Case 1*) **then**

    for $i=1$ to N

        if $r_i=+1$ compute $b_{i,0} = \phi(x_i)/\phi(x_0)$

        if $r_i=-1$ compute $a_0\alpha_{i,0} = a_0/\phi(x_i) = 1/\phi(x_0)\phi(x_i)$

$$r_z = +1; \quad c_0 = 1 + \sum_{r_i=+1} s_i b_{i,0} + \sum_{r_i=-1} s_i a_0 \alpha_{i,0}$$

  If $r_0=-1$ (*Case 2*) **then**

    for $i=1$ to N compute $\beta_{i,0} = \phi(x_0)/\phi(x_i)$

$$r_z = -1; \quad c = 1 + \sum_{i=1}^{N} s_i \beta_{i,0}$$

  If $c=0$ then $Z=0$; return

    else $s_z = \mathrm{sgn}(c); \quad c = |c|, \; l_z = 1$

Case 1: if $c<a_0$ then $r_z = -1; \; h = -\ln(c/a_0)$

      compute h-sequence:

        repeat

          $h=\ln h; \; l_z = l_z+1$

        until $h<1$

      $z = l_z + h$; return

Case 2: if $ca_0>1$ then $r_z = +1; \quad h = \ln ca_0$

      compute h-sequence;

      $z = l_z + h$; return

If $[x]=1$ then $h = f_0 + r_0 \ln c$;

      compute h-sequence;

      $z = l_z + h$; return

$c = 1 + r_0 a_1 \ln c$

For $i=1$ to $[x]-2$

    if $c<a_i$ then $z = l_z + c/a_i$; return

      else $l_z = l_z+1; \quad c = 1 + a_{l_z} \ln c$

If $c_{[x]-1} < a_{[x]-1}$ then $z = l_z + c/a_{[x]-1}$; return

    else $l_z=[x]; \; h = f_0 + \ln c$

      compute h-sequence

      $z = l_z + h$; return.

This is the algorithm which is implemented in the software implementation of SLI arithmetic described in [16] although it was convenient there to implement separate c-sequences for the two cases. Our primary concern here though is with potential hardware implementations of the SLI system including the extended summation operation.

The implementation described here is an extension of the ideas of [14] using modified CORDIC algorithms for the computation of the various sequences.

The preliminary step is the determination of the maximal element in the sum. We recall from [16] that the electronic representation of the SLI system is such that the order relations are precisely those of the same bit-strings compared as 2's complement integers.

It follows that this preliminary step can be achieved with a simple tree of "integer-discriminators". This tree will have $\log_2(N+1)$ stages assuming sufficient parallelism that at least $(N+1)/2$ such discriminators can operate simultaneously.

It is clearly important here that an efficient integer discrimination algorithm is used here. The comparison of integers is essentially the same operation as the determination of the sign of a "double number" which was a critical operation in the modified CORDIC approach to SLI arithmetic outlined in [14]. We digress briefly to consider this operation and to present an efficient low-level algorithm for integer comparison.

Consider then the operation of determining which is the larger of two 2's complement integers p and q. The algorithm proceeds as though both representations were standard binary integers and then reverses the outcome if the sign bit of p is a 1. It suffices therefore that our algorithm determines the larger of two aligned **positive** binary numbers of $n+1$ bits.

The first stage of the algorithm replaces each pair of bits (p_i,q_i) with the pair (d_i,p_i) $(i=0,1,...,n)$ where

$$d_i = p_i \text{ XOR } q_i = p_i+q_i \bmod 2 = \begin{cases} 1 & p_i \neq q_i \\ 0 & p_i = q_i \end{cases} \quad (10)$$

Now if $d_i=1$ and $d_j=0$ for $j<i$ then $p>q$ if $p_i=1$ while $p<q$ if $p_i=0$. (If there is no $d_i=1$ then $p=q$.) The task is therefore to find the first such d_i and the corresponding p_i. This is achieved through a tree of single bit tests as follows.

For simplicity, we assume here that $n=2^k-1$ so that the word length is a power of 2. (Clearly any shorter words can be extended with 0's.) The second phase is a k-stage process in each stage of which we consider successive pairs (d_{2i}, p_{2i}) and (d_{2i+1}, p_{2i+1}). The first of these is retained if $d_{2i}=1$ while the second is selected otherwise. These single-bit tests can be effected simultaneously for each pair. The number of such pairs is thus halved at each step so that, after k such steps, there is only one pair (D,P) say remaining. If $D=0$ then $p=q$, whereas if $D=1$, then $p>q$ if and only if $P=1$. For 2's complement representations, the final decision is reversed if $p_0=1$.

The algorithm is summarized for positive numbers below. Clearly the arguments could be any aligned representations though we describe the process for integers.

Algorithm 2   Fast discriminator
Input: Positive binary integers p, q of 2^k bits.
Initialize:

For $i=0$ to 2^k-1 set $(d_i^{(0)}, p_i^{(0)}) = (p_i \lor q_i, p_i)$ (Parallel)
Loop: For $j=1$ to k (Serial)
   for $i=0$ to $2^{k-j}-1$ (Parallel)

$$(d_i^{(j+1)}, p_i^{(j+1)}) = \begin{cases} (d_{2i}^{(j)}, p_{2i}^{(j)}) & \text{if } d_{2i}^{(j)} = 1 \\ (d_{2i+1}^{(j)}, p_{2i+1}^{(j)}) & \text{if } d_{2i}^{(j)} = 0 \end{cases}$$

Output: **If $d_0^{(k)} = 0$ then $p=q$**

else $p>q \Leftrightarrow p_0^{(k)} = 1$

The process is illustrated in the following examples.

Example 1

We consider two examples using 8-bit words.
(a) $p = 1\,1\,0\,0\,1\,0\,1\,0$ (b) $p = 1\,0\,1\,0\,1\,0\,1\,0$
    $q = 1\,0\,1\,1\,1\,1\,1\,0$ $q = 1\,0\,1\,1\,1\,1\,1\,0$
Initialization:
    $d = 0\,1\,1\,1\,0\,1\,0\,0$ $d = 0\,0\,0\,1\,0\,1\,0\,0$
    $p = 1\,1\,0\,0\,1\,0\,1\,0$ $p = 1\,0\,1\,0\,1\,0\,1\,0$
$j{=}1$ \/ \/ \ /\ /   \/ \ / \/ \/
    $d = 1\;\;1\;\;\;\;1\;\;0$ $d = 0\;\;1\;\;\;\;1\;\;0$
    $p = 1\;\;0\;\;\;\;0\;\;0$ $p = 0\;\;0\;\;\;\;0\;\;0$
$j{=}2$ \ /   \ /   \ /   \ /
    $d = \;\;\;\;1\;\;\;\;\;\;1$ $d = \;\;\;\;1\;\;\;\;\;\;1$
    $p = \;\;\;\;1\;\;\;\;\;\;0$ $p = \;\;\;\;0\;\;\;\;\;\;0$
$j{=}3$ \   /   \   /
    $d = \;\;\;\;\;\;\;\;1$ $d = \;\;\;\;\;\;\;\;1$
    $p = \;\;\;\;\;\;\;\;1$ $p = \;\;\;\;\;\;\;\;0$
Output: $p > q$ $p < q$

In both cases, $p_0=1$ and so the conclusion would be reversed if these were 2's complement integers. Note how the first (i.e left most) pair in $(d^{(0)}, p^{(0)})$ for which $d_i^{(0)} = 1$ is preserved throughout the process to yield the required result. If $d_{2i}^{(j)} = d_{2i+1}^{(j)} = 0$ then there is no difference

between p and q in the appropriate positions. The fact that it is the second pair which are transmitted to the next phase in this situation is an arbitrary decision - but this is choice which allows the simple one-bit tests to be performed in the (parallel) inner loop.

It should be observed that this integer discriminator could easily be incorporated into floating-point hardware too.

The motivation for this discriminator in the present context was the need to detect the sign of a "double number" quickly. That is we have a quantity x represented by the sum of two aligned 2's complement words p and q; we wish to identify the sign of x. This can be achieved using the same idea. If the sign bits of the two components are the same then clearly x has this same sign. Otherwise we must identify the value of the first bit of p for which the corresponding bit of q is the **same**. This is achieved by simply reversing the test in the inner loop in Algorithm 1. If no such bit exists then p and q are 1's complements and so their sum is negative; otherwise x is positive if and only if the appropriate $p_i=1$.

It follows that the determination of the sign of a double number can be achieved in $\lceil \log_2 n \rceil$ one-bit discrimination steps for a wordlength of n bits.

Returning to the main objective of this section, we are now in position to start Algorithm 1. The a-sequence for this largest component can be computed exactly as in [14] which is based on the CORDIC-like algorithm for $e^{-1/x}$ defined by

$$\begin{aligned} v_1 &= 1, \; u_1 = K \\ \delta_k &= -\text{sgn}(v_k) \\ u_{k+1} &= u_k + u_k \delta_k 2^{-k} \\ v_{k+1} &= v_k + \delta_k \xi_k \end{aligned} \tag{11}$$

where

$$\begin{aligned} \xi_k &= x \varepsilon_k, \; \varepsilon_k = \tanh^{-1} 2^{-k}, \\ K &= \prod \cosh \varepsilon_k \end{aligned} \tag{12}$$

with the usual repetitions in the sequence ε_k and the product K. Two additional steps using $\varepsilon_1 = \tanh^{-1} 1/2$ are also included in order to extend the range of convergence of the algorithm. Of course, the same algorithm is used to generate any α-sequences which are needed in Case 1.

The b- and β-sequences are computed using similar relations:

$$\begin{aligned} b_j &= \exp((b_{j+1}-1)/a_{j+1}) \\ \beta_j &= \exp((\beta_{j+1}-1)/a_{j+1}\beta_{j+1}) \end{aligned} \tag{13}$$

with appropriate starting values. Minor modifications of

the above algorithm can be used to compute these terms.

One important observation here is that all of these sequences can be computed simultaneously which has an obviously beneficial effect on the speed of extended summation. We shall also see later that compressing the operation in this manner reduces the overall rounding error inherent in the operation to a significant extent.

The next stage is the accumulation of c_0. This is the result of summation of 1 and N fixed-point fractions. (Recall that all the internal arithmetic required for the SLI system can be performed to **fixed** absolute precisions.)

One aspect of the implementation proposed in [14] was the use of a "double-number" format throughout the internal arithmetic. This facility is retained by the extended operation so that the formation of c_0 is reduced to the summation of $2N+1$ fixed point fractions using a tree of Carry Save Adders to obtain the double-number representation of c_0.

The remainder of Algorithm 1 can be implemented just as it would be for a single SLI arithmetic operation. The terms of the c-sequence can again be computed using a modified CORDIC algorithm for the function $a \ln c$ in which the constants used are

$$\alpha_k = a\,\varepsilon_k = a\tanh^{-1}2^{-k}. \tag{14}$$

These, like the ξ_k of (12), can be computed in parallel. The standard CORDIC logarithm algorithm can be used for any terms of the h-sequence which are needed. The reader is referred to [14] for a full description of these modified algorithms in which again "double number" representations are used throughout the internal computation.

3. Timing Estimates and Comparisons

We begin this section with estimated timings for the extended SLI summation algorithm. We shall also make comparisons of these estimates with serial and parallel floating-point implementations. Throughout this section we shall assume that the prospective SLI hardware unit has "sufficient" parallelism that any operations which can, in principle, be performed simultaneously, can be so performed in practice.

We list below our notation and assumptions regarding the relative timings of the underlying basic operations measured in some base "time unit", t.u. The basis of our comparisons will be single precision SLI against single precision floating-point operations, which we shall assume

take approximately 2c t.u.

Operation	Time in t.u.s
3 to 2 Carry Save Adder operation	a
Shift left or right	$b \approx a/3$
32-bit Carry propagate adder	$c \approx 32a$
32-bit aligned discrimination	$d \approx 5e \approx a$
One-bit logical test or negation	$e \approx a/5$

On a serial machine therefore, the floating-point summation of the terms $X_0,...,X_N$ will require approximately $2Nc$ t.u. For a machine with sufficient parallelism to perform this summation using a typical reduction algorithm, the number of separate reduction stages required is $\lceil \log_2 N + 1 \rceil$ so that the overall timing is then approximately $2c\,\log_2 N + 1$. These are the two quantities with which we make our comparisons later.

We note that the time taken for the SLI operation will vary according to the levels of the maximum element and the final result. As in [14] we shall make a "worst case" estimate in which both of these quantities are at level 5 and a "typical" estimate using level 3. It should be noted at the outset that the worst case comparison is especially unfair to the SLI system since the **corresponding floating-point computation could NOT be performed** by any straightforward means.

With the use of the algorithm of the previous section for the signs of the double number representations (and a slightly improved "squarer") it follows that the timings given in [14] for single SLI operations are overestimated. If we also modify the computation of the starting value for the b-sequence so that this sequence is computed entirely in parallel with the a-sequence, the "typical" and "worst case" times quoted there can be reduced to 30c and 49c t.u. respectively.

This represents a slowdown by a factor of about 15 compared to the same algorithm executed in floating-point on a fast serial machine. Much of this loss may well be recovered by the fact SLI arithmetic allows much simpler program structures to be used since it does not need protecting from possible overflow or underflow problems.

The principal finding of this section is that by using the extended Algorithm 1, this loss of speed is significantly reduced or even removed.

The precomputation of the largest component requires $\log_2(N+1)$ steps each of which is an integer discriminator implemented according to Algorithm 2 in 5e t.u.

The remainder of the algorithm splits into two principal cases depending on whether $X_0 \geq 1$. We shall concentrate first on the "large" case, that is $X_0 \geq 1$ or, equivalently, $r_0 = +1$. We shall subdivide this case further but the computation proceeds the same way at least as far as the computation of c_0.

The computation of the a-sequence for X_0 is common to all cases and is achieved as in [14] in a total of

$$78a+39b+39d + ([x]-1)(132a+50b+41d) \text{ t.u.}$$

It was observed in [14] that the b-sequence can be computed in parallel with the a-sequence provided that the starting value is defined appropriately. The less severe precision requirement for terms of this sequence also implies that it can be computed alongside the a-sequence at no extra cost in time.

For terms of magnitude less than 1, the α-sequence is required. The computation is just the same as for the a-sequence and so can also be computed in the same time. For each such term we then need the multiplication of the two representations of a_0 and α_0. These are both 37-bit double numbers and so 148 terms need to be summed with a CSA-tree. This multiplication requires $12a+b$ t.u.

It follows that the total time required to produce all the components of the sum which forms c_0 is

$$90a+40b+39d + (l_{max}-1)(132a+50b+41d) \text{ t.u.}$$

where l_{max} is the maximum level of any component of the sum. (This could be greater than $[x]$ if very small quantities with reciprocals greater than $|X_0|$ are present.)

The next stage is the summation of N double numbers and 1. This can be achieved with another CSA-tree in approximately $\log_{3/2}(2N+1) \approx 1.7(1+\log_2 N)$ steps. This requires $1.7(1+\log_2 N)a$ t.u.

There are now three possibilities: namely $|Z| \geq |X_0|$, $1 \leq |Z| < |X_0|$ or the "flip-down" case where $|Z| < 1$ for which the final result must be represented in reciprocal form. In each of the first two of these cases, the remaining parts of the algorithm are performed as in the proposed implementation in [14]. The only difference of any significance is the possibility that a longer h-sequence is needed. In ordinary SLI addition at most one term of this sequence is needed. Here there is a possibility of needing up to 4 such terms if the series is **very** long and all terms are close to unity. However the total length of the c-sequence and h-sequence will still not exceed 5 for nontrivial summation. Since the terms of the h-sequence are computed by a straightforward logarithm routine, which is quicker than the modified CORDIC algorithm for the c-sequence, we obtain an upper bound for the time by taking each step (except possibly the final one) to be a computation of the form $1 + a \ln c$. The time obtained for this operation in [14] is $83a+37b+34d$ t.u. per step. The final step in the case where $[z] \geq [x_0]$ is necessarily just a logarithm (of a double number) which can be achieved in $70a+35b+34d$ t.u. Finally, we must convert the final result to "single" form using the Carry Propagate (or Carry Look-Ahead) adder.

For the "flip-down" case where $|X_0| \geq 1 > |Z|$, the computation of a c-sequence is replaced with an h-sequence which as we have already observed is a quicker calculation.

Summarizing, in the case of "large" extended summation the total time for the "worst case" where $[x_0]=[z]=5$ is given by

Find X_0	$\log_2(N+1)*5e$
a-,b-,α-sequences	$90a+40b+39d + 4(132a+50b+41d)$
c_0	$1.7(1+\log_2 N)a$
c-,h-sequences	$70a+35b+34d + 4(83a+37b+34d)$
Final CPA	c
TOTAL	$1022a+423b+c+373d+\lceil \log_2(N+1) \rceil(5e+1.7a)$

$$\approx 49c + 2.7a\lceil \log_2(N+1) \rceil$$
$$\leq 50c \text{ t.u for } N \leq 2^{11}.$$

We see immediately that this operation can be expected to be faster than the floating-point calculation would be for $N \geq 25$ on a serial machine - but of course the floating-point computation would be impossible.

For a more typical estimate, we consider the case where no component of the sum or the result is at a level exceeding 3. The corresponding total is now $592a+249b+c+223d+2.7a\lceil \log_2(N+1) \rceil \leq 30c$ t.u. for the same range of N. In this case the extended summation - even of 2000 terms - compares favorably with the serial floating-point summation of just 15 terms.

We see that the fact that the summation is compressed into the formation of c_0 has the effect of making the extended aspect of the operation virtually free. This statement remains valid for the case of the extended summation of small quantities. The only important difference from the point of view of timing the operation is the replacement of the b-sequence with the β-sequence. Comparing the two equations in (13), we see that the essential difference is the extra multiplication $a_{j+1}\beta_{j+1}$.

This operation entails forming the product of two double numbers of which β has the shorter wordlength. We find that we must sum 128 terms; using a CSA tree this is reduced to just two in 11a+b t.u. This together with the standard *b*-sequence calculation still implies that the computation of the next term of the β-sequence is quicker than that of the *a*-sequence. It follows that the bound on the time taken to produce c_0 for the large case remains valid here.

The final stages of the "Case 2" algorithm are identical with those of the large case above. The only time loss is derived from the slightly slower test of $ca_0>1$ rather than $c<a_0$. This entails a further "double × double" multiplication which adds 11a+b t.u. to the overall time. The approximate timings quoted above remain valid.

The comparisons made above were not entirely fair since we assumed considerable parallelism in the SLI computation but only a serial architecture for the floating-point summation. Of course this is not entirely unfair either as it points to a possible trade-off to be considered in deciding how faster technology is best utilized.

We conclude this section with a comparison of projected times for our parallel SLI implementation with floating-point computation using similar parallelism. The summation of $N+1$ terms on a sufficiently parallel floating-point architecture will require $\lceil \log_2(N+1) \rceil$ steps of a typical reduction algorithm which takes, under our assumptions, $2c\lceil \log_2(N+1) \rceil$ t.u. For the "typical" range calculation, the SLI summation takes $29c + 2.7a\lceil \log_2(N+1) \rceil$ t.u. compared with $49c + 1.7a\lceil \log_2(N+1) \rceil$ t.u. for the "worst case". In Table 1, below, we show the relative timings for different vector lengths. Again recall that floating-point would fail for the worst case computation.

Table 1 *Comparison of extended summation times for floating-point and SLI arithmetic on parallel architectures*

# terms	Flp time	"Typical" SLI time	SLI:flp	"Worst case" SLI time
16	8c	30c	4:1	50c
64	12c	30c	2.5:1	50c
256	16c	30c	2:1	50c
1024	20c	30c	1.5:1	50c
4096	24c	30c	1.25:1	50c

It is clear from the table that for even moderate length sums, the time-penalty incurred by the more robust SLI arithmetic is very small while for massively parallel machines it may be almost undetectable.

From the entries in Table 1, we can also deduce relative timings for scalar products of two vectors. On our "sufficiently parallel" machine, the floating-point calculation requires just one more parallel multiplication increasing each of the quoted times by 2c t.u. The corresponding SLI multiplications can also be performed simultaneously - but more slowly.

Since the multiplication of two SLI quantities is equivalent to the addition of their logarithms the effective levels of all quantities are reduced by one. It follows that the "worst case" and "typical" timings are reduced to 40c and 20c t.u. respectively. The corresponding overall times for the scalar product are therefore 90c and 50c t.u. The slowdown ratios for scalar products of vectors of the same lengths as in Table 1 thus vary from 5:1 down to 2:1.

4. Error Analysis and Control

In much the same way as for the consideration of timings, in order to obtain error bounds for these extended operations, we must consider separate cases. We shall again treat one of the main cases in some detail and then describe the differences and summarize the results for the others. In the spirit of [4], the aim of the analysis is to demonstrate that the rounding errors involved in Algorithm 1 can be restricted to the order of the inherent error in the operation. As in [4], we find that working to fixed absolute precisions in the various stages of the internal computation achieves the desired control. We begin by studying the case of "large" arithmetic with a "large" result; that is, $|X_0|, |Z| \geq 1$. Note here that we do **NOT** assume that all terms are of the same sign, nor even that the signs of X_0 and Z are the same.

In this case, the inherent error can be approximated by linear perturbation theory to obtain

$$|\delta z| \leq \left(\left| \frac{\phi'(x_0)}{\phi'(z)} \right| + \sum_{r_i=+1} \left| \frac{\phi'(x_i)}{\phi'(z)} \right| + \sum_{r_i=-1} \left| \frac{\phi'(x_i)}{\phi^2(x_i)\phi'(z)} \right| \right) \gamma_0 \quad (15)$$

where each component is correct to the accuracy, γ_0, of the representation, so that $|\delta x_i| \leq \gamma_0$. Also using the facts that ϕ' is an increasing function, $\phi^2(x)/\phi'(x) = \phi(x)/\phi'(x-1)$ and $|\phi(x)/\phi'(x-1)| \geq 1$ for $x \geq 1$ (see Lemma 3.1 of [4]) it follows for the case where $z \geq x_0$ that this inherent error is bounded by $(N+1)\gamma_0$. If, on the other hand, $z < x_0$, then the appropriate bound is $(N+1)\gamma_0\phi'(x_0)/\phi'(z)$. These are the bounds which our algorithm should be designed to achieve by choosing appropriate working precisions.

Many of the details are similar to the analysis used in [4]. The error bounds for the computation of individual *a*-, *b*-

or α-sequences are unchanged for the extended algorithm from those obtained in [2]. Thus, we have

$$|\delta a_j|, \ |\delta \alpha_j| \ \le \ \lambda \gamma_1 \qquad (16)$$
$$|\delta b_{i,0}| \ \le \ \phi'(x-1)\rho(\gamma_2 + \lambda \gamma_1/e)$$

where γ_1, γ_2 are the working precisions to which terms of the a- and b-sequences are computed and

$$\lambda \ = \ \frac{4+e^2}{e^e}+1 \ \approx \ 1.75, \quad \rho \ = \ \sum_{j=0}^{\infty} \frac{1}{\phi'(j)} \ < \ 2.4. \quad (17)$$

For the extended sum under present consideration, this allows us to bound the error in c_0:

$$|\delta c_0| \ \le \ N\rho(\gamma_2 + e^{-1}\lambda \gamma_1)\phi'(x-1) \qquad (18)$$

which is just N times the error bound at this stage in the simple large addition algorithm.

The improvement in the error control for the extended sum derives from the fact that only one c- or h-sequence is needed so that the propagation of this error is minimized. In the situation where $|Z| \ge |X_0|$, the analysis continues precisely as for the addition case in [2] except that additional terms of the h-sequence may be needed. The propagation of the error through the calculation of the remainder of the c-sequence (with working precision γ_2) yields the bound

$$|\delta c| \ \le \ \rho\big(N(\gamma_2 + \lambda \gamma_1/e) + \gamma_2 + \lambda \gamma_1 \ln(1/\gamma_2)\big) \qquad (19)$$

The computation of the necessary terms of the h-sequence adds at most a further $\rho\gamma_2$ to the overall error. (See [4], for details.) It follows that the final error is bounded by

$$|\delta z| \ \le \ \rho\big(\gamma_2(N+2) + \lambda \gamma_1(N/e + \ln(1/\gamma_2))\big) \qquad (20)$$

The main term in the bound for a single addition is $2(\rho+1)\gamma_2$ since $\gamma_1 \ll \gamma_2$. It follows that the error in this extended sum of $N+1$ terms is less than $N/2$ times that for the single operation.

For the situation where $z<x_0$, the inherent error is bounded by $(N+1)\gamma_0\phi'(x_0)/\phi'(z)$. This is the situation where some cancellation has occurred but it has not been so severe as to cause the result to flip down to reciprocal form. The algorithm, and its analysis, are unchanged for the computation of c_0. Since $|Z| \ge 1$, it follows that

$$c_0 \ = \ \phi(z)/\phi(x_0) \ \ge \ 1/\phi(x_0) \ = \ a_0 \ \ge \ \gamma_1. \qquad (21)$$

Just as with the large subtraction algorithm considered in [2], the c-sequence is increasing and so we have

$$|\ln c_j| \ \le \ \ln 1/\gamma_1$$

for every j for which the sequence is to be computed.

The obvious modifications to the analysis of [2] can now be used to obtain the final error bound

$$|\delta z| \ \le \ \frac{\phi'(x)}{\phi'(z)}\{\gamma_2(1 + \rho(N+1)) + \lambda \gamma_1(1 + \rho(N/e + \ln 1/\gamma_1))\}$$

$$(22)$$

in which we again see that the dominant term is less than $N/2$ times that for the single subtraction algorithm.

The last case we consider in any detail is the one which might be expected to be most troublesome. This is the situation where $|X_0| \ge 1 > |Z|$ which is the case of severe cancellation. Specifically then we have

$$s_z\phi(z)^{-1} \ = \ \phi(x_0) + \sum_{i=1}^{N} s_i\phi(x_i)^{r_i} \qquad (23)$$

which has the inherent error

$$|\delta z| \ \le \ \phi^2(z)\left\{\left|\frac{\phi'(x_0)}{\phi'(z)}\right| + \sum_{r_i=+1}\left|\frac{\phi'(x_i)}{\phi'(z)}\right| + \sum_{r_i=-1}\left|\frac{\phi'(x_i)}{\phi^2(x_i)\phi'(z)}\right|\right\}\gamma_0$$

which, by a similar argument, we may bound by

$$|\delta z| \ \le \ (N+1)\gamma_0\frac{\phi'(x_0)\phi(z)}{\phi'(z-1)}. \qquad (24)$$

Yet again the analysis is unchanged as far as the calculation of c_0. At this stage, for this case we have $c_0<a_0$ and we form $h_1 = -\ln(c_0/a_0)$ and proceed to compute further terms of the h-sequence as necessary. The analysis of [4] for the case of large subtraction with cancellation can now be used to obtain

$$|\delta z| \ \le \ \frac{\phi'(x_0)\phi(z)}{\phi'(z-1)}\{\gamma_2(1 + \rho(N+1)) + \lambda \gamma_1(1 + N\rho/e)\}. \quad (25)$$

Again the rounding error is magnified by less than $N/2$. In the various cases of extended "small" sums, in which $r_i=-1$ for every i, the inherent error is bounded by

$$|\delta z| \ \le \ (N+1)\gamma_0\frac{\phi'(x-1)\phi(z)}{\phi(x)\phi'(z-1)} \qquad (26)$$

for $|Z| \le 1$ and by

$$|\delta z| \ \le \ (N+1)\gamma_0\frac{\phi'(x-1)}{\phi(x)\phi'(z)} \qquad (27)$$

for the "flip-up" case where $|Z|>1$. Extensions of the analysis of [4] similar to those used above yields an error bound for the cancellation case (which for small arithmetic corresponds to $z>x_0$) which shows an increase by a factor of about $7N/12$ over the single small subtraction algorithm. Similar results apply to the remaining cases so that in all cases the roundoff error of the extended operations is increased by only about half the expected factor. It follows that with the same working precisions as suggested in [4] - and used in the software implementation in [16] - the roundoff errors of extended summation are only about half of the inherent error bounds.

The corresponding bounds in [4] were used to find working precisions which control the roundoff error to be of the order of the inherent error. The same reasoning here allows us the opportunity to control the error of extended summation to give the effect of only a single rounding error - at least for the cases where severe cancellation does not occur. The extra precision needed is clearly dependent on the maximum vector length available in the extended algorithm. For illustrative purposes we shall take this maximum to be $N+1=1024$. Since in [4], we had $\gamma_1 \leq 2^{-5}\gamma_2$ it follows that, in the case where $|Z|>|X_0|\geq 1$, we must choose γ_2 so that $2.5\times 2^{10}\gamma_2 \leq \gamma_0$. For single length SLI arithmetic, $\gamma_0=2^{-28}$ and so using $\gamma_2=2^{-12}2^{-28}=2^{-40}$ will suffice.

This amounts to adding just 9 bits of precision to the internal computation of the SLI algorithm. This is a very low cost in order to achieve the often sought after goal of a single roundoff error for extended summation.

5. CONCLUSIONS

In this paper, we have described an algorithm for extended arithmetic operations in SLI arithmetic and discussed its possible hardware implementation and error analysis.

The implementation details suggest that any time-penalty associated with the use of SLI arithmetic can be kept to a very small factor on highly parallel computers - perhaps of the order of just 2 or 3 for typical scientific computing programs. Not only do we see that extended operations can be executed in times comparable with single SLI arithmetic operations, the use of this extended algorithm also results in a significant relative reduction in the roundoff error bound. We see too that the error analysis of such extended operations with elements of mixed sign is more straightforward than has sometimes been suggested.

Putting these results together provides us with alternatives for the utilization of improved technology and massive parallelism. One possibility is that any such improvements be used to provide more raw speed. Perhaps a better solution would be to compromise by accepting about half the potential speed-up and using SLI arithmetic instead. A speed-up by a factor of 5 rather than 10, for example, may well be an acceptable "price" for freedom from

scaling problems in order to avoid overflow or underflow. We might even conclude that a smaller speed-up would be acceptable in order to perform the internal arithmetic to greater accuracy and so compute vector sums and scalar products with just a single roundoff error.

References

[1] C.W.Clenshaw and F.W.J.Olver, *Beyond floating point*, J. ACM 31 (1984) 319-328.

[2] C.W.Clenshaw and F.W.J.Olver, *Level-index arithmetic operations*, SIAM J Num Anal 24 (1987) 470-485.

[3] C.W.Clenshaw, F.W.J.Olver and P.R.Turner, *Level-index arithmetic: An introductory survey*, Numerical Analysis and Parallel Processing (P.R.Turner Ed.) LNM 1397, Springer Verlag, 1989, pp. 95-168.

[4] C.W.Clenshaw and P.R.Turner, *The symmetric level-index system*, IMA J Num Anal 8 (1988) 517-526.

[5] C.W.Clenshaw and P.R.Turner, *Root-squaring using level-index arithmetic*, Computing 43 (1989) 171-185.

[6] H.Hamada *URR: Universal representation of real numbers*, New Generation Computing, 1 (1983) 205-209.

[7] U.W.Kulisch and W.L.Miranker, *The arithmetic of the digital computer: A new approach*, SIAM Review 28 (1986) 1-40.

[8] D.W.Lozier and F.W.J.Olver, *Closure and precision in level-index arithmetic*, SIAM J Num. Anal, to appear.

[9] D.W.Lozier and P.R.Turner, *Supercomputers need super arithmetic*, NIST Tech Rep, NISTIR 89-4135.

[10] S.Matsui and M.Iri *An overflow/underflow-free floating-point representation of numbers*, J.Inf.Proc. 4 (1981) 123-133

[11] F.W.J.Olver, *A new approach to error arithmetic*, SIAM J Num Anal. 15 (1978) 368-393

[12] F.W.J.Olver, *Rounding errors in algebraic processes - in level-index arithmetic*, Proc. Reliable Numerical Computation (M.G.Cox and S.Hammarling, eds.) Oxford, 1990, pp.197-205.

[13] F.W.J.Olver and P.R.Turner, *Implementation of level-index arithmetic using partial table look-up*, Proc. ARITH8, (M.J.Irwin and R.Stefanelli, Eds.) IEEE Computer Society, Washington, DC, 1987, 144-147.

[14] P.R.Turner, *Towards a fast implementation of level-index arithmetic*, Bull IMA 22 (1986) 188-191.

[15] P.R.Turner, *Algorithms for the elementary functions in level-index arithmetic*, Scientific Software Systems (M.G.Cox and J.C.Mason Eds.) Chapman and Hall, 1990, pp. 123-134.

[16] P.R.Turner, *A software implementation of sli arithmetic*, pp. 18-24, Proc.ARITH9, (M.D.Ercegovac and E.Swartzlander, Eds) IEEE Computer Society, Washington DC, 1989.

Specifications for a
Variable-Precision Arithmetic Coprocessor

T.E. Hull, M.S. Cohen* and C.B. Hall**

Department of Computer Science
University of Toronto
Toronto, Ontario, Canada M5S 1A4

Abstract

The authors have been developing a programming system which is intended to be especially convenient for scientific computing. Its main features are variable precision (decimal) floating-point arithmetic and convenient exception handling. The software implementation of the system has evolved over a number of years, and a partial hardware implementation of the arithmetic itself was constructed and used during the early stages of the project. Based on this experience, the authors have developed a set of specifications for an arithmetic coprocessor to support such a system. The main purpose of this paper is to describe these specifications. An outline of the language features and how they can be used is also provided, to help justify our particular choice of coprocessor specifications.

Introduction

The general purpose of our project is to provide better programming language facilities for scientific computing, especially in terms of precision control and exception handling. Early efforts were concerned primarily with precision control and led to the development of preprocessors, both for Algol and Fortran, which made it easy to change precision dynamically between single, double, triple, etc., precisions. Then an attached processor, called CADAC, which carried out variable precision decimal arithmetic was built and attached to a Vax [3,4]. Finally, the present software system, which implements the programming language Numerical Turing (NT) and which provides for variable-precision decimal arithmetic and exception handling, was developed [7,8,9]; special arithmetic capabilities are also included, such as directed roundings and exponent

This work was supported by the Natural Sciences and Engineering Research Council of Canada.

* Present address: 141 North Meadow Crescent, Thornhill, Ontario, Canada L4J 3C4.
** Present address: IBM Canada Limited, 844 Don Mills Road, North York, Ontario, Canada M3C 1V7.

manipulations. It runs under Unix on either a Vax or a Sun 3.

Based on experience with these developments, the language specifications have of course evolved. But they are now at a stage where we have some confidence in their appropriateness, and believe that it would be useful to design and build the necessary hardware support. The main purpose of this paper is to describe the corresponding specifications, for the arithmetic unit, and also to indicate what other hardware features would be most helpful to the systems programmer, especially for implementation of the exception handling.

Precision control

We first describe the floating-point arithmetic (the real arithmetic) and related functions. In this section we do so in terms of the programming language, that is, from the user's point of view. Implications for the hardware will be considered in the next section.

Real values are p-digit, decimal, normalized floating-point numbers with exponents in the range $[-10p, 10p]$, where $1 \leq p \leq maxprecision$. (In NT the parameter $maxprecision$ is only 200, but it should be much larger, at least 1000.)

A programmer specifies what precision is to prevail by means of a precision statement. In NT the precision statement is of the form

precision *intexp*

where *intexp* is any integer expression. The specified precision is to prevail throughout the scope of the precision statement. For example, in Figure 1, the value of p determines the precision in effect from immediately after the precision statement to the end of the loop.

All declarations and floating-point operations (arithmetic, elementary functions, procedures, etc.) within this scope are carried out in precision p. In this example

```
----
----
var x : real
var p := currentprecision
loop
   precision p
   var y : real
   ----
   ----
   solve( ----, y)
   if y ---- then
     x := y
     exit
   end if
   p := p+10
end loop
----
```

Figure 1. All declarations and floating-point operations (arithmetic, elementary functions, procedures, etc.) within the scope of the precision statement are carried out in precision p.

p is first set equal to the value of the current precision (in NT *currentprecision* is a function which returns this value). The value of p is then increased by 10 each time the loop is executed until the condition in the *if* statement is *true*. Then y is assigned to x, and, if the precision of y is higher than that of x, the value of y will have to be rounded to the precision of x before being assigned. There must be a default precision in case no precision statement is yet in effect (in NT it is 16).

Any value in any expression that appears within the scope of a precision statement must be rounded or, in effect, extended to the prevailing precision before it can be used, if its precision differs from that of the prevailing precision.

This example illustrates one way in which precision control can be used. The procedure *solve* is executed in higher and higher precision until some criterion, presumably an accuracy criterion, is satisfied.

Basically this same idea can be implemented in Aberth's system [1], where the calculations inside the loop would be done in interval arithmetic and the decision to exit is then based on whether the interval associated with the final result is small enough. This is possible also in NT with the help of the directed roundings. (In a future version of NT, it is intended to have interval arithmetic "built-in", but in either case the implications for the coprocessor are the same: directed roundings must be supported.)

A second example is shown in Figure 2. In this example the increase in precision is used primarily to make sure that no intermediate overflow or underflow can

```
---- % n and the array a must
---- % be known at this point
var x : real
begin
   precision 2*currentprecision + getexp(n) + 1
   var sum := 0.0
   for i : 1..n
     sum := sum + a(i)*a(i)
   end for
   x := sqrt(sum)
end
----
```

Figure 2. This program fragment computes the Euclidean norm of the array a. The precision is increased just enough so that no intermediate overflow or undrflow can occur. Overflow may however occur in the final assignment to x.

take place. (The exponent range increases with the precision.) In fact, the precision is increased by the minimum amount that is needed to guarantee that no intermediate overflow or underflow can occur. (The function *getexp* returns the exponent of its argument, which, for the purpose of this example, means that $getexp(n) + 1 = ceil(\log_{10}n)$.) Of course, overflow might occur in the final assignment, because the calculated value to be assigned must, at that point, be rounded to the precision of x before being assigned to x.

In a third example one might use precision control to run a calculation in two different precisions and subtract the final results to measure the cumulative effect of rounding errors. (Of course, care must be taken in the use of any such technique; otherwise the results could be misleading.)

In another example, the "exact dot product", which plays an important role in ACRITH [12], can be implemented in NT. The maximum precision must of course be higher than it is in the current implementation of NT, and this should be kept in mind in developing specifications for the coprocessor.

The available floating-point operations in all these and other examples must of course include addition, subtraction, multiplication and division. The rounding in these operations is important and it should be unbiased. (In NT it is "round to nearest, or nearest even in case of a tie", which is what is done in IEEE arithmetic [10,11], but we are now considering a simpler rule, as described in the next section.) The same rounding rule should hold for the implicit rounding that occurs when a higher precision value must be rounded down (coerced) to some prevailing current precision, or when an assignment to some lower precision variable is about to be made.

As already indicated, directed roundings (round up and round down) should also be available with the four basic arithmetic operations. The functions *floor*, *ceil*, and *round* (the latter being to nearest integer in case of a tie) should also be available.

Other functions are required to determine quotients and remainders, and also to determine and set exponents, and to convert from integer to real. Of course comparison operators are also needed.

The operations and functions just described are sufficient for the construction of other functions which might be required, especially the elementary functions. In fact, the elementary functions can be programmed using these operations and functions in a quite straightforward and easily understood manner, especially when it is possible to increase the precision at appropriate stages in the calculations [5,6].

Other helpful functions, such as the *nextafter* function recommended in the IEEE standard, can also be programmed quite easily with those described above, but it may be that it is convenient to build some of them directly in the hardware.

Besides the functionality of the language, the user's primary concern is efficiency. Most of the computation will be in the default precision, which of course must therefore be as efficient as possible. But the other most frequently used precisions are only at most a few digits more than the default precision. For example, when dot products are accumulated in higher precision they will usually need only about two extra digits of precision [6]. The requirements of the elementary functions are more variable, but most of the time only two or three extra digits are required to deliver results to within an error of less than one unit in the last place [5,6 for example]. It would therefore be desirable that precisions slightly more than the default precision be the second consideration in terms of efficiency.

The next most frequently used precisions are double precision, and occasionally slightly more than double. An example was shown in Figure 2. These precisions should therefore be given third priority.

To support the ACRITH "exact dot product" facility, some consideration should also be given to making efficient the accumulation of dot products in a very long "accumulator".

It should be acknowledged again that the higher precisions will not be used very much of the time and the overall efficiency of a program will therefore usually not be seriously affected if efficiency in these precisions is not very great.

In our experience, high precision calculations are quite often used only as test programs. They are used to find the "true" solutions of some problems (matrix calculations, or elementary function evaluations, for example) in order to test the accuracy of programs running in some standard precision. In such cases, the high precision program needs to be run only once for each test problem, whereas there may be many test programs or variations on a test program to be run on the same problem.

From another point of view, we have also found that the availability of higher precision has often made our programming more efficient. In such cases using higher precision has enabled us to handle a problem in a direct and relatively simple way, rather than in a roundabout and relatively convoluted manner.

Arithmetic specifications

The functionalities described in the previous section, provide all the requirements to be met by a hardware floating-point unit, apart from what happens when exceptions occur. Some further details about the arithmetic, especially with regard to rounding, will be discussed in this section, and exception handling will be considered in the next section.

It might be helpful to consider a possible representation of floating-point numbers. We are not committed to the following representation, but we did use it recently in some preliminary design investigations. We used 4 bytes to store the following:

sign	1 bit	
exponent	$\geq$ 15 bits	(for exponents at least in [-10000,10000])
precision	$\geq$ 10 bits	(for precisions at least in [1,1000])
extended	1 bit	(to indicate whether or not the format is extended)
uninitialized	1 bit	(to indicate whether or not the value has not yet been assigned)

and this was followed by a sufficient number of 4-byte words to represent the normalized significand (4 bits for each decimal digit).

This representation is more compact than what is used by the software implementation in NT, but is otherwise equivalent. (The "extended" bit is there to allow for the possibility of having an alternative format for even higher precisions, which could be handled entirely in software, but we have not yet tried to take advantage of this idea, either in the original CADAC design or in the present software implementation.)

Precision specifications for the arithmetic operations, directed roundings, quotient and remainder, etc., have already been spelled out in detail elsewhere [6], so they will not be repeated here. The only new possibility has to do with rounding, which up until now has always followed the IEEE rule: ''round to nearest, or nearest even in case of a tie.'' We are now considering another rule, like one suggested by von Neumann [2, pp 57-58], which is easier to implement and is, at the same time, also ''unbiased''. The rule is:

> ''if there is something non-zero following the least significant digit in the true result, set the last bit to 1 (so that the last digit will be odd) and throw away the ''something'', otherwise just throw away the zeros following the least significant digit''.

With this rule, there is no possibility of a carry operation, or any subsequent renormalization. The only disadvantage is that the maximum error can be almost 1 in the last place, compared to a maximum of 1/2 in the last place with the earlier rule. But this disadvantage seems to us to be one we need not worry about in a variable precision environment where it is such a simple matter to increase precision whenever one wishes. In fact, the elementary functions can also be made accurate to within an error of less than 1 unit in the last place (just as they are in the current implementation of NT), so that the elementary functions *and* the arithmetic operations would then both meet the same accuracy requirement.

Note that the requirement is *less than* 1 unit, not *less than or equal to* 1 unit. This is important because we can then deduce some convenient results, such as, for example, that the square root of a perfect square will be exact, or that sines and cosines cannot exceed 1 in value.

Whatever the rounding rule, the arithmetic unit should be designed so that the rounding operation is easily implemented for the implicit rounding that can arise with coercion or assignment, as mentioned earlier, as well as with the usual arithmetic operations. Note that, because of the coercion rule, any two numbers involved in an arithmetic operation will always be, at least in effect, in the same precision, the precision prevailing at the time the operation is carried out.

Exception handling

The exception handling facilities are described in detail elsewhere [8]. For the purposes of this paper, the key feature is that handlers are attached to operators (arithmetic operators, functions, procedures − including assignments because a precision change can take place with assignment, and this could result in overflow or underflow). An example in NT is shown in Figure 3.

```
function norm (a : array 1..* of real): real
   var sum := 0.0
   for i:1..upper(a)
      sum := sum + a(i)**2
   end for
   result sqrt(sum)
end norm
---- % n and the array b are determined in
---- % these statements, where n is
---- % the number of elements in b
handler h
   on failure:
      precision 2*currentprecision + getexp(n) + 1
      x := norm(b)
      nextstatement
end h
var x := norm(b)@h
```

Figure 3. The function computes the Euclidean norm. It is first invoked in the current precision. Any intermediate overflow or underflow that might occur is not handled, and this causes the function to raise the *failure* exception. This in turn invokes the handler called *h* which causes the calculation to be repeated in high enough precision so that no intermediate overflow or underflow can occur. An overflow might still occur on assignment to *x*. For this to be handled, another handler would have to be attached to the assignment operator inside the handler *h*.

The advantage in this example, compared to Figure 2, is that the norm is almost always done in the more efficient lower precision, and the extra work in higher precision is done only when necessary, and presumably only rarely. This illustrates an exception handling technique that can be applied quite generally.

There are only 4 built-in exceptions associated with arithmetic operations − *overflow, underflow, domainerror,* and *uninitialized* − plus one other, namely *failure*, which is raised by a function or procedure in which an unhandled exception occurs. (There are only 2 other built-in exceptions, but they are associated with input.) Users can also explicitly raise *user-defined* exceptions. (For example, a user could decide to declare *toonearsingular* to be an exception, and *raise* it in a procedure for solving linear equations.)

The *overflow* exception can be raised by a floating-point add, or an integer multiply, or the exponential function, and so on. The user knows which operation raises the *overflow* exception because the handler is associated with the operation. (This is one of the advantages of attaching the handler to an operator. Another is that anyone reading the program knows which operators have been modified by a handler, and which have not.)

Based on our experience in implementing these exception handling facilities on both Vax and Sun 3 systems, we have been led to the following conclusions:

(1) The hardware should be designed so that it is easy for the system to identify the location and type of an exception. This means it must be possible to provide precise interrupts and to provide for a software mechanism equivalent to UNIX signals.

(2) The system also needs to be able to restart computations following an exception (that is, return to a machine instruction following the location where the exception occurred, as opposed to restarting the instruction which raised the exception). Besides being able to provide precise interrupts, this means that it must also be possible to preserve at least part of the machine state.

(3) Tracing back from an unhandled exception through what might be a sequence of *failure* exceptions, due to a nesting of subprograms, requires that the system be able to 'unroll' stack frames and return to an arbitrary point after an exception. This is mainly a software problem, but the hardware can make it easier − for example, the register mask saved by the VAX 'calls' instruction simplifies restoring the registers when a stack frame is released.

Concluding remarks

On the basis of our experience with a software system that supports variable-precision floating-point arithmetic and exception handling which have been especially designed to facilitate scientific computing, as well as experience with an earlier hardware unit with some aspects of these features, we have described the specifications we believe are most suitable for an arithmetic coprocessor that could support such a system.

Bibliography

[1] Aberth, O. and Schaefer, M., Precise Computation using Range Arithmetic, via C++, private communication.

[2] Burks, A.W., Goldstine, H.H. and von Neumann, J., Preliminary Discussion of the Logical Design of an Electronic Computing Instrument, Part I, Vol.1. Report prepared for U.S. Army Ord. Dept. (1946). Reprinted in John von Neumann Collected Works, Vol.V (MacMillan Coy., New York, 1963), 34-79.

[3] Cohen, M.S., Hamacher, V.C. and Hull, T.E., CADAC: An Arithmetic Unit for Clean Decimal Arithmetic and Controlled Precision, Proceedings 5th Symposium on Computer Arithmetic (IEEE Computer Society, Ann Arbor, Michigan, 1981), 106-112.

[4] Cohen, M.S., Hull, T.E. and Hamacher, V.C., CADAC: A Controlled-Precision Decimal Arithmetic Unit, IEEE Transactions on Computers, vol. C-32, 4 (1983), 370-377.

[5] Hull, T.E. and Abrham, A., Properly Rounded Variable Precision Square Root, ACM TOMS 11, 3 (1985), 229-237.

[6] Hull, T.E. and Abrham, A., Variable Precision Exponential Function, ACM TOMS 12, 2 (1986), 79-91.

[7] Hull, T.E. and Cohen, M.S., Toward an Ideal Computer Arithmetic, Proceedings 8th Symposium on Computer Arithmetic (IEEE Computer Society, Como, Italy, 1987), 131-138.

[8] Hull, T.E., Cohen, M.S., Sawchuk, J.T.M. and Wortman, D.B., Exception Handling in Scientific Computing, ACM TOMS, 14, 3 (1988), 201-217.

[9] Hull, T.E. and Hall, C.B., Precision Control and Exception Handling in Scientific Computing, Proc. Symp. Scientific Software (ed. D.Y. Cai, L.D. Fosdick and H.C. Huang), May 31-June 3, 1989 (China University of Science and Technology Press, Beijing, PRC, 1989), 118-131.

[10] IEEE Standard for Binary Floating-Point Arithmetic, ANSI/IEEE Standard 754-1985 (IEEE Computer Society, 345 East 47th Street, New York, NY 10017, USA, 1985).

[11] IEEE Standard for Radix-Independent Floating-Point Arithmetic, ANSI/IEEE Standard 854-1987 (IEEE Computer Society, 345 East 47th Street, New York, NY 10017, USA, 1987).

[12] Kulisch, U.W. and Miranker, W.L., The Arithmetic of the Digital Computer: a New Approach, SIAM Review 28, 1 (1986), 1-40.

Algorithms for Arbitrary Precision Floating Point Arithmetic

Douglas M. Priest
Department of Mathematics
University of California
Berkeley, California 94720

Abstract

We present techniques which may be used to perform computations of very high accuracy using only straightforward floating point arithmetic operations of limited precision, and we prove the validity of these techniques under very general hypotheses satisfied by most implementations of floating point arithmetic.

To illustrate the application of these techniques, we present an algorithm which computes the intersection of a line and a line segment. The algorithm is guaranteed to correctly decide whether an intersection exists and, if so, to produce the coordinates of the intersection point accurate to full precision. Moreover, the algorithm is usually quite efficient; only in a few cases does guaranteed accuracy necessitate an expensive computation.

1   Introduction

"How accurate is a computed result if each intermediate quantity is computed using floating point arithmetic of a given precision?" The casual reader of Wilkinson's famous treatise [21] and similar roundoff error analyses might conclude that the most one can hope to say about the accuracy of a computation carried out in fixed precision floating point arithmetic is that the computed solution is close to the exact solution of a problem close to the given problem, where the precise meaning of "close" depends on the precision of the arithmetic. He might further surmise that if he wishes to compute a result with more accuracy than such an analysis can guarantee based on the widest precision of arithmetic supported in whatever computing environment is available, then he must instead resort to a subroutine library such as Brent's MP package [3] in order to compute with higher precision arithmetic.

That both conclusions are wrong follows from the existence of techniques which allow a program to compute to arbitrarily high accuracy using only fixed precision floating point arithmetic operations. These techniques can be used in virtually any computing environment which incorporates a "reasonable" floating point arithmetic, and they do not rely on dirty tricks such as accessing a floating point number as though it were an integer value, nor on inefficient conversions such as extracting the exponent field of a floating point number. Instead they use only floating point additions, comparisons, multiplications, and divisions, which are commonly supported in hardware and often greatly optimized. Armed with

such techniques, a programmer can write code which performs none but straightforward floating point operations, looks entirely ordinary both to the eye and to the compiler, runs almost as efficiently as any other floating point computation, and produces a result whose accuracy is limited only by the overflow and underflow threshholds.

Of course, extra accuracy is not free: a program guaranteed to produce an accurate answer must be more expensive than one which is allowed to emit erroneous results. How much more expensive must such a program be? In this paper, we present algorithms which suggest that the cost of extra accuracy may be quite reasonable for many problems—at least, reasonable enough that we are obliged to consider more carefully the trade-offs between cost and accuracy. Our algorithms are based on an approach pioneered by Møller [17], Kahan [7], Dekker [4], Pichat [19], Linnainmaa [13, 14], and several others [10, 12], but our hypotheses are slightly more general than theirs. Moreover, rather than simply extending the accuracy to approximately twice the working precision, as do most of the aforementioned references, our algorithms expand upon methods developed by Bohlender [1] and Kahan [9] which compute to arbitrarily high accuracy.

Below, we give algorithms for exact addition and multiplication and arbitrarily accurate division of extended precision numbers using only fixed precision floating point arithmetic operations. We express the cost of these algorithms in terms of the number of fixed precision operations required. Section 2 describes floating point arithmetics and defines the criteria which an arithmetic must satisfy for our results to be valid. Lemma 1 in this section generalizes results from [4, 9, 10, 13, 17, 19]. Sections 3 and 4 focus on addition algorithms for extended precision numbers: section 3 considers the exact addition of two such numbers, while section 4 presents an algorithm for the problem Kahan [9] has called "distillation", namely, expressing the exact sum of n arbitrary fixed precision floating point numbers as a single extended precision number. Kahan gives an algorithm which requires at most $O(n \log n)$ fixed precision operations; we present another, very different algorithm which nevertheless has the same cost bound. Sections 5 and 6 present algorithms for multiplication and division, respectively. Although the techniques motivating these algorithms are well known, we present the algorithms both for completeness and to obtain explicit cost

132

bounds. Section 7 gives an example of an algorithm which uses several of our extended precision arithmetic routines to compute to full accuracy the point of intersection of a line and a line segment. Finally, section 8 summarizes our results and discusses some related topics.

With the exception of section 7 and several remarks in section 8, we ignore the possibility of overflow and underflow. Therefore, unless otherwise noted, the phrase "provided no overflow or underflow occurs" should be appended to each result stated below. We stress, however, that we are not advocating a complete, self-contained software package upon which a user must rely if he wishes to perform extended precision arithmetic, but rather a paradigm for numerical computation in which each intermediate quantity is computed to such precision as is necessary to guarantee the desired accuracy in the final result. Our intent is merely to demonstrate the feasibility of such a paradigm up to the limits imposed by overflow and underflow; in most cases, those limits are not particularly constraining, as our example in section 7 illustrates.

2 Properties of Floating Point Arithmetics

We begin by defining floating point arithmetic and noting several important properties. For integers t and β both greater than 1, let $R_{\beta,t}$ denote the set of all rational numbers of the form $m\beta^k$ where m and k are integers and $|m| < \beta^t$. These numbers are called the *t-digit, radix β floating point numbers*. A *floating point arithmetic* of radix β and t-digit precision is one which, given any $a, b \in R_{\beta,t}$ and $\circ \in \{+, -, \times, /\}$, determines a quantity $c \in R_{\beta,t}$ (assuming $b \neq 0$ if $\circ = /$, and subject to the usual caveats regarding overflow and underflow). We write $c = \mathrm{fl}(a \circ b)$ to denote the result of computing $a \circ b$ in floating point arithmetic.

We view arbitrary precision arithmetic in the following context. Let R_β denote the subring of $\mathbf{Q}$ generated by $R_{\beta,t}$. Note that each $x \in R_\beta$ may be written as a finite sum $x = \sum x_i$ with $x_i \in R_{\beta,t}$. In particular, if $x \neq 0$ we can choose the x_i satisfying a *non-overlapping condition*: writing $x_i = m_i\beta^{k_i}$ with $\beta^{t-1} \leq |m_i| < \beta^t$, we require that $k_i - k_j \geq t$ for $i < j$. (In other words, if $i < j$ then the most significant digit of x_j is at least one order of magnitude smaller than the least significant digit of x_i.) Such an expression is called a *t-digit expansion* for x; each term of the sum is a *component* of the expansion. A t-digit expansion for zero consists of a single zero component. Our goal is to show that all arithmetic operations over the ring R_β may be computed using only t-digit floating point arithmetic on components of t-digit expansions. (Here the "arithmetic" operations over R_β consist of the usual addition and multiplication as well as an approximate division which produces a quotient as an expansion accurate to a specified number of components.)

To guarantee that we can reduce arithmetic over R_β to fixed precision floating point arithmetic, we must make some assumptions about the nature of the floating point arithmetic we are using. The most important assumption we shall make is embodied in the following definition.

Definition: For t-digit numbers a and b and $\circ \in \{+, -, \times, /\}$, let $c = a \circ b$ exactly (assuming $b \neq 0$ if $\circ = /$). Suppose x and y are consecutive t-digit floating point numbers with the same sign as c such that $|x| \leq |c| < |y|$. Then the floating point arithmetic is called *faithful* if $\mathrm{fl}(a \circ b) = x$ whenever $c = x$ and $\mathrm{fl}(a \circ b)$ is either x or y whenever $c \neq x$.

All of the results which follow assume that the floating point arithmetic is faithful. In fact, most results assume nothing else; only the multiplication and division algorithms of sections 5 and 6 require additional hypotheses in the form of modest lower limits on the precision of the arithmetic. As we shall see, faithfulness is a powerful property, yet our assumption does not severely limit the applicability of our algorithms. Any arithmetic which conforms to the IEEE 754 or 854 standard is faithful, as are DEC VAX and IBM 370 arithmetics. In fact, any arithmetic in which subtraction is performed using at least one guard digit possesses faithful addition; faithful multiplication and division are not much more difficult to achieve. (In fairness, however, we must warn the reader that a number of machines still lack faithful arithmetic: notable exceptions include Crays and CDC Cybers. Of course, with somewhat more complicated algorithms and substantially more complicated proofs, our results could be extended to those machines as well.)

We now prove a crucial lemma which says that we can compute exactly the error incurred in the fixed precision addition of two t-digit numbers. This technique forms the basis of the algorithms presented in subsequent sections. As we will note below, the proof given here generalizes various special cases which are considered in [4, 9, 10, 13, 17, 19]. In addition to the hypothesis of faithfulness, we rely heavily on the following well-known property of floating point numbers (see Sterbenz [20]): if a and b are t-digit floating point numbers such that $1/2 \leq a/b \leq 2$ then $a - b$ is also a t-digit floating point number. In particular, if the floating point arithmetic is faithful, $\mathrm{fl}(a - b) = a - b$ exactly.

Lemma 1: Let a and b be any t-digit numbers. If the floating point arithmetic is faithful then the t-digit numbers c and d calculated by the following algorithm satisfy $c + d = a + b$ and either $d = 0$ or $c + d$ is an expansion for $a + b$ (i.e., $c = m_c\beta^{k_c}$ and $d = m_d\beta^{k_d}$ with $\beta^{t-1} \leq |m_c|, |m_d| < \beta^t$ and $k_c - k_d \geq t$).

Algorithm 1: (Addition of two t-digit numbers with explicit error term)

```
1 procedure sum_err( a, b )
2 begin
3 if |a| < |b|
4     swap( a, b )
5 c := fl(a + b), e := fl(c - a)
6 g := fl(c - e), h := fl(g - a), f := fl(b - h)
7 d := fl(f - e)
8 if fl(d + e) ≠ f
9     c := a, d := b
```

10 return c, d
11 end

Proof: Without loss, assume $|a| \geq |b|$. If $b = 0$ then clearly the algorithm produces $c = a$, $d = 0$, so the lemma holds. For the general case, assume $a > 0$; the proof when $a < 0$ is essentially identical. Let $c = \mathrm{fl}(a+b)$ as above and define $r = a + b - c$, so r is the error in computing c. We prove the lemma by establishing three claims.

Claim 1: $d = r$ if and only if r is a t-digit floating point number. Of course, if r is not a t-digit number, then we cannot have $d = r$. For the converse, suppose first that $-a \leq b \leq -a/2$. Then $1 \leq a/(-b) \leq 2$ so $a + b = a - (-b)$ is a t-digit number; hence, by faithfulness, $c = \mathrm{fl}(a + b) = a + b$ exactly and $r = 0$. Again by faithfulness, we must have $e = \mathrm{fl}(c - a) = b$, $g = a$, $h = 0$, $f = b$, and $d = 0 = r$, a t-digit number.

Now suppose $-a/2 < b < 0$. Let a' denote the largest t-digit number not greater than $a/2$, and note that $a - a'$ is the smallest t-digit number not less than $a/2$. Since $|b| \leq a'$, we must have $a + b \geq a - a'$, so by faithfulness, $c = \mathrm{fl}(a + b) \geq a - a' \geq a/2$. But then $1/2 \leq c/a \leq 1$, so $e = \mathrm{fl}(c - a) = c - a = b - r$ exactly. Again we have $g = a$, $h = 0$, and $f = b$, so if r is a t-digit number then $d = \mathrm{fl}(f - e) = r$.

Finally, suppose $b > 0$. If $c \leq 2a$ then $1 \leq c/a \leq 2$ so $e = \mathrm{fl}(c - a) = c - a = b - r$ exactly. As before, $g = a$, $h = 0$, $f = b$, and thus $d = r$ if r is a t-digit number. If instead $c > 2a$, then $e = \mathrm{fl}(c - a)$ may not be computed exactly; let $s = c - a - e$, so s is the roundoff error incurred in computing e. By arguments similar to those of the preceding paragraph, we have $e \geq c/2$, so $1 \leq c/e \leq 2$ and $g = \mathrm{fl}(c - e) = c - e = a + s$ exactly. Now write $a = m\beta^k$ and $b = n\beta^j$ with $\beta^{t-1} \leq |m| < \beta^t$ and likewise for n. We can obtain $c > 2a$ only when $j = k$, so all computed quantities must be integer multiples of β^k. Likewise r is a multiple of β^k, but we must have $|r| < \beta^{k+1}$, so r is a t-digit number. (In fact, r is a one-digit multiple of β^k.) We also see that $|s| < \beta^{k+1}$, hence $h = \mathrm{fl}(g - a) = s$ exactly, and clearly $f = \mathrm{fl}(b - h) = b - s$ is also exact. Thus $d = \mathrm{fl}(f - e) = r$, and the claim holds.

Claim 2: $\mathrm{fl}(d + e)$ is always computed exactly. Recall that $\mathrm{fl}(d+e) = f$ exactly whenever r is a t-digit number. Suppose r is not a t-digit number. Then we must have $e = \mathrm{fl}(c - a) = c - a = b - r$, $g = a$, $h = 0$ and $f = b$, so $d = \mathrm{fl}(b - e)$. Write $b = n\beta^j$ with $\beta^{t-1} \leq |n| < \beta^t$ and note that a, b, and c are all integer multiples of β^j, so r is an integer multiple of β^j. If $br \geq 0$ then clearly $|r| \leq |b|$, so r would be a t-digit number; therefore suppose $br < 0$ and also $|r| > |b|$. Then $de < 0$ and $1/2 \leq d/(-e) \leq 1$ so $\mathrm{fl}(d + e) = \mathrm{fl}(d - (-e)) = d + e$ exactly (although we need not have $d + e = f$).

Claim 3: If r is not a t-digit number then a and b satisfy the non-overlapping condition. To see this, assume that r is not a t-digit number, and write $c = m\beta^k$, $a = n\beta^j$ with $\beta^{t-1} \leq m < \beta^t$ and likewise for n; we must show $|b| < \beta^j$. Suppose $|b| \geq \beta^j$ and recall that $d \neq r$

only if $|b| < |r|$. By faithfulness, $|r| < \beta^k$, so we must have $j < k$, but since $|b| \leq |a|$ this can happen only if $j = k - 1$. Now a, b, and c are all integer multiples of $\beta^{j-t+1} = \beta^{k-t}$, so r is an integer multiple of β^{k-t} and $|r| < \beta^k$. Thus r is a t-digit number, a contradiction.

From the preceding claims, the proof of the lemma proceeds as follows. If r is a t-digit number then $d = r$ and the condition in line 8 must fail because $\mathrm{fl}(d + e) = f$ exactly; in this case, the algorithm outputs the computed sum and the roundoff error which, by faithfulness, must satisfy the non-overlapping condition. Otherwise, the condition in line 8 must succeed because $\mathrm{fl}(d + e)$ is computed exactly but $d + e \neq f$, so the algorithm simply returns a and b; but in this case a and b already satisfy the non-overlapping condition. ∎

By examining the preceding proof, we can easily obtain several corollaries which show that the algorithm may be simplified substantially when the floating point arithmetic possesses additional properties besides faithfulness. Many of these special cases have been considered separately in some of the references; to clarify the relationship to our results, we introduce several definitions. Let a, b be any t-digit floating point numbers and $\circ$ one of $\{+, -, \times, /\}$ (with $b \neq 0$ if $\circ = /$). Let $c = a \circ b$ and x and y two consecutive t-digit numbers with $|x| \leq |c| < |y|$. Then the floating point arithmetic is *correctly rounding* if it is faithful and also $\mathrm{fl}(a \circ b) = x$ whenever $|c - x| < |c - y|$ and $\mathrm{fl}(a \circ b) = y$ whenever $|c - y| < |c - x|$. The arithmetic's addition is *properly truncating* if $\mathrm{fl}(a + b) = x$ whenever $c = x$ or $ab > 0$, but $\mathrm{fl}(a + b) = y$ whenever $c \neq x$ and $ab < 0$. We have borrowed the last definition from Dekker [4]; the reader should not confuse this term, as Linnainmaa did [13], with the notion of *correctly chopping*, in which $\mathrm{fl}(a + b) = x$ always.

The following simplification was established by Knuth [10] for correctly rounding arithmetic; here we prove a slightly more general result which includes properly truncating arithmetics. (Møller [17] obtained a similar but weaker result using still different hypotheses.)

Corollary 1: If the arithmetic has the property that the roundoff error of a sum is always a t-digit number (as it is in correctly rounding and properly truncating arithmetics), then lines 8 and 9 may be eliminated from the above algorithm.

Proof: Returning to the proof of the lemma, we are assuming that r is always a t-digit number; hence the test in line 8 never succeeds. ∎

Dekker [4] and Linnainmaa [13] consider a different variation, which also appears in Pichat [19]. Again, we state a slightly more general result which follows easily from the preceding lemma.

Corollary 2: If the arithmetic has the property that $|\mathrm{fl}(a + a)| \leq 2|a|$ for all a (as it is in properly truncating arithmetics, correctly chopping arithmetics, and all faithful binary arithmetics), then line 6 may be eliminated and b substituted for f in lines 7 and 8 of the above algorithm. In particular, if the arithmetic also has the property of the preceding corollary, that the roundoff error of a sum is a t-digit number (e.g., in a properly truncating arithmetic or a binary correctly rounding arithmetic), then lines 6, 8, and 9 may all be eliminated and b substituted for f in line 7.

Proof: For the first statement, note that if $|\mathrm{fl}(a+a)| \leq 2|a|$ for all a then $|c| \leq 2|a|$, so we always have $g = a$, $h = 0$, and $f = b$. The second statement follows by combining the first with the previous corollary. $\blacksquare$

Although the original algorithm is an interesting theoretical result, in practice virtually every computer which provides a faithful floating point arithmetic can take advantage of one of the simplifications described in the corollaries. For example, arithmetic which conforms to the IEEE 754 standard is binary and correctly rounding, so the simplest version suggested in corollary 2 may be used. For different reasons, the same version may be used on DEC VAX, IBM 370, and similar computers. The only faithful arithmetics which arise in practice and for which neither simplification applies are decimal correctly rounding arithmetics found on some hand calculators.

In the results which follow, we state the cost of our algorithms as the maximum number of t-digit floating point arithmetic operations performed; note that the number of operations required in the preceding algorithms is at most eleven, but may be reduced to as few as six if a simplified version may be used. (Here and below we count a comparison of two numbers as one floating point operation, since comparison implies subtraction. We also count the absolute value operation as one t-digit floating point operation, namely a comparison against zero. One may regard this as pessimistic, since the sign of a floating point number can be, and often is, represented in one bit. On the other hand, separate counts for each type of operation may easily be obtained from the algorithms presented; we have chosen to combine the counts of different operations only for simplicity.)

3   Addition

The addition algorithm we present is a simple variation of the classical algorithm in which the radix points are aligned and digits are summed pairwise from right to left. The difference is that we are summing components rather than digits, and we regard the error of the computed sum as the "sum" of the components and the computed sum itself as a "carry". Since we can only add two components at once, we always add the carry of the last addition to the next smaller component not yet added. (In effect, we are merge-sorting the components of the two expansions by increasing magnitude and adding in this order.) The exact sum is then the sum of the final carry and the error terms from each preceding addition. Since the error terms need not satisfy the non-overlapping condition for expansions, we must renormalize their sum to obtain the result as a t-digit expansion.

We present the addition and renormalization algorithms separately since we will reuse the renormalization process later. To facilitate the proofs we need one more definition: for $0 \leq d < t$, a finite sequence $x_1, x_2, \ldots, x_n$ of t-digit floating point numbers is said to *overlap by at most d digits* if for each $j = 1, 2, \ldots, n-1$ there exist

$i \leq j$ and k such that $x_1, \ldots, x_{i-1}$ are integer multiples of β^k, x_i is an integer multiple of β^{k-d}, $x_{i+1}, \ldots, x_j$ are all zero, and $|x_{j+1}| < \beta^k$. (Loosely speaking, this condition says that if the x_i were written in positional notation, the significant digits of any two successive non-zero terms would coincide in at most d digit positions, and moreover, no three terms would mutually coincide in any position. For example, in four digit decimal floating point, the sequence $12340, 5678, 9.123$ overlaps by at most three digits, but the sequence $12340, 5678, 91.23$ does not because all three numbers coincide in the tens place.) Note in particular that if $\beta^{k-1} \leq |x_1| < \beta^k$ and the sequence $x_1, \ldots, x_n$ overlaps by at most d digits then $|\sum x_j| < \beta^{k+1}$; that is, the sum can carry over to at most one larger place than the largest term. Note also that the components of a t-digit expansion do not overlap at all (i.e., they overlap by at most zero digits).

Proposition 2.1: Let $x = \sum_{i=1}^{n} x_i$ be a t-digit expansion with n components and $y = \sum_{i=1}^{m} y_i$ a t-digit expansion with m components. If the arithmetic is faithful then the following algorithm computes a sequence $e_1, \ldots, e_{n+m}$ which overlaps by at most one digit and satisfies $x + y = \sum e_j$.

Algorithm 2.1: (Addition)

```
procedure add( n, x₁, ..., xₙ, m, y₁, ..., yₘ )
begin
i := n, j := m
if |xᵢ| < |yⱼ|
      while i > 1 and |xᵢ₋₁| ≤ |yⱼ|
          e₍ᵢ₊ⱼ₎ := xᵢ, i := i − 1
else if |xᵢ| > |yⱼ|
      while j > 1 and |yⱼ₋₁| ≤ |xᵢ|
          e₍ᵢ₊ⱼ₎ := yⱼ, j := j − 1
a := xᵢ, b := yⱼ
while i > 1 or j > 1
    ( c, e₍ᵢ₊ⱼ₎ ) := sum_err( a, b )
    a := c
    if i = 1 or ( j > 1 and |yⱼ₋₁| < |xᵢ₋₁| )
        b := yⱼ₋₁, j := j − 1
    else
        b := xᵢ₋₁, i := i − 1
( c, e₂ ) := sum_err( a, b )
e₁ := c
return n + m, e₁, ..., e₍ₙ₊ₘ₎
end
```

Proof: Clearly $x + y = \sum_{j=1}^{n+m} e_j$. To prove that the e_j overlap by at most one digit, we proceed by induction on the number of times the algorithm calls sum_err. In particular, we show that after each execution of sum_err there exists k such that $|e_{i+j}| < \beta^k$ and $c, x_1, \ldots, x_{i-1}, y_1, \ldots, y_{j-1}$ are all integer multiples of β^{k-1} with all but one being multiples of β^k. The proposition will follow easily.

First observe that prior to the first call to sum_err, we have $i = n$ or $j = m$ or both. If $i = n$ but $j < m$ then the final loop starts with $e_{n+j+1}, \ldots, e_{n+m}$ equal to the last $m - j$ components of y, which do not overlap by assumption. Since this case can occur only if $y \neq 0$, we may write $y_j = p\beta^u$ with $\beta^{t-1} \leq |p| < \beta^t$. Then $|x_n| \geq |y_j|$, so $x_1, x_2, \ldots, x_n$ and $y_1, y_2, \ldots, y_j$ are all integer multiples of β^u, while $|e_{n+j+1}| = |y_{j+1}| < \beta^u$. A similar argument applies if $i < n$ and $j = m$. Hence in any case we may begin the induction assuming that the algorithm has entered the final loop; in particular, if $i > 1$ then $|x_{i-1}| > |y_j|$, whereas if $j > 1$ then $|y_{j-1}| > |x_i|$.

To start the induction, let $a = x_i$, $b = y_j$, and (c, e_{i+j}) denote the result returned from the first call to sum_err. If $c = 0$ then $e_{i+j} = 0$ and the initial case holds. Otherwise write $c = q\beta^v$ with $\beta^{t-1} \leq |q| < \beta^t$ and note that $|e_{i+j}| < \beta^v$. Suppose $|a| \geq |b|$; the case $|a| < |b|$ is similar. Write $a = x_i = p\beta^u$ with $\beta^{t-1} \leq |p| < \beta^t$. Then $x_1, \ldots, x_{i-1}$ are multiples of β^{u+t}, and since $|y_{j-1}| > |x_i|$ then y_{j-1} is a multiple of β^u so $y_1, \ldots, y_{j-2}$ are also multiples of β^{u+t}. Since $u \geq v - 1$, we find that $c, x_1, \ldots, x_{i-1}, y_1, \ldots, y_{j-1}$ are all multiples of β^{v-1} and all but y_{j-1} are multiples of $\beta^{v-1+t} \geq \beta^v$ so again the initial case holds. Thus the proposition is established for the first call to sum_err.

Now suppose we have returned from the end of the final loop. Then the current value of a is the previous value of c and the current value of b is one of $\{x_i, y_j\}$. Let c and e_{i+j} be the quantities calculated in this call to sum_err and note that $c = \sum_{k=i}^{n} x_k + \sum_{k=j}^{m} y_k - \sum_{k=i+j}^{n+m} e_k$. If $c = 0$ then $e_{i+j} = 0$ and we need only appeal to the induction hypothesis. Otherwise write $c = q\beta^v$ with $\beta^{t-1} \leq |q| < \beta^t$ and again note $|e_{i+j}| < \beta^v$. Assume $b = x_i$; the case $b = y_j$ is similar. Note that $|b| \leq |y_{j-1}|$ (otherwise we would have chosen $b = y_{j-1}$ in the previous iteration of the loop) and that $|b| \geq |y_j|$ (otherwise we would have chosen $b = x_i$ prior to the previous iteration). Write $b = p\beta^u$ with $\beta^{t-1} \leq |p| < \beta^t$. Then

$$\beta^{v+t-1} \leq |c| \leq \left| \sum_{k=i}^{n} x_k \right| + \left| \sum_{k=j}^{m} y_k \right| + \left| \sum_{k=i+j}^{n+m} e_k \right|$$

$$< \beta^{u+t} + \beta^{u+t} + \beta^{v+1}$$

where the bounds on the first two terms follow from the non-overlapping conditions on x and y and the inequality $|y_j| \leq |b| = |x_i|$, and the bound for the last term follows from the induction hypothesis. Consequently, $\beta^{v+t-1} < \beta^{u+t+1} + \beta^{v+1}$, so $v + t - 1 < \max(u + t + 2, v + 2)$. Since $t \geq 3$ we must have $v + t - 1 < u + t + 2$ so $v < u + 3$ and hence $v \leq u + 2$. Now $x_1, \ldots, x_{i-1}$ are multiples of β^{u+t} and thus of β^v. To complete the induction it suffices to show that y_{j-1} is a multiple of β^{v-1}; then $y_1, \ldots, y_{j-2}$ will be multiples of $\beta^{v-1+t} \geq \beta^v$. Write $y_{j-1} = r\beta^s$ with $\beta^{t-1} \leq |r| < \beta^t$ so that $|\sum_{k=j}^{m} y_k| < \beta^s$. Then the previous inequality

becomes $\beta^{v+t-1} < \beta^{u+t} + \beta^s + \beta^{v+1}$. Since $v \leq u+2$, if $s < u+t$ then we have $v+t-1 < u+t+1$ so $v < u+2$ and hence $v \leq u+1$, and since $|y_{j-1}| \geq |b|$ this implies y_{j-1} is a multiple of $\beta^u = \beta^{v-1}$ and the induction step follows. If instead $s \geq u+t$ then clearly $s \geq v-1$, so again y_{j-1} is a multiple of β^{v-1}.

From the above argument, we clearly must have either $c, x_1, \ldots, x_{i-1}, y_1, \ldots, y_{j-1}$ all multiples of β^v or else exactly one of $\{x_{i-1}, y_{j-1}\}$ is a multiple of β^{v-1} but not of β^v. Suppose as above that y_{j-1} is not a multiple of β^v. Then clearly $|y_{j-1}| < |x_{i-1}|$ so the next iteration of the loop will choose $b = y_{j-1}$. Consequently, either all of the quantities $e_1, \ldots, e_{i+j-1}$ are multiples of β^v or there exists $l \leq i + j - 1$ such that $e_1, \ldots, e_{l-1}$ are multiples of β^v, e_l is a multiple of β^{v-1}, and $e_{l+1}, \ldots, e_{i+j-1}$ are zero. This completes the proof. ∎

Proposition 2.2: Let $e_1, \ldots, e_n$ overlap by at most $t - 2$ digits. If the arithmetic is faithful then the following algorithm computes an expansion $s = \sum_{j=1}^{m} s_j$ with $m \leq n$ such that $s = \sum e_j$.

Algorithm 2.2: (Renormalization)

```
procedure renorm( n, e_1, ..., e_n )
begin
    c := e_n
    for i := n - 1, n - 2, ..., 1
        ( c, f_{i+1} ) := sum_err( c, e_i )
    f_1 := c
    s_1 := f_1, k := 1
    for j := 2, 3, ..., n
        ( c, d ) := sum_err( s_k, f_j )
        s_k := c
        if d ≠ 0
            l := k - 1, k := k + 1
            while l ≥ 1
                ( c', d' ) := sum_err( s_l, s_{l+1} )
                if d' = 0
                    s_l := c', l := l - 1, k := k - 1
                else
                    l := 1
            s_k := d
    return k, s_1, ..., s_k
end
```

Proof: After the first for loop terminates, we clearly have $\sum f_j = \sum e_j$. Furthermore, we claim that the f_j do not overlap at all. We again argue by induction. Specifically, let c and f_{i+1} be the return values from a call to sum_err in the first loop. If $c = 0$ then $f_{i+1} = 0$ so the induction holds; otherwise write $c = q\beta^v$ with $\beta^{t-1} \leq |q| < \beta^t$ so that $|f_{i+1}| < \beta^v$. Also write $e_i = p\beta^u$ with $\beta^{t-1} \leq |p| < \beta^t$. Then since the e_j overlap by at most $t - 2$ digits, we have $v \leq u + 1$. Consequently, $e_1, \ldots, e_{i-1}$ are all multiples of $\beta^{u+2} \geq \beta^v$, so $f_1, \ldots, f_i$ must be multiples of β^v while $|f_{i+1}| < \beta^v$. This establishes the claim.

Proceeding to the rest of the algorithm, we note that the output $s_1, \ldots, s_k$ clearly satisfies $\sum_{j=1}^{k} s_j = \sum_{j=1}^{n} f_j$.

We need only show that the output qualifies as an expansion. Induct on k: let c and d be the return values from a call to sum_err in the outer for loop. If $d = 0$ there is nothing to prove, so assume $d \neq 0$. Then also $s_k = c \neq 0$ so we may write $s_k = q\beta^v$ with $\beta^{t-1} \leq |q| < \beta^t$ and we have $|d| < \beta^v$. Since the f_i do not overlap, we also have $|d + \sum_{i>j} f_i| < \beta^v$. Therefore each subsequent call to sum_err in the outer for loop must yield $|c| \leq \beta^v$ with equality only if c is rounded (so that again $d \neq 0$). If we always have $|c| < \beta^v$ then clearly $|s_i| < \beta^v$ for $i > k$ so the induction holds. If instead we obtain $|c| = \beta^v$ at some later step, then by the induction hypothesis, upon completion of the predicate of the outer if statement, we still find $s_1, \ldots, s_k$ (for the value of k then in effect) is an expansion. In addition, we then have $|d| < \beta^{v-t}$, so again the induction holds, and the proof is complete. ∎

It follows from the proof above that sum_err is never called from the innermost while loop more than $n - 1$ times. An easy operation count yields the following.

Corollary: Let x and y be t-digit expansions having n and m components, respectively. Then the composition of algorithms 2.1 and 2.2 computes a t-digit expansion $s = x + y = \sum_{i=1}^{k} s_i$ with $k \leq n + m$ using at most $49(n + m - 1) + 9$ t-digit floating point operations.

4 Distillation

In this section we consider the following problem: given any t-digit floating point numbers $x_1, x_2, \ldots, x_n$, compute a t-digit expansion y such that $y = \sum_{i=1}^{n} x_i$. Kahan has coined the term "distillation" to descibe the process by which the components of y may be obtained from the x_i.

The first algorithm to implement distillation was given by Pichat [19], who was only interested in obtaining the first component of y. Pichat's algorithm simply passes through each of the x_i in turn applying the basic sum_err procedure. (In fact, the first loop in our renormalization procedure above constitutes one pass of Pichat's algorithm.) Pichat showed that this method converges; i.e., that after a sufficient number of passes, one obtains the first component of y. Bohlender [1] then showed that Pichat's algorithm can be used to obtain the entire expansion for y with at most $n - 1$ passes, and he added a stopping criterion for the case where only a given number of leading components of y are desired.

Many other algorithms which evaluate a sum or an inner product to working precision (i.e., yielding the first component of an expansion) have been developed; see Bohlender [2] for a survey of several approaches. Unfortunately, many of these techniques rely on special features, such as an extra wide fixed point accumulator, directed roundings, or interval arithmetic, which must be implemented in a combination of hardware and low-level software to be efficient. Other techniques, such as one proposed by Kulisch and Miranker [11], require direct access to the exponent and significand fields of floating point numbers, and obtaining these quantities, even when the programming environment provides a convenient way to do so, is much too time-consuming compared to the highly optimized and streamlined floating point additions and comparisons of Pichat's algorithm, at least for the modest values of n one typically encounters.

On the other hand, Pichat's algorithm is not the only way to implement distillation using only faithful floating point addition and comparison. Kahan [9] gives a distillation algorithm which first sorts the x_i by decreasing magnitude, then uses successive passes like Pichat's algorithm, but alternating in order: first smallest to largest, then largest to smallest. Whereas Pichat's algorithm requires $O(n^2)$ t-digit floating point operations in the worst case (i.e., $n - 1$ passes through n numbers), Kahan claims his algorithm requires at most $O(n \log n)$ operations. We now present yet another algorithm which is easily seen to have the same worst-case bound. Our method, a straightforward divide-and-conquer approach using the algorithms developed above, recursively adds successive pairs of partially distilled sums in a binary tree-like reduction. A similar approach was proposed for some parallel computer architectures by Leuprecht and Oberaigner [12].

Proposition 3: Given t-digit floating point numbers $x_1, x_2, \ldots, x_n$. Then the following algorithm carried out with faithful arithmetic computes a t-digit expansion $y = \sum_{i=1}^{m} y_i$ with $m \leq n$ such that $y = \sum x_i$. The algorithm requires at most $49n\lceil \log n \rceil$ t-digit floating point operations.

Algorithm 3: (Distillation)

```
procedure distill( n, x_1, ..., x_n )
begin
if n = 1
    l := 1, w_1 := x_1
else
    m := ⌊n/2⌋
    ( j, y_1, ..., y_j ) := distill( m, x_1, ..., x_m )
    ( k, z_1, ..., z_k ) := distill( n - m, x_{m+1}, ..., x_n
    )
    ( l, w_1, ..., w_l ) := renorm( add( j, y_1, ..., y_j,
    k, z_1, ..., z_k ) )
return l, w_1, ..., w_l
end
```

Proof: That the output is an expansion for $\sum x_i$ is obvious. As for the cost, note that the number of components of a sum as computed by the renormalize and add algorithms does not exceed the sum of the numbers of components of the summands; hence for each level of the binary tree, the total number of components of the expansions added at that level is at most n. Therefore the cost of each level is at most $49n$, and since the number of such levels is clearly $\lceil \log n \rceil$, the total cost is at most $49n\lceil \log n \rceil$. ∎

Kahan's algorithm in [9], though very different from the algorithm above, also incurs a cost at most proportional to $n \log n$, presumably with a modest constant, as opposed to a worst case cost of $11n(n - 1)$ for Pichat's algorithm (assuming the first version of the sum_err procedure is used). Although algorithms are known which have asymptotic cost at most proportional to n, they are impractical for most purposes since they explicitly refer to the exponent and significand fields of the numbers to be added; as noted above, these quantities are very expensive to obtain compared with the cost of a few additions and comparisons. Moreover, Kahan notes that the actual cost of these algorithms in practice is usually far less than the worst case bound, so we are not likely to gain anything by continuing to look for asymptotically cheaper distillation methods.

5 Multiplication

To compute the exact product of two t-digit expansions, we must guarantee that no significant digits are lost when the product of two components is computed in t-digit arithmetic. We accomplish this by splitting each component into a sum of either two or three numbers, each with fewer nonzero digits than the original components. We may then multiply these split components with no error, finally adding all the partial products and renormalizing to obtain a t-digit expansion. The method of splitting components was first proposed by Dekker [4], and we shall rely on the proof given by Linnainmaa [14]. To clarify the statements and proofs, we define the number of *leading nonzero digits* of a t-digit number x to be the smallest integer d such that $x = m\beta^k$ with $\beta^{d-1} \leq |m| < \beta^d$, or zero if $x = 0$.

Proposition 4.1: Let x be a t-digit floating point number. For $1 < k < t$, define $a_k = \beta^{t-k} + 1$. Provided the floating point arithmetic is faithful, the following algorithm computes t-digit numbers x' and x'' such that $x = x' + x''$, x' has at most k leading nonzero digits, and x'' has at most $t - k + 1$ leading nonzero digits.

Algorithm 4.1 (Splitting of a t-digit number)
```
procedure split( x, k )
begin
  y := fl(a_k × x), z := fl(y − x)
  x' := fl(y − z), x'' := fl(x − x')
  return x', x''
end
```

Proof: This is Theorem 1 in [14]. Linnainmaa actually states the theorem with an additional hypothesis which guarantees that in fact x'' has at most $t - k$ leading nonzero digits; in a remark following his proof, however, he shows that even if the additional hypothesis is not satisfied, x' will be computed with $k-1$ leading nonzero digits and x'' will have $t - k + 1$ leading nonzero digits.■

We are now ready to give the multiplication algorithm. Note that we choose to split each component of the first expansion into two parts and each component of the second expansion into three parts. Dekker [4] shows that with binary correctly rounding arithmetic it suffices to split each factor into just two parts; Linnainmaa [14] gives other criteria under which splitting into two parts is sufficient. Consequently, the following algorithm can be improved in many cases. Note that we also do not sum the partial products into one net accumulator but instead separate them into groups (denoted $a^{(1)}, \ldots, a^{(6)}$) which are guaranteed to overlap by at

most two digits so that they may be renormalized without first being processed through the add algorithm. We then accumulate these renormalized expansions into a smaller accumulator (b) which in turn is added to the overall product accumulator. This nesting of additions significantly reduces the coefficient of the largest order term in the cost estimate.

Proposition 4.2: Let $x = \sum_{i=1}^{n} x_i$ be a t-digit expansion with n components and $y = \sum_{i=1}^{m} y_i$ a t-digit expansion with m components. Assume the arithmetic is faithful and that $t \geq 7$, and set $k_2 = \lfloor t/2 \rfloor + 1$ and $k_3 = \lfloor t/3 \rfloor + 1$. Then the following algorithm computes the product xy expressed as a t-digit expansion with at most $2nm$ components. The algorithm requires at most $98n^2m + 1049nm + 8m - 446n$ t-digit floating point operations.

Algorithm 4.2: (Multiplication)
```
procedure multiply( n, x_1, ..., x_n, m, y_1, ..., y_m )
begin
for i := 1, 2, ..., n
    ( x_i', x_i'' ) := split( x_i, k_2 )
for i := 1, 2, ..., m
    ( y_i', z ) := split( y_i, k_3 )
    ( y_i'', y_i''' ) := split( z, k_3 )
p_1 := 0, k := 1
for i := 1, 2, ..., n
    for j := 1, 2, ..., m
        a_j^(1) := fl(x_i' × y_j')
        a_j^(2) := fl(x_i' × y_j'')
        a_j^(3) := fl(x_i' × y_j''')
        a_j^(4) := fl(x_i'' × y_j')
        a_j^(5) := fl(x_i'' × y_j'')
        a_j^(6) := fl(x_i'' × y_j''')
    ( j, b_1, ..., b_j ) := renorm( m, a_1^(1), ..., a_m^(1) )
    ( l, c_1, ..., c_l ) := renorm( m, a_1^(2), ..., a_m^(2) )
    ( j, b_1, ..., b_j ) := renorm( add( l, c_1, ..., c_l,
        j, b_1, ..., b_j ) )
    ( l, c_1, ..., c_l ) := renorm( m, a_1^(3), ..., a_m^(3) )
    ( j, b_1, ..., b_j ) := renorm( add( l, c_1, ..., c_l,
        j, b_1, ..., b_j ) )
    ( l, c_1, ..., c_l ) := renorm( m, a_1^(4), ..., a_m^(4) )
    ( j, b_1, ..., b_j ) := renorm( add( l, c_1, ..., c_l,
        j, b_1, ..., b_j ) )
    ( l, c_1, ..., c_l ) := renorm( m, a_1^(5), ..., a_m^(5) )
    ( j, b_1, ..., b_j ) := renorm( add( l, c_1, ..., c_l,
        j, b_1, ..., b_j ) )
    ( l, c_1, ..., c_l ) := renorm( m, a_1^(6), ..., a_m^(6) )
    ( j, b_1, ..., b_j ) := renorm( add( l, c_1, ..., c_l,
        j, b_1, ..., b_j ) )
```

$$(\, k, p_1, \ldots, p_k \,) := \text{renorm}(\text{ add}(\, j, b_1, \ldots, b_j,$$
$$k, p_1, \ldots, p_k \,) \,)$$
$$\text{return } k, p_1, \ldots, p_k$$
$$\text{end}$$

Proof: Note that for each i both x_i', x_i'' have at most $\lfloor t/2 \rfloor + 1$ leading nonzero digits, and for each j, y_j', y_j'', y_j''' have at most $\lfloor t/3 \rfloor + 1$ leading nonzero digits. Since $t \geq 7$, each of the six products in the innermost for loop is computed exactly. By the non-overlapping condition for expansions together with the bounds on the number of leading nonzero digits of split components, we see that the six sequences $a_j^{(1)}, \ldots, a_j^{(6)}, j = 1, \ldots, m$ computed in the inner loop overlap by at most two digits. Hence the renormalization and addition procedures in the outer loop succeed without error, so the output satisfies $p = xy$. Note that for each i the sums accumulated in b form expansions with at most $2m$ components. Therefore p never exceeds $2nm$ components. A straightforward operation count completes the proof. ∎

The above multiplication algorithm is based on the classical algorithm of multiplying digits pairwise; hence, the number of multiplications is proportional to the product of the numbers of components in each expansion. In particular, if both expansions have n components, the number of basic multiply-and-add steps is $O(n^2)$. Algorithms which multiply n-digit numbers in fewer than $O(n \log^2 n)$ steps are known (see Knuth [10]), but we have elected not to use them for several reasons. Specifically, most of these algorithms improve upon the classical method only for very large n, while in practice we would expect to apply the above algorithm only to expansions with relatively few components (e.g., at most 39 for a typical implementation of binary floating point arithmetic using 64 bit storage; see the remarks on underflow in section 8 below). In addition, some of these algorithms are based on integer arithmetic, so that the input is assumed to consist of a contiguous string of n digits representing each multiplicand. Since we are given input in the form of t-digit expansions, the digits of which need not be contiguous, we cannot always benefit from the application of the techniques suggested.

Notice that although the total number of multiplications is only proportional to nm, the total cost is proportional to n^2m. This extra factor of n in the total cost arises from the cost of adding the partial products. Using slightly more intermediate storage, we can rearrange the order of the additions in the form of a binary tree, much like the arrangement in algorithm 3 above, and reduce this term to $nm \log n$. The remark following algorithm 3 suggests, however, that reducing the cost to $O(nm)$ is not practical.

6 Division

Our division algorithm is analogous to the usual method of long division. In that algorithm, we take the quotient of the most significant digits of the current dividend and divisor as a guess for the next digit of the overall quotient. If the product of this guess with the divisor is larger than the dividend, however, we must choose a smaller digit. For our algorithm, however, since we do not require the components of an expansion to have the same sign, we do not need to adjust the guess in this way. We simply take the quotient of the most significant components (as computed by t-digit arithmetic) as the next term of the quotient, and we show that once a sufficient number of terms have been computed, we can renormalize to express the quotient as an expansion.

Proposition 5: Let $x = \sum_{i=1}^{n} x_i$ be a t-digit expansion with n components and $y = \sum_{i=1}^{m} y_i$ a t-digit expansion with m components, and suppose $y \neq 0$. Assume the arithmetic is faithful and $t \geq 9$; for $d > 0$ let $k = \lfloor (t-4)d/t \rfloor$. Then the following algorithm computes a t-digit expansion $q = \sum q_i$ with at most d components such that $|q - x/y| < \beta^{1-k} |x/y|$. The algorithm requires at most $98(d-1)^2 m + (d-1)(1255m + 49n - 479) + 7$ t-digit floating point operations.

Algorithm 5: (Division)

```
procedure divide( d, n, x_1, ..., x_n, m, y_1, ..., y_m )
begin
    j^(1) := n
    for i := 1, 2, ..., n
        e_i^(1) := x_i
    for i := 1, 2, ..., d
        q_i := fl(e_1^(i)/y_1)
        if i < d
            ( k, f_1, ..., f_k )  :=  multiply( 1, q_i,
                m, y_1, ..., y_m )
            for l := 1, 2, ..., k
                f_l := -f_l
            ( j^(i+1), e_1^(i+1), ... )  :=  renorm( add(
                k, f_1, ..., f_k, j^(i), e_1^(i), ... ) )
    ( k, s_1, ..., s_k ) := renorm( i, q_1, ..., q_i )
    return k, s_1, ..., s_k
end
```

Proof: The result is trivial if $x = 0$; for $x \neq 0$ it suffices to show that the sequence $q_1, \ldots, q_i$ computed in the main loop overlaps by at most four digits. Proceeding by induction on i, let q_i be the quantity computed in the i-th iteration of the loop for some i and write $q_i = p\beta^u$ with $\beta^{t-1} \leq |p| < \beta^t$. Note that $q_i = (e_1^{(i)}/y_1)(1 + \epsilon)$, $f = q_i y = q_i y_1(1 + \delta)$ and $e_1^{(i)} = e^{(i)}(1 + \eta)$ where $|\epsilon|, |\delta|, |\eta| < \beta^{1-t}$. Hence writing $\alpha = 1 - (1 + \epsilon)(1 + \delta)(1 + \eta)$ we have $e^{(i+1)} = e^{(i)} - f = e^{(i)}\alpha$ and $|\alpha| < \beta^{3-t}$. Similarly, $e_1^{(i+1)} = e^{(i+1)}(1 + \mu) = e^{(i)}\alpha(1 + \mu) = e_1^{(i)}\alpha(1 + \mu)(1 + \eta)^{-1}$ and hence $q_{i+1} = q_i\alpha(1 + \mu)(1 + \nu)(1 + \eta)^{-1}(1 + \epsilon)^{-1}$ with $|\mu|, |\nu| < \beta^{1-t}$. Consequently, $|q_{i+1}| < |q_i|\beta^{4-t} < \beta^{u+4}$. Since $t \geq 9$, the induction hypothesis implies that $q_1, \ldots, q_{i-1}$ are multiples of $\beta^{u+t-4} > \beta^{u+4}$ and the proof is complete, save for an easy operation count. ∎

7 Example: Intersecting a Line Segment and a Line

We illustrate the application of our algorithms to a well-studied problem in computational geometry, that of computing the intersection of a line segment and a line. For the purpose of this section, we assume that we have an arithmetic conforming to the IEEE Standard for Binary Floating Point Arithmetic. In particular, the arithmetic is binary and correctly rounding, so the simplest form of sum_err may be used throughout; in addition, we assume that two different precisions are available, corresponding to two different values of t: *single precision*, for which the precision is t_1 ($t_1 = 24$ in the IEEE standard), and *double precision*, with a precision t_2 (the standard specifies $t_2 = 53$). The arithmetic includes operations between single precision operands yielding a single precision result, operations between double precision operands yielding a double precision result, and two conversion operations: one which converts a single precision number to double precision (exactly), and one which correctly rounds a double precision number to single precision provided no overflow or underflow occurs. Most importantly, we suppose that double precision, as the name implies, carries at least twice as many digits as single precision, i.e., $t_2 \geq 2t_1$; this assumption guarantees that we can multiply single precision numbers exactly by converting to and multiplying with double precision.

The general problem is this: given the coordinates of four points in the plane, P_1, P_2, P_3, and P_4, decide whether the line segment $P_1 P_2$ intersects the line containing P_3 and P_4, and if so, compute the coordinates of the intersection point. This problem arises in many computer graphics and computer-aided design applications; numerous authors have recognized both the importance and difficulty of solving the problem accurately (see [6, 15, 16, 18] and references therein). Milenkovic [16], in particular, has shown that if the single and double precision arithmetics satisfy $t_2 \geq 2t_1 + 1$ then line and line segment intersection calculations may be computed to full accuracy provided lines and line segments are defined by the coefficients of line equations given as single precision numbers. Unfortunately, his approach requires that line segments defined by endpoint coordinates be perturbed and replaced by new line segments defined by line equation coefficients; he apparently overlooks the fact that perturbing two nearly parallel lines may move their intersection point a large distance. Another practical drawback of Milenkovic's method stems from the fact that intersections are characterized exactly, but each intersection point, once computed, is rounded to single precision, so that subsequent tests fail to indicate that the computed intersection point actually belongs to the lines determining the intersection. Milenkovic sidesteps this problem by arguing that any efficient algorithm would not make such a test because it would be redundant (i.e., the point has already been found to be the intersection of the two lines, so we need not test for its inclusion in either line). In practice, however, many otherwise simple applications of line and segment intersection calculations become much more complicated when non-redundancy is required, due to the overhead of maintaining all accumulated topological knowledge.

We propose instead to compute line and line segment intersections to full accuracy—i.e., correct to single precision—directly from endpoint coordinates using the algorithms presented above as necessary. Unlike Milenkovic's method, our algorithm may easily be modified to allow approximate endpoint intersections to be recognized; in this way, we obtain a consistent set of calculations and no longer need to assume non-redundancy. Ottman, et al, [18] propose essentially the same approach for the intersection problem and other geometric problems. In fact, they show how to compute intersection points to full accuracy using only single precision arithmetic and a means of computing inner products to full precision. Their method, however, relies on interval arithmetic, which in turn requires directed roundings, to obtain the final result to full accuracy. In contrast, our method uses only the faithfulness of the arithmetic and could easily be implemented entirely in single precision by using the multiplication and division algorithms of sections 5 and 6. Before presenting the algorithm, we formulate a concise statement of the problem.

Problem: Given single precision numbers (x_1, y_1), (x_2, y_2), (x_3, y_3), and (x_4, y_4). Decide whether the line segment $(x_1, y_1)(x_2, y_2)$ intersects the line determined by (x_3, y_3) and (x_4, y_4) at a unique point, and if so, compute the coordinates (x, y) of the intersection point accurate to single precision. (That is, if the exact intersection point is (x', y'), then return a point (x, y) such that x is the nearest single precision number to x' and y is the nearest single precision number to y'.)

We first devise a straightforward but somewhat costly algorithm which is easily seen to solve the problem. Writing the intersection problem as a system of two linear equations, we see that the line segment intersects the line at a unique point if and only if $d \neq 0$, where $d = (y_4 - y_3)(x_1 - x_2) + (x_3 - x_4)(y_1 - y_2)$. In this case, the coordinates of the intersection point are given by

$$x = \frac{1}{d}\left[(x_1 - x_2)t + (x_3 - x_4)s\right]$$

$$y = \frac{1}{d}\left[(y_1 - y_2)t + (y_3 - y_4)s\right]$$

where $s = (x_2 y_1 - x_1 y_2)$ and $t = (x_3 y_4 - x_4 y_3)$. We can compute d by rewriting the previous expression as a sum of products, multiplying each term using double precision, then distilling the terms to obtain an expansion for d. (This is precisely the manner in which Ottman, et al, use the accurate inner product evaluation.) Then d vanishes if and only if the first component of the expansion vanishes, and otherwise this component is accurate to double precision. Similarly, we can express the numerators of the expressions for x and y as sums of products, then multiply these terms (although since each term has degree three, we need to use the Dekker/Linnainmaa splitting method), distill, and use the leading components, which are again accurate to double precision. Taking the quotient of these leading components with the leading component of d gives coordinates which are accurate to near double precision, and at any rate sufficiently accurate to round to the correct single precision quantities, provided no underflow occurs. We formalize these ideas in the following proposition; we omit the straightforward proof. (Since the number of terms being distilled is never larger than sixteen, we have assumed for the operation count that Pichat's distillation algorithm is used.)

Proposition: The following algorithm solves the above problem using at most 3301 IEEE 754-compatible double precision floating point arithmetic operations including precision conversions. Moreover, no computation overflows, and the only computation which can underflow is the final conversion to single precision. (Here fl denotes double precision computation.)

Algorithm: (Intersect line and segment, version 1)

    procedure intersect1($x_1, y_1, \ldots, x_4, y_4$)

    begin

    convert $x_1, y_1, \ldots, x_4, y_4$ to double precision numbers $x'_1, y'_1, \ldots, x'_4, y'_4$

    $e_1 := \mathrm{fl}(x'_1 \times y'_4)$, $e_2 := \mathrm{fl}(-x'_1 \times y'_3)$

    $e_3 := \mathrm{fl}(-x'_2 \times y'_4)$, $e_4 := \mathrm{fl}(x'_2 \times y'_3)$

    $e_5 := \mathrm{fl}(x'_3 \times y'_1)$, $e_6 := \mathrm{fl}(-x'_3 \times y'_2)$

    $e_7 := \mathrm{fl}(-x'_4 \times y'_1)$, $e_8 := \mathrm{fl}(x'_4 \times y'_2)$

    $(n, d_1, \ldots, d_n) := \mathrm{distill}(8, e_1, \ldots, e_8)$

    if $d_1 = 0$

        return "No unique intersection"

    else

        $(e'_1, e''_1) := \mathrm{split}(e_1, t_1), \ldots, (e'_8, e''_8) := \mathrm{split}(e_8, t_1)$

        $b_1 := \mathrm{fl}(e'_1 \times x'_3), \ldots, b_{16} := \mathrm{fl}(e''_8 \times x'_1)$

        $(j, b_1, \ldots, b_j) := \mathrm{distill}(16, b_1, \ldots, b_{16})$

        $c_1 := \mathrm{fl}(e'_5 \times y'_4), \ldots, c_{16} := \mathrm{fl}(e''_1 \times y'_2)$

        $(k, c_1, \ldots, c_k) := \mathrm{distill}(16, c_1, \ldots, c_{16})$

        $x' := \mathrm{fl}(b_1/d_1)$, $y' := \mathrm{fl}(c_1/d_1)$

        convert x', y' to single precision numbers x, y

        return x, y

    end

The preceding algorithm is easy to deduce, but seems rather costly. Can we do better? For a global algorithm, the answer seems to be "no", since we must avoid destructive cancellation in computing sums. Nevertheless, we can rearrange the computation slightly to guarantee that only two sums can incur destructive cancellation, and then only rarely, by proceeding as follows. Note that we may write the coordinates of the intersection point as $x = x_1 + a(x_2 - x_1)$ and $y = y_1 + a(y_2 - y_1)$ where $a = u/(u-v)$, $u = (y_4 - y_3)(x_1 - x_3) + (x_3 - x_4)(y_1 - y_3)$, and $v = (y_4 - y_3)(x_2 - x_3) + (x_3 - x_4)(y_2 - y_3)$. In terms of u and v, we have the following possibilities: if $uv > 0$, the line segment does not intersect the line; if $u = 0$ and $v = 0$, the line segment coincides with the line; if $u = 0$ but $v \neq 0$, the line segment intersects the line at P_1, and similarly if $u \neq 0$ but $v = 0$; finally, if $uv < 0$, the line segment intersects the line at some point between P_1 and P_2.

Suppose we compute u and v accurate to double precision, by means described above. We can test for each of the aforementioned cases, and we need only perform additional computations if $uv < 0$. In this case, however, no cancellation occurs when we compute $u - v$, so a can be computed to near double precision. Likewise, we can compute $x_2 - x_1$ and $y_2 - y_1$ to near double precision, and therefore the products $a(x_2 - x_1)$ and $a(y_2 - y_1)$ may be computed to near double precision. Specifically, each of these quantities will be accurate to all but the last two significant bits. Thus if neither sum $x_1 + a(x_2 - x_1)$ and $y_1 + a(y_2 - y_1)$ cancels in more than $t_2 - 2 - t_1$ leading bits, then both will be sufficiently accurate to round to the correct single precision values. Moreover, we can test whether this much cancellation has occurred, and if so, we can always resort to the lengthy algorithm presented above. To summarize, we state our results (again without proof).

Proposition: Let $c = 2^{t_2 - 2 - t_1}$. The following algorithm solves the above problem using at most 721 IEEE 754-compatible double precision floating point operations including conversions, *provided the condition in the last if statement fails.* Moreover, no computation overflows, and the only computation which can underflow is the final conversion to single precision. (Again, fl denotes computation in double precision.)

Algorithm: (Intersect line and segment, version 2)

    procedure intersect2($x_1, y_1, \ldots, x_4, y_4$)

    begin

    convert $x_1, y_1, \ldots, x_4, y_4$ to double precision numbers $x'_1, y'_1, \ldots, x'_4, y'_4$

    $u_1 := \mathrm{fl}(y'_4 \times x'_1), \ldots, u_8 = \mathrm{fl}(x'_4 \times y'_3)$

    $(j, u_1, \ldots, u_j) := \mathrm{distill}(8, u_1, \ldots, u_8)$

    $v_1 := \mathrm{fl}(y'_4 \times x'_1), \ldots, v_8 = \mathrm{fl}(x'_4 \times y'_3)$

    $(k, v_1, \ldots, v_k) := \mathrm{distill}(8, v_1, \ldots, v_8)$

    if $u_1 = 0$

        if $v_1 = 0$

            return "Intersection not unique"

        else

            return x_1, y_1

    else if $v_1 = 0$

        return x_2, y_2

    else if ($u_1 < 0$ and $v_1 < 0$) or ($u_1 > 0$ and $v_1 > 0$)

        return "No intersection"

    $a := \mathrm{fl}(u_1/\mathrm{fl}(u_1 - v_1))$

    $x' := \mathrm{fl}(x'_1 + \mathrm{fl}(a \times \mathrm{fl}(x'_2 - x'_1)))$, $y' := \mathrm{fl}(y'_1 + \mathrm{fl}(a \times \mathrm{fl}(y'_2 - y'_1)))$

    if $|x'| \leq \mathrm{fl}(c \times |x'_1|)$ or $|y'| \leq \mathrm{fl}(c \times |y'_1|)$

        $(x, y) := \mathrm{intersect1}(x_1, y_1, \ldots, x_4, y_4)$

    else

        convert x', y' to single precision numbers x, y

    return x, y

    end

The last if test fails only when either the x or y coordinate of the intersection point is much closer to zero than

x_1, respectively y_1. We cannot expect to give a meaningful estimate of the probability with which the test fails, but we can say loosely that it will happen "only rarely". Even then, we simply resort to the longer computation given as version 1, and here we could exercise greater care and reuse some of the quantities already computed in version 2, so that the overall cost will be slightly less than the sum of the costs of each version alone. (In this respect, our approach is similar to that suggested by Dobkin and Silver [6].) Moreover, the most pessimistic quantities in the cost estimates arise from the distillation procedure, which, as Kahan [9] notes, often takes far less time than we can predict. Therefore, we suggest that the preceding algorithm is not only provably accurate and robust, but usually very efficient as well.

8 Further Remarks

We have exhibited algorithms which perform exact addition and multiplication and arbitrarily precise division using only fixed precision floating point arithmetic operations by expressing extended precision numbers as unevaluated sums of t-digit floating point numbers. The algorithms are based on a simple property possessed by most "reasonable" implementations of floating point arithmetic. As an application of these algorithms, we have given a provably robust, accurate, and reasonably efficient algorithm for computing the intersection of a line and a line segment. Unlike many algorithms which are *backwards stable* in that they actually compute the intersection of a nearby line and line segment, our algorithm is *forwards stable*: it computes the point nearest the exact intersection of the given line and segment.

We express the cost of our algorithms as the number of t-digit floating point arithmetic operations required as a function of the number of components of the input expansions. Note that a typical number in R_β does not have a unique t-digit expansion; in fact, even the number of components is not uniquely determined. Since the cost of each operation depends on the number of components, we could in principle demand that we always compute *minimal* t-digit expansions. We choose not to do so, however, since this would require the development of an algorithm which finds a minimal expansion given an arbitrary one, and such algorithms are difficult to formulate in a machine-independent way. Instead we only require that an expansion satisfy the non-overlapping condition; i.e., that significant digits of successive components do not overlap. This requirement incurs additional cost in the form of the renormalization steps following each operation, but it provides a convenient form for subsequent rounding and usually leads to "nearly minimal" expansions. We conjecture that for many applications, the cost of producing minimal expansions outweighs the cost of working with nearly minimal ones. (In fact, for some applications we may save time by renormalizing even less frequently; our algorithms could easily be made more efficient in this respect, although some of the proofs would need to be refined.)

In designing our algorithms, we have tacitly assumed that the exponent range of the arithmetic is unlimited, so that no overflow or underflow occurs. In fact, although it is possible for our distillation and multiplication algorithms to overflow when the exact results lie within representable range, such overflow can occur only when the inputs are already very close to the overflow threshhold and is therefore not likely. On the other hand, since the trailing components of expansions quickly become very small, underflow imposes a certain limitation on the accuracy of these algorithms. (For example, IEEE standard double precision with a 64 bit storage format contains eleven bits of exponent and 53 bits of fraction. Hence the net exponent range is $2^{11} = 2048$, so the maximum number of representable components in an expansion is $\lceil 2048/53 \rceil = 39$. Exact calculations can easily exceed this limit after several operations.) Nevertheless, we believe that in most applications, intermediate results may be rounded or rescaled so that underflow does not pose a serious problem. Even when underflow cannot be avoided, Demmel [5] points out that the cumulative effect of the resulting errors can often be estimated by extending the error analysis to include an underflow term whose magnitude is bounded by the underflow threshhold, if underflow is gradual, or by the ratio of this threshhold to the unit roundoff otherwise.

In conclusion we note that the techniques we have proposed for calculating to arbitrary precision can be used equally well to simulate either floating point or fixed point arithmetic. Specifically, if after each calculation we truncate the resulting expansion to a fixed number of components, then the relative error of each intermediate result will be bounded by a corresponding constant, as is true of single precision floating point arithmetic. Alternatively, if we drop only trailing components whose magnitude is below a given threshhold (as with underflow), then we can instead bound the absolute error of each operation, thereby emulating fixed point arithmetic. More importantly, in either case we may adjust the precision dynamically, calculating certain intermediate results to very high precision and others to lower precision, to obtain the desired accuracy in the solution; the cost of the computation, not the precision of the arithmetic, determines the size of acceptable errors. We should then rephrase the traditional question of roundoff error analysis, "How accurate is a computed result if we calculate each intermediate quantity using fixed precision arithmetic?", as the more natural question, "What is the cheapest way to compute a result to a specified accuracy?" We believe further study in this direction will provide both practical and theoretically useful results.

References

[1] Bohlender, G., Floating-Point Computation of Functions with Maximum Accuracy, *IEEE Trans. Comput.* **C-26** (1977), 621–632.

[2] Bohlender, G., What Do We Need Beyond IEEE Arithmetic?, in Ullrich, C. (Ed.), *Computer Arithmetic and Self-Validating Numerical Methods*, Academic Press, Boston, 1990.

[3] Brent, R., A Fortran Multiple Precision Arithmetic Package, *ACM Trans. Math. Soft.* **4** (1978), 57–70.

[4] Dekker, T., A Floating-Point Technique for Extending the Available Precision, *Numer. Math.* **18** (1971), 224–242.

[5] Demmel, J., Underflow and the Reliability of Numerical Software, *SIAM J. Sci. Stat. Comput.* **5** (1984), 887–919.

[6] Dobkin, D., and D. Silver, Recipes for Geometry and Numerical Analysis, in *Proc. Fourth Annual Symposium on Computational Geometry*, Association for Computing Machinery, New York, New York, 1988.

[7] Kahan, W., Further Remarks on Reducing Truncation Errors, *Comm. ACM* **8** (1965), 40.

[8] ——, A Survey of Error Analysis, in Freeman, C. V. (Ed.), *Proc. IFIP Cong. 1971*, vol. 2, North-Holland, Amsterdam, 1972.

[9] ——, Paradoxes in Concepts of Accuracy, lecture notes from Joint Seminar on Issues and Directions in Scientific Computation, U. C. Berkeley, 1989.

[10] Knuth, D., *The Art of Computer Programming*, vol. 2 (2nd ed.), Addison-Wesley, Reading, Mass., 1981.

[11] Kulisch, U., and W. Miranker, *Computer Arithmetic in Theory and Practice*, Academic Press, New York, New York, 1981.

[12] Leuprecht, H., and W. Oberaigner, Parallel Algorithms for the Rounding-Exact Summation of Floating-Point Numbers, *Computing* **28** (1982), 89–104.

[13] Linnainmaa, S., Analysis of Some Known Methods of Improving the Accuracy of Floating-Point Sums, *BIT* **14** (1974), 167–202.

[14] ——, Software for Doubled-Precision Floating-Point Computations, *ACM Trans. Math. Soft.* **7** (1981), 272–283.

[15] Milenkovic, V., Verifiable Implementations of Geometric Algorithms Using Finite Precision Arithmetic, *Artificial Intelligence* **37** (1988), 377–401.

[16] ——, Double Precision Geometry: A General Technique for Calculating Line and Segment Intersections Using Rounded Arithmetic, in *Proc. Thirtieth Annual Symposium on Foundations of Computer Science*, IEEE Computer Society Press, Los Alamitos, Calif., 1989.

[17] Møller, O., Quasi Double Precision in Floating-Point Addition, *BIT* **5** (1965), 37–50.

[18] Ottman, T., G. Thiemt, and C. Ullrich, Numerical Stability of Simple Geometric Algorithms in the Plane, in Børger, E. (Ed.), *Computation Theory and Logic*, Springer-Verlag, Berlin, 1987.

[19] Pichat, M., Correction d'une Somme en Arithmetique à Virgule Flottante, *Numer. Math.* **19** (1972), 400–406.

[20] Sterbenz, P., *Floating-Point Computation*, Prentice-Hall, Englewood Cliffs, New Jersey, 1974.

[21] Wilkinson, J., *Rounding Errors in Algebraic Processes*, Prentice-Hall, Englewood Cliffs, New Jersey, 1964.

Session 6:

Adders I

Chair:
Tony Carter
University of Utah

Designing Optimum Carry-Skip Adders

Vitit Kantabutra

Department of Mathematics and Computer Science
Drexel University
Philadelphia, Pennsylvania 19104

1 Introduction

In designing binary adders, the computer architect has
many criteria to consider. Circuit speed, power consumption, and cost are of prime importance. For VLSI
implementation, chip area must be kept low. In recent
years, *carry-skip adders* have often been the adders
of choice, since the best carry-skip adders (even the
simple *one- or two-level* ones) can be comparable in
speed to the fastest adders such as carry-lookahead
adders [4]; yet unlike the carry-lookahead adders, the
carry-skip adders remain quite low in cost, chip area,
and power consumption [10][1]. We must mention,
though, that for extremely large operand sizes, Ofman's adders are faster than carry-skip adders[2].

The present paper describes a method for designing optimum-speed one-level carry-skip adders. This
method always yields the fastest adders if the assumptions of Guyot, et al. [4] hold – that is, if the *ripple
time* (a circuit parameter) of a carry signal is a linear function of the number of bit positions that the
carry signal propagates through, and if the *skip time*
(another circuit parameter) of a carry signal is a linear function of the number of blocks of bit positions
skipped by the signal, or if these two parameter are
such mildly nonlinear functions that they can be mod-
eled by a linear function without any effect on any of
the results obtained. The circuit design method to be
described in this paper is useful because Guyot, et al.
have shown that in device technologies such as 2-Alu
CMOS the nonlinearities are often insignificant.

The skip time / ripple time ratio is called ρ. The
method presented in this paper gives the fastest circuit, no matter what ρ is. The method given by
Guyot, et al. [4] is only guaranteed to yield circuits
whose speeds are within twice the skip time of the
optimum. The author has programmed his own algorithm, as well as the algorithm of Guyot, et al., and
has performed a comparison study of the resulting circuits by these two algorithms, which is presented in
Section 4 of this paper. There it is shown that Guyot,
et al.'s circuits can have total carry propagation time
that is more than 11% slower than those produced by
the algorithm of the present paper. Our algorithm
runs in polynomial time and takes about 8 seconds
to run on a MacIntosh II for a sample input with an
operand size of 128 bits and a ρ value of 2.0001. The
circuits obtained are always optimum-speed. Additionally, our algorithm yields optimum-speed circuits
for both integer and noninteger values of ρ, whereas
Oklobdzija and Barnes' algorithm [9] only yields optimum circuits for integer values of ρ. If ρ is, in fact, not
an integer, then their algorithm could still be run with
the nearest integer input as an approximate value for
ρ, but doing this would typically yield a circuit that
is far from optimum. In Section 4 we give an example in which Oklobdzija and Barnes' algorithm yields
a circuit that has a total carry propagation time over
16% slower than the carry propagation time of our
optimum-speed circuit.

Two relevant papers appeared in the previous IEEE
arithmetic circuit symposium. One of these papers is
by Chan and Schlag. Chan and Schlag's work [3] allows the ripple time to be a wide range of functions,
including any quadratic function of the number of bits
rippled through by the carry signal. For certain technologies, this assumption would be an improved model
for the ripple time, and could yield faster circuits.
However, their algorithm appears to be slower than
the one in this paper for practical operand sizes, although the exact analysis of the speeds of both algorithms seems difficult. Algorithm speed is important in the cases where the circuit designer wants to
try different values of ρ to see which value gives the
best circuit for the technology and the application
at hand. It would be interesting to test the speed
of all the carry-skip circuit design algorithms for the
operand sizes for which carry-skip adders are fast.
Asymptotic algorithm analysis doesn't seem very useful, since the $O(\log n)$-delay adders, such as Ofman's
adder [8], would beat the speed of carry-skip adders for
extremely large operand sizes n. Guyot, et al.'s algo-

---

[1] One-level carry-skip adders may be as fast as carry-
lookahead adders, but are more likely to be 20%-30% slower for
operand sizes of 64 bits or more. However, this is offset by the
fact that these carry-skip adders are *many times* more efficient
in terms of power consumption and design and implementation
costs.

[2] If n is the operand size, then Ofman's adder exhibit a total delay of $O(\log n)$, whereas carry-skip adders exhibit a total
delay of $O(\sqrt{n})$. Carry-lookahead adders are often also faster
than carry-skip adders, though that isn't clear, since technological limitations currently dictate that carry-lookahead adders
for operands of sizes larger than a few bits be in a modified
(multi-level) form. And in this form, those adders are slowed
down somewhat.

146

rithm seems to be the fastest algorithm, even though it doesn't always yield the fastest circuits. The other paper is by Silvio Turrini. Turrini's paper [10] suggests, without any convincing analysis, an algorithm that supposedly computes an optimum circuit for any combination of skip time and ripple time, with no restrictions on the number of skip levels. However, Turrini's work only covers a limited case of Guyot, et al.'s Model 1 (a description of Model 1 exists in the next subsection). The present paper, however, covers the full range of Model 1.

Carry-skip adders were invented for decimal arithmetic operations by Charles Babbage in the 1800's, and became quite popular in mechanical adding machines later that century. Modern interest in carry-skip adders only began in the early 1960's with the publication of Lehman and Burla's paper [6]. In the next section, we will present a technical introduction to carry-skip adders and then we will go on to survey the modern literature pertaining to carry-skip adders.

The author hopes that this paper will help carry-skip adders to gain the recognition they deserve for their high speed, energy efficiency, cost efficiency, and ease of implementation.

2   Technical Details

2.1   A Detailed Description of Carry-Skip Adders

Let us now develop a detailed description of carry-skip adders. Let the two binary operands be called $X = x_{n-1}x_{n-2}\ldots x_2 x_1 x_0$ and $Y = y_{n-1}y_{n-2}\ldots y_2 y_1 y_0$. Call the carry into and the carry out of bit position i respectively c_i and c_{i+1}. The result bit at position i is to be called z_i. The result can have at most $n+1$ bits; z_0 up to z_n. The following relations for z_i and c_{i+1} are well-known:

$$z_i = x_i \oplus y_i \oplus c_i \qquad (1)$$
$$c_{i+1} = x_i y_i + y_i c_i + c_i x_i \qquad (2)$$

where $\oplus$ and $+$ respectively denote the EXCLUSIVE OR and the (INCLUSIVE) OR operations, and when two variables are written next to each other they are to be ANDed together. The carry input c_0 to the circuit is 0 when the adder is used for adding two numbers in the usual way, but this input can be connected to the c_n output of another adder to create a multiple-precision adder.

Carry-skip adders are best understood by first looking at the carry-ripple adder. The carry-ripple adder is obtained by a straightforward implementation of an adder circuit according to the logical relations (1) and (2) for z_i and c_{i+1} given above. In this adder a carry signal may have to propagate all the way from the least significant position to the most significant position. An example would be the case where $\forall i$ except 0, $x_i = 0, x_0 = 1$, and $\forall i, y_i = 1$. Note that in every bit position i through which the carry signal travels, x_i and y_i are different. This is a significant fact, because in general, a carry signal will propagate through a bit

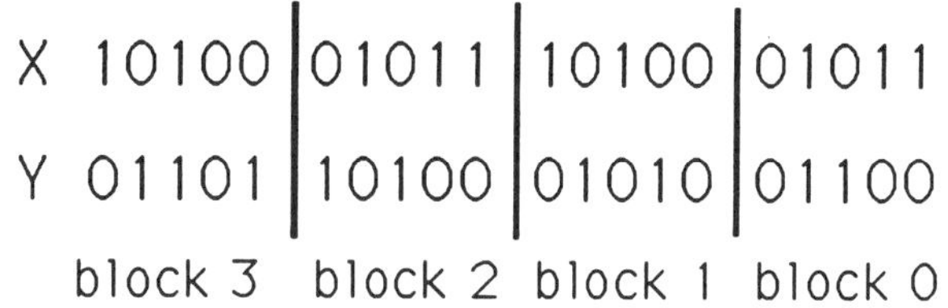

Figure 1: Dividing the bit positions into blocks

position i if and only if $x_i \neq y_i$. (To make this clearer, note that if $x_i = y_i$, then all carry signals to the left of position i are independent of all inputs in positions 0 through $i-1$.) Thus, the chances that a carry signal travels a long way is small, because the probability of having a long string of bit positions, all with $x_i \neq y_i$, is small. In fact, early electronic computer researchers [1] have shown that the average time delay of a carry-ripple adder is only $O(\log n)$. (See [1].)

In a carry-skip adder, we take advantage of the observations just discussed. The bit positions are to be divided into *blocks* of contiguous positions[3], as shown in Figure 1. Let us number these blocks $0, 1, 2, \ldots$ from right to left. When a block has only positions i with $x_i \neq y_i$, like block 2 in Figure 1, then we know that the carry out of that block is the same as the carry into the block[4]. We also know that the carry input to the bit positions in the block are either all 0's or all 1's, depending on whether the carry input to the least significant bit of the block is a 0 or a 1, respectively. Thus, no carry signal has to propagate through this block in a position-by-position fashion. Finding out if all positions i in the block have $x_i \neq y_i$ is an easy $O(\log n)$ process, and will even be regarded instead as a constant-time process for practical block sizes. (For example, even when a block is as large as 16 bits, which is unusual, we can use a layer of EXCLUSIVE OR gates and 2 layers of 4-input AND gates.) Similarly, it is a quick process to force the carry input bits in the block to a uniform value. Aside from these two types of time delays, there is also the delay caused by the final addition step, which is constant, and the delay due to carry propagation. Thus, in practical circuits this very last type of delay is the only type that causes nonconstant total circuit delay, and is the quantity to be minimized in order to obtain an optimum circuit.

There are two kinds of carry propagation delays. First there is the *ripple time*, r, which is the time it takes for a carry signal to pass through a single adder cell

---

[3] We consider $n+1$, not n, bit positions because the result of the addition operation can have up to $n+1$ bits.

[4] On the contrary, if for some position i in the block, $x_i = y_i$, then we may correctly think of the carry signal into the block (no matter it is 0 of 1) as stopping at the **rightmost** such bit position inside the block. This is because the carry signals to the left of that point is independent of the carry into the block. The carry out of the block is completely determined by the value of x_i at the **leftmost** bit position at which $x_i = y_i$. In fact, the carry out of the block is equal to x_i.

(a single bit position). Then there is the *skip time, s*, which is the time a carry signal needs to skip a block, i.e.; the time interval between the moment it is known to the circuit that the carry into a particular block is of a certain value and the moment when the carry out of that block gets set to the same value. The ripple time per bit is certainly constant, and usually the skip time for each block doesn't depend on the block size enough to be regarded as anything other than a constant[5].

Thus, minimizing the total circuit delay involves partitioning the bit positions into blocks in the best possible way. Lehman and Burla [6] found the best partitioning scheme, if one is to try only blocks of equal size. The authors suggest, though, that unequal block sizes may yield a faster circuit. This suggestion is, indeed, correct. Majerski [7] studied how to minimize carry propagation time and the number of skips when unequal blocks are allowed. In both these papers, the ratio s/r (henceforth called ρ) was assumed to be 1, which is often unrealistic. (This ratio ρ depends on the particular circuit technology and the particular implementation involved.) Oklobdzija and Barnes [9] gave a method for computing optimum block sizes (block sizes that yield the fastest circuit), but only for $\rho = 2, 3, 4, 5, 6$, or 7. Yet in practice ρ is usually a non-integer, and sometimes less than 2. Recently Guyot, Hochet, and Muller [4] showed how to compute near-optimum block sizes for all rational values of ρ. The block sizes computed by their method are guaranteed to be such that the total circuit delay falls within $2s$ of the best possible. In the present paper we compute an *optimum* (guaranteed fastest possible, not just nearly so) circuit for any given ρ. Also, ρ can be any real number.

2.2 Deriving the Optimum-Circuit Algorithm

We will study a mathematical model for carry-skip adders and compute the delay based on that model. Two models for carry-skip adders are considered by Guyot, et al. [4] – Model 1 was developed based on the assumption that the skip time doesn't depend on the length of the block being skipped, whereas Model 2 was based on the opposite assumption. However, the dependence of the skip time on the block size was often found to be so small that Model 1 yields results that are no different from those obtained by using the more complicated Model 2. So in the present work we will use only Model 1, which can be described as follows:

A Description of Model 1:

We have mentioned before that the only nonconstant delay is that which is caused by carry propagation. Suppose that in position i, $x_i = y_i$. Then there is a carry out of either 0 or 1 from position i. The value of this carry depends solely on the values of x_i and y_i, and not on the input bits at any other positions. Because of this independence on other positions, the carry is said to be *generated* at position i. Now let j be the first position to the left (left = more significant) of i such that $x_j = y_j$. Then the carry is said to *propagate* through positions $i + 1, i + 2, \ldots, j - 1$, and finally to be *terminated* at position j. We must model and minimize the total carry propagation delay, which is equal to the longest delay (taken over all carry signals) caused by the propagation of a carry signal from the moment after it is generated up until when it is terminated. The delay caused by a carry signal from generation to termination is called the *life* of the signal. The life of a carry signal is divided into 3 parts:

1. Right after generation, the carry has to travel left to the nearest boundary of a block, unless the termination position j is reached first, and,

2. then the carry has to skip over as many blocks as it takes to reach the closest block boundary lying to the right of position j, and finally,

3. the carry travels to its termination at position j.

The delay involved in parts 1. and 3. combined is equal to the number of positions traveled multiplied by the ripple delay, r, whereas the delay caused by part 2. is given by the number of blocks skipped multiplied by the skip time, s.

To achieve the objective of minimizing the longest possible carry signal life, we have to partition the $n + 1$ bit positions into blocks in the best way. (Recall that we consider the total number of bit positions to be $n + 1$.) To see what "best" means, let us introduce a method of representation of how the bit positions are partitioned. This representation is equivalent to the one invented by Guyot, et al. [4].

Imagine an X-Y plane such that the Y axis points upwards as usual but the X axis points to the left instead of to the right. Again, let the blocks be numbered $0, 1, 2, \ldots$ from right to left as in Figure 1. Now, if block b contains $m(b)$ bit positions, then put a special marker at the point $(b, m(b))$ on the X-Y plane. See Figure 2. Let M be the set of all such markers, and let μ be the set of all points b for which there is a marker with X coordinate b. Note that $\sum_{\forall b \in \mu} m(b) = n + 1$. Note also that the time needed for a carry signal to ripple from the first to the last bit in a block of size $m(b)$ is $r(m(b) - 1)$. Guyot, et al., however, used $rm(b)$. The difference doesn't affect the results in any way.

Theorem 1 *The longest possible carry life in a circuit is given by the expression*

$$\max_{\substack{\forall b_2, b_1 \in \mu \\ b_2 > b_1}} \{r(m(b_2) + m(b_1) - 2) + s(b_2 - b_1 - 1)\}.$$

Proof. This theorem follows immediately from the previous discussion on the life of a carry. ∎

Some important isosceles triangles will now be introduced, because we'll find the best partitioning of the

---

[5] See [4] for a study of the effect of block size on skip time.

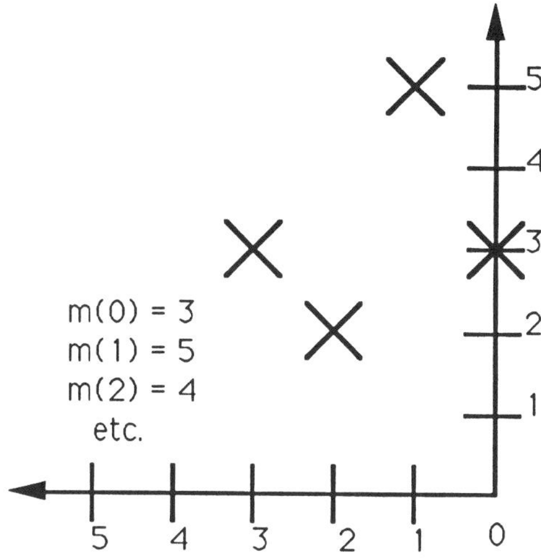

Figure 2: A representation of blocks

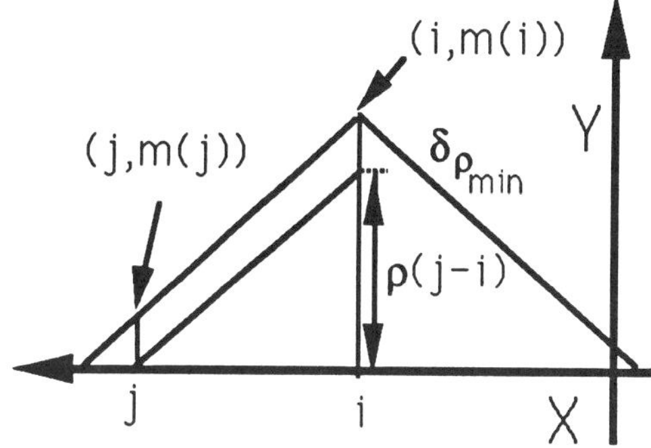

Figure 3: Two markers, one on each side of $\delta_{\rho\min}$

bit positions into blocks by finding the lowest height isosceles triangle that contains all the markers in the set M.[6] Let $\rho = s/r$, and let Δ_ρ be the class of all isosceles triangles such that the two sloped sides have slopes ρ and $-\rho$, and such that the horizontal side (the base) lies on the X axis. Let $\delta_{\rho\min}$ be the minimum-height member of Δ_ρ that contains all the markers of M. It is clear that $\delta_{\rho\min}$ exists. That $\delta_{\rho\min}$ is unique will be shown in Lemma 1, below. Let $H(\delta)$ denote the height of any isosceles triangle δ.

Lemma 1 *The triangle $\delta_{\rho\min}$ is unique.*

Proof. Assume the contrary – that there are at least two distinct triangles of minimum height that contain all the markers of M. Let two of them be called δ_1 and δ_2. Let $\delta_3 = \delta_1 \cap \delta_2$. Note that δ_3 is a triangle of smaller height than δ_1 and δ_2, has sides of slope ρ and $-\rho$, and contains all the markers of M. This observation contradicts the assumption made at the beginning of the proof. ∎

The following theorem is important because, as we will later see, it implies that in most cases, $\delta_{\rho\min}$ completely determines an optimum-speed circuit.

Theorem 2 *If the sloped sides of $\delta_{\rho\min}$ each touch at least one marker, and if at least one of these sides touches a marker somewhere other than at the apex of the triangle, then the worst carry signal life is given by $2rH(\delta_{\rho\min}) - 2r - s$.*

Proof. Assume that the hypothesis of the theorem is true. Then there must be natural numbers i and

---

[6] This statement isn't quite accurate, but serves as a good intuitive explanation of the reason of existence of these isosceles triangles. As the paper progresses, the reader will see more accurately how these triangles are used for obtaining the best partitioning of the bit positions.

j such that the point $(i, m(i))$ lies on the right side (right sloped boundary) of $\delta_{\rho\min}$ and such that the point $(j, m(j))$ lies on the left side of the same triangle. In addition, at least one of those two points lies off the apex. Without loss of generality assume that $(j, m(j))$ does, as shown in Figure 3. Let us look at the case in which $(i, m(i))$ is at the apex. Consider the situation where a carry signal has to travel from the rightmost bit of block i to the leftmost bit of block j. This situation obviously can occur. By Theorem 1, the life of this carry signal is equal to $r(m(j) + m(i) - 2) + s(j - i - 1)$, which is the same as $r(m(j) + m(i) + \rho(j - i)) - 2r - s$. From Figure 5 we see that $m(j) + \rho(j - i) = H(\delta_{\rho\min})$. It follows immediately that the carry signal life is $2rH(\delta_{\rho\min}) - 2r - s$. The case in which both $(i, m(i))$ and $(j, m(j))$ are off $\delta_{\rho\min}$'s apex can be dealt with similarly. Note finally that no carry signal can have a life longer than does a carry signal discussed above. ∎

If $\delta_{\rho\min}$ touches no markers other than one (which would be at the apex), then some definitions are needed, as follows: (Refer to Figure 4.) Let l_1 and l_2, respectively, be lines parallel to the left and the right sides of $\delta_{\rho\min}$ so that l_1 and l_2 both touch some markers, and the shaded triangle sided by l_1, l_2, and part of $\delta_{\rho\min}$'s base contains all markers excepts the one at the apex of $\delta_{\rho\min}$. Assume that l_1 is farther from the left side of $\delta_{\rho\min}$ than l_2 is from the right side of the same triangle. (The other cases can be treated similarly.) Draw a vertical line V through the apex of $\delta_{\rho\min}$, and let $A, B, C, D,$ and G be points of intersections among previously-defined lines, as shown in Figure 4. E and F will be explained in the proof of Theorem 3, coming up next. Let $x(P)$ and $y(P)$, respectively, denote the X and the Y coordinates of a point P.

Theorem 3 *If $\delta_{\rho\min}$ touches a marker at the apex (D) but touches no marker at any other place, then the worst possible total carry propagation delay of the circuit is given by the expression $r(y(C) + y(D)) - 2r - s$.*

Proof. By the definition of l_1 and l_2, there must be some marker on l_2. Let E be the location of this

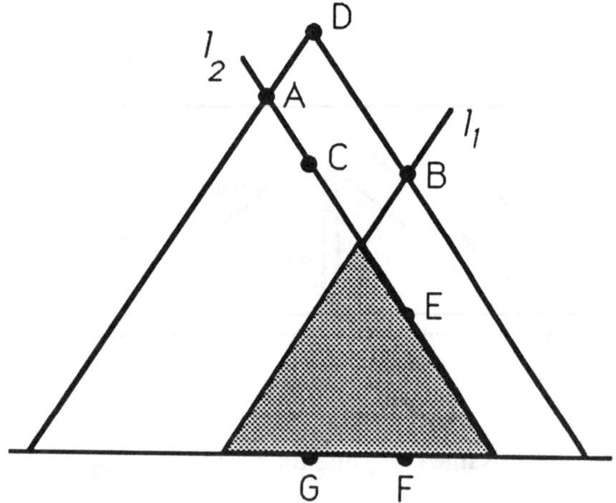

Figure 4: The case where $\delta_{\rho_{\min}}$ touches the markers only at the apex

marker, and let F be the point on the X axis vertically aligned with E. It is easy to see that the longest carry life in the entire circuit is the life of a carry signal that is generated at the rightmost bit position in the block numbered $x(E)$ and is terminated at the leftmost bit position in the block numbered $x(D)$. Thus, the longest carry signal life is equal to $r(y(C) + y(E) - 2) + s(\|\mathbf{FG}\| - 1)$. Using similar geometric reasoning as in the proof of Theorem 2, we can see that this expression is equal to the expression $r(y(C) + y(D)) - 2r - s$. ∎

Now define Δ_ρ^h to be the set of all members of height h of Δ_ρ.

Lemma 2 *For every circuit with delay no more than $2rh - 2r - s$, there exists a member δ of Δ_ρ^h such that at most one marker of the circuit lies strictly outside δ. In case that for every member of Δ_ρ^h, there is at least one marker outside the member, there exists a $\delta \in \Delta_\rho^h$ such that there is exactly one marker outside δ, and such that the marker outside δ is vertically aligned with the apex.*

Proof. Pick any circuit with no more than $2rh - 2r - s$ units of delay. Draw an arbitrary member of Δ_ρ^h — call it δ. If the circuit doesn't have more than one marker outside δ, and the marker outside δ is vertically aligned with the apex, then the lemma is already proven. Otherwise there are two cases to consider:

(a) There are two markers at positions i and j such that these markers are vertically above the right and the left sides of δ, respectively. (There might be more than 2 markers outside δ.) One of these markers may be vertically aligned with δ's apex.

(b) All markers are above the same side of δ and no marker is vertically aligned with δ's apex.

In case (a) we can see, using a similar geometric argument as in the proofs of Theorems 2 and 3, that the longest possible carry life is longer than $2rh - 2r - s$. So case (a) can't occur.

In case (b), assume without loss of generality that all markers are above the left side. Slide δ to the left. Eventually one of the following events must occur:

1. All markers are in δ.

2. Some marker, say $(i, m(i))$, which has always remained outside δ, is now aligned with the apex of δ.

In case event (1) occurs, the lemma is clearly proven. In case event (2) occurs, it is easy to see that at the time of such occurrence, no other marker than $(i, m(i))$ can be outside of δ, or else the circuit's maximum delay would be greater than $2rh - 2r - s$. Hence the lemma is also proven in this case. ∎

Theorem 4 *Let h be a positive real number such that the members of Δ_ρ^h has base length at least 2. Then, among all circuits with largest n (with $s/r = \rho$) of delay up to $2rh - 2r - s$, there is at least one with one of the following properties:*

1. *All the markers are contained in[7] some member δ of Δ_ρ^h, or,*

2. *For some member δ of Δ_ρ^h, only 1 marker of the circuit is strictly outside δ. This marker is vertically aligned with the apex of δ, and is less than 1 unit above it.*

Proof. Pick any largest circuit c (largest = largest number n of operand bits) with delay up to $2rh - 2r - s$, and suppose that there is *no* member of Δ_ρ^h that will contain all of c's markers. Then by Lemma 2, there must be a $\delta \in \Delta_\rho^h$ that contains all but one marker, and this marker is vertically aligned with the apex of δ. We will modify c to obtain a new circuit, c', such that c' has the same size and delay upper bound as mentioned above for c, but either has all the markers in δ or has only one marker outside δ, such that this marker is vertically aligned with the apex and is less than 1 unit above it. If we can do this then we will have proven the theorem. Understanding how to modify c involves looking at Figure 5. Figure 5 is just Figure 4 with δ and its apex H added in, and with unneeded labels G, F, and E removed. By Theorem 3, the given circuit has delay $r(y(C) + y(D)) - 2r - s$. Since the delay is at most $2rh - 2r - s$, we have $h \geq \frac{y(C)+y(D)}{2}$. The strategy now is to shorten the block numbered by $x(H)$ (the X coordinate of the point H)

---

[7]Not strictly.

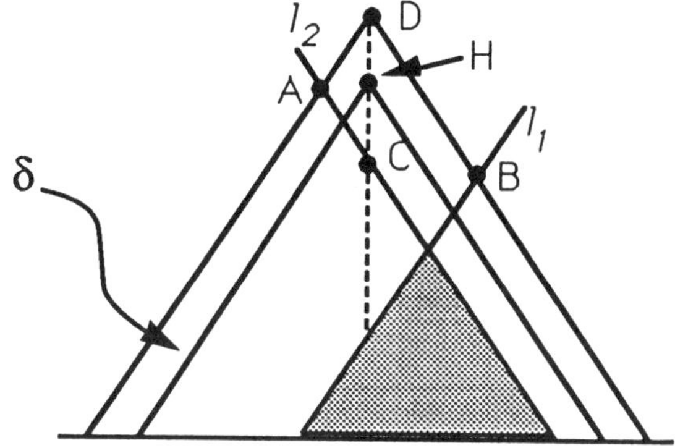

Figure 5: Related to obtaining a new circuit with all the markers or all but one in δ.

just enough so that $\lfloor m(x(H)) \rfloor \leq h$, and then to put the bit positions removed from that block elsewhere. Since $h \geq \frac{y(C)+y(D)}{2}$, it follows that $y(H) - y(C) \geq y(D) - y(H)$. This last relation is very important, because it means that roughly, we can remove $y(D) - y(H)$ bit positions from block $x(H)$ and insert them in any other block. Exactly speaking, though, this is not always the case, because $y(D) - y(H)$ and $y(H) - y(C)$ are not necessarily integers.

There are 2 cases to consider. The first case is when $y(D) - y(H) < 1$. Then there is nothing to do, since the conclusion of the theorem is already true. So let us consider the second case, in which $y(D) - y(H) \geq 1$. In this case we can move $\lfloor y(D) - y(H) \rfloor$ bit positions from block $x(H)$ to some other block. From the geometry of Figure 5, it is clear that if this other block exists, then this move can be done. We have thus assumed the existence of 2 blocks in all. This is fine, since we have assumed that the base of δ is of length at least 2, and so the base covers at least 2 integer points on the X axis. ∎

Our main goal is that when n is given, we would like to design a minimum-delay n-bit circuit. But it follows from Theorem 4 that for any delay d, some maximum-n circuits of delay up to d have all or all but one of their markers in some $\delta \in \Delta_\rho^h$ (provided that the base of δ is larger than 2 units). This will now be stated formally as follows:

Theorem 5 *Let a real number d be given. If $d \geq s - 2r$, then among all the maximum-n circuits with the skip time/ripple time ratio equal to the given ρ, there is one such that either*

1. *all the circuit's markers fit into some member of Δ_ρ^h, where $h = (d + 2r + s)/2r$, or,*

2. *there exists a member δ of Δ_ρ^h (again, where $h = (d + 2r + s)/2r$) such that only 1 marker is outside δ. Moreover, this marker is vertically aligned with the apex of δ, and is less than 1 unit above it.*

Such a circuit will be of delay at most d.

Proof. If $d \geq s - 2r \geq 2\rho r - 2r - s$ then $h = (d + 2r + s)/2r \geq 2\rho r/2r = \rho$. The base length b of each member of Δ_ρ^h is given by the expression $\frac{h}{b/2} = \rho$, and so $b = 2h/\rho \geq 2$. Thus the corollary follows from Theorem 4. ∎

In the next section, we will use Theorem 5 and the other theorems to design an algorithm that outputs optimum-speed carry-skip adders.

3 The Procedure

We will now describe the algorithm such that, when given the number n of operand bit positions, will compute a partitioning of the positions into blocks[8] so as to yield the minimum worst-case circuit delay. If the number of blocks in all optimum circuits is less than 2 (the trivial and unrealistic case), then the optimum circuit is simply the only existing circuit. Otherwise, an optimum circuit can be found by using the results of the previous section, which culminated in Theorem 5. The trick is simply to find the minimum delay $d_{\min}$ such that there exists a circuit of at least n bit positions with delay $d_{\min}$. The technique used will be a "binary search" on the amount of delay. That is, we will first start with a range of delays (real numbers) $[d_{\text{low}}, d_{\text{high}}]$ such that we know that $d_{\min}$ is within this range. Then compute what the size (size = maximum number of result bits) of the largest circuit is that will possess a worst-case delay of up to d_{mid}, where $d_{\text{mid}} = \frac{d_{\text{low}} + d_{\text{high}}}{2}$. Then halve the range of delays at d_{mid} in such a way that the new range will still contain $d_{\min}$. It remains only to show how to determine the size of the largest circuit when a delay figure is given. This will now be explained.

Let us suppose that we have drawn the picture of some member δ of Δ_ρ^h for some arbitrarily chosen h. Then the size of the largest circuit whose markers are all contained in δ is equal to the number of points in the set $\delta \cap S$ where $S = \{(x,y) | x \in \mathbf{Z} \text{ and } y \in \mathbf{Z}_+\}$. Moreover, we can read off the design of one such circuit from the picture by inspection! The circuit we mean is the one whose markers correspond to the maximal points[9] of S in δ.

So if there is a largest circuit for the given delay figure such that this circuit's markers are all contained in some triangle in Δ_ρ^h for the appropriate h, then to compute this circuit we just have to find which triangle contains the most points of S. This can be done by starting with some arbitrary triangle $\delta \in \Delta_\rho^h$, and then sliding the base of δ along the X axis. During the slide, points in S enter and leave δ. Is this slide a finite process? To answer this question, notice two things:

---

[8]As noted earlier, we actually partition $n + 1$, not n, bit positions into blocks, because we must accommodate the most significant bit of the result.

[9]Points with maximum Y coordinate.

First, $|\delta \cap S|$ is a periodic function of the position[10] of δ, with period 1; and second, the events that points of S enter and leave δ are discrete events. The first observation means that we only need to slide δ by 1 unit, in either direction we choose. Because of this and the second observation, the slide is indeed a finite process.

Now there is a possibility that the largest circuits are such that the markers are not all contained in the relevant member of Δ_ρ^h. Then it is not hard to see that in this case, there is only one largest circuit. This circuit can be computed by drawing a member δ of Δ_ρ whose apex is vertically aligned with an integer point on the X axis. Suppose the apex lies at the point which we will call (i, y). By Theorem 4, the apex is less than 1 unit below the marker of block i. Since the marker must be at an integer point, y can't be an integer. Let $j = \lceil y \rceil$. Then the circuit is simply the one whose markers are the maximal points of S in δ, except that the marker for block i is taken to be at (i, j). Compute a circuit by this latter precedure anyway whether its delay is within the amount specified or not, and whether it is a largest circuit or not. If the delay is too great, or if the circuit is not a largest one, then just discard it.

The sliding part of the process is detailed in Table 1.

4   Test Results

4.1   Comparison of This Paper's Circuits with Guyot, et al.'s Circuits

As mentioned before, the author has programmed both his algorithm and Guyot, et al.'s, and has run them on a variety of inputs. Table 2 compares the circuits resulting from some test runs. It is not really understood when Guyot, et al.'s algorithm performs well, and when it does relatively poorly, except that it seems to perform poorly when ρ is only a small amount away from a round number, like 1.01 or 1.0001. There are times when the circuits produced by their algorithm are as good as our results, and there are times when their circuits possess carry propagation delays that are more than 11% worse than ours, which are guaranteed to be optimum-speed circuits. All circuit delays are all calculated according to the formulas presented in this paper.

4.2   Comparison with Oklobdzija and Barnes' Result

As mentioned in the introduction, Oklobdzija and Barnes' algorithm [9] can be used for finding optimum-speed carry-skip adders, but only for integer values of ρ. If ρ isn't an integer, then their algorithm can still find a circuit, but such a circuit can be far from optimum. In the following example, their algorithm produces a circuit which has a carry propagation delay that is 16.08% slower than the carry propagation delay of the optimum circuit, produced by the algorithm in our paper:

<hr>

[10] The X coordinate of the left end of the base.

```
begin
place δ so that its left vertex is at 0;
pool := number of markers in δ;
left1 := number of markers on left side of δ;
left2 := 0; right := 0; oldmax := 0;
deltaposition := 0; {Initial position of δ}
flag := true;
while flag do begin
{evaluate number of markers in triangle}
        pool := pool − left2;
        pool := pool + right;
        if pool > oldmax then
            begin
                oldmax := pool;
                deltaposition := X coordinate of
                            left vertex of δ;
            end;
{slide triangle}
        slide δ to the right until a marker is hit;
        if the left corner of δ is at or
        beyond x = 1 then flag := false;
        left2 := left1;
        left1 := no. of markers on left side of δ;
        right := no. of markers on right side of δ;
end; {while}
output the circuit determined by deltaposition;
end;
```

Table 1: Triangle sliding program

Suppose $\rho = 1.334$ and the number of bits is 32. Then our algorithm produces the circuit 1 2 3 5 6 6 4 3 2, with a carry propagation delay of 10.3380r units. They, on the other hand, must approximate ρ by either 1 or 2. If they were to use $\rho = 1$, then the circuit they would obtain is 1 2 3 4 5 6 5 4 3 2 1, with a carry delay of 12.006r units. On the other hand if they were to use $\rho = 2$, then the circuit they would obtain is 2 4 6 8 6 4 2, with a carry delay of 12.000r units. So they should use $\rho = 2$, which gives a deterioration of 16.08% over our optimum circuit's speed. (We calculate all carry delays according to our theorems. When calculating the delays of Oklobdzija and Barnes' circuits, we must remember that ρ is 1.334, *not* 1 or 2.)

5   Conclusion

In this paper, we have presented a method of designing the fastest one-level carry-skip adders. These adders are almost the fastest known adders, and are of clean design. It is possible, however, to apply the carry-skip principle once or several more times to obtain even faster, but somewhat more complicated, circuits. In [4] a procedure for finding near-optimum two-level carry-skip circuits was presented. An important next step in this research area would be to find a way to compute optimum two-level carry-skip adders. It shouldn't be too difficult to see that the method presented in this paper can be extended for that purpose.

bits	ρ	This paper's circuit's carry delay	Guyot, et al.'s circuit's carry delay	Guyot, et al.'s % deviation from optimum	This paper's circuit	Guyot, et al.'s circuit
32	1.0001	9.0005r	10.0000r	11.105%	1 2 3 4 5 6 5 3 2 1	1 2 3 4 6 6 4 3 2 1
32	1.025	9.1250r	10.0000r	9.589%	1 2 3 4 5 6 5 3 2 1	1 2 3 4 6 6 4 3 2 1
32	1.05	9.2500r	10.0000r	8.108%	1 2 3 4 5 6 5 3 2 1	1 2 3 4 6 6 4 3 2 1
32	5.5	19.0000r	20.0000r	5.2632%	2 8 13 7 2	5 10 11 6
64	2.0001	19.0007r	20.0001r	5.2598%	2 4 6 8 10 11 9 7 5 2	1 3 5 7 9 12 10 8 5 3 1
128	2.0001	28.0013r	29.0006r	3.5688%	2 4 6 8 10 12 14 16 14 12 10 8 6 4 2	2 4 6 9 11 13 15 16 14 12 10 7 5 3 1
128	.85	18.8500r	19.0000r	0.7958%	1 2 3 4 4 5 6 7 8 9 10 10 10 9 8 7 6 5 4 4 3 2 1	1 2 2 3 4 5 6 7 7 8 9 10 10 9 8 7 7 6 5 4 3 2 2 1
64	.85	12.8r	12.8r	0.0000%	1 2 3 4 5 5 6 7 7 6 5 4 3 3 2 1	1 2 3 4 5 5 6 7 7 6 5 4 3 3 2 1

Table 2: Comparing the circuits produced by the algorithms of Kantabutra and of Guyot, et al.

However, it seems that such an extension would be very complicated.

6 Acknowledgments

The author would like to thank Deborah L. Arnold for technical assistance in this project.

References

[1] A. Burks, H. H. Goldstine, and J. von Neumann, "Preliminary discussion of the logic design of an electronic computing instrument," Instit. Adv. Study, princeton, NJ, 1946 (reprinted in C. G. Bell, *Computer Structures: Readings and Examples,* New York: McGraw-Hill, 1971).

[2] J. J. F. Cavanagh, *Digital Computer Arithmetic, Design and Implementation.* New York: McGraw-Hill, 1984.

[3] Pak K. Chan and Martine D.F. Schlag, "Analysis and design of CMOS manchester adders with variable carry-skip," in *Proc. 9th IEEE Symp. on Comput. Arithmetic,* 1989.

[4] A. Guyot, B. Hochet, and J.-M. Muller, "A way to build efficient carry-skip adders," *IEEE Transactions on Computers,* October 1987.

[5] J. P. Hayes, *Computer Architecture and Organization, 2nd ed.* New York: McGraw-Hill, 1988.

[6] M. Lehman and N. Burla, "Skip techniques for high-speed carry propagation in binary arithmetic units," *IRE Trans. Electron. Comput.,* p. 691, Dec 1961.

[7] S. Majerski, "On determination of optimal distributions of carry skips in adders," *IEEE Trans. Electron. Comput.,* vol. EC-16, Feb. 1967.

[8] Yu. Ofman, "On the algorithmic complexity of discrete functions," *Soviet Physics –Doklady,* vol. 7, no. 7, January 1963.

[9] V. G. Oklobdzija and E. R. Barnes, "Some optimal schemes for ALU implementation in VLSI technology," in *Proc. 7th Symp. Comput. Arithmetic,* 1985.

[10] Silvio Turrini, "Optimal group distribution in carry-skip adders," in *Proc. 9th IEEE Symp. on Comput. Arithmetic,* 1989.

Delay Optimization of Carry-Skip Adders
and Block Carry-Lookahead Adders

Pak K. Chan, Martine D.F. Schlag*
Computer Engineering
University of California, Santa Cruz
Santa Cruz, California 95064

Clark D. Thomborson
Department of Computer Science
University of Minnesota
Duluth, Minnesota 55812

Vojin G. Oklobdzija
IBM T.J. Watson Research Center
Yorktown Heights, NY 10598

Abstract

The worst-case carry propagation delays in carry-skip adders and block carry-lookahead adders depend on how the full adders are grouped structurally together into blocks as well as the number of levels.

We report a multidimensional dynamic programming paradigm for configuring these two adders to attain minimum latency. Previous methods are applicable only to very limited delay models that do not guarantee a minimum latency configuration. Under our delay model, critical path delay is calculated not only taking into account the intrinsic gate delays, but also the fanin and fanout contributions.

1 Introduction

The worst-case carry propagation delays in carry-skip adders depend on how the full adders are grouped together (into blocks). The problem of configuring carry-skip adders to minimize the carry propagation delay has been the subject of several papers. Lehman has shown that carry-skip adders with variable-size blocks are faster than adders with fixed-size blocks [1]. Later, Majerski suggested that multilevel implementation of the variable-block-size carry-skip adders would provide further improvement in speed [2]. The optimization technique developed for the choice of block sizes by Majerski is limited to a specific ratio between the carry-generate and carry-skip propagation delay. Almost two decades later, Oklobdzija and Barnes developed algorithms for determining *near-optimal* block sizes for one-level and two-level implementations, and a generalization of their method was given by Guyot *et al* [3, 4]. Their algorithms have very elegant geometrical interpretations but do not guarantee optimality of the design. Moreover, their algorithms work only if the carry-skip propagation delay is a constant. In [5] it is noted that in CMOS Manchester adders with carry-skip, the carry-skip propagation delay is not necessarily a constant, but depends on the number of bits in the adder. Chan and Schlag developed a polynomial time algorithm to configure block sizes to attain minimum latency for one-level carry-skip adders under a linear carry-skip delay model. Simultaneously, an indirect enumeration approach was taken by Turrini to generate (multilevel) block distributions containing the maximum number of bits under a specified delay constraint [6]. Unfortunately, this approach is applicable only to constant carry-skip delay models, since it constructs a configuration from the top down by calculating delay constraints for lower-level blocks without knowing the number of bits encompassed in these blocks; when the lowest level is reached each block is filled with the maximum number of bits satisfying its delay constraints.

The idea of varying block sizes to further reduce delays was also suggested in [7], where an exhaustive search was employed to search for an optimum block carry-lookahead adder. Much earlier, Montoye and Cook used an analytical delay model to guide an iterative search for area-time optimal parallel prefix adders generated by a binary recursion [8]. They supplied no runtime analysis of their search technique, although they did indicate that an optimal 34-bit adder could be found in 30 minutes of IBM 3033 time. Wei and Thomborson [9] devised a dynamic programming technique that found, in $O(n^2 h^2)$ time, *all* area-time optimal parallel-prefix adders in a class generated by a binary recursion similar to Montoye's. Here, h is the height of the minimum-delay adder of data width n. They found optimal 66-bit adders in a few seconds of SUN-3 CPU time.

In this paper, we formulate the problems of configuring carry-skip adders and variable-block-size block carry-lookahead adders as dynamic programs. The resulting dynamic programs have multidimensional objective functions. It is thus necessary to carry forward a list of optimal structures from each stage of the dynamic program. In the traditional (unidimensional)

---

*Supported in part by NSF Presidential Young Investigator Grant MIP-8896276

154

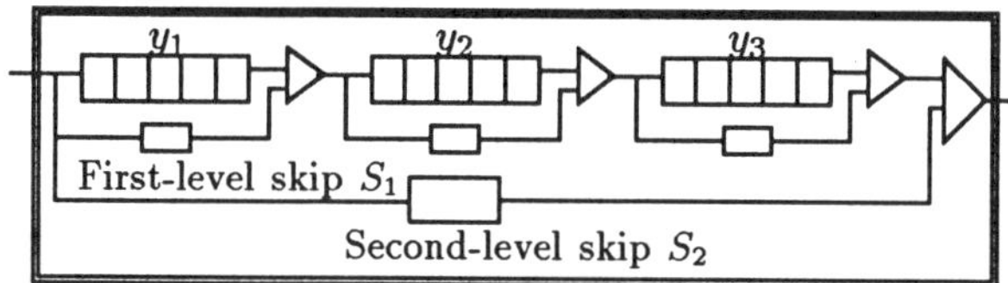

Figure 1: Forming a stage from blocks

dynamic program, only a single optimum structure is carried forward. The existence of multidimensional dynamic programs was noted in an early paper by Dantzig [10]. Weingartner [11] was apparently the first to suggest that this would be an effective method of solving multidimensional knapsack problems. Subsequent researchers [12, 13, 14] refined Weingartner's algorithm, adding more sophisticated data structures and list-pruning strategies.

The multidimensional dynamic programming formulations to configure carry-skip and block carry-lookahead adders, in this article have not been reported elsewhere in form, scope, or generality [1]. In contrast to previously-published optimization techniques, our method immediately generalizes to a wide class of gate delay models and is guaranteed to find minimum latency circuits.

2 Carry-skip adders

2.1 Constructing a stage from blocks

We group several full adders together to form an adder block. Each block has a block-level carry skip mechanism S_1, which can be implemented with a multiplexor selected by the group propagate. The basic structure of a *stage* of a 2-level carry-skip adder is illustrated in Fig. 1. Each stage encompasses several blocks, and contains a second-level carry skip mechanism. The pertinent components of carry-propagation delays in a block are shown in Fig. 2.

2.2 Glossary of terms

The basic notations used in this section are listed below. The meanings of the notations are illustrated in Figs. 1 and 2 .

1. $I(y)$ — internal-carry delay, the maximum delay it takes a carry to generate within a block of y full-adder units and assimilate within the block.

2. $G(y)$ — carry-generate delay, the maximum delay it takes a carry to generate within a block of y full-adder units. This also includes the time it takes a carry to propagate through the buffer

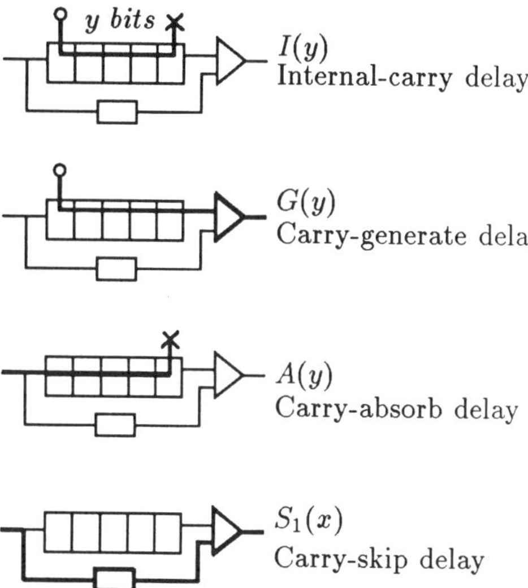

Figure 2: Characterization of delays at the block level

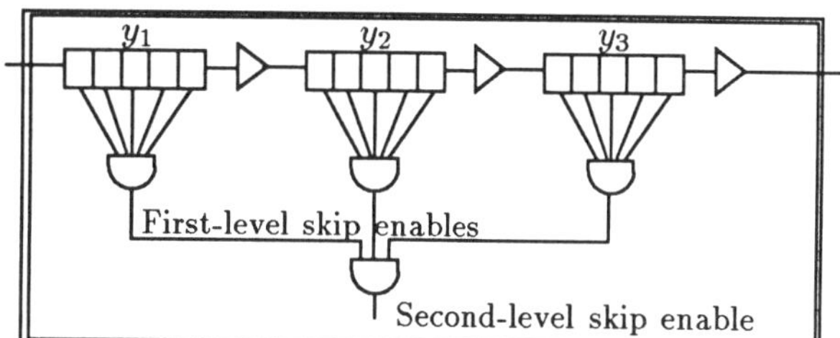

Figure 3: Skip enable generation, inputs to the first level AND gates are the carry propagates p_i

(the triangle). Typically, the buffer computes the logical 'or' of its two carry-input signals.

3. $A(y)$ — carry-assimilate delay, the maximum delay it takes a carry to enter a block of y full-adders and assimilate within the block.

4. $S_l(y)$ — lth level carry-skip delay, the time it takes a carry to skip through y full-adder units using the lth-level carry skip mechanism. For $l = 0$, this is the time for a carry to propagate through a block consisting of y full-adder units. For $l \geq 1$, this is the time to compute the logical 'and' of a carry-in signal with the skip-enable signal of the block. This also includes the time it takes a carry to propagate through the buffer.

5. $Set\_up_l(y)$ — lth level setup time, the amount of time it takes to enable the skip circuitry at level l, see Fig. 3. This reflects the delay to generate a group propagate for y bits ($\Pi_{i=1}^{y} p_i$, where p_i is the carry propagate of the ith full adder).

---

[1] The difficulty of using the dynamic programming technique to solve optimization problems, as noted by Dreyfus [15], lies in the formulation.

3 A 2-D dynamic programming formulation for finding minimum latency configurations

In order to present the idea in a readily-understandable form, we start the discussion with a two-dimensional optimization problem based on one-level carry-skip adders. The method we derive in this section delivers the same results as a previously-published algorithm [5], but at a much higher computational cost. However, this section's method can be easily generalized to more complicated timing models and to higher-dimensional optimizations.

3.1 Problem statement: one-level carry-skip adder

Let y_k denote the number of bits in block k. We say that a vector $\vec{y} = (y_1, y_2, ..., y_m)$ is an m-block *configuration* of a one-level n-bit adder if $\sum_{i=k}^{m} y_k = n$ and all y_k are positive integers. Let $\mathcal{C}_n$ be the set of all configurations of one-level n-bit adders. We shall assume that all skip circuitries are setup at time zero. The effect of nonzero setup time is treated in Section 3.5. The minimum-latency design problem for a one-level carry-skip adder can now be stated as

Given timing models for internal-carry $I()$, carry-generate $G()$, carry-assimilate $A()$, and carry-skips $S_0()$ and $S_1()$, find the configuration $\vec{y}^* \in \mathcal{C}_n$ with minimum latency.

The carry-propagation delay between blocks i and j of a configuration $\vec{y}$ is

$$D(\vec{y}, \alpha, \beta) = G(y_\alpha) + \sum_{k=\alpha+1}^{\beta-1} S_1(y_k) + A(y_\beta) \; ; \; \text{for } \alpha < \beta \; ,$$

$$D(\vec{y}, \alpha, \alpha) = I(y_\alpha) \; .$$

The worst-case carry-propagation delay of a configuration $\vec{y}$ is therefore $D(\vec{y}) = \max_{1 \leq \alpha \leq \beta \leq m} D(\vec{y}, \alpha, \beta)$. Then our problem is to find a minimum worst-case delay configuration $\vec{y}^*$ for given carry-generate, carry-assimilate and skip delay functions,

$$D_n \equiv D(\vec{y}^*) = \min_{\vec{y} \in \mathcal{C}_n} D(\vec{y}). \qquad (1)$$

3.2 Algorithm: one-level carry-skip adder

We refer to i-bit, j-block carry-skip adders as (i, j)-adders. Note that $j \leq i$, since each block must have at least one bit. Also note that, for small blocks, rippling through a single block may be faster than using a one-level skip. For this reason we amend the problem formulation of section 3.1 to allow the possibility of having an initial and/or final block without a skip. During our construction of an optimal configuration we

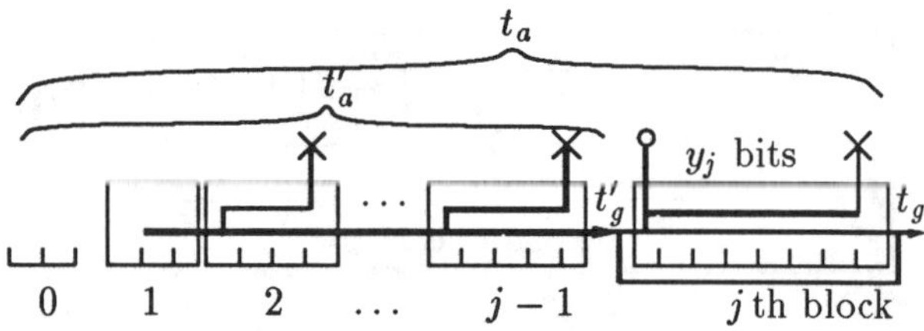

Figure 4: Appending a new block

shall consider only (i, j)-adders consisting of an initial (possibly empty) block with no skip, followed by j non-empty consecutive blocks. Given (i, j) there are $\binom{i}{j}$ such adder configurations, since we have the freedom to distribute $i - j$ bits among $j + 1$ blocks. A final step will consider adding a block to the end with no skip. We use a pair (t_a, t_g) to characterize the worst-case carry propagation delays of an (i, j)-adder, where

- t_a is the worst-case delay of any "carry chain" that terminates *before* or *at* block j, and

- t_g is the worst-case delay of any "carry chain" that emerges from block j.

We shall construct for each i and j, a list $t(i, j)$ of pairs (t_a, t_g) for all (i, j)-adders. The basis for the dynamic programming is

$$t(i, 0) = (\max\{A(i), I(i)\}, \max\{G(i), S_0(i)\}); \text{ for } 0 \leq i \leq n,$$

and for $i \geq j \geq 1$, $i - j + 1 \geq y_j \geq 1$, the minimal worst-case delays of (i, j)-adders are formed by composing $(i - y_j, j - 1)$ adders and a new jth block with y_j bits. For each (t'_a, t'_g) in $t(i - y_j, j - 1)$, we construct a pair (t_a, t_g) by:

$$t_a = \max\{t'_a, t'_g + A(y_j), I(y_j)\} \qquad (2)$$

$$t_g = \max\{G(y_j), t'_g + S_1(y_j)\} \qquad (3)$$

We then solve for $D_n = \min_{\vec{y} \in \mathcal{C}_n} D(\vec{y})$ by 2–D dynamic programming in a tableau that retains, for each $t(i, j)$, a list of the minimal (t_a, t_g) pairs for all (i, j)-adders. The list in tableau cell (i, j), for $j \geq 1$, is obtained by using the recursion above to process the lists in cells $(i - y_j, j - 1)$ for all "last block" sizes $i - j + 1 \geq y_j \geq 1$. Once the entire tableau for $1 \leq i \leq n$ and $0 \leq j \leq i + 1$ has been computed, the lists in column i are concatenated into one list $T(i)$, and a final block of $n - i$ bits without a skip is added. D_n is the minimum of the set

$$\cup_{i=0}^{n} \{\max(t_a, t_g + A(n-i), G(n-i), t_g + S_0(n-i)) | (t_a, t_g) \in T(i)\}.$$

This algorithm delivers the correct minimum for any non-negative $G()$, $A()$, $S_0()$ and $S_1()$ functions, but it potentially requires exponential time and space. The next section addresses this issue by presenting techniques to prune the search and limit the number of configurations generated.

There is a reason to expect good performance, however. If the t_a and t_g values in the retained lists are independently distributed, then each list will have $O(\log n)$ elements with high probability [16, 17]. In this case, the optimization algorithm for an n-bit adder will run in $O(n^3 \log^3 n)$ time, with high probability. Positive correlation among the t_a and t_g values would shorten the lists and hence the runtimes; any negative correlation would lengthen the lists. We expect to see a slight positive correlation for any reasonable delay model, so we believe this method will prove feasible for optimizing adders with hundreds of bits.

3.3 Number of configurations in the tableau

The maximum number of configurations for cell (i, j) in the tableau is the binomial coefficient $\binom{i}{j}$. Fig. 5 shows the potential number of configurations for any 10-bit carry-skip adder. There are ten possible configurations for a 10-bit one-block adder because of the possibility of an initial block (0th block) with no skip. This initial block can hold zero to nine bits. Fortunately, many configurations can be thrown away using the following pruning techniques.

- In each tableau cell $t(i, j)$ the non-dominated pairs must be retained. For example, let (t_a, t_g) and (t'_a, t'_g) be pairs in cell $t(i, j)$ of the tableau. We say (t_a, t_g) is dominated if either ($t_a > t'_a$ and $t_g \geq t'_g$) or ($t_a \geq t'_a$ and $t_g > t'_g$). From equations (2) and (3) it is clear that any pair constructed by adding another block to (t_a, t_g) will be sub-optimal to the corresponding pair generated by adding the same block to (t'_a, t'_g), and hence the former can be discarded. If only one optimal configuration is desired, further pruning can be achieved by breaking ties arbitrarily and discarding all but one of the pairs involved (pruning by domination [11]).
- Once the $t(n, j)$ cell is filled in we can examine its entries and determine the minimum worst-case delay of an (n, j)-adder. This delay, $D_n(j)$, is an upper bound on the final delay, D_n, and can be used to discard any pair generated with either $t_a \geq D_n(j)$ or $t_g \geq D_n(j)$. In order to take full advantage of this bound, we fill each row of the tableau from right to left so that $D_n(j)$ can be used in filling the rest of the row (pruning by fathoming [13]).
- Finally, it is not necessary to fill in the entire tableau. If each pair in $t(i, j)$ is sub-optimal or equal to a pair in $t(i, j-1)$ then $j-1$ is the optimal number of blocks for i bits. There is no point in computing column i above row $j - 1$.

Fig. 5 shows the potential number of configurations for any 10-bit carry-skip adder. However, upon using the aforementioned pruning techniques the number of configurations can be drastically reduced. Fig. 6 shows the actual number of configurations generated in each cell of the tableau for the delay model

$$G(y) = y \quad A(y) = y \quad S_0(y) = y \quad S_1(y) = 1.$$

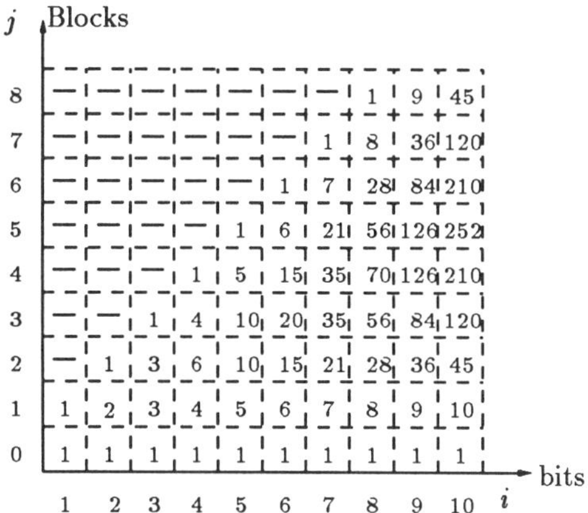

Figure 5: Potential # of entries in the tableau

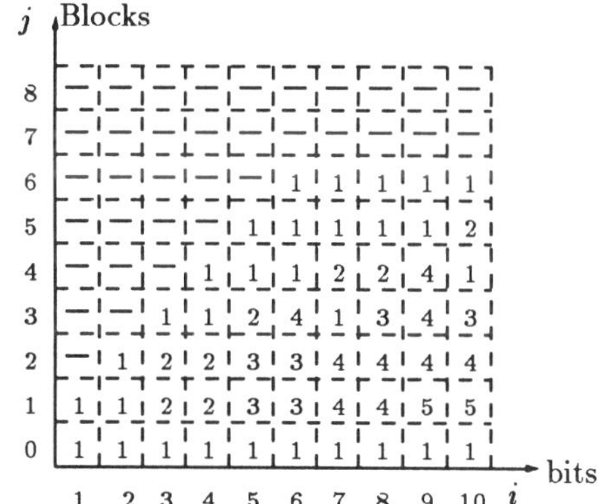

Figure 6: Actual # of entries in the tableau

3.4 Algorithm: *l*-level carry-skip adder

In this section we generalize the one-level skip algorithm to multiple levels. We shall assume that all skip circuitries are setup at time zero. The effect of nonzero setup time is treated in Section 3.5.

We shall construct carry-skip adders having a total of i bits and j 'stages' *at* level l and denote these as (i, j, l)-*adders*. Again we shall consider only (i, j, l)-adders where the j non-empty stages are consecutive and follow an initial number of bits (possibly none) forming an adder with only lower-level skips. If we were going to apply the algorithm from the one-level case, we would need to have available, $G_{l-1}(y)$, $A_{l-1}(y)$, and $I_{l-1}(y)$ functions. Unfortunately, these

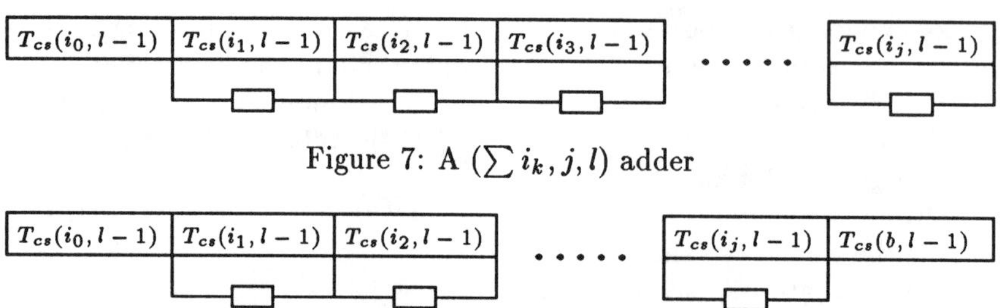

Figure 7: A $(\sum i_k, j, l)$ adder

Figure 8: A $(b + \sum i_k, j, l)$ adder

delays are configuration-sensitive and cannot solely be characterized by y. This difficulty is surmounted by determining the values of these delays for all $(i, j, l-1)$-adders. We use a 4-tuple (t_i, t_g, t_a, t_s) to characterize the worst-case carry propagation delays of an (i, j, l)-adder, where

- t_i is the worst-case delay of any "carry chain" that generates *at* or *before* stage j and terminates *at* or *before* stage j (at level l),
- t_g is the worst-case delay of any "carry chain" that generates *at* or *before* stage j and continues through stage j,
- t_a is the worst-case delay of any "carry chain" that enters the adder and terminates *at* or *before* stage j,
- t_s is the worst-case delay of any "carry chain" that enters the adder and continues through stage j.

Again, we shall compute a tableau in which $t_{cs}(i, j, l)$ contains the minimal 4-tuples for all (i, j, l)-adders. Fig. 7 shows a $(\sum i_k, j, l)$-adder. We also characterize the worst-case delays of a 'stage' of a carry-skip adder having i number of bits and l levels *regardless* of the number of stages it contains, with a (possibly zero) number of bits in lower-level blocks at the end. We call these $(i, *, l)$-*adders*, as illustrated in Fig. 8. Their 4-tuples can be obtained from those of the (i, j, l)-adders using the following equations:

$$
\begin{aligned}
T_{cs}(i, l) &= \{(T_i, T_g, T_a, T_s) \mid \exists\, b, 0 \le b \le i, \\
&\quad \exists\, (T_i', T_g', T_a', T_s') \in T_{cs}(b, l-1), \\
&\quad \exists\, j, \exists\, (t_i, t_g, t_a, t_s) \in t_{cs}(i-b, j, l), \\
&\quad \text{such that} \\
T_i &= \max(t_i, T_i', t_g + T_a') \\
T_g &= \max(t_g + T_s', T_g') \\
T_a &= \max(t_a, t_s + T_a') \\
T_s &= t_s + T_s'\}.
\end{aligned} \tag{4}
$$

The recurrence relationship for the set $t_{cs}(i, j, l) = \{(t_i, t_g, t_a, t_s)\}$ is defined below. The recurrence formula (5) expresses the worst-case propagation delays of (i, j, l)-adders composed of $(i-b, j-1, l)$-adders and $(b, *, l-1)$-adders.

$$
\begin{aligned}
t_{cs}(i, j, l) &= \{(t_i, t_g, t_a, t_s) \mid \exists\, b, 0 \le b \le i, \\
&\quad \exists\, (T_i', T_g', T_a', T_s') \in T_{cs}(b, l-1), \\
&\quad \exists\, (t_i', t_g', t_a', t_s') \in t_{cs}(i-b, j-1, l), \\
&\quad \text{such that} \\
t_i &= \max(t_i', T_i', t_g' + T_a') \\
t_g &= \max(t_g' + S_l(b), T_g') \\
t_a &= \max(t_a', t_s + T_a') \\
t_s &= t_s' + S_l(b)\}.
\end{aligned} \tag{5}
$$

The basis for the dynamic programming is $(l = 0)$

$$
T_{cs}(i, 0) = \{(I(i), G(i), A(i), S_0(i))\} \;;\; \text{for } 0 \le i \le n,
$$

and for $l \ge 1$, $t_{cs}(i, 0, l)$ is defined as

$$
t_{cs}(i, 0, l) = T_{cs}(i, l-1); \qquad \text{for } 0 \le i \le n.
$$

In this expression, the worst-case delay of an n-bit adder using at most l-level skips is the minimum T_i appearing in the sets $T_{cs}(n, k)$, for $1 \le k \le l$.

We control the number of configurations in each set by adopting pruning techniques similar to those described in the previous section. In addition, once an additional skip-level produces only sub-optimal 4-tuples for a given number of bits i, no more new skip levels are considered for i bits.

3.5 Incorporation of setup time for the skip gates

The setup time $Set\_up_l(y)$ is the amount of time needed to enable the skip circuitry at level l. This reflects the delay to generate a group propagate of y bits. Our dynamic programming formulation cannot be easily adapted to take care of the effects of setup time. The problem is that the worst-case absorb and skip times computed for $(i, *, l)$-adders can no longer be used in generating $(i, j, l+1)$-adders since the setup times have been incorporated assuming that carries arrive to the adder at time 0. A compromise is to charge the setup time only to the carry generate, subsequently the formulation is modified as:

$$
\begin{aligned}
t_{cs}(i,j,l) \quad &= \{(t_i, t_g, t_a, t_s) \mid \exists\, b, 0 \le b \le i, \\
&\quad \exists\, (T_i{}', T_g{}', T_a{}', T_s{}') \in T_{cs}(b, l-1), \\
&\quad \exists\, (t_i{}', t_g{}', t_a{}', t_s{}') \in t_{cs}(i-b, j-1, l), \\
&\quad \text{such that} \\
&\quad t_i = \max(t_i{}', T_i{}', t_g{}' + T_a{}') \\
&\quad t_g = \max(\max(t_g{}', Set\_up_l(b)) + S_l(b), T_g{}') \\
&\quad t_a = \max(t_a{}', t_s + T_a{}') \\
&\quad t_s = t_s{}' + S_l(b)\}.
\end{aligned}
\tag{6}
$$

In this formulation, the generate delay t_g is exact, while the other three components may be under-estimated. Care should be taken during the pruning to verify the actual delays of the current best delay for an $(n, *, l)$-adder which will be used to discard configurations.

3.6 Results

We coded our dynamic programming formulation in the "T" programming language [18], and used Turrini's [6] delay model and results to validate our algorithm. Turrini's delay model is

$$
\begin{aligned}
I(y) &= y \quad & G(y) &= y + 1 \quad & A(y) &= y - 1 \\
S_l(y) &= 1 \quad & Set\_up_l(y) &= l + 1.
\end{aligned}
$$

Despite the under-estimation of the delay resulting from ignoring the setup time for skips in all but the generate delays, our algorithm was able to generate the same optimal size adders as Turrini [6]. However, we emphasize that our approach is applicable to *any* delay model. Turrini's analysis is limited to models with a constant value for the skip delay, regardless of the number of blocks being skipped.

4 Delay optimization of block carry-lookahead adders

This work was motivated by Wei and Thomborson [9] who used dynamic programming techniques to optimize parallel-prefix adders, as well as a study carried out by Lee [7]. In his paper, Lee discusses the possibility of varying the block sizes in a block carry-lookahead adder (**BCLA**) to further optimize the carry propagation delay. We begin by recalling the structure of block carry-lookahead adders.

Fig. 9 shows a 16-bit 2-level equal-block-size block carry-lookahead adder. Each box is a 4-bit carry-lookahead generator as shown in Fig. 10. These two figures illustrate the notation that we shall be using in this section. We use small letters to denote *global* signal names, e.g., g_0, c_1, and capital letters to denote signal names relative to a block, e.g., G_0, C_1. The goal of our optimization is to minimize the worst-case

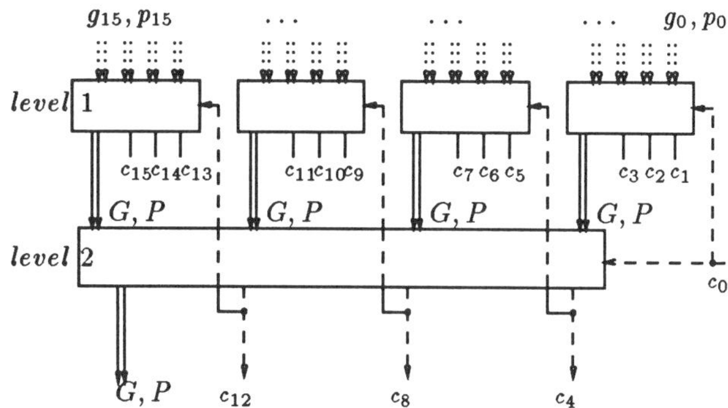

Figure 9: A 16-bit two-level equal-block-size BCLA

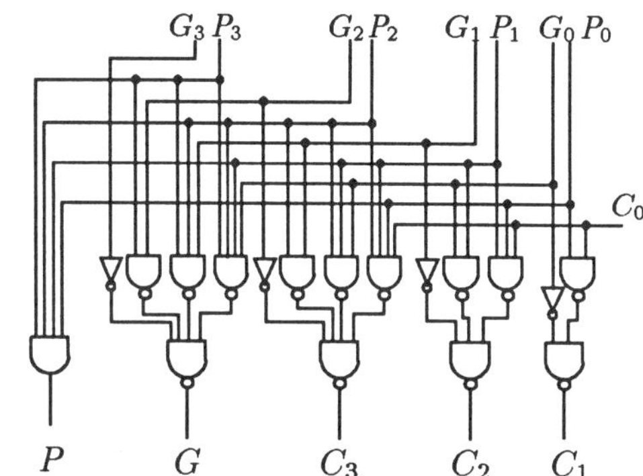

Figure 10: A 4-bit carry-lookahead generator

delay of carries c_1 to c_{n-1} of an n-bit adder. Notice in Fig. 10 that there is no connection from the carry input C_0 of the block to the carry propagate P and generate G outputs. In terms of the structure of the BCLA, this means that P and G are the only signals which travel down the carry-lookahead tree; the carry outputs at the lower levels travel back up to determine the carry outputs of some of their ancestor blocks.

An equal-block-size BCLA minimizes the height of the tree. The latency would be minimal if delays were measured merely by summing unit gate delays along paths. However, in practice the delay of a gate depends on fanin and fanout. The interior of a "block" of a BCLA is a two-level network. Hence the delay of a block is a function primarily of the size of the gates (fanin) as well as the fanout of the signals feeding these gates. Each pair of generate and propagate signals G, P fanout to only one block, however within the block their fanout is linear and quadratic in the size of the block, respectively. These factors tend to limit the block size. In contrast, the carries fanout to multiple blocks (to each of their rightmost ancestor blocks) and hence their delay minimization is improved by decreasing the height of the tree. Smaller blocks are faster and their increased speed may offset additional levels of logic on interior paths, if the sizes of blocks can be varied to balance path delays.

Fig. 11 shows an 8-bit variable-block-size BCLA. Lee shows that the (3-level, 8-bit) BCLA as shown in Fig. 12(a) has the minimum latency according to a gate delay model which considers fanouts and fanins; the next best adder has the configuration of Fig. 12(b). However, Lee found neither exact algorithms nor heuristics to size a BCLA to attain minimum latency [7].

Here, we formulate a multidimensional dynamic program to solve the problem for a particular class of gate delay models, in which gate delay depends *linearly* on fanout and fanin. We refer to BCLA adders having i bits as *i-adders*. For a given m, we construct a BCLA i-adder by selecting m (smaller) BCLA adders of sizes $i_0, i_1, \ldots i_{m-1}$ respectively and combining them to form an $(i_0 + i_1 + \ldots + i_{m-1})$-adder with an m-bit carry-lookahead generator.

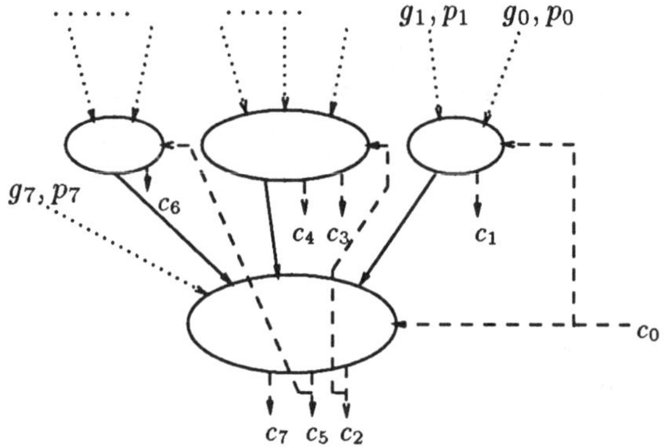

Figure 11: An 8-bit variable-block-size BCLA

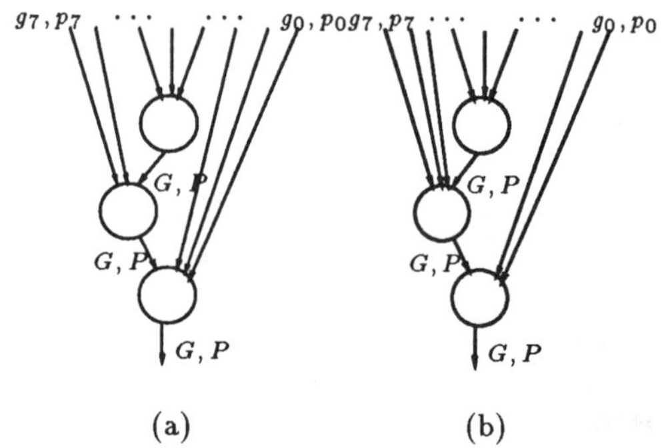

(a)   (b)

Figure 12: Optimal 8-bit variable-block-size BCLAs

Instead of trying all possible combinations of $i_0, i_1, \ldots i_{m-1}$ which total to i bits, we construct a BCLA adder incrementally, by filling in the positions in our m-bit carry-lookahead generator starting with the least significant position.

A partially completed adder with $i = i_0 + i_1 + \ldots + i_{j-1}$ bits and an m-bit block having only positions $0, 1, \ldots, j-1$ filled is called an (m, i, j)-adder, as shown in Fig. 13. Clearly this is only defined for $1 \leq j < m$. Since not all inputs of the m-bit block are provided, we assume temporarily that these are constants and compute worst case delays of the adder only from the inputs to positions 0 through $j - 1$ of the m-bit block.

Since our goal is to minimize the worst-case carry of an i-adder we must maintain enough delay information in our i-adders and (m, i, j)-adders to compute accurately the worst-case carry delay and guarantee the minimum latency. Because of the structure of the carry-lookahead generator and the gate delay model, we shall be able to compute the delay of an $(m, i, j+1)$-adder without retaining complete information about the arrival time of the inputs to the (m, i, j)-adder. (In fact, the only arrival time that must be retained is that of the most significant generate.) One complication with this construction is that the fanout of the carry-in to an i-adder increases when the i-adder is connected to another block; this may further increase its worst-case carry delay. Fortunately, since the dependence on fanout is linear we can account for the extra delay by maintaining two versions of the worst-case carry delay of an i-adder; one for paths originating from the carry-in and the other for the overall worst-case. Before discussing the delay components which will characterize our i-adders and (m, i, j)-adders in any more detail, we first present our gate delay model.

4.1 Gate delay model

We assume that the input arrival time $(t_{in,j})$ and the output available time (t_{out}) of a gate are related by

$$t_{out} = \max_j \{t_{in,j}\} + FO \cdot \tau + \Delta(FI) \qquad (7)$$

where FO is the fanout of the output signal, FI is the fanin of the gate, τ is the delay per unit fanout, and $\Delta(FI)$ is the delay of a gate of fanin FI under zero load. We define specific delay functions $\Delta_{nand}(FI)$, $\Delta_{and}(FI)$, and Δ_{inv} to model the behavior of 'NAND' gates, 'AND' gates, and inverters under zero load. The functions Δ_{nand}, Δ_{and}, and Δ_{inv} must be monotone nondecreasing, but may take infinite values beyond a certain point in their domain. This ensures that our designs will not contain 17-input-NANDs if an 8-input-NAND is the widest one available.

For simplicity of presentation, we assume that NAND gates, AND gates, and inverters have the same τ value, although this is not a limitation of our formulation. We must, however, require that all Δ functions take nonnegative values over their domains.

We also define δ to express the incremental change of delay per unit fanin: $\delta(FI) = \Delta(FI) - \Delta(FI - 1)$. When considering different gates, we add a suffix to identify the gate in question, for example, δ_{and} and

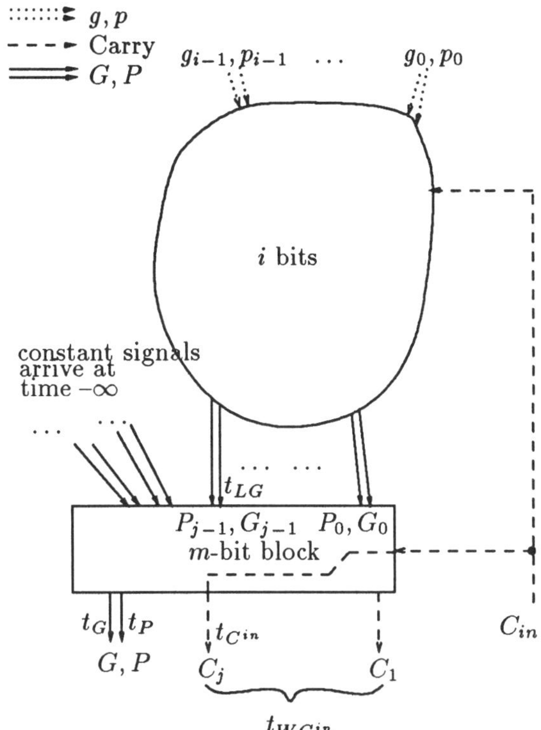

Figure 13: An (m, i, j)-adder

δ_{nand}. Under our linearity assumption on $\Delta()$, δ_{and} and δ_{nand} are nonnegative constants.

The loading on the input signals of a k-bit BCLA adder connected to the jth input of an m-bit BCLA adder [19] can be expressed as

- G_j of the k-bit BCLA adder has fanout $m - j$, i.e., $f_G(m, j) = m - j$. Notice that the fanout is largest at the 0th input position.
- P_j of the k-bit BCLA adder has fanout $f_P(m, j) = (m - j)(j + 1)$.
- The carry-in C_0 to the m-bit carry-lookahead generator has fanout $m - 1$.

4.2 Constructing BCLA adders

For this construction we need to characterize an i-adder with a 5-tuple, $(T_G, T_P, T_{WC}, T_{WC^{in}}, F_{C^{in}})$, where

- T_G is the worst-case delay of the group *generate* output,
- T_P is the worst-case delay of the group *propagate* output,
- $T_{WC^{in}}$ is the worst-case delay of any path from the carry input to any carry output, and
- T_{WC} is the worst-case delay of any carry output,
- $F_{C^{in}}$ is the fanout of the carry input inside the adder.

All the delay values above are calculated under the assumption of zero fanout. When we use these adders as building blocks of larger adders, we shall add appropriate multiples of τ to the delay. Note the two versions of the worst-case carry delay. As discussed, these are necessary in order to account for additional loading on the carry-in when the i-adder is connected to other blocks.

Recall that an (m, i, j)-adder has a partially completed m-bit block with $i = i_0 + i_1 + \ldots + i_{j-1}$ bits having only positions $0, 1, 2, \ldots, j - 1$ filled. We shall characterize an (m, i, j)-adder by an 8-tuple $(t_G, t_P, t_{LG}, t_C, t_{C^{in}}, t_{WC}, t_{WC^{in}}, f_{C^{in}})$, where [2]

- t_G is the worst-case delay of the group generate output,
- t_P is the worst-case delay of the group propagate output,
- t_{LG} is the arrival time of the group generate G_{j-1},
- $t_{C^{in}}$ is the worst-case path delay from the carry input to the currently last carry output C_j of the m-bit carry-lookahead generator,
- t_C is the overall worst-case delay of the currently last carry output C_j of the m-bit carry-lookahead generator,
- $t_{WC^{in}}$ is the worst-case path delay from the carry input to any carry output,
- t_{WC} is the overall worst-case delay of any carry output, and
- $f_{C^{in}}$ is the fanout of the carry input inside the adder.

The arrival time of the input G_{j-1} (t_{LG}) at the m-bit block is the only input arrival time retained. We shall be able to compute all the components of an $(m, i, j + 1)$-adder from those of an $(m, i - b, j)$-adder and a b-adder.

As in the algorithm for carry-skip adders, we retain a tableau of lists for constructed adders:

$$
\begin{aligned}
T(i) &= \{(T_G, T_P, T_{WC}, T_{WC^{in}}, F_{C^{in}}) \\
&\quad \mid \text{for all } i\text{-adders}\} \\
t(m, i, j) &= \{(t_G, t_P, t_{LG}, t_C, t_{C^{in}}, t_{WC}, t_{WC^{in}}, f_{C^{in}}) \\
&\quad \mid \text{for all } (m, i, j)\text{-adders}\}. \quad (8)
\end{aligned}
$$

Three sets of equations in our dynamic programming formulation cover respectively, filling in the first position of an m-bit block, an intermediate position, and the last position.

---

[2] The $t_{C^{in}}$ term is redundant since it will always be equivalent to $f_{C^{in}}\tau + \tau$, but is made explicit here to elucidate the delay equations.

An $(m, i, 1)$-adder is generated from a 5-tuple $(T_G, T_P, T_{WC}, T_{WC^{in}}, F_{C^{in}})$ in $T(i)$, by

$$
\begin{aligned}
t_G &= T_G + f_G(m, 0)\tau + \Delta_{nand}(m) + \tau + \Delta_{nand}(m) \\
t_P &= T_P + f_P(m, 0)\tau + \Delta_{and}(m) \\
t_{LG} &= T_G \\
t_{C^{in}} &= \tau(F_{C^{in}} + m - 1) + \Delta_{nand}(2) + \Delta_{nand}(2) + \tau \\
t_C &= \max\{t_{C^{in}}, \\
&\qquad T_G + f_G(m, 0)\tau + \Delta_{inv} + \tau + \Delta_{nand}(2), \\
&\qquad T_P + f_P(m, 0)\tau + \Delta_{nand}(2) + \tau + \Delta_{nand}(2)\} \\
t_{WC^{in}} &= \max\{t_{C^{in}}, T_{WC^{in}} + (m - 1)\tau\} \\
t_{WC} &= \max\{t_{WC^{in}}, T_{WC}, t_C\} \\
f_{C^{in}} &= F_{C^{in}} + m - 1.
\end{aligned}
$$

For $j > 1$ (and $j < m$), an (m, i, j)-adder is generated by connecting the G_{j-1}, P_{j-1} inputs of an $(m, i - b, j - 1)$-adder with the G, P outputs of a b-adder. As discussed, the worst-case carry of the b-adder is adjusted by the increase in fanout of its carry-in. Notice that the two-level network computing the new carry is similar to the network for the previous network; it differs only in that the fanin of the gates have increased and the inputs P_{j-1}, G_{j-1} and G_{j-2} must be incorporated. This is the reason for the retaining the t_{LG} component. As a result, the delay of C_j can be computed based on the arrival times of P_{j-1}, G_{j-1} (provided by the b-adder) and the t_{LG} and t_C components of the $(m, i-b, j-1)$-adder. Specifically, if $(T_G, T_P, T_{WC}, T_{WC^{in}}, F_{C^{in}})$ is the 5-tuple which characterizes the b-adder and $(t'_G, t'_P, t'_{LG}, t'_C, t'_{C^{in}}, t'_{WC}, t'_{WC^{in}}, f'_{C^{in}})$ is the 8-tuple which characterizes the $(m, i - b, j - 1)$-adder, then the 8-tuple for our new (m, i, j)-adder is :

$$
\begin{aligned}
t_G &= \max\{t'_G, \\
&\quad T_G + f_G(m, j - 1)\tau + \Delta_{nand}(m - j - 1) + \tau \\
&\quad + \Delta_{nand}(m), \\
&\quad T_P + f_P(m, j - 1)\tau + \Delta_{nand}(m) + \tau + \Delta_{nand}(m)\} \\
t_P &= \max\{t'_P, T_P + f_P(m, j - 1)\tau + \Delta_{and}(m)\} \\
t_{LG} &= T_G \\
t_{C^{in}} &= t'_{C^{in}} + \delta_{nand} + \delta_{nand} \\
t_C &= \max\{T_G + f_G(m, j - 1)\tau + \Delta_{inv} + \tau + \Delta_{nand}(j + 1), \\
&\quad T_P + f_P(m, j - 1)\tau + \Delta_{nand}(j + 1) + \tau + \Delta_{nand}(j + 1), \\
&\quad t'_{LG} + f_G(m, j - 2)\tau + \Delta_{nand}(2) + \tau + \Delta_{nand}(j + 1) \\
&\quad + \delta_{nand} + \delta_{nand}, \; t'_C + \delta_{nand} + \delta_{nand}\} \\
t_{WC^{in}} &= \max\{t'_{WC^{in}}, t_{C^{in}}, T_{WC^{in}} + t'_{C^{in}}\} \\
t_{WC} &= \max\{t'_{WC}, T_{WC}, T_{WC^{in}} + t'_C, t_C\} \\
f_{C^{in}} &= f'_{C^{in}}.
\end{aligned}
$$

When the mth position of an $(m, i - b, m - 1)$-adder is filled in with a b-adder, we obtain an i-adder . If $(T'_G, T'_P, T'_{WC}, T'_{WC^{in}}, F'_{C^{in}})$ is the 5-tuple which characterizes the b-adder and $(t_G, t_P, t_{LG}, t_C, t_{C^{in}}, t_{WC}, t_{WC^{in}}, f_{C^{in}})$ is the 8-tuple which characterizes the $(m, i - b, m - 1)$-adder, then the 5-tuple for our new i-adder is :

$$
\begin{aligned}
T_G &= \max\{t_G, \\
&\quad T'_G + f_G(m, m - 1)\tau + \Delta_{inv} + \tau + \Delta_{nand}(m), \\
&\quad T'_P + f_P(m, m - 1)\tau + \Delta_{nand}(m) + \tau \\
&\quad + \Delta_{nand}(m)\} \\
T_P &= \max\{t_P, T'_P + f_P(m, m - 1)\tau + \Delta_{nand}(m)\} \\
T_{WC^{in}} &= \max\{t_{WC^{in}}, T'_{WC^{in}} + t_{C^{in}}\} \\
T_{WC} &= \max\{t_{WC}, T'_{WC}, T'_{WC^{in}} + t_C\} \\
F_{C^{in}} &= f_{C^{in}}.
\end{aligned}
\tag{9}
$$

The basis for the dynamic programming is $T(1) = \{(\Delta_{and}(2), \Delta_{or}(2), 0, 0, 0)\}$. The first two components are the delays to generate the g_i and p_i, respectively.

The minimum worst-case delay of an n-bit BCLA adder is the minimum T_{WC} appearing in the set $T(n)$.

4.3 Implementation

Instead of a 2-D tableau to fill as in the case of the carry-skip adders, we must fill the 3-dimensional volume as depicted in Fig. 14. To construct an (m, i, j)-adder for $j > 1$, we must have already constructed all $(m, b, j - 1)$-adders and all b-adders, for $1 \le b < i$. To construct all i-adders we must have constructed all $(m, i - b, m - 1)$-adders for $1 \le b < i$ and $2 \le m \le i$. Finally, to construct an $(m, i, 1)$-adder, we must have constructed all i-adders. These dependencies lead us to the following steps depicted in Fig. 14 which are repeated for $z = 2, 3, \ldots, n$.

1. Construct all $(z, b, 1)$-adders for $b = 1, 2, \ldots, z - 1$.
2. Construct all (z, b, j)-adders for $j = 2, 3, \ldots, z - 1$ and $b = j, j + 1, \ldots, z - 1$.
3. Construct all (m, z, j)-adders for $m = 2, 3, \ldots, z$ and $j = 2, 3, \ldots, m$. (We now have all z-adders, since they are in fact the (m, z, m)-adders for $m = 2, 3, \ldots, z$.)
4. Construct all $(m, z, 1)$-adders from the z-adders, for $m = 2, 3, \ldots, z$.

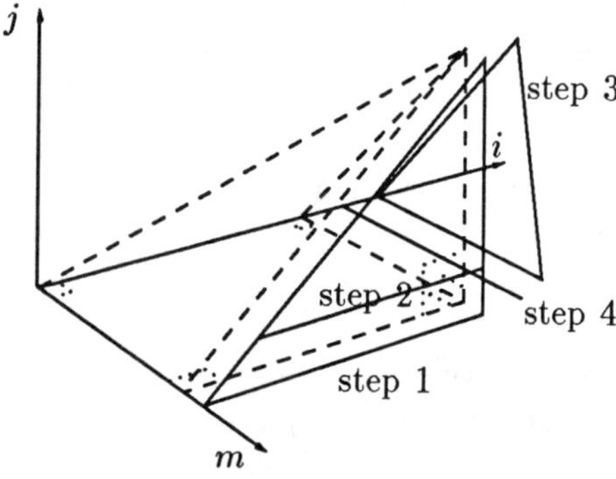

Figure 14: Filling the volume

4.4 Pruning techniques

The maximum number of configurations in the volume even for a small number of bits is prohibitively high. We have employed the following pruning techniques to reduce the number of configurations.

- Set an upper bound on m, the maximum number of inputs to a block, based on a known technological constraint.

- Compute the worst-case delays of equal-block-size adders using the gate delay model. This sets an upper bound on any delay component of the variable-block-size adders that we are building in the "volume." Hence any configuration which has a delay component greater than this upper bound can be thrown away.

- Since the worst-case delays of equal-block-size adders are typically 20% higher than the minimal latency ones, a tighter upper bound can be obtained by temporarily disregarding some delay components during the ranking of the configurations (e.g., t_P or T_P), and running the algorithm to obtain a suboptimal configuration. In effect, this is optimization by a lower-order dynamic program. We then use the maximum delay component of this suboptimal configuration as the new upper bound for a new trial (after reinstating the deleted delay components). This iterative improvement scheme turns out to be very effective in pruning infeasible configurations and reaching the minimum latency configurations.

4.5 Results

We use a gate delay model obtained by fitting data from an ASIC–CMOS standard cell library [20] to linear Δ functions [21]. We select τ to be 5 so that all the parameters in the equations are scaled to integers. Note that an unloaded inverter has delay of 12 units.

$$\tau = 5$$
$$t_{NAND,out} = t_{in} + FO \cdot \tau + FI \cdot 20$$
$$t_{OR,out} = t_{in} + FO \cdot \tau + FI \cdot 20 + 17$$
$$t_{AND,out} = t_{in} + FO \cdot \tau + FI \cdot 20 + 17$$
$$t_{INV,out} = t_{in} + FO \cdot \tau + 12. \tag{10}$$

We shall represent the carry-lookahead adder tree in parentheses notation. For example, the adder structures shown in Figs. 12a and b are represented as ((1 1 3) 1 1 1) and ((1 1 1 3) 1 1), respectively; and the equal-block-size 16-bit BCLA of Fig. 9 appears as (4 4 4 4). The numbers in the expression represent the block sizes at the top level.

For an n-bit adder, T_{WC} in Table 1 indicates the worst-case delay to generate the carries c_1 to

n	T_G T_P T_{WC}	configuration
2	152 124 152	2
3	202 154 202	3
4	309 258 247	((1 2) 1)
	252 184 252	4
6	424 350 314	((1 (1 2)) 1 1)
	337 251 319	(1 1 2 2)
7	399 318 347	((1 1 2 2) 1)
	362 281 364	(1 1 3 2)
8	424 355 369	((1 (1 2) 3) 1)
	382 286 379	(1 1 2 2 2)
	317 251 434	(3 3 2)
	347 251 499	(4 4)
16	559 477 489	((1 (1 (1 2)) (1 1 2) (3 2)) 1 1)
	489 410 506	(((1 2) (1 1 2) 4 (2 1)) 2)
	407 350 569	((2 (2 1)) (2 2 2) (3 2))
	437 311 597	(4 4 4 4)
24	479 385 609	((1 3) (2 (2 2)) (2 3 2) (3 2 2))
	532 378 789	((4 4) (4 4 4 4))
32	621 544 627	(((1 (1 (1 2))) ((1 2) ((1 2) 3)) (2 3 2) (3 2 2)) 2 2)
	532 378 879	((4 4 4 4) (4 4 4 4))
48	716 599 716	(((1 2) ((1 (1 2) (2 2)) ((1 2) (1 2) (3 2)) ((1 2) 3 3 2) (4 3 3))) 3 2)
	577 408 932	((4 4 4 4) (4 4 4 4) (4 4 4 4))
64	789 639 797	((2 3 ((1 2 3 3) (2 3 4 (2 2)) ((1 2 2) (2 2 2) (3 2)) ((2 2 2) (3 2) 3)) (3 2)) 2)
	622 438 977	((4 4 4 4) (4 4 4 4) (4 4 4 4) (4 4 4 4))
66	794 649 802	((2 3 ((1 2 3 3) (2 3 4 (2 2)) ((1 2 2) (2 2 2) (3 2)) ((3 2 2) (3 2) (2 2))) (3 2)) 2)
	717 505 977	(2 ((4 4 4 4) (4 4 4 4) (4 4 4 4) (4 4 4 4)))
84	803 674 856	((3 4 ((2 3 (1 2 2) (3 2 2)) ((1 2 2) (2 3 2) (3 2 2)) ((2 3 2) (3 2 2) (3 2)) ((3 2 2) (3 2) 3)) (3 2)) 2)
	762 530 1069	(4 (4 4 4 4) ((4 4 4 4) (4 4 4 4) (4 4 4 4) (4 4 4 4)))

Table 1: Delays of equal-block-size vs variable-block-size n-bit BCLAs. Configurations for 2-bit to 16-bit adders are optimal (in latency)

c_{n-1}. The overflow condition is indicated by the final carry c_n which depends on the carry generate and propagate (T_G and T_P). Table 1 shows the delays of variable-block-size BCLAs versus their equal-block-size counterparts. These results are generated by restricting the maximum fan-in of any CMOS gate to 4. The delay of an inverter in a typical 1.5μm CMOS technology is roughly 0.3 ns, so we can convert our integer delay values in Table 1 to nanoseconds in such a technology by multiplying by (0.3/12) ns. For 8-bit adders, we have 9.225 ns and for 16-bit adders we have 12.225 ns.

Except for the $n \leq 8$ cases, T_{WC} is the dominant delay component. This experiment demonstrates that variable-block-size BCLAs out-perform their equal-block-size counterparts by 15-25%, in terms of their T_{WC}. However, variable-block-size adders are not as modular as the equal-block-size adders. The best variable-block-size BCLAs tend to have more levels but less fan-ins than their equal-block-size counterparts. This suggests that the number of levels is not a good measure of latency for VLSI technology.

5 Conclusion

We have formulated the problems of minimizing the latencies in carry-skip and block carry-lookahead adders as multidimensional dynamic programs. Based on these formulations, we implemented programs to carry out the minimization. The dynamic programming formulations are appealing because of their generality. On the other hand, the computational requirement of the optimization process is also high. All the algorithms presented are coded in the **"T"** language [18]. The program requires 60 Megabyte of swap space and ran for over 3 hours before completion on a SPARC station. The algorithms generate adder configurations that are not modular, but the adders' latencies are 15-25% less than their modular counterparts.

The delay model that we have established considers fanin and fanout, and is therefore more realistic than counting the number of levels. However, we do not account for the effect of wire lengths in the model, this will be considered in future work.

References

[1] M. Lehman and N. Burla, "Skip Techniques for High-Speed Carry-Propagation in Binary Arithmetic Units," *IRE Transactions on Electronic Computers*, vol. EC-10, pp. 691–698, Dec. 1961.

[2] S. Majerski, "On determination of optimal distribution of carry skips in adders," *IEEE Transactions on Electronic Computers*, vol. EC-16, pp. 45–48, Feb. 1967.

[3] A. Guyot, B. Hochet, and J.-M. Muller, "A Way to Build Efficient Carry-Skip Adders," *IEEE Transactions on Computers*, vol. C-36, pp. 1144–1151, Oct. 1987.

[4] V. G. Oklobdzija and E. R. Barnes, "Some optimal schemes for ALU implementation in VLSI technology," in 7th *Computer Arithmetic Symposium*, pp. 2–8, 1985.

[5] P. K. Chan and M. Schlag, "Analysis and Design of CMOS Manchester Adders with Variable Carry-Skip," *IEEE Transactions on Computers*, vol. C-39, pp. 983–992, Aug. 1990.

[6] S. Turrini, "Optimal group distribution in carry-skip adders," in 9th *Computer Arithmetic Symposium*, (Santa Monica, Los Angeles), pp. 96–103, Sept 1989.

[7] B. Lee, "VLSI Implementation of Fast Arithmetic Algorithms: Optimizing Delays in Carry Lookahead Adders." CS292I: Class project report, UC Berkeley, December 1989.

[8] R. K. Montoye, "Area-time efficient addition in charge based technology," in *ACM IEEE* 18th *Design Automation Conference Proceedings*, pp. 862–872, 1981.

[9] B. W. Wei and C. D. Thompson, "Area-Time Optimal Adder Design," *IEEE Transactions on Computers*, vol. C-39, pp. 666–675, May 1990.

[10] G. Dantzig, "Discrete-Variable Extremum Problems," *Operations Research*, vol. 5, pp. 266–277, 1957.

[11] H. Weingartner, "Capital Budgeting of Interrelated Projects: Survey and Synthesis," *Management Science*, vol. 12, pp. 485–516, Mar. 1966.

[12] H. Weingartner and D. Ness, "Methods for the solution of the multidimensional 0/1 knapsack problem," *Operations Research*, vol. 15, pp. 83–103, 1967.

[13] T. Morin and R. Marsten, "Branch-and-bound strategies for dynamic programming," *Operations Research*, vol. 24, pp. 611–627, July-August 1976.

[14] R. Marsten and T. Morin, "A hybrid approach to discrete mathematical programming," *Mathematical Programming*, vol. 15, no. 1, pp. 21–40, 1978.

[15] S. E. Dreyfus and A. M. Law, *The Art and Theory of Dynamic Programming*. 111 Fifth Avenue, New York, New York 10003: Academic Press, Inc., 1977.

[16] J. Bentley, H. Kung, M. Schkolnick, and C. Thompson, "On the average number of maxima in a set of vectors and applications," *Journal of ACM*, vol. 11, no. 1, pp. 536–543, 1978.

[17] L. Devroye, "A note on finding convex hulls via maximal vectors," *Information Processing Letters*, vol. 11, pp. 53–56, Aug. 1980.

[18] J. A. Rees, N. I. Adams, and J. R. Meehan, *The T Manual*. New Haven, Connecticut: Yale University, March 1983.

[19] T. Rhyne, "Limitations on Carry Lookahead Networks," *IEEE Transactions on Computers*, vol. C-33, pp. 373–373, Apr. 1984.

[20] LSI Logic Corporation, 1551 McCarthy Boulevard, Milpitas, CA 95035, *Compacted Array Technology Data Book*, July 1987.

[21] V. G. Oklobdzija and E. R. Barnes, "On implementing addition in VLSI technology," *Journal of Parallel and Distributed Computing*, vol. 5, 1988.

The Redundant Cell Adder

Tom Lynch and Earl Swartzlander
Department of Electrical and Computer Engineering
University of Texas at Austin
Austin, Texas 78712

Abstract

This paper describes the design of the 56-bit significand adder for the Advanced Micro Devices Am29050[1] microprocessor. This is a 1 μm design rule CMOS realization of a high performance RISC microprocessor that implements IEEE Standard 754 floating point arithmetic. To achieve an add time of under 4 ns for the 56-bit significand and to avoid multistage pipelines which significantly impair compiler efficiency, a redundant cell adder has been developed. This new design is a key to realizing the high performance for floating point arithmetic that is achieved by the Am29050 microprocessor.

1 Introduction

Classic high speed adders include carry lookahead, carry skip, carry select, and conditional sum adders. A variant of the binary carry lookahead adder was analyzed by Brent and Kung [1]. Another variant, based on reformulating the carry equations was developed by Ling [2]. Fast implementations of carry lookahead adders have been developed by Hwang and Fisher [3] in CMOS and Bewick, Song, and Flynn [4] in bipolar ECL. For large word sizes, the carry lookahead adder is generally considered to be the fastest. For gate level designs where each gate has unit delay, the carry lookahead adder greatly reduces delay with a modest increase in complexity relative to the ripple carry adder. In dynamic CMOS implementations where the delay depends on the gate size and loading, the use of Manchester carry chains [5] to realize the carry lookahead logic significantly reduces the gate loading, which produces a substantial speed increase.

This paper presents the design of a redundant cell adder implemented in dynamic CMOS which combines carry lookahead adders realized with Manchester carry chains and

the carry select adder concept to achieve approximately twice the speed of the traditional carry lookahead adder. Because this is a cellular adder that performs redundant calculations to generate the control signals for a carry select adder, it is referred to as the "redundant cell adder".

2 Logic Design of the Redundant Cell Adder

Notation: a_i and b_i denote the i-th bits of the words to be added, p_i denotes that a carry will propagate across bit position i (i.e., $p_i = a_i + b_i$), and g_j denotes that a carry is generated at bit position j (i.e., $g_j = a_j b_j$). $p_{i:j}$ signifies that a carry will propagate from bit j (LSB) to bit i (MSB). Similarly $g_{i:j}$ denotes that a carry is generated in each of the bit positions from j to i inclusive. The fundamental carry operation, "fco," introduced by Brent and Kung [1] is used:

$$(p_{i:j}, g_{i:j}) = (p_{i:k+1}, g_{i:k+1}) \, fco \, (p_{k:j}, g_{k:j}) \qquad (1)$$

which is defined as:

$$p_{i:j} = p_{i:k+1} p_{k:j} \text{ and } g_{i:j} = g_{i:k+1} + p_{i:k+1} g_{k:j}$$

Brent and Kung have shown the associativity of the fco. For: $i > m > k > j$.

$$[(p_{i:m+1}, g_{i:m+1}) \, fco \, (p_{m:k+1}, g_{m:k+1})] \, fco \, (p_{k:j}, g_{k:j})$$
$$= (p_{i:m+1}, g_{i:m+1}) \, fco \, [(p_{m:k+1}, g_{m:k+1}) \, fco \, (p_{k:j}, g_{k:j})] \qquad (2)$$

A block diagram of a 64 bit redundant cell adder is shown in Figure 1. The adder uses a tree of Manchester carry chain carry lookahead modules (on the left) to calculate the carries for each of the eight bit carry select adders (on the right). On the left, the Manchester carry chain carry lookahead modules (Mcc) produces group generate and propagate signals for four bit groups, except for the least significant group, where carry c_4 is produced. On the second level of the tree, the first level signals are

165

---

[1] Am29050 is a trademark of Advanced Micro Devices

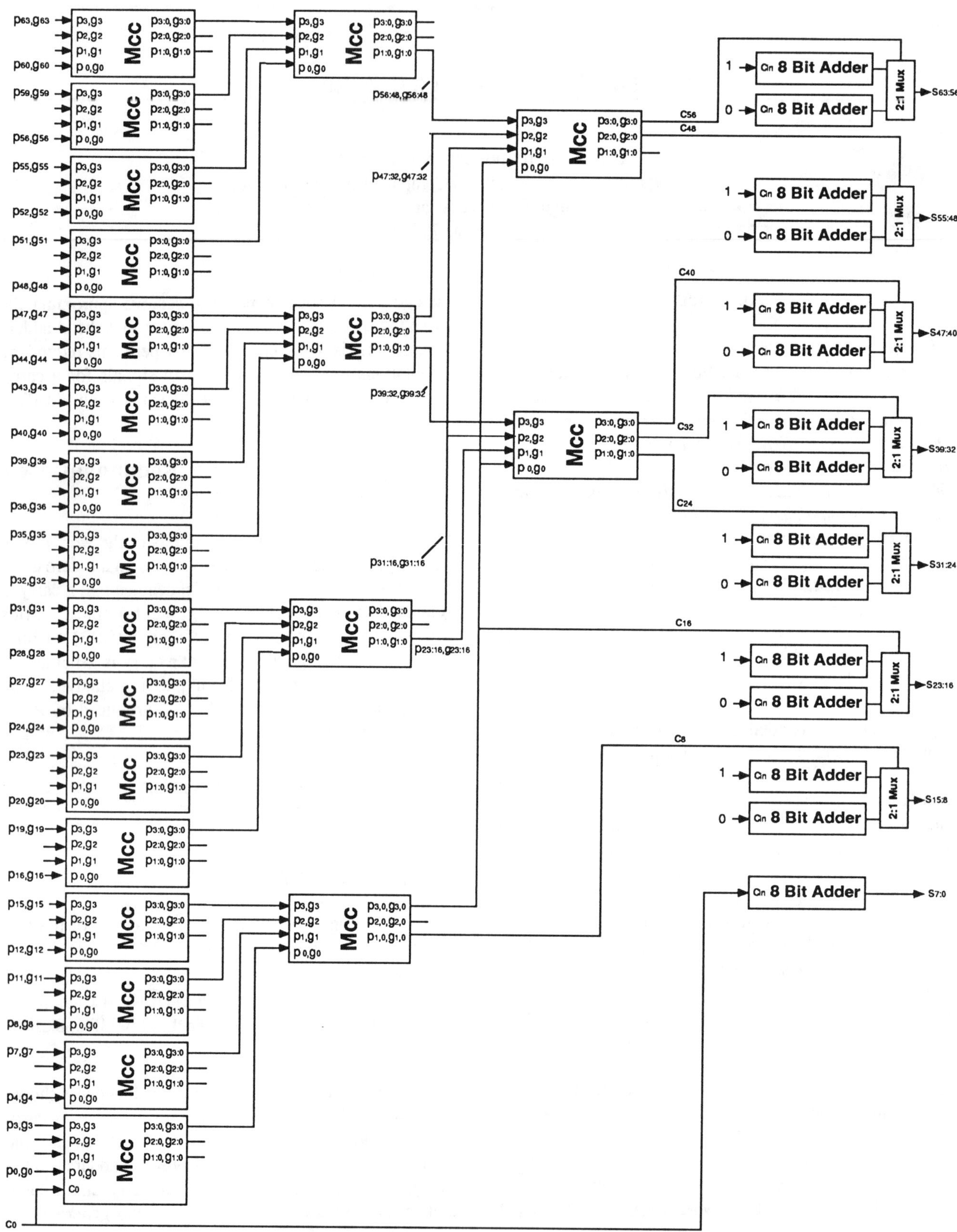

Figure 1. Redundant Cell Adder Block Diagram.

combined to produce group propagate and generate signals for 16 bit boundaries, and overlapping eight bit boundaries. For example, $p_{47:32}, g_{47:32}$ and the overlapping eight bit boundary signals $p_{39:32}, g_{39:32}$ are produced. Carries c_8 and c_{16} are available from the least significant group on the second level.

On the third level two Manchester carry chain modules are used. The most significant third level module combines the three less significant 16 bit boundary group propagate and generate signals, $p_{15:0}, g_{15:0}$; $p_{31:16}, g_{31:16}$; and $p_{47:32}, g_{47:32}$, and the most significant eight bit boundary group propagate and generate signals $p_{55:48}, g_{55:48}$ to produce carries c_{48} and c_{56} (note that carry c_{32} is generated both here and at the lower module in this column).

The idempotency of the fundamental carry operation must be shown to explain the operation of the least significant third level carry lookahead module. As shown in Equation (2), combining group p and g signals with themselves produces correct results. The result for propagate follows trivially since ANDing any signal, say x, with itself returns x, while the result for generate follows by factoring $g_{i:j} + p_{i:j}g_{i:j} = g_{i:j} (1 + p_{i:j}) = g_{i:j}$. Thus:

$$(p_{i:j}, g_{i:j}) = (p_{i:j}, g_{i:j}) \text{ fco } (p_{i:j}, g_{i:j}) \tag{3}$$

Because of this idempotency property, the fundamental carry operation can be applied to overlapping regions:

Theorem:
Given $i > m \geq k > j$, then $(p_{i:j}, g_{i:j})$ can be derived from $(p_{i:k}, g_{i:k}) \text{ fco } (p_{m:j}, g_{m:j})$

Proof:
The proof begins by applying Equation (1) to expand both terms in $(p_{i:k}, g_{i:k}) \text{ fco } (p_{m:j}, g_{m:j})$:

$$(p_{i:k}, g_{i:k}) = (p_{i:m+1}, g_{i:m+1}) \text{ fco } (p_{m:k}, g_{m:k}) \tag{4}$$
and
$$(p_{m:j}, g_{m:j}) = (p_{m:k}, g_{m:k}) \text{ fco } (p_{k-1:j}, g_{k-1:j}) \tag{5}$$

Substituting these for $(p_{i:k}, g_{i:k})$ and $(p_{m:j}, g_{m:j})$ in $(p_{i:k}, g_{i:k}) \text{ fco } (p_{m:j}, g_{m:j})$ yields:

$$(p_{i:k}, g_{i:k}) \text{ fco } (p_{m:j}, g_{m:j}) = [(p_{i:m+1}, g_{i:m+1}) \text{ fco } (p_{m:k}, g_{m:k})] \text{ fco } [(p_{m:k}, g_{m:k}) \text{ fco } (p_{k-1:j}, g_{k-1:j})]$$

Associativity from Equation (2) is applied to move the brackets:

$$(p_{i:k}, g_{i:k}) \text{ fco } (p_{m:j}, g_{m:j}) = [(p_{i:m+1}, g_{i:m+1}) \text{ fco } [(p_{m:k}, g_{m:k}) \text{ fco } (p_{m:k}, g_{m:k})]] \text{ fco } (p_{k-1:j}, g_{k-1:j})$$

By the idempotency of the fco operation from Equation (3) $[(p_{m:k}, g_{m:k}) \text{ fco } (p_{m:k}, g_{m:k})]$ reduces to $(p_{m:k}, g_{m:k})$:

$$(p_{i:k}, g_{i:k}) \text{ fco } (p_{m:j}, g_{m:j}) = [(p_{i:m+1}, g_{i:m+1}) \text{ fco } (p_{m:k}, g_{m:k})] \text{ fco } (p_{k-1:j}, g_{k-1:j})$$

Applying Equation (1) to the bracketed quantity:

$$(p_{i:k}, g_{i:k}) \text{ fco } (p_{m:j}, g_{m:j}) = (p_{i:k}, g_{i:k}) \text{ fco } (p_{k-1:j}, g_{k-1:j})$$

Applying Equation (1) again:

$$(p_{i:k}, g_{i:k}) \text{ fco } (p_{m:j}, g_{m:j}) = (p_{i:j}, g_{i:j}) \qquad \text{QED}$$

Since overlapping regions can be combined with the fundamental carry operation, the overlapping group propagate and generate signals $p_{31:16}, g_{31:16}$; $p_{23:16}, g_{23:16}$; and c_{16} can be applied to the least significant carry lookahead module on the third level to produce carries for c_{24}, c_{32}, and c_{40}. Since carries on all eight bit boundaries are now known, these are used in standard carry select fashion to select the correct result from the eight bit adders.

3 Implementation of the Redundant Cell Adder

The redundant cell adder was implemented for the Am29050 microprocessor using AMD's 1 μm minimum drawn feature size, two layer metal CMOS process. The basic Manchester carry chain cell is shown in Figure 2. Outputs are tapped from the appropriate points in the chain. In order to make the Manchester carry chain as fast as possible, each series transistor is sized to approximately fill the bit cell where it is placed. The tree is laid out by placing the Mcc modules in roughly the same position as they are shown in the block diagram of Figure 1. The third level (and higher level if necessary) Mcc modules are placed in holes left in the column of second level Mcc modules to reduce the width of the layout.

The adder floor plan is shown in Figure 3. Metal two is used for long horizontal runs carrying the input operands, the bit propagate and generate signals, the calculated carries, and the results. Metal one runs vertically, and is used for local interconnect. The LSB of the adder is at the bottom. The inputs come from the left, while the outputs leave at the right. The first block in the floor plan (going from left to right) is the propagate and generate logic. The second block is a stack of four bit Mcc modules connected in pairs to form eight bit ripple carry sections. A ZERO carry is input to each section in this column.

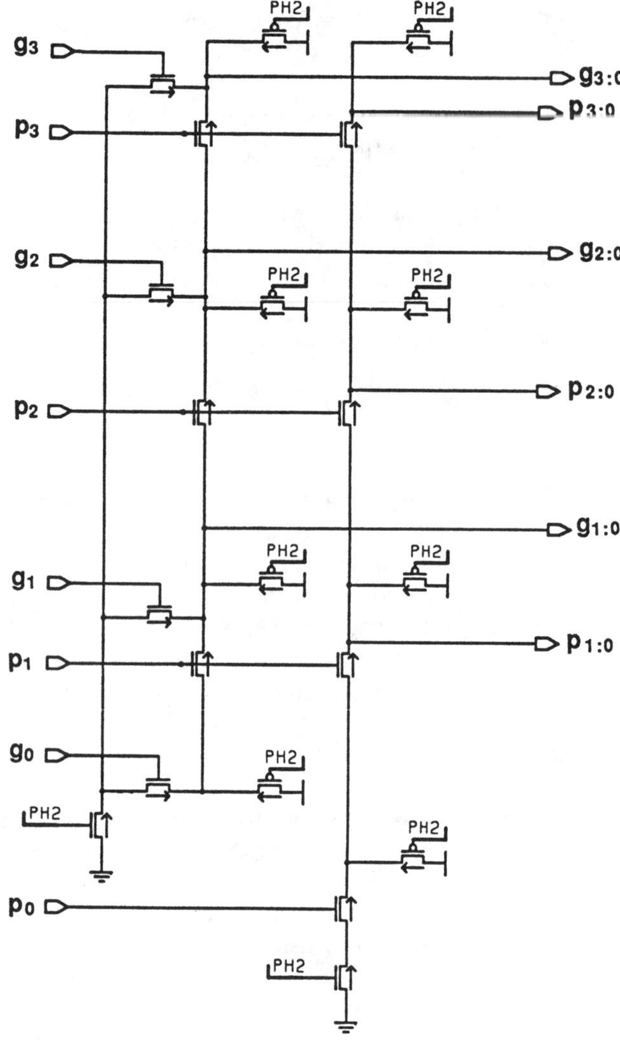

Figure 2.   Manchester Carry Chain Cell.

Figure 3.   Redundant Cell Adder Floor Plan.

The third column consists of ripple carry sections with a carry in of ONE. Next are the exclusive-OR gates for calculating the sums from the ripple carries and the bit propagates. The fifth column is the carry tree. The last column consists of multiplexers which select between the eight bit add results.

For the IEEE 754 floating point standard, double precision significand calculation, only 56 bits are required, so carry c_{56} is the carry out, and the top carry select adder (bits 56 through 63) and its multiplexer are omitted. Also, the associated carry tree logic may be omitted, which includes the top four Manchester carry chain blocks on the first level and the top Manchester carry chain block on the second level.

4   Performance

The width of the adder is roughly proportional to log n. The second level of the adder has holes large enough to contain three Manchester carry chains, so the third (and fourth if required) levels of the tree can be packed into the second level for adders of up to 256 bits. The total width is the sum of the widths of the blocks in the floor plan:

$$W = W_{pg} + (1 + \log_{16}N) W_{Mcc} + W_{csa8} \qquad (6)$$

Where: W is the total width of the N bit adder, W_{pg} is the width of the propagate/generate logic, W_{Mcc} is the width of the Manchester carry chain logic, and W_{csa8} is the width of the eight bit carry select adder. W is 450 µm wide (for $24 \leq N < 256$), as implemented in the AMD 1µm CMOS technology. The total area is the product of the width times the height:

$$A = W N H_c \qquad (7)$$

Where: H_c is the height of a bit slice adder cell. Since each bit cell is 71.6 µm high, the 56 bit adder shown on Figure 4 is 4010 µm high for a total area of 1.8×10^6 square µm. The height and area of adders for word sizes from 24 to 256 bits scales in direct proportion to the wordsize.

As with any carry select adder, the delay is determined by the slower of the sum computation and the carry computation. In this implementation, the sum computation is faster. For the carry computation, propagate and generate signals are already set up when the clock asserts. First, the least significant Manchester carry chain in the first level fires. The critical delay path signal travels through three levels of Manchester carry chains. The worst case load is seen by c_{16} since it drives two

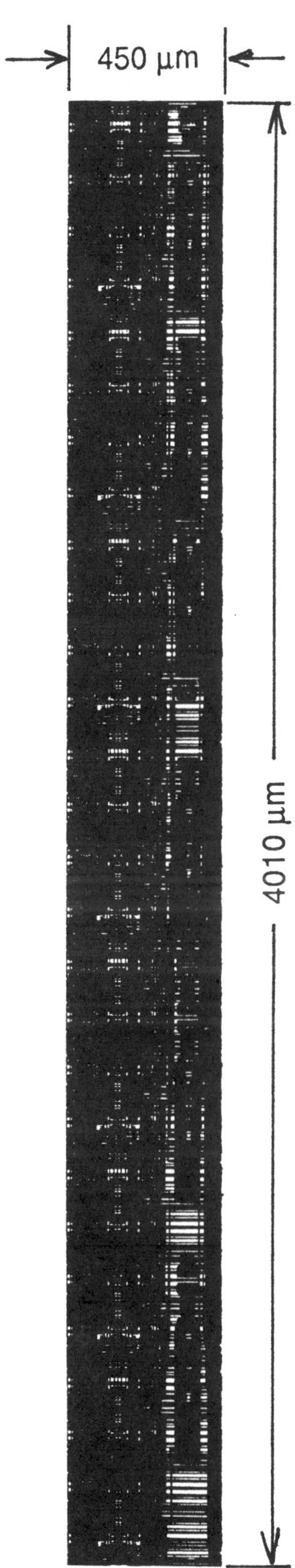

Figure 4. Redundant Cell Adder Layout.

Manchester carry chain inputs; the multiplexers are well buffered, so they present a small load. Finally the critical carry signal arrives at the multiplexer for the most significant eight bit adder.

$$D = (\log_4 N)\, D_{Mcc} + D_{mux} \tag{8}$$

Where: D is the delay of the N bit adder, D_{Mcc} is the delay of the Manchester carry chain, and D_{mux} is the delay of a 2:1 multiplexer. In the Am29050 microprocessor, the time from the rising edge of the clock to the sum is approximately 3.2 ns as shown in the electron beam timing waveforms on Figure 5. With comparable CMOS technologies the speed should be nearly constant for $24 \leq N < 64$.

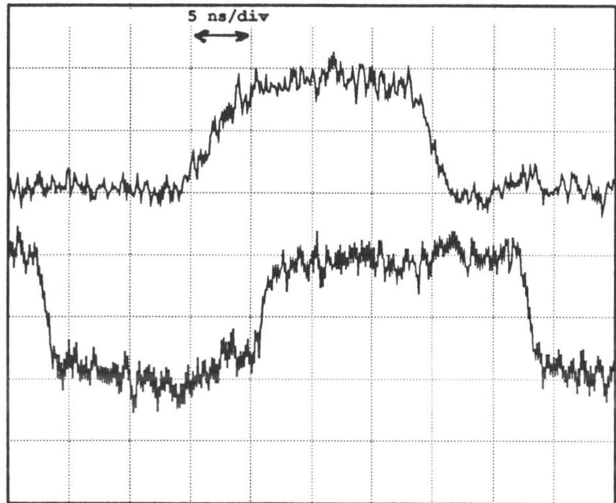

Figure 5. Measured Performance of the Redundant Cell Adder. The Top Waveform is the Clock and the Lower Waveform is the Most Significant Sum Bit.

5 Conclusion

The idempotence of the fundamental carry operation allows the construction of a fast adder that has small performance degradation as the add width grows. This adder is very fast, achieving a 3.2 ns measured add time for 56 bit operands and is of reasonable size, as shown by the Am29050 microprocessor implementation.

Acknowledgement

The design and initial implementation of the redundant cell adder was performed by the first author while he was employed by Advanced Micro Devices. Special thanks are due Stephen McIntyre, Danny English, and Tom Burghart for their help in implementing this adder.

References

[1] R. P. Brent and H. T. Kung, "A Regular Layout for Parallel Adders," *IEEE Transactions on Computers*, Vol. C-31, 1982, pp. 260-264.

[2] H. Ling, "High-Speed Binary Adder," *IBM Journal of Research and Development*, Vol. 25, pp. 156-166, May 1981.

[3] I. S. Hwang and A. L. Fisher, "A 3.2 ns 32-bit CMOS Adder in Multiple Output Domino Logic," *1988 IEEE International Solid-State Circuits Conference Digest of Technical Papers*, pp. 140, 141, 332, and 333.

[4] G. Bewick, P. Song, G. DeMichel, and M. J. Flynn, "Approaching a Nanosecond; a 32-bit Adder," *Proceedings of the 1988 IEEE International Conference on Computer Design: VLSI in Computers and Processors*, pp. 221-226.

[5] T. Kilburn, D. B. G. Edwards, and D. Aspinall, "Parallel Addition in Digital Computers: A New Fast "Carry" Circuit," *IEE Proceedings*, Vol. 106, pt. B, 1959, pp. 464-466.

Session 7:

Adders II

Chair:
Mary-Jane Irwin
Pennsylvania State University

Optimal Purely Systolic Addition

Lars Kühnel
Institut für Informatik und Praktische Mathematik
Christian-Albrechts-Universität
W–2300 Kiel 1, Germany

Abstract

We introduce a purely systolic hardware algorithm for addition which is based on a mesh connected arrangement of cells. The proposed $\mathcal{FASTA}$-algorithm is well-suited for realization in integrated technologies. Its area, computation time, and period are satisfying $A(n) = O(n)$, $T(n) = O(\sqrt{n})$, $P(n) = O(\sqrt{n})$, respectively, where n denotes the operand length. Therefore, this adder is T-, APT-, and AT^2-optimal in the linear model for signal propagation delays. In the class of $\Theta(\sqrt{n})$ time adders it is optimal with respect to A, P, T, AP, AT, APT, AP^2, and AT^2. The suggested algorithm essentially is a solution to the general problem of "parallel prefix computation". Therefore, it can serve as a paradigm for the design of optimal purely systolic hardware algorithms in a wide range of application domains.

1 Introduction

We consider the addition of non–negative integers in conventional positional binary n–bit representation. Given the $2n$ bits of two operands a and b, the task is to compute the $(n+1)$–bit representation of the sum s.

$$s = \sum_{i=0}^{n} s_i 2^i = a + b = \sum_{i=0}^{n-1} a_i 2^i + \sum_{i=0}^{n-1} b_i 2^i$$

For each $n \in \mathbb{N}$ (the *problem size*) this formulation defines a mapping $ADD_n : \mathbb{B}^n \times \mathbb{B}^n \rightarrow \mathbb{B}^{n+1}$ that maps each pair of n–bit operands to the corresponding $(n+1)$–bit tuple of the sum, leading to the sequence of mappings $ADD = (ADD_n)_{n \in \mathbb{N}}$.

In this paper we introduce a novel globally clocked hardware algorithm $\mathcal{FASTA}$ for the computation of ADD. $\mathcal{FASTA}$ actually is a sequence $(\mathcal{FASTA}_n)_{n \in \mathbb{N}}$ of hardware algorithms with $\mathcal{FASTA}_n$ computing ADD_n. This new adder is well–suited for realization in integrated technologies like VLSI, ULSI, and WSI. The algorithm can be easily extended in order to cover two's complement addition and subtraction, too.

According to the model introduced in [12], a hardware algorithm H is a tuple (G, L, S). G is a finite directed graph, the *computation graph* of H. The nodes correspond to processing elements (PEs) resp. input and output nodes and the edges correspond to the wires connecting the cooperating PEs. The *layout* L describes an embedding of G into the plane $\mathbb{R}^2$, thus

suggesting a two–dimensional realization of G. S is the *I/O–scheme* of H and specifies the times (i.e. clock ticks) and locations (i.e. input resp. output nodes) at which values for $\{a_0, \ldots, a_{n-1}, b_0, \ldots, b_{n-1}\}$ are read resp. for $\{s_0, \ldots, s_n\}$ are written.

Our complexity analysis refers to the usual VLSI complexity measures. Let $\mathcal{H} = (H_n)_{n \in \mathbb{N}}$ be a sequence of hardware algorithms. The *area* $A_{\mathcal{H}}(n)$ of H_n is the area of the smallest convex subset of $\mathbb{R}^2$ that contains the layout of H_n. The *time* (or *latency*) $T_{\mathcal{H}}(n)$ of H_n is the time elapsed between the application of the first input bit and the generation of the last output bit with respect to a single computation. Since hardware algorithms are working on infinite streams of arguments, we are additionally considering the *period* $P_{\mathcal{H}}(n)$ of H_n. The period measures the time elapsed between the application of the first input bits of two successive computations (i.e. the time between two successive sets of inputs). Note that the reciprocal of $P_{\mathcal{H}}(n)$ corresponds to the *throughput* of H_n. Moreover, we investigate the behaviour with respect to the usual product measures $AP_{\mathcal{H}}(n)$, $AT_{\mathcal{H}}(n)$, $APT_{\mathcal{H}}(n)$, $AT_{\mathcal{H}}^2(n)$, and $AP_{\mathcal{H}}^2(n)$.

In [5, 9] Foster and Kung have identified a set of properties that make a hardware algorithm especially well-suited for realization in integrated technologies. These *systolic* algorithms employ a high degree of parallelism and pipelining for the efficient solution of a given problem. Based on the above papers, Schmeck [12] suggests the following concise definition of systolic algorithms.

- The computation graph uses only few types of simple cells.
- The layout suggests a realization with simple and regular data and control flows.
- For each node, the number of direct neighbours is small and does not depend on the problem size.
- The design uses extensive concurrency.
- For each node, the area occupied by its realization does not depend on the problem size.
- The suggested realization uses only physically local communication for the exchange of data between PEs. In particular, the length of the longest wire used for data communication does not depend on the problem size.
- The realization can be operated at a clock frequency that does not depend on the problem size.
- Average and maximum data rates have the same magnitude.

It is crucial to note that *purely systolic* hardware algorithms totally avoid long–distance or irregular wires for data communication (cf. [9]).

Purely systolic hardware algorithms are in particular promising low costs for the design of realizations, high throughput, modular expansibility without slowing down the system clock, and steady use of the interface to their environment. Moreover, some of the above properties prove to be advantageous with respect to *testability* aspects. Mesh connected array–like hardware algorithms with simple cells are important examples for purely systolic systems and are thus ideal candidates for the realization in integrated technologies.

Note that the well–known fast adders (see e.g. [1, 2, 13]) are based on tree–like hardware structures and thus don't belong to the class of purely systolic systems. A first hint concerning fast *and* purely systolic addition can be found in [3]. Unfortunately, Chazelle and Monier's description is not very detailed . Moreover, their algorithm is not *purely* systolic in the above strong sense.

In this paper we propose a new hardware algorithm for f̲ast purely s̲ystolic a̲ddition (*FASTA*). The *FAST* adder is T-, APT-, and AT^2–optimal in the linear model for signal propagation delays. Moreover, it is optimal with respect to *all* usual measures in the class of time optimal (i.e. $\Theta(\sqrt{n})$ time) adders. Since the algorithm is based on the well–known carry lookahead technique, it suggests a purely systolic solution to the general problem of *parallel prefix computation* (see [10]). Thus the *FASTA* hardware algorithm is of significant relevance for a wide range of application domains.

The paper is organized as follows. Section 2 contains a concise summary of lower bounds in the linear model and includes a special treatment of time optimal hardware algorithms for addition. Based on the list of lower bounds, we develop the *FASTA* algorithm in Section 3. Section 4 contains some remarks concerning additional properties and possible extensions of the *FAST* adder.

2 Lower Bounds for Purely Systolic Addition

Obviously, the value of the mapping ADD_n depends on the operands. Thus a hardware algorithm $\mathcal{H} = (H_n)_{n \in \mathbb{N}}$ for addition has to read the argument bits and to perform a calculation. Since there exists a minimum feature width (cf. e.g. [4]), this results in

$$A_\mathcal{H}(n) = \Omega(1) .$$

A simple counting argument, involving the number of bits to be produced and the number of output nodes that are used for these output purposes, leads to (see e.g. [4, 8])

$$AT_\mathcal{H}(n) = \Omega(n) \text{ and } AP_\mathcal{H}(n) = \Omega(n) . \tag{1}$$

If the number of output nodes used by H_n is $O(n)$, then

$$P_\mathcal{H}(n) = \Omega(1) \text{ and thus } AP_\mathcal{H}^2(n) = \Omega(n) . \tag{2}$$

Clocked versions of a ripple carry adder (see e.g. [6]) with a constant number of full–adder cells resp. of an n–cell ripple carry adder show that the above lower bounds are tight.

The lower bound for the computation time heavily depends on the underlying model for signal propagation delays. The widely used *constant model* assumes that the time for propagating a signal across a wire of length L does not depend on L. Thus this model should not be used as a basis for the derivation of purely systolic hardware algorithms since it does not punish the designer for using long wires. In a purely systolic algorithm, every datapath of length L crosses $\Theta(L)$ processing elements with each of these elements contributing at least one clock tick to the overall delay of a signal that has to be propagated across this datapath. Thus it is appropriate to base the analysis on the *linear model* advocated in [4]. In this model, the time for the propagation of a signal across a wire of length L is assumed to be $\Omega(L)$. Chazelle and Monier [4] have shown that this assumption leads to the following lower bounds for addition.

$$T_\mathcal{H}(n) = \Omega(\sqrt{n}) \tag{3}$$
$$APT_\mathcal{H}(n) = \Omega(n^2) \tag{4}$$
$$AT_\mathcal{H}^2(n) = \Omega(n^2) . \tag{5}$$

Note that the "conventional" fast adders are not time optimal in the linear model since they are based on tree structures and therefore employ wires whose length depend on n. Due to the latter fact, the length of a basic clock cycle depends on n, too. In general, this leads to a computation time in $\omega(\sqrt{n})$. For instance, the carry lookahead adder suggested in [2] has a time complexity $\Omega(n)$ in the linear model.

Section 3 shows that there exists a purely systolic adder whose time is $\Theta(\sqrt{n})$. Thus the lower bound 3 is tight. We derive from 5 that the area of a time optimal adder Opt satisfies

$$A_{Opt}(n) = \Omega(n) . \tag{6}$$

Moreover, "reasonable" time optimal adders satisfy

$$P_{Opt}(n) = \Omega(\sqrt{n}) \tag{7}$$

(this bound is shown in [8]). A combination of 3, 6, and 7 yields

$$AP_{Opt}(n) = \Omega(n\sqrt{n}) , \; AT_{Opt}(n) = \Omega(n\sqrt{n}) ,$$

$$AP_{Opt}^2(n) = \Omega(n^2) .$$

Table 1 contains a compilation of the above bounds. Complete proofs can be found in [8].

3 *FASTA*: A hardware algorithm for fast purely systolic addition

The *FAST* adder is based on the well–known technique of carry lookahead addition. The following subsection outlines the basic ideas of this technique.

3.1 Carry Lookahead Technique

The bits of the sum s can be computed through the use of carry signals c_i, $i \in [-1, n-1]$. (Given two integers i, j, let $[i, j]$ denote the set of integers $\{k \mid i \leq k \text{ and } k < j\}$. Furthermore, let $\wedge, \vee, \oplus$ denote the logical AND, OR, and exclusive OR, respectively.)

$$
\begin{aligned}
c_{-1} &= 0 \\
c_i &= (a_i \wedge b_i) \vee (a_i \wedge c_{i-1}) \vee (b_i \wedge c_{i-1}) \\
s_i &= a_i \oplus b_i \oplus c_{i-1} \\
s_n &= c_{n-1}
\end{aligned}
$$

The carry lookahead technique involves assigning a pair of bits $(G[i,j], P[i,j])$ to each group of bit positions $[i,j] \subseteq [0, n-1]$ with G and P satisfying

$(G[i,j] = 1) \iff$ The group $[i,j]$ generates a carry into position $(j+1)$ regardless of an incoming carry.

$(P[i,j] = 1) \iff$ The group $[i,j]$ propagates an incoming carry from position $(i-1)$ into position $(j+1)$.

Obviously, we have $G[0,i] = c_i$ for all $i \in [0, n-1]$. The (G, P) pair of a single bit position i can be computed according to

$$ G[i,i] = a_i \wedge b_i \; ; \; P[i,i] = a_i \oplus b_i \; . $$

Introducing the operator $\circ : \mathbb{B}^2 \times \mathbb{B}^2 \to \mathbb{B}^2$ with

$$ (G, P) \circ (G', P') = (G' \vee (P' \wedge G), P \wedge P') \, , $$

we derive the property

$$ (G[i,j], P[i,j]) = (G[i,k-1], P[i,k-1]) \circ (G[k,j], P[k,j]) $$

for all $k \in [i+1, j]$.

The concept of carry lookahead addition is based on the observation that the operator 'o' is associative. Due to the latter fact, the carries $G[0,i]$ can be computed from the single bit (G, P) pairs in any order as long as only associativity is exploited.

3.2 Design Considerations

The derivation of a purely systolic time optimal adder was guided by the following crucial ideas. Throughout the rest of the paper, let $n = m^2, m \in \mathbb{N}$.

Firstly, the n bit positions are divided into $\sqrt{n}$ groups of $\sqrt{n}$ bit positions each. This leads to the sequence of *blocks*

$$ ([q\sqrt{n}, (q+1)\sqrt{n}-1])_{q \in [0, \sqrt{n}-1]} \; . $$

The (G, P)–pair of each of these blocks is computed in a ripple carry adder like fashion using the property

$$
\begin{aligned}
&(G[q\sqrt{n}, k], P[q\sqrt{n}, k]) = \\
&\quad (G[q\sqrt{n}, k-1], P[q\sqrt{n}, k-1]) \circ (a_k \wedge b_k, a_k \oplus b_k) \, .
\end{aligned}
$$

Thus each of these block computations takes $O(\sqrt{n})$ time. It is possible to pipeline the block computations and therefore to complete the computation of *all* $\sqrt{n}$ block–(G, P)s in $O(\sqrt{n})$ time.

The sequence of *cumulative block carries*

$$ (G[0, q\sqrt{n} - 1])_{q \in [1, \sqrt{n}]} $$

can be computed iteratively from the block–(G, P)s using the equality

$$
\begin{aligned}
G[0, q\sqrt{n}-1] = \; &G[(q-1)\sqrt{n}, q\sqrt{n}-1] \vee \\
&\vee (P[(q-1)\sqrt{n}, q\sqrt{n}-1] \wedge G[0, (q-1)\sqrt{n}-1]) . \quad (8)
\end{aligned}
$$

The generation of the sum bits s_i is divided into two stages. Let $i \in [q\sqrt{n}, (q+1)\sqrt{n}-1]$, $i = q\sqrt{n}+k$, i.e. i belongs to the qth block. s_i satisfies the equation

$$ s_i = a_i \oplus b_i \oplus c_{i-1} = a_i \oplus b_i \oplus G[0, i-1] $$

and thus

$$
s_i = a_i \oplus b_i \oplus \\
\oplus (\underbrace{G[q\sqrt{n}, i-1]}_{\substack{\text{intra–block} \\ \text{carry}}} \vee (P[q\sqrt{n}, i-1] \wedge \underbrace{G[0, q\sqrt{n} - 1]}_{\substack{\text{carry into} \\ q\text{th block}}})).
$$

Note that $(G[q\sqrt{n}, i-1], P[q\sqrt{n}, i-1])$ are intra–block (G, P) signals, whereas $G[0, q\sqrt{n}-1]$ is the cumulative block carry entering the qth block.

Stage 1 of the sum bit generation process performs the computation of *preliminary sum bits* $s[q\sqrt{n}, i]$ from the operands a_i, b_i and the intra–block carries:

$$ s[q\sqrt{n}, i] = a_i \oplus b_i \oplus G[q\sqrt{n}, i-1] \; . $$

Exploiting the property

$$ s_i = s[q\sqrt{n}, i] \oplus (P[q\sqrt{n}, i-1] \wedge G[0, q\sqrt{n}-1]) \, , $$

the second stage of the sum bit generation computes the s_is from the preliminary sum bits, the intra–block propagate signals, and the sequence of cumulative block carries.

The generation of the preliminary sum bits and of the cumulative block carries can be integrated into the ripple carry adder like computation mentioned above. Stage 2 of the sum generation is performed by using an additional linear array of identical PEs with a time skewed input.

The following list summarizes the basic ideas that have led to the $\mathcal{FASTA}$ hardware algorithm.

- Carry lookahead addition, using $\sqrt{n}$ blocks of $\sqrt{n}$ successive bit positions.
- Ripple carry adder-like (G, P) computation within blocks, using a linear array of $\sqrt{n}$ PEs.
- Pipelining of block computations on the latter array.

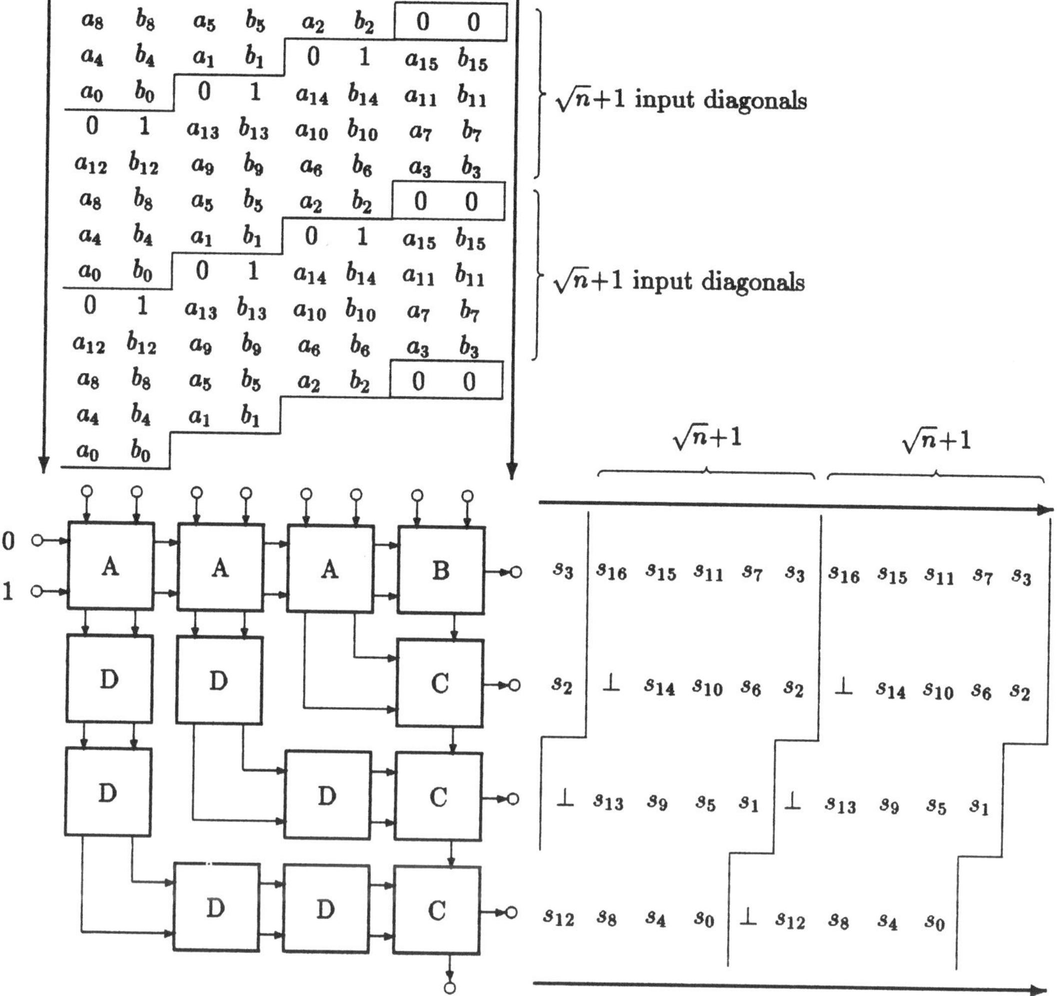

Figure 1: Layout and I/O–scheme of $\mathcal{FASTA}_{16}$.

- Iterative computation of the cumulative block carries by the rightmost cell.
- Two-stage generation of sum bits. Computations of Stage 1 incorporated into the above ripple carry adder-like computation. Computations of Stage 2 performed by a second linear array of $\sqrt{n}-1$ cells. Intermediate results of first stage are delayed appropriately, using a sequence of $\sqrt{n}-2$ shift registers of decreasing length.

3.3 An Overview

Figure 1 gives the layout of the computation graph and sketches the I/O–scheme of $\mathcal{FASTA}_n$ for $n = 16$. The layout uses a mesh connected array like arrangement of $n - (\sqrt{n} - 1)$ cells with the linear array of $\sqrt{n} - 1$ A–cells and the B–cell in the top row performing the ripple carry adder like computation mentioned above. This linear array produces the preliminary sum bits and provides the sequence of cumulative block carries via the bottom output line of the B–cell. The latter sequence is read by the top cell of a linear array of $\sqrt{n}-1$ C–cells in the rightmost column, which implements the second stage of sum bit generation. Accordingly, the intermediate results of the linear array of A–cells have to be delayed appropriately in order to meet the corresponding cumulative block carries in the array of

C–cells. This can be achieved by using the indicated arrangement of $n - 3\sqrt{n} + 2$ D–cells with each D–cell simply delaying its two input signals by one clock tick. The input is read in the form of $\sqrt{n}$ diagonals of $2\sqrt{n}$ operand bits, each input diagonal corresponding to one of the blocks introduced in 3.2. An additional diagonal of constant pairs separates the inputs of successive computations. The output is generated in the form of $\sqrt{n} + 1$ distorted diagonals of $\sqrt{n}$ sum bits each. Again, each diagonal corresponds to one of the blocks of 3.2. The last output diagonal of a computation contains the most significant sum bit s_n and $\sqrt{n}-1$ "don't cares".

3.4 Components

First of all, we will describe the linear array of A– and B–cells. Let $i = q\sqrt{n} + k < (q + 1)\sqrt{n}$. The A–cells are used for computing the intra–block signals $(G[q\sqrt{n}, i], P[q\sqrt{n}, i])$ and the preliminary sum bits $s[q\sqrt{n}, i]$ from the inputs $(G[q\sqrt{n}, i-1], P[q\sqrt{n}, i-1])$, a_i, and b_i (see Fig. 2). Moreover, an A–cell propagates the incoming intra–block P–signal via the second bottom output line. Note that the outputs are produced on the basis of a global clock. Thus the output signals indicated in Fig. 2 are available at clock tick $t + 1$, provided that the shown inputs are applied to the cell

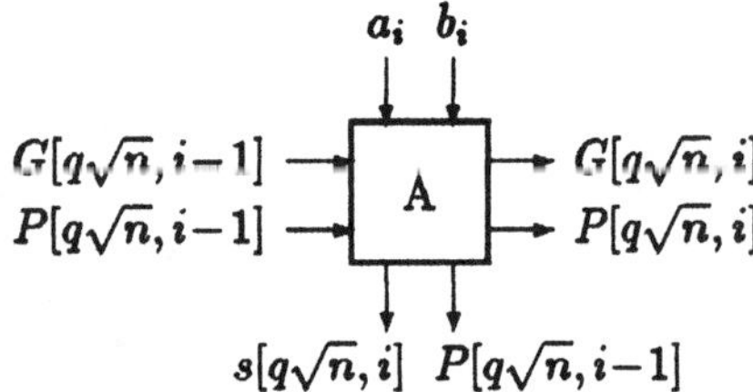

Figure 2: I/O–behaviour of the A–cells

input lines at clock tick t. Obviously, the A–cells can be realized by using a small number of logical gates and flipflops.

A linear array of $\sqrt{n}$ A–cells together with a time skewed input (as shown in Fig. 1) could be used for generating all the preliminary sum bits and their corresponding intra–block propagate signals. It is crucial to note that the rightmost A–cell produces the signals that are needed for computing the sequence of cumu-

$$G[q\sqrt{n},(q+1)\sqrt{n}-2] \to \boxed{A} \to G[q\sqrt{n},(q+1)\sqrt{n}-1]$$
$$P[q\sqrt{n},(q+1)\sqrt{n}-2] \to \quad \to P[q\sqrt{n},(q+1)\sqrt{n}-1]$$

Figure 3: I/O of a hypothetical $\sqrt{n}$th A–cell

lative block carries (cf. 8 and Figure 3). Moreover, since

$$s_{(q+1)\sqrt{n}-1} = s[q\sqrt{n},(q+1)\sqrt{n}-1] \vee$$
$$\vee \left(P[q\sqrt{n},(q+1)\sqrt{n}-2] \wedge G[0,q\sqrt{n}-1]\right) ,$$

the same holds for the sequence of sum bits $(s_{(q+1)\sqrt{n}-1})_{q\in[0,\sqrt{n}-1]}$, too. Therefore, we incorporate the corresponding computations into the hypothetical $\sqrt{n}$th A–cell, thus introducing a new cell type B. Figure 5 outlines a possible realization of the B–cell. The symbols used for standard logical components are described in Fig. 4.

The I/O–behaviour of the complete linear array is given in Fig. 6. At the beginning of a computation, the D–flipflop controlling the O2 output line of the B–cell has to be cleared. Therefore, the input pair $(0,0)$ precedes the first pair of operand bits in the sequence of inputs for the B–cell. At the end of a computation, this D–flipflop contains the value $G[0,n-1] = s_n$. In Fig. 6, the last input diagonal of $\sqrt{n}-1$ $(0,1)$–pairs and a single $(0,0)$–pair generates the input $I1 = I2 = I4 = 0$ and $I3 = 1$ with respect to the B–cell. It can be easily verified that this input combination makes the content of the O2–flipflop (i.e. s_n) appear on the O1 output line of the B–cell.

Since the input pair $(0,0)$ of the last input diagonal clears the O2–flipflop of the B–cell, successive computations can be pipelined as indicated in Fig. 1.

The sequences of values available at the bottom output lines of the above linear array are comprising ex-

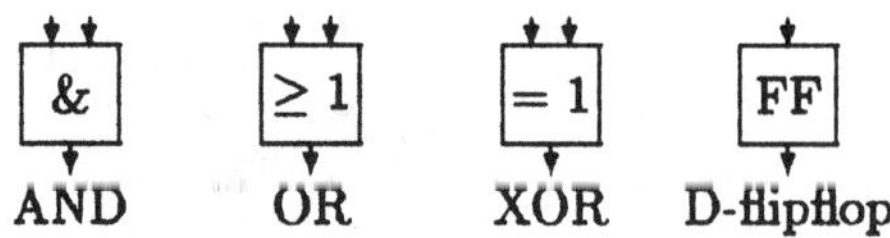

Figure 4: Symbols for logical components

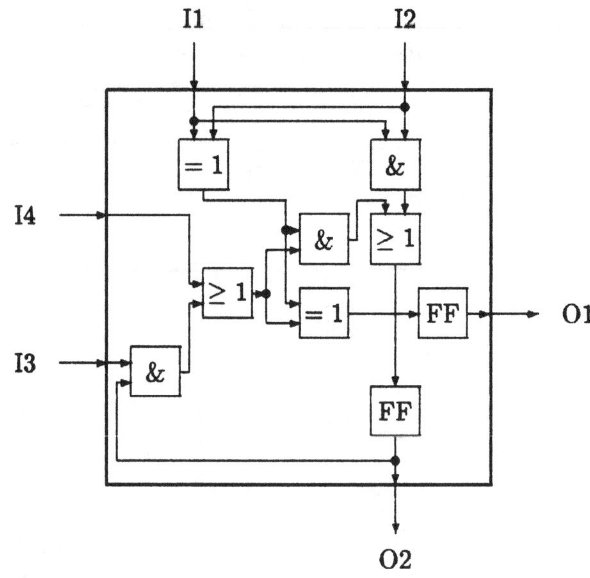

Figure 5: Realization of the B–cell

actly the inputs needed for the second stage of sum bit generation. Thus a linear array of $\sqrt{n}-1$ cells can be used to perform the corresponding computations. The latter array uses cells of type C, whose I/O–behaviour is given by Fig. 7. Figure 8 outlines the I/O–behaviour of the complete linear array of C–cells. The time skewed input format for this array can be generated by employing the arrangement of D–cells given in Fig. 1.

3.5 Analysis

Each cell of the $\mathcal{FAST}$ adder occupies only constant area. Thus the complete layout can be embedded into an $O(\sqrt{n}) \times O(\sqrt{n})$ array of constant area rectangles. Thus

$$A_{\mathcal{FASTA}}(n) = O(n) .$$

$$G[0,q\sqrt{n}-1]$$
$$P[q\sqrt{n},i-1] \to \boxed{C} \to s_i$$
$$s[q\sqrt{n},i] \to$$
$$G[0,q\sqrt{n}-1]$$

Figure 7: I/O–behaviour of the C–cells

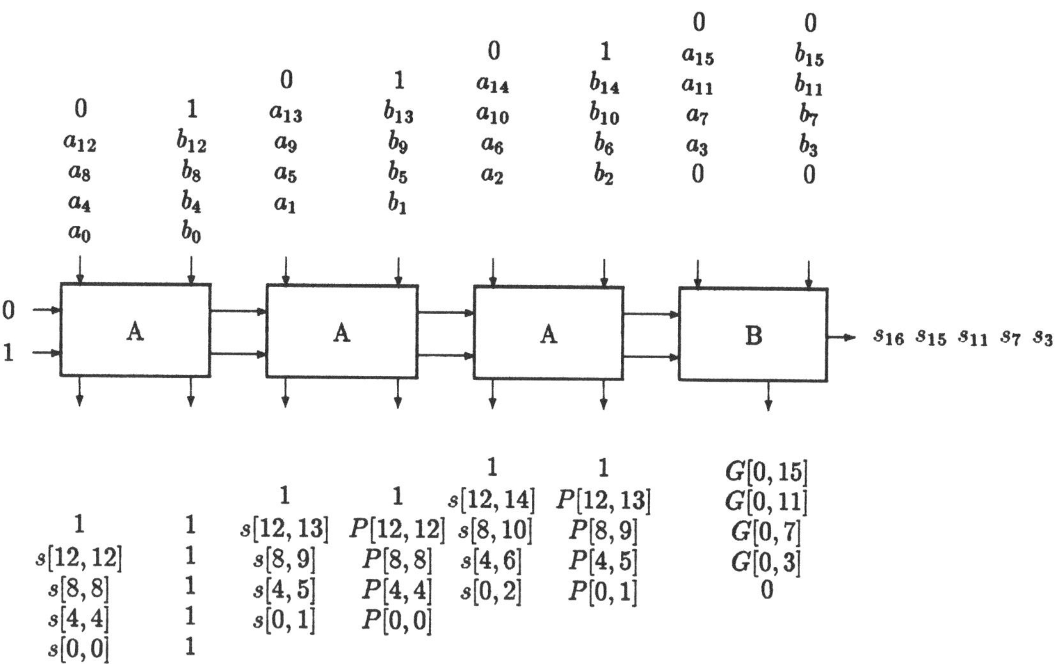

Figure 6: I/O–behaviour of the top row of cells

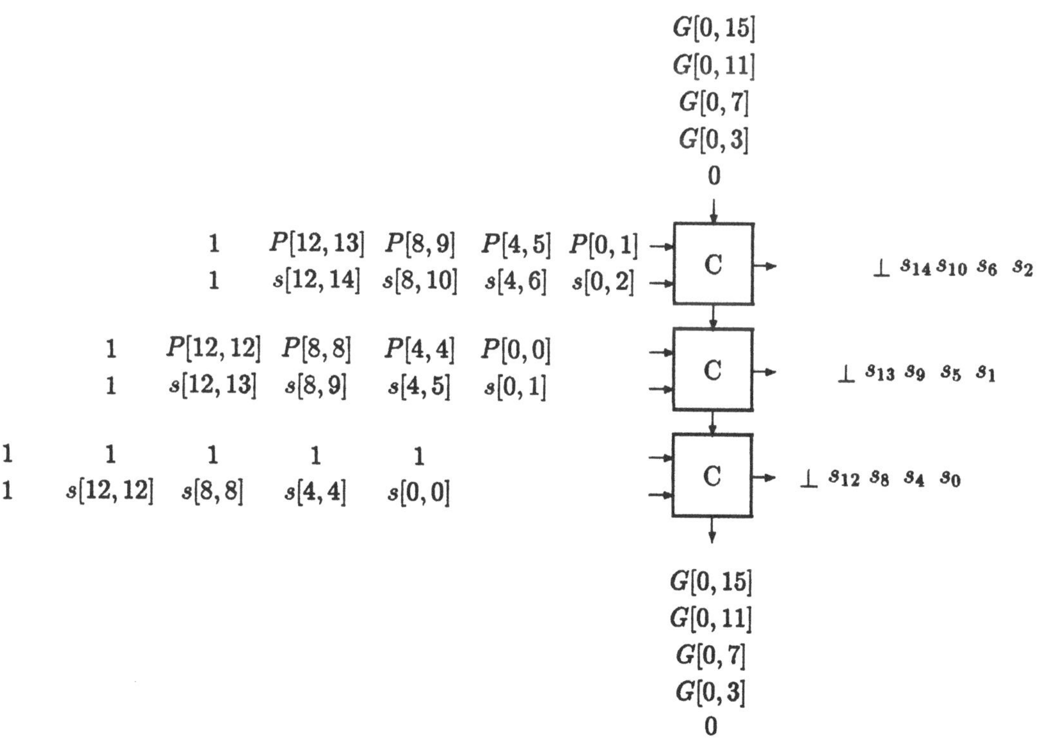

Figure 8: I/O–behaviour of the array of C–cells

	General lower bound	Lower bound for time-optimal addition	$\mathcal{FASTA}_n$
$A(n)$	$\Omega(1)$	$\Omega(n)$	$O(n)$
$P(n)$	$\Omega(1)$	$\Omega(\sqrt{n})$	$O(\sqrt{n})$
$T(n)$	$\Omega(\sqrt{n})$	$\Theta(\sqrt{n})$	$O(\sqrt{n})$
$AP(n)$	$\Omega(n)$	$\Omega(n^{3/2})$	$O(n^{3/2})$
$AT(n)$	$\Omega(n)$	$\Omega(n^{3/2})$	$O(n^{3/2})$
$APT(n)$	$\Omega(n^2)$	$\Omega(n^2)$	$O(n^2)$
$AP^2(n)$	$\Omega(n)$	$\Omega(n^2)$	$O(n^2)$
$AT^2(n)$	$\Omega(n^2)$	$\Omega(n^2)$	$O(n^2)$

Table 1: Tight bounds for addition

Let $\tau_{\mathcal{FASTA}}(n)$ denote the minimum length of a basic clock cycle of $\mathcal{FASTA}_n$. The switching time of each cell is $O(1)$. Therefore, we have $\tau_{\mathcal{FASTA}}(n) = O(1)$. One can easily verify that

$$T_{\mathcal{FASTA}}(n) = (3\sqrt{n} - 3) \cdot \tau_{\mathcal{FASTA}}(n)$$
$$P_{\mathcal{FASTA}}(n) = (\sqrt{n} + 1) \cdot \tau_{\mathcal{FASTA}}(n) .$$

Thus we deduce

$$T_{\mathcal{FASTA}}(n) = O(\sqrt{n})$$
$$P_{\mathcal{FASTA}}(n) = O(\sqrt{n})$$

for the time and period of the $\mathcal{FASTA}$ hardware algorithm.

Table 1 summarizes the asymptotic behaviour of $\mathcal{FASTA}_n$ with respect to A, P, T, and the usual compound measures. Moreover, it is confronting these upper bounds with the corresponding lower bounds given in Section 2.

We derive from the entries of this table that the $\mathcal{FAST}$ adder is T-, APT-, and AT^2-optimal. In the class of time optimal adders the $\mathcal{FASTA}$ algorithm is optimal with respect to *all* usual measures.

Note that the computation graph uses only four types of simple cells. Due to the mesh connected array-like structure, the data flow is *particularly* simple and regular. The maximum number of direct neighbours of a node is determined by the leftmost A–cell and thus equals 6 (the input nodes have to be considered, too). The chosen arrangement results in *very* short data communication lines, the maximum wire length being bounded from above by a constant. The adder is extensively exploiting concurrency mainly through a high degree of parallelism. Moreover, the average and maximum data rates have the same magnitude. Thus we conclude that the $\mathcal{FASTA}$ hardware algorithm is purely systolic.

4 Additional Properties and Extensions

Employing a standard technique outlined in [6, p. 71], the $\mathcal{FAST}$ adders can be easily extended in order to be capable of performing addition/subtraction of integers in two's complement representation.

Each single adder $\mathcal{FASTA}_n$ is quite adaptable to varying operand lengths. Note that the computation graph of $\mathcal{FASTA}_N$, N satisfying $N < n$, appears as a subgraph of the computation graph of $\mathcal{FASTA}_n$ (see Fig. 1). Therefore, ADD_N can be computed by the computation graph of $\mathcal{FASTA}_n$ at no increase in computation time and period, when compared to the performance of $\mathcal{FASTA}_N$. Since the underlying computation principle essentially does not depend on the number of input diagonals, the computation graph of $\mathcal{FASTA}_n$ can tackle the case $N > n$, too. There exists a straightforward modification of the $\mathcal{FASTA}_n$-I/O-scheme which leads to a latency $(N/\sqrt{n} + 2\sqrt{n} - 3) \cdot \tau_{\mathcal{FASTA}}(n)$ and period $(N/\sqrt{n} + 1) \cdot \tau_{\mathcal{FASTA}}(n)$ for the computation of ADD_N.

Besides from the criteria considered above, there exists another criterion of increasing importance for the valuation of the practical relevance of a hardware algorithm: the *testability* with respect to a given fault model. The testability of a hardware algorithm relates to the minimum number of input patterns needed to detect all possible faults considered by the underlying fault model. There exists a slightly modified version of the $\mathcal{FASTA}$ hardware algorithm which is *C–testable* with respect to the widely used *single stuck-at* fault model (see e.g. [11]). This means that the minimum number of test patterns needed to detect all such faults does not depend on n. It suffices to furnish each of the A–cells with an additional OR-gate and an additional flipflop in order to allow an 11 pattern test sequence to detect each possible stuck-at fault (see [8]).

The application of the carry lookahead technique does not depend on the radix r of the underlying number representation. Thus, choosing $r = 2^m$ and employing a conventional m–bit representation for the r–digits leads to a class of coarse–grained $\mathcal{FAST}$ adders in which each cell handles m–bit blocks of bit positions. Note that the A–, B–, and C–cells of these adders essentially have to perform an m–bit addition. Thus it may be appropriate to generate *hybrid* adders which use the purely systolic $\mathcal{FASTA}$ structure for the high level inter-node communication and a binary tree based addition scheme within the PEs.

It is well–known that carry lookahead adders are essentially a solution to the general problem of *parallel prefix computation* (PPC, see [10]). PPCs occur in the solution of many other problems such as the simulation of finite-state machines, linear recurrences, digital filtering, various graph problems, sorting, and others. Therefore, the $\mathcal{FAST}$ adder suggests a general paradigm for the derivation of purely systolic hardware algorithms in a wide range of application domains.

5 Conclusion

In this paper we have introduced a novel purely systolic hardware algorithm for integer addition which is well-suited for realization in integrated technologies like VLSI, ULSI, WSI, and discrete technologies as well. It is shown in [7] that the use of few types of simple cells and interconnection patterns together with a rigorous specification of the I/O-behaviour lead to a straightforward and comparatively simple proof for this parallel algorithm.

If the linear model for signal propagation delays is assumed, the $\mathcal{FASTA}$ hardware algorithm turns out to be T-, APT-, and AT^2-optimal. In the class of time optimal adders it is optimal with respect to all usual complexity measures for VLSI-algorithms. Its concrete area requirements are essentially determined by $(2\sqrt{n}-1)$ logical nodes. The overall $\mathcal{FASTA}$ area of $O(n)$ has to be compared with the lower bound $\Omega(n^2)$ for the area of very fast non-purely systolic adders (cf. [14]).

The $\mathcal{FAST}$ adder can be easily adapted to varying operand lengths. Moreover, there exist straightforward augmentations of the computation graph which render possible the processing resp. generation of non-diagonal input and output formats without changing the asymptotical properties of the $\mathcal{FASTA}$ algorithm. Besides from these properties, this new class of adders is especially advantageous with respect to testability aspects.

Since the suggested algorithm allows the generation of optimal adders at low costs and, on the other hand, gives a basic principle for the purely systolic solution of many other problems, the $\mathcal{FASTA}$ hardware algorithm seems to be of significant practical and theoretical relevance.

Future work should include the practical realization of the $\mathcal{FAST}$ adder. Moreover, the testability analysis should be extended to more general fault models. The incorporation of fault tolerance mechanisms is another important topic for future research.

Due to lack of space, this paper merely outlines the concept of the $\mathcal{FAST}$ adder. For a complete description of all details and a proof of correctness see [7]. [8] contains a detailed and systematic treatment of purely systolic hardware algorithms for parallel prefix computations.

Acknowledgements

I am especially indebted to Dr. H. Schmeck for many valuable discussions and suggestions. Moreover, I wish to thank Prof. Dr. W. Thomas and my colleagues of the VLSI group in Kiel for their support. My thanks also go to the anonymous referees. Their comments have helped to improve the presentation of the paper.

References

[1] B. Becker, R. Kolla. On the construction of optimal time adders. In *Proc. STACS 88*, pages 18–28, Springer-Verlag 1988. LNCS 294.

[2] R. P. Brent, H. T. Kung. A regular layout for parallel adders. *IEEE Trans. Computers*, C-31(3):260–264, 1982.

[3] B. Chazelle, L. Monier. Optimality in VLSI. In J. P. Gray, ed., *VLSI 81*, pages 269–278, London, 1981. Academic Press.

[4] B. Chazelle, L. Monier. A model of computation for VLSI with related complexity results. *J. ACM*, 32(3):573–588, 1985.

[5] M. J. Foster, H. T. Kung. The design of special-purpose VLSI chips. *IEEE Computer*, 13(1):26–40, 1980.

[6] K. Hwang. *Computer Arithmetic. Principles, Architecture, and Design.* Wiley, New York, 1979.

[7] L. Kühnel. Optimal purely systolic addition. Technical Report 9002, Inst. f. Informatik, Universität Kiel, W–2300 Kiel 1. Germany, April 1990.

[8] L. Kühnel. Optimale systolische Präfixberechnungen. Dissertation. In preparation, 1991.

[9] H. T. Kung. Why systolic architectures? *IEEE Computer*, 15(1):37–46, 1982.

[10] R. E. Ladner, M. J. Fischer. Parallel prefix computation. *J. ACM*, 27(4):831–838, 1980.

[11] P. K. Lala. *Fault Tolerant and Fault Testable Hardware Design.* Prentice-Hall, London, 1985.

[12] H. Schmeck. Modellierung und Bewertung von VLSI-Algorithmen. Habilitationsschrift. Institut für Informatik. Universität Kiel, 1989.

[13] J. Sklansky. Conditional-sum addition logic. *IRE Trans. Electr. Comp.*, EC-9(2):226–231, 1960.

[14] B. Sugla, D. A. Carlson. Extreme area-time trade-offs in VLSI. *IEEE Trans. Computers*, 39(2):251–257, 1990.

<h1 style="text-align:center">Constant Time Arbitrary Length
Synchronous Binary Counters</h1>

J. E. Vuillemin

Digital Equipment Corp.

Paris Research Laboratory

85 Av. Victor Hugo

92500 Rueil-Malmaison, France

Abstract

We introduce a synchronous binary counter which can be operated under a high clock frequency, independent of the counter's length n: all signals traverse at most two 3-inputs logic gates during each clock phase. The proposed design is simple enough to have practical implications, as illustrated by a cMOS programmable gate array implementation which has counted up to 2^{40} with a 40MHz clock. The area required for laying out our design is no larger than that of the (much slower) carry-ripple counter.

1 Introduction

Binary counters are found in nearly all digital systems, and their design is extensively covered in all hardware courses and textbooks.

- Since their implementation is so simple, binary counters are seldom in the *critical path* of a synchronous digital design. Whatever device the counter is controlling is typically more complex, hence slower than our small friend.

- Even in technologies where a synchronous counter can be operated with a 1GHz (1ns period) clock, one can only count up to about 2^{50} during the course of one day. Practical uses for longer counters are thus doubtful.

We have nevertheless two reasons for being interested in fast counters.

1. Counters provide the most basic mean for testing, debugging and measuring digital systems. In order for such an instrumentation to be effective, it needs to operate *faster* than the system under observation. In this context, counters thus naturally need to be the *fastest possible designs* in any given technology.

2. The theory of binary arithmetic circuits has reached a very clean conceptual state: time $log_2(n)$ is necessary and sufficient for both n bits addition ([2],[7],[4]) and multiplication ([6],[8],[5]).

While the $log_2(n)$ lower bound of [7] also applies to circuits performing combinatorial incrementation, it breaks down for *synchronous* counters. This is shown by [3], who construct a synchronous binary counter whose clock period is independent of the counter size, thus showing that incrementation can be performed substantially faster than addition.

3. The counter presented here is similar in nature to that of [3]; it is, however, somewhat simpler, using half as many flip-flops, with the same clock speed. In a comparative 64 bits test implementation (see below), our counter's area is twice smaller than that of [3].

We exclude from our discussion carry-save, pipelined and asynchronous counters, which share some, but not all features in our design. Our timing model is the simple gate-depth count, with limited fan-in. It ignores fan-out and far-away signal distribution problems, which have to be delt with specifically for each implementation technology.

2 Linear time binary counters

Let us review some classical designs, starting with the *carry ripple counter*:

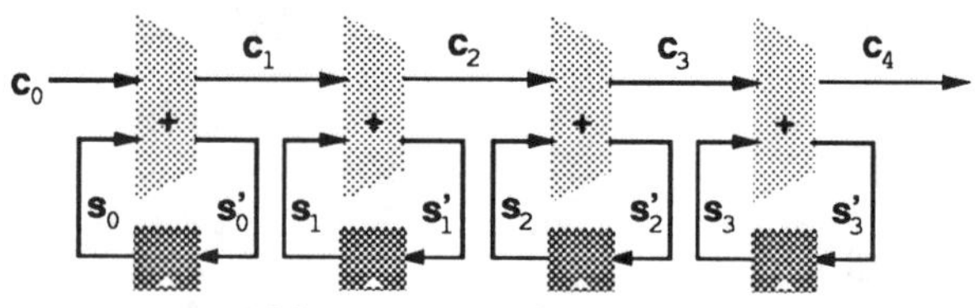

4 bits carry ripple counter.

The counter's only input (besides the clock) is the *increment* signal c_0. Each sum bit $S = {}_2[s_0 s_1 \cdots] = \sum_{i \geq 0} s_i 2^i$ is the output $s_i = \rho_t(s_i')$ of a synchronous register whose input is the corresponding bit of the next sum $S' = {}_2[s_0' s_1' \cdots] = \sum_{i \geq 0} s_i' 2^i = S + c_0$. Each bit of S' is the *exclusive or* $s_i' = s_i \oplus c_i = s_i + c_i - 2 s_i \times c_i$

of the corresponding sum and carry bits. Each consecutive carry $c_{i+1} = s_i \wedge c_i = s_i \times c_i$ is the *logical and* between the previous carry and sum bit. The carry ripple counter is therefore completely specified by the following equations:

$$s_i = \rho_t(s_i'), \quad s_i' = s_i \oplus c_i, \quad c_{i+1} = s_i \wedge c_i \text{ for } i \geq 0.$$

Let $c_0^{(0)} c_0^{(1)} \cdots c_0^{(t)} \cdots$ denote the boolean values of the counter's increment at times $0\ 1 \cdots t \cdots$; the corresponding counter's values $S^{(0)} S^{(1)} \cdots S^{(t)} \cdots$ sum up in binary the increments seen up to time t:

$$S^{(t)} = {}_2[s_0^{(t)} s_1^{(t)} \cdots] = \sum_{i \geq 0} s_i^{(t)} 2^i = \sum_{0 \leq k < t} c_0^{(k)}.$$

The value of S' at time t is that of S at time $t+1$, so $S'^{(t)} = S^{(t+1)} = S^{(t)} + c_0^{(t)}$. The carry vector $C^{(t)} = S^{(t)} \oplus S'^{(t)} = {}_2[c_0^{(t)} c_1^{(t)} \cdots] = \sum_{i \geq 0} c_i^{(t)} 2^i$ is the bit-wise exclusive or of S and S'. A small simulation will refresh our memory of the binary number system:

t	$s_0^{(t)}$	$s_1^{(t)}$	$s_2^{(t)}$	$s_3^{(t)}$	$c_0^{(t)}$	$c_1^{(t)}$	$c_2^{(t)}$	$c_3^{(t)}$	$c_4^{(t)}$
0	0	0	0	0	1	0	0	0	0
1	1	0	0	0	1	1	0	0	0
2	0	1	0	0	1	0	0	0	0
3	1	1	0	0	1	1	1	0	0
4	0	0	1	0	1	0	0	0	0
5	1	0	1	0	1	1	0	0	0
6	0	1	1	0	1	0	0	0	0
7	1	1	1	0	1	1	1	1	0
8	0	0	0	1	1	0	0	0	0

Let us assume that any logic gate with up to 3 inputs has a *unit delay* τ. The minimal *clock period* under which we may correctly operate a n bit carry ripple counter (crc) is:

$$T_{crc}(n) = n \times \tau. \tag{1}$$

Attempts to operate the *crc* under a lower clock period will fail: consider a time t when the counter's value is $S = 2^n - 1 = {}_2[11 \cdots]$, with $c_0^{(t)} = 0$, so $C^{(t)} = 0 = {}_2[00 \cdots]$. An increment $c_0^{(t+1)} = 1$ in that state causes all carries and sum bits to change; this requires a $n \times \tau$ combinatorial delay for the increment signal to ripple through up to the n-th bit.

Consider now the following *carry anticipate counter*:

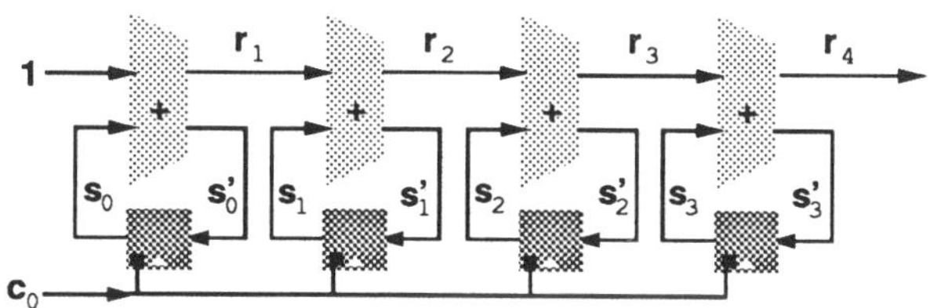

4 bits carry anticipate counter.

It is derived from the carry ripple counter by setting the initial carry to 1, and by enabling the sum registers through the increment signal c_0. Its defining equations are thus:

$$\begin{aligned} s_i &= \rho_t(if\ c_0 = 1\ then\ s_i'\ else\ s_i), \\ s_i' &= s_i \oplus r_i, \text{ for } i \geq 0, \\ r_{i+1} &= s_i \wedge r_i, \text{ for } i \geq 0 \text{ with } r_0 = 1. \end{aligned}$$

Using the same notations as before, we have: $S^{(t)} = \sum_{0 \leq k < t} c_0^{(k)}$, $S' = S+1$, $R = S \oplus S'$. Our interest for the carry anticipate counter (cas) is not immediate, since

$$T_{cac}(n) = n \times \tau, \tag{2}$$

which gets established through the very same argument as (1). We can however operate a n bits carry anticipate counter at full clock speed $T_{crc} = \tau$ provided that we restrict the non-zero increments to only occur during clocks phases which are at least n cycles apart. Indeed in this case, two time instants t_1 and t_2, for which $c_0^{(t_1)} = c_0^{(t_2)} = 1$ and $c_0^{(t)} = 0$ for $t_1 < t < t_2$, are such that $t_2 - t_1 \geq n \times \tau$. It follows that the length $n-1$ carry chain $R^{(t_1)} = S^{(t_1)} \oplus (S^{(t_1)} + 1)$ reaches a *stable state* no later than time $t_2 - \tau$. Any subsequent non-zero increment $c_0^{(t)}$ at time $t \geq t_2$ will make the input multiplexer to each sum bit switch within delay τ after the clock pulse, so the correct value of S' will be latched. This observation, together with an (easy) analysis of the frequency of non-zero carries c_k is the key to designing a *constant time* binary counter.

3   Constant time binary counter

The constant time $T(n) = 2 \times \tau$ (independent of n) counter is organized as a sequence of carry anticipate counter blocks. In an actual implementation, choices for the blocks' lengths may be different from the ones given below. Our choices match the clock period $2 \times \tau$, under the assumption that any logic gate up to 3 inputs has internal delay less than τ. The first block is a one bit *crc*:

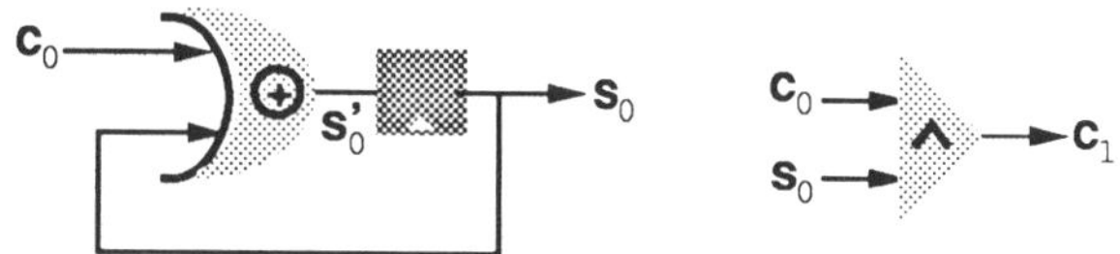

with defining equations: $s_0 = \rho_t(c_0 \oplus s_0)$, $c_1 = c_0 \wedge s_0$. The second block is a 4 bits carry anticipate counter:

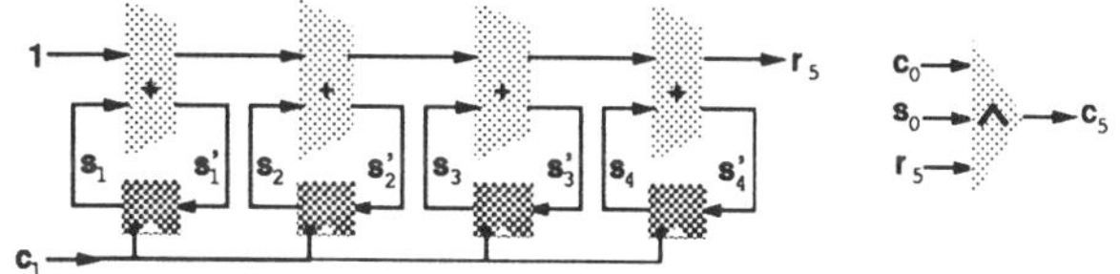

with defining sum equations:

$$s_1 = \rho_t(if\ c_1 = 1\ then\ s_1 \oplus 1\ else\ s_1),$$

$$s_2 = \rho_t(if\ c_1 = 1\ then\ s_2 \oplus r_2\ else\ s_2),$$

$$s_3 = \rho_t(if\ c_1 = 1\ then\ s_3 \oplus r_3\ else\ s_3),$$

$$s_4 = \rho_t(if\ c_1 = 1\ then\ s_4 \oplus r_4\ else\ s_4).$$

The carry equations for this 4 bits cac block are:

$$r_2 = s_1,\ r_3 = r_2 \wedge s_2,\ r_4 = r_3 \wedge s_3,$$
$$r_5 = r_4 \wedge s_4,\ c_5 = r_5 \wedge s_0 \wedge c_0. \tag{3}$$

In order to verify that this 5 bits counter correctly operates within clock period $2 \times \tau$, consider a time t when the carry chain $R_{2,5} = {}_2[r_2 r_3 r_4 r_5]$ changes state, i.e. $s_1^{t-1} \neq s_1^t$. Since the value of s_1 changes at most every second clock tick, the earliest time t' when state $R_{2,5}$ may change again is such that $t' - t \geq 4 \times \tau$. The gate depth of equations (3) being less than $3 \times \tau$ for each r_i ($2 \leq i \leq 5$), we see that r_2, r_3, r_4 and r_5 reach their stable value no later than time $t' - \tau$; it follows that ${}_2[s_1' s_2' s_3' s_4']$ has the correct value $1 + {}_2[s_1 s_2 s_3 s_4]$ no later than time t'. By the same reasoning carry c_5 has its correct value at all times.

For most practical purposes, we are through with our counter design. When properly adapted to the characteristics of a given technology, the first counter block will typically have $k \sim 3$ to 8 bits, and the second block more than 2^k bits, so there is no need for a third block. In order to implement such a counter in a given technology, one has to:

1. Derive from the technology parameters an *a-priori* estimate for gate, signal distribution and register delays, in order to determine a structure for the first k bits counter block. The logic gate computing the enabling carry $c_k = c_0 \wedge r_k$ to the next block must be carefully implemented so as to be *glitch-free*.

2. Design and optimize for the given technology the second level cac block.

3. Implement, test and measure the resulting counter. If measurement does not validates the initial assumptions, another design iteration is called for.

With help from Alan Skea, we have carried out this task for an existing $\tau = 10ns$ (100MHz) cMOS process, based upon Xilinx's programmable gate array [9] and our own PAM technology [BRV89]. The resulting counter, with a $k = 3$ bits first block, has been tested and measured. It runs at 40MHz (slightly under the theoretical 50MHz limit), for over 40 bits, which is as far as we could test it within a work-day. It has proved over 20 % faster than a similarly inspired counter, designed by Peter Alfke, which is part of the standard library in that technology [9].

For the sole benefit of our theoretically minded reader, let us pursue our counter's construction. The 3-rd block is a carry anticipate counter of length $64 = 2 \times 2^5$, with defining sum equations:

$$s_i = \rho_t(if\ c_5 = 1\ then\ s_i \oplus r_i\ else\ s_i)\ for\ 5 \leq i \leq 64.$$

The corresponding carry equations are: $r_6 = s_6$, $r_{i+1} = r_i \wedge s_i$ for $5 < i \leq 64$. This 63 long carry chain has enough ($63 \times \tau$) time to settle, since state changes in bits $s_5 s_6 \cdots$ of the counter are at least 32 cycles ($\geq 64\tau$) apart. By now, the reader has presumably infered the general pattern for our counter. The 4-th block is an intimidating 2^{69} bits long cac[1], with enable signal

$$c_{69} = (c_0 \wedge s_0 \wedge (r_5 \wedge r_{69})).$$

An observer of gate c_{69} and r_{69} better be patient, since not much happens until clock ticks T_1 and T_2, determined by $\sum_{k < T_1} c_0^{(k)} = 2^{69} - 32$ and $\sum_{k < T_2} c_0^{(k)} = 2^{69}$. By clock tick $T_2 - 1$, enough cycles have gone by so r_{69} has settled to 1, no later than time $(2T_2 - 1)\tau$, and no earlier than time $2T_1\tau$. At this point, $c_0 = r_5 = r_{69} = 1$, but $s_0 = 0$ so $c_{69} = 0$. During the next clock cycle T_2, signals $c_0 = r_5 = r_{69} = 1$ keep their value 1, but our counter has just changed its parity $s_0 = 1$; so c_{69} becomes 1 no later than one gate delay τ after the clock tick; just in time to enable the multiplexer controlling the next value of bit s_{69}. Bit s_{69} is thus set to 1 at the next clock tick while all 69 previous bits *and* carries are reset to 0.

The 5-th block[2] is a cac of length $l(5) = 2^{T_2} - 1$. The carry enable $c_{L(4)}$, with $L(4) = T_2 = 2^{69}$ into this block gets computed by :

$$c_{L(4)} = (c_0 \wedge s_0 \wedge (r_5 \wedge r_{69} \wedge r_{L(4)})).$$

In general, the k-th cac block has length $l(k) = 2^{L(k)} - k/2$, where $L(k) = \sum_{i < k} l(i)$ is the sum of the lengths of the preceding blocks. The incoming carry to block k is:

$$c_{L(k)} = (c_0 \wedge s_0 \wedge (r_{L(2)} \wedge r_{L(3)}(\wedge r_{L(4)} \cdots \wedge r_{L(k)}) \cdots)).$$

In picture:

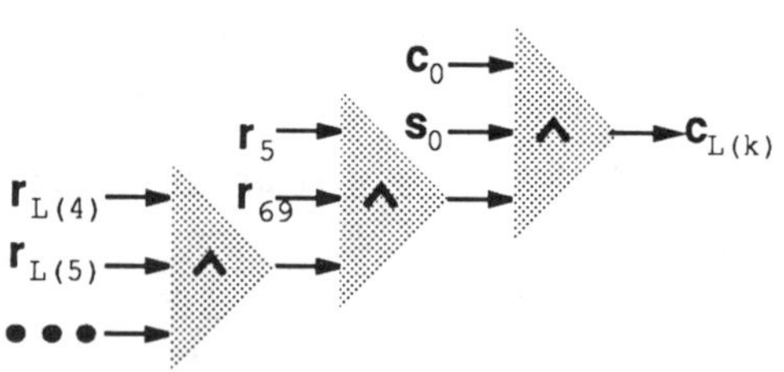

While the depth of such gates grows as $k/2$, this (tiny) delay gets absorbed by our choice $l(k) = 2^{L(k)} - k/2$ in the block length.

---

[1] over a billion Tera bits!

[2] by now, we have far exceeded the time and space limits of the known Universe!

4   Conclusion

What makes this design work is the very structure of the binary number system. Let the *2adic valuation* $v_2(n)$ of integer $n > 0$ be the exponent of the largest power of 2 which divides n, so $n = 2^{v_2(n)}(2p + 1)$ for some natural number $p \geq 0$. The number of non-zero carries in the n-th line of the binary table is precisely $1 + v_2(n + 1)$. The total number of carries $C(n)$ effectively propagated when we increment consecutive integers from 1 up to n is $C(n) = \sum_{0 < i \leq n}(1 + v_2(n + 1)) = 2n - \nu_1(n)$, where $\nu_1(n) = \sum_{i \geq 0} n_i$ is the number of non zero bits when one writes $n = {}_2[n_0 n_1 \cdots]$ in binary. It follows that $n < C(n) < 2n$, and we can say that the *average carry chain* in a counter has length (almost) 2. What our counter effectively does is to *amortize* throughout all cycles the bursts of long yet rare carry chains which plague the worst case behaviour of naive implementations. From the above analysis $(C(n)/n \sim 2)$, we are tempted to conjecture that the clock period $T = 2 \times \tau$ cannot be reduced to $T = \tau$, under our limited 3 fan-in assumption.

To conclude, we leave the following as an open question: is it possible to design a synchronous, arbitrary length, constant time up-down[3] counter?

Acknowledgements

Thanks to Alan Skea for a fine implementation and challenging test of the counter. Thanks to Patrice Bertin for stimulating observations.

References

[1] Xilinx, *The Programmable Gate Array Data Book,* Product Briefs, Xilinx Inc., 1987.

[2] V. S. Burtsev, "Accelerating Multiplication and Division Operations in High Speed Digital Computers," *Exact Mechanics and Computing Techniques,* Moscow Academy of Sciences, 1958.

[3] M. Ercegovac, T. Lang, "Binary Counter with Counting Period of One Half Adder Independent of Counter Size," *IEEE Trans. on Circuits and Systems,* Vol. 36, No 6, pp. 924-926, 1989.

[4] L. Guibas, J. E. Vuillemin, "On fast binary addition in n-MOS technologies," *Proc. of ICCC 82*, IEEE New York, pp. 147-151, 1982.

[5] J. E. Vuillemin, *A very fast multiplication algorithm for VLSI implementation,* INTEGRATION, the VLSI Journal, Vol. 1, No 1, pp. 39-52 1983.

[6] C. S. Wallace, *A suggestion for a Fast Multiplier,* IEEE Trans. El. Comp., Vol. EC-13, No 1, pp. 14-17, 1964.

[7] S. Winograd, *On the time required to perform addition,* J. ACM, 12,2, pp. 277-285, 1965.

[8] S. Winograd, *On the time required to perform multiplication,* J. ACM, 14,4, pp. 793-802, 1967.

[9] Xilinx, *The Programmable Gate Array Data Book,* Product Briefs, Xilinx Inc., 1987.

---

[3] a counter which may be incremented or decremented at each cycle.

Session 8:

Division

Chair:
George Taylor
MIPS Computer Systems

Integer Division Using Reciprocals

Robert Alverson
Tera Computer Company
400 North 34th St., Suite 300
Seattle, WA 98103

Abstract

As logic density increases, more and more functionality
is moving into hardware. Several years ago, it was uncommon to find more than minimal support in a processor for integer multiplication and division. Now, several
processors have multipliers included within the central
processing unit on one integrated circuit [8, 12]. Integer
division, due to its iterative nature, benefits much less
when implemented directly in hardware and is difficult
to pipeline. By using a reciprocal approximation, integer division can be synthesized from a multiply followed
by a shift. Without carefully selecting the reciprocal,
however, the quotient obtained often suffers from off-
by-one errors, requiring a correction step. This paper
describes the design decisions we made when architecting integer division for a new 64 bit machine. The result
is a fast and economical scheme for computing both unsigned and signed integer quotients that guarantees an
exact answer without any correction. The reciprocal
computation is fast enough, with one table lookup and
five multiplies, that this scheme is competitive with a
dedicated divider while requiring much less hardware
specific to division.

1   Introduction

Not too long ago, the cost of integer multiplication was
sufficiently high that compiler writers often found it
useful to reduce the strength of a multiplication by a
constant into a series of additions. Lately, such "optimizations" must be done more carefully, since several
current microprocessors include fast integer multipliers.

Integer division has remained an enigma. If a processor has any support for integer division, it is usually in
the form of a simple iterative divider producing a single
quotient bit per clock. While the dynamic frequency
of integer divides is typically small, the time weighted
impact on path length can be significant. One investigation found nearly 10 percent of all cycles for a spread-
sheet application were devoted to integer division[1]. To

---

*This research was supported by the United States Defense
Advanced Research Projects Agency under Contract MDA972-
89-C-0002. The views and conclusions contained in this document are those of Tera Computer Company and should not be
interpreted as representing the official policies, either expressed
or implied, of DARPA or the U. S. Government.

mitigate this effect, various solutions have been used.
The straightforward iterative divider is area-efficient
but not pipelineable, complicating the instruction issue logic in a pipelined processor. A second option is a
non-iterative array divider, which is pipelineable but not
area-efficient. Both solutions result in divide operations
which take about four times as many cycles as multiplies, for 32 bit integers. This is fundamental, since the
partial products in a multiply may be summed in parallel while the remainders for divide must be calculated
serially. With 64 bit integers, the divide latency worsens
to around six times longer than that for multiply.

Modern compilers have been effective at reducing the
cost of integer division, as some researchers have devised
methods for dividing by small constants using scaled
reciprocals[3, 9, 10]. This approach is quite effective,
since constant divisors are quite frequent in practice.
However, computing the reciprocal at run-time has been
too costly for integer division in general until now. This
paper presents an integer division scheme which supports fast division by a constant, is still efficient for nonconstant divisors, requires little extra hardware, and is
pipelinable. The ideas presented reflect the design decisions we made when architecting integer division for
the Tera Computer[2], a new 64 bit machine.

Section 2 refines the integer division problem and informally describes its solution using reciprocals. Section 3
proves that the outlined division scheme produces the
correct result for unsigned division in all cases. These
results are extended to the division of signed integers in
section 4.

2   Reciprocal Division

Integer division has several forms. Unsigned division
gives the positive quotient and positive remainder from
the division of two positive numbers. Extended precision division is the same as unsigned division except
the dividend is a double word rather than a single word.
For division of signed integers, two ways of defining the
desired result are common, round toward zero (chopped
division) and round toward negative infinity (floored division). Thus, there are four common variants of integer
division: unsigned, extended precision, signed chopped,

and signed floored.

unsigned	$q = \lfloor x/y \rfloor$
extended-precision	$q = \lfloor (x_h 2^w + x_l)/y \rfloor$
signed chopped	$q = \lfloor abs(x/y) \rfloor * sign(x/y)$
signed floored	$q = \lfloor x/y \rfloor$

In all thses cases, the remainder is $x - q*y$ (for extended-precision division, $x = x_h 2^w + x_l$). The remainder takes the sign of the dividend in signed chopped division, while the sign of the remainder matches the sign of the divisor in signed floored division. FORTRAN requires signed chopped division, while C leaves the choice to the implementor. In addition, integer division is implicitly used in calculating C pointer differences and converting binary values to decimal for output.

The basic strategy for reciprocal division is to find values a and sh such that $x/y = x*a/2^{sh}$. For the moment, consider only the unsigned case $0 \le x < 2^w, 0 < y < 2^w$, where w is the wordlength (our target machine has a 64 bit wordlength). Given a shift sh, the obvious choice for the scaled reciprocal is $a = round(2^{sh}/y)$. Thus, reciprocal division has the dubious advantage of computing a single division using a division, a multiply, and a shift. Fortunately, the reciprocal need not be calculated using division, but may be computed using a Newton-Raphson iterative approximation [13].

To guarantee quick convergence, the reciprocal iteration must start with an accurate initial guess. The standard solution, which works well for floating point reciprocals, is to use a table lookup[4]. The table index is taken from the k most significant mantissa bits after removing the leading one (for a normalized binary floating point number). The easiest way to look up an initial guess for the reciprocal of an integer divisor is to normalize the integer by converting it to float and applying the same lookup method.

While it is possible to convert the initial reciprocal estimate back to an integer immediately, we chose to perform the intermediate iteration steps with floating point. Since we use reciprocals for floating point divide as well, this choice allows the instruction that updates the reciprocal to be used by both integer and floating point divide.

The other solution is to use an integer iteration. While this option is conceptually simpler, the floating point iteration is adequate and reduces the overall architectural impact of divide support. Because the standard 53 bit floating point mantissa does not provide enough precision for 64 bit division, the result eventually must be converted to an integer.

When the reciprocal is converted to an integer, the shift sh must be selected. Essentially, the shift sh must be large enough for the final shift to truncate away the rounding error in the reciprocal. This guarantees that the quotient is either correct or low by one. Consequently, the remainder must be calculated and the quotient conditionally adjusted by one. Alternately, the shift sh can be made even larger to always yield a correct quotient. Of course, a larger sh leads to a larger

scaled reciprocal, with which it is harder to compute. Thus the shift sh should be as small as possible while guaranteeing the desired level of accuracy.

Consider the division $(2^w - 3)/(2^w - 2)$. Here, a shift sh of $2w$ is necessary to compute the correct quotient, even though a shift of w yields an answer within one.

$$\begin{aligned} q &= \lfloor (2^w - 3) * \lceil 2^{sh}/(2^w - 2) \rceil / 2^{sh} \rfloor \\ &= 2^{w-sh} - 1 \text{ if } 2 \le sh < w \\ &= 1 \text{ if } w \le sh < 2w \\ &= 0 \text{ if } 2w \le sh \end{aligned}$$

For this division to be handled without exception, the shift of $2w$ must yield the scaled reciprocal $2^w + 3$. To represent this quantity requires $w + 1$ bits, which is a challenge. For machines with 32 bit words, the reciprocal could be expressed using double precison IEEE floating point numbers[6]. With our wordsize of 64 bits, a double-extended format with 65 significand bits is needed to be useful. This width is more than the de facto standard for double-extended (64 bits in the 8087[7]). In addition, we have no other reason to support a double-extended format for our machine.

With the above constraints in mind, we present our integer division procedure. Each line represents one machine instruction. All float variable names start with f; the others are integers. First, we convert the integer divisor y to float and compute an initial reciprocal approximation which is correct to 17 bits. If our machine had floating point divide in hardware, we could use that to compute the initial reciprocal. Like the IBM RS/6000[11], we use reciprocals for floating point division. The shift sh is computed by counting zeros at the left of the divisor (powers of two are handled as special cases).

```
fy ← double(y);
fa ← approx(1/fy);
sh ← w + ceil(log2(y));
```

Next, we iterate to improve the accuracy of the reciprocal. This iteration mirrors the float reciprocal iteration, except we use the integer divisor to maximize the precision of the error term e. As a final step, the reciprocal is multiplied by 2^{sh} and rounded to ceiling.

```
fe ← nearest(1.0 - fa*y);
fa ← nearest(fa + fa*fe);
fe ← nearest(1.0 - fa*y);
a ← ceil((fa + fa*fe)*2^sh);
```

Unless the unrounded result of a reciprocal iteration has at least $2w$ correct bits, the reciprocal may still need a correction for proper rounding. For example, consider $y = 2^w - 1$. On conversion to float, this number rounds to 2^w. The float reciprocal will come to 2^{-w} and the scaled reciprocal is initially calculated as $2^w + 1$. Since the properly rounded answer is $2^w + 2$, a correction must be made.

This correction could be accomplished by another reciprocal iteration. Instead, we detect the cases where the reciprocal is low by one with a trial division of y/y, which should give one. If the result is zero, then increment the reciprocal. While this correction is an annoyance, its cost may be amortized over the iterations of a loop or eliminated during compilation. The cost of a correction to the quotient, however, must be paid with each divide.

```
q ← floor(y*a/2^sh);
a ← a + (1-q);
```

Finally, do the division. The dividend x is multiplied by the scaled reciprocal a and shifted right by the amount sh. This multiply-shift logic can be shared with the logic needed to implement a floating point multiply add with a single round[5].

```
q ← floor(x*a/2^sh);
```

Notably, the compiler need only load a and sh for division by a constant, leaving only one operation to be computed at run-time. In the general case, only five more multiplies are needed to compute the scaled reciprocal. The only logic specifically needed for implementing integer division is the initial reciprocal lookup table. This same table can be used for computing floating point reciprocals, so its cost is also amortized over both integer and floating point divides.

3   Unsigned Division

While the preceding informal discussion identified some pitfalls in performing reciprocal division, it did not attempt to prove that we avoid them in our procedure. Here, we formally show that our division procedure determines the correct quotient when applied to unsigned integers. Let the wordlength be w bits. For a given divisor $y \neq 0$, let $sh = w + \lceil log_2 y \rceil$. Then $a = \lceil 2^{sh}/y \rceil$ implies $2^w \leq a < 2^{w+1}$. The ceiling of the reciprocal is the most useful rounding since the error introduced is truncated away when the final quotient is converted to an integer. The ceiling never rounds up to 2^{w+1}, since $2^{sh}/y$ is never more than $2^{w+1} - 1$:

$$
\begin{aligned}
\lceil log_2 y \rceil &< log_2 y + 1 \\
2^{\lceil log_2 y \rceil} &< 2y \\
2^{\lceil log_2 y \rceil} &\leq 2y - 1 \\
2^{sh} &\leq 2^w(2y - 1) \\
\lceil 2^{sh}/y \rceil &\leq \lceil 2^w(2y - 1)/y \rceil \\
a &\leq 2^{w+1} - \lfloor 2^w/y \rfloor \\
a &\leq 2^{w+1} - 1 \qquad \text{(as long as } y \leq 2^w)
\end{aligned}
$$

Now the quotient can be computed by multiplying by a and dividing by 2^{sh}:

$$
\begin{aligned}
q &= \lfloor x/y \rfloor \\
&= \lfloor (x/2^{sh}) * (2^{sh}/y) \rfloor \\
&\approx \lfloor (x/2^{sh}) * a \rfloor \\
&= \lfloor (x * a)/2^{sh} \rfloor \\
&= Q
\end{aligned}
$$

For what range of values of the dividend x is the quotient q guaranteed to be correct? The boundary cases are $x = q * y$ and $x = q * y + y - 1$. In the former case, a small negative error in the scaled reciprocal a might make the quotient one too small. In the latter, a small positive error in a may cause the quotient to be one too large. Since a is the ceiling of the true value and both x and y are positive, the rounding of a can only cause the unrounded product $x * a$ to be too large. To show this error never propagates to the quotient, compare the computed quotient Q with the true quotient $q = \lfloor x/y \rfloor$:

$$
\begin{aligned}
\text{Let } r &= a * y - 2^{sh} \\
x &= q * y + s \\
0 &\leq s < y
\end{aligned}
$$

$$
\begin{aligned}
Q &= \lfloor x * a/2^{sh} \rfloor \\
&= \lfloor (q * y + s) * a/2^{sh} \rfloor \\
&= \lfloor (q * (a * y) + a * s)/2^{sh} \rfloor \\
&= \lfloor (q * (2^{sh} + r) + a * s)/2^{sh} \rfloor \\
&= q + \lfloor (q * r + a * s)/2^{sh} \rfloor
\end{aligned}
$$

For Q to equal q, $q * r + a * s$ must be less than 2^{sh}. While $q * r + a * s$ also must not be less than zero, it is clear that $Q \geq q$ from the rounding of the scaled reciprocal. As long as there is a bound on the size of the dividend, we can choose a scaled reciprocal which guarantees a correct quotient. The first step is to bound the term $q * r$, taking advantage of the fact that r is the remainder of 2^{sh} divided by y:

$$
\begin{aligned}
r &\leq y - 1 \\
q * r &\leq q * (y - 1) \\
q * (y - 1) &= x - s - q \\
q * r &\leq x - q - s
\end{aligned}
$$

When the dividend x is a maximum 2^w, q must be at least one since y cannot be greater than a maximum x. Thus, the difference $x - q$ cannot be as large as 2^w.

$$
\begin{aligned}
x - q &< 2^w \\
q * r &< 2^w - s \\
q * r + a * s &< 2^w + (a - 1) * s
\end{aligned}
$$

Similarly, the term $a * s$ can be bounded by recalling that s is the remainder of x/y and thus is between zero and $y - 1$.

$$
\begin{aligned}
s &\leq y - 1 \\
q * r + a * s &< 2^w + (a - 1)(y - 1) \\
q * r + a * s &< 2^w + a * y - a - y + 1 \\
r &\leq y - 1 \\
a * y &\leq 2^{sh} + y - 1 \\
q * r + a * s &< 2^{sh} + 2^w - a
\end{aligned}
$$

Finally, the lower bound on the scaled reciprocal a leaves the desired relation which guarantees the computed quotient is correct.

$$
\begin{aligned}
a &\geq 2^w \\
q * r + a * s &< 2^{sh} \\
Q &= q
\end{aligned}
$$

Hence, Q is correct if $x \leq 2^w$. While the dividend x is never 2^w for unsigned division with wordlength w, the stronger result will prove useful for signed division where two's complement allows the value -2^{w-1}.

We have shown that a scaled reciprocal can be used to compute a quotient of positive integers exactly. To achieve this accuracy, we need a reciprocal with one more bit than will fit in a word. Like a floating point mantissa, the scaled reciprocal has a leading one which need not be explicitly represented. With this improvement, the scaled reciprocal nicely fits in one machine word. To maintain the hidden bit, the shift sh is constrained to yield a scaled reciprocal $2^w \leq a < 2^{w+1}$, even though other values of a and sh might work just as well.

4 Signed Division

In the realm of signed integers, the two most common definitions of division round the infinitely precise quotient towards zero or towards negative infinity. Rounding towards zero (chopped rounding) results in a division algorithm in which the operands are separated into sign and magnitude parts. Except for the extra cases where the dividend or divisor is -2^{w-1}, chopped division is exactly the same as unsigned division with a wordlength of $w-1$ and one sign bit. However, the previous proof already allowed for a dividend of 2^{w-1}, so there is no mathematical problem. The hardware still must allow w bits for the magnitude of the dividend in this case. Thus, the signed integer chopped division of x by y is simply performed:

$$a = \lceil abs(2^{sh}/y) \rceil * sign(y)$$
$$q = \lfloor abs(x * a)/2^{sh} \rfloor * sign(x * a)$$

Note that the scaled reciprocal computation effectively rounds away from zero, a round mode that is not part of the IEEE floating point standard.

For rounding towards negative infinity (floored rounding), the handling depends on the sign of the quotient. For a non-negative quotient, floored division may be computed exactly as chopped division, since the results will always match. Complication arises when the signs of the dividend and divisor do not match. To compute the quotient as

$$q = \lfloor x * a/2^{sh} \rfloor$$

the scaled product $x * a/2^{sh}$ must not be less than x/y. When the quotient is negative, this constraint requires a to be rounded towards zero. As it stands, we must round a away from zero when the signs of x and y match and toward zero when their signs are different. Rounding the reciprocal depending on the sign of the dividend x is inconvenient and reduces the performance improvement possible when the divisor is constant. Fortunately, there is another solution. We may use the same reciprocal as with chopped division by biasing the product so that any error is one-sided, allowing the exact quotient to be calculated without any correction. While the same reciprocal is computed as always, the quotient

calculation requires the bias b:

$$a = \lceil abs(2^{sh}/y) \rceil * sign(y)$$
$$b = \text{if } x * a \geq 0 \text{ then } 0 \text{ else } 2^{w-1} - 1$$
$$q = \lfloor (x * a + b)/2^{sh} \rfloor$$

The bias must be large enough to cancel out negative error when the product of the dividend and the scaled reciprocal is negative. This error is most significant when the dividend is an exact multiple of the divisor: $x = qy$. In this case, the slightest negative error will cause the quotient to be low by one. To get the correct quotient from the floor, its argument must not be less than the true quotient q:

$$
\begin{aligned}
(x * a + b)/2^{sh} &\geq q \\
x * a + b &\geq q * 2^{sh} \\
b &\geq q * 2^{sh} - x * a \\
b &\geq q * 2^{sh} - q * y * a \\
b &\geq q * (2^{sh} - a * y) \\
b &\geq -q * r \qquad \text{(By definition of } r\text{)}
\end{aligned}
$$

Note that r is zero when y is a power of two. Consequently, the bias is constrained by divisors which are not powers of two. Since $x = -2^{w-1}$ can only be exactly divided by powers of two, the minimum dividend to consider is $-2^{w-1} + 1$.

$$
\begin{aligned}
r &\leq y - 1 \\
-q * r &\leq -q * y + q \\
-q * r &\leq -x + q \\
-x &\leq 2^{w-1} - 1 \\
q &\leq -1 \\
-q * r &\leq 2^{w-1} - 2 \\
b &\geq 2^{w-1} - 2 \qquad \text{(For a correct quotient)}
\end{aligned}
$$

As an example, consider dividing $(-2^{w-1} + 1)/(2^{w-1} - 1)$. We select the shift $sh = 2w - 2$, yielding the scaled reciprocal $a = 2^{w-1} + 2$. This requires a bias $b \geq 2^{w-1} - 2$ to get the correct quotient of negative one.

The bias must not be so large that a correct answer is incremented to become too large (i.e. smaller in magnitude). When the quotient is negative, the biased product $x * a + b$ is independent of the signs of x and y. Thus, only the case of negative x and positive y must be considered. The answer is most sensitive to the bias when $x = q * y + y - 1$. As before, the critical constraint is to compute a value which rounds to the correct quotient:

$$
\begin{aligned}
(x * a + b)/2^{sh} &< q + 1 \\
x * a + b &< (q + 1) * 2^{sh} \\
b &< (q + 1) * 2^{sh} - x * a \\
b &< (q + 1) * 2^{sh} - ((q + 1) * y - 1) * a \\
b &< (q + 1) * (2^{sh} - a * y) + a \\
b &< -r * (q + 1) + a
\end{aligned}
$$

If b is less than the minimum value of $-r * (q + 1) + a$, then correctness is assured. To find the minimum value, consider each component. Recall that we assume x is negative and y is positive, yielding a negative quotient after floored rounding:

$$
\begin{aligned}
q + 1 &\leq 0 \\
r &\geq 0 \\
-r * (q + 1) &\geq 0 \\
a &\geq 2^{w-1} \\
b &< 2^{w-1}
\end{aligned}
$$

We see that the bound on the scaled reciprocal a leads to a sufficient condition on b for the correctness of the quotient. In fact, this upper limit for b is also necessary, since the division $-1/1$ will not give the correct answer if $b \geq 2^{w-1}$. Thus, the bias of $b = 2^{w-1} - 1$ is correct for all cases with a negative quotient.

Unsigned integer division readily generalizes to signed integer division. While floored division does entail extra complication, our carefully chosen bias allows the extra complexity to be hidden from the compiler, so that constant divisors may be converted to scaled reciprocals without regard to the rounding desired for the integer quotient.

5 Conclusion

While reciprocal division has been used for floating point for many years, until now the ideas have only been applied to integer arithmetic when the divisor is constant. The division scheme described here is fast in the general case, requiring only six multiplies and a table lookup. Our reciprocal division supports unsigned integer, signed chopped, and signed floored division. The real strength of our method is division by a constant, which takes only a single multiply and shift, one operation on our machine. The analysis we presented shows that the computed quotient is always exact — no adjustment or correction is necessary. There is very little extra hardware required, primarily rounding logic and an initial reciprocal lookup table. Thus, we believe this design strikes a balance between hardware and software support for 64 bit integer division.

Further research is needed to determine how best to implement extended precision division using reciprocals. Besides generating the correct quotient in the normal case, extended precision division may result in overflow if the quotient is larger than will fit in one word. A simple change to the unsigned division code sequence seems to allow computing the extended-precision division $(x_h 2^w + x_l)/y$ correct to $w - 2$ bits, although overflow is not detected. We are currently investigating the best way to increase the accuracy to guarantee the full w bits.

Acknowledgements

The author is indebted to Burton Smith, who wanted to do integer division just a little bit better. Thanks also to Gail Alverson and Jon Bertoni for their comments on an early version of this paper.

References

[1] Thomas L. Adams and Richard E. Zimmerman. An Analysis of 8086 Instruction Set Usage in MS DOS Programs. In *Proceedings of the Third International Conference on Architectural Support for Programming Languages and Operating Systems*, pages 152–160, April 1989.

[2] R. Alverson, D. Callahan, D. Cummings, B. Koblenz, A. Porterfield, and B. Smith. The Tera Computer System. In *International Conference on Supercomputing*, pages 1–6, June 1990.

[3] E. Artzy, J. A. Hinds, and H. J. Saal. A Fast Division Technique for Constant Divisors. *Communications of the ACM*, 19(2):98–101, February 1976.

[4] D. L. Fowler and J. E. Smith. An Accurate, High Speed Implementation of Division by Reciprocal Approximation. In *Proceedings of the Ninth Symposium on Computer Arithmetic*, pages 60–67, September 1989.

[5] E. Hokenek, R. K. Montoye, and P. W. Cook. Second-Generation RISC Floating Point with Multiply-Add Fused. *IEEE Journal of Solid-State Circuits*, 25(5):1207–1213, October 1990.

[6] *IEEE Standard for Binary Floating Point Arithmetic*. IEEE, New York, NY, 1985.

[7] *Microprocessor and Peripheral Handbook, Volume I - Microporcessor*. Intel Corporation, Santa Clara, CA, 1989.

[8] Gerry Kane. *MIPS R2000 RISC Architecture*. Pentice Hall, Englewood Cliffs, NJ, 1987.

[9] S-Y. R. Li. Fast Constant Division Routines. *IEEE Transactions on Computers*, C-34(9):866–869, September 1985.

[10] Daniel J. Magenheimer, Liz Peters, Karl Pettis, and Dan Zuras. Integer Multiplication and Division on the HP Precision Architecture. In *Proceedings of the Second International Conference on Architectural Support for Programming Languages and Operating Systems*, pages 90–99, April 1987.

[11] P. W. Markstein. Computation of elementary functions on the IBM RISC System/6000 processor. *IBM Journal of Research and Development*, 34(1):111–119, January 1990.

[12] Brett Olsson, Robert Montoye, Peter Markstein, and MyHong NguyenPhu. RISC System/6000 Floating-Point Unit. In Mamata Misra, editor, *IBM RISC System/6000 Technology*, pages 34–42. IBM Corporation, Austin, TX, 1990.

[13] Shlomo Waser and Michael J. Flynn. *Introduction to Arithmetic for Digital Systems Designers*. Holt, Rinehart and Winston, New York, NY, 1982.

Fast Division Using Accurate Quotient Approximations to Reduce the Number of Iterations

Derek C. Wong and Michael J. Flynn
Electrical Engineering Department
Stanford University
Stanford, CA 94305

Abstract

A class of iterative integer division algorithms is presented based on look-up table and Taylor-series approximations to the reciprocal. The algorithm iterates by using the reciprocal to find an approximate quotient and then subtracting the quotient multiplied by the divisor from the dividend to find a remaining dividend. Fast implementations can produce an average of either 14 or 27 bits per iteration, depending on whether the basic or advanced version of this method is implemented. Detailed analyses are presented to support the claimed accuracy per iteration. Speed estimates using state-of-the-art ECL components show that this method is faster than the Newton-Raphson technique and can produce 53-bit quotients of 53-bit numbers in about 28 or 22 ns for the basic and advanced versions. [1]

1 Introduction

This is a description of a high-speed divide algorithm. The algorithm operates on positive, binary, non-redundant integers. Since an exact remainder can be produced, rounding is straightforward. A reduced number of iterations is used to produce the result, and the algorithm is general and can be applied to any size numbers. With a sufficient number of iterations, any accuracy of quotient can be produced, with a well-defined remainder (where remainder is defined relative to the significance of the last digit of the quotient). For the purposes of this paper, we concentrate our examples on 53-bit integers (same number of bits as the mantissa of the IEEE floating-point standard).

Unlike other subtractive algorithms which produce 1-3 bits of quotient per iteration, a reasonable implementation of the basic algorithm takes 4 iterations to divide 53-bit numbers to produce a 53-bit quotient or about 14 bits/iteration. Implementations of the advanced version of this algorithm can produce a 53-bit result in 2 or even 1 iteration.

Some comparative work may be found in the following references. In references [1] and [5], Atkins and Fandrianto describe higher-radix methods using SRT. Krishnamurthy [9] describes a related idea for iterative

[1] This work was performed with support from the Center for Integrated Systems at Stanford University, and from NSF Contract No. MIP88-22961 using equipment provided by NASA under contract NAGW 419.

division by transforming the range of the divisor to a number close to one. In this case, the leading bits of the dividend can be used as bits of the quotient. The Cyrix co-processor's division algorithm [11] is based on an algorithm similar to the one discussed in section 2.2 (more details later in this paper).

As RAM density and speed increases, the use of larger look-up tables becomes practical and advantageous. This algorithm accurately estimates the reciprocal using large look-up tables. Since the algorithm can then estimate quotients accurately, subtraction can reduce the dividend by many bits per iteration.

This paper first outlines our method. This method is then compared to other methods using reciprocal approximations. Then speed estimates are made for fast implementations of this scheme. This is compared to the performance of other algorithms using the same technology assumptions.

2 The Algorithm

This algorithm uses a fixed look-up table to get adjustments to the current quotient. Each adjustment, appropriately shifted, is added to the previous cumulative estimate of the quotient. Thus, each new estimate is an adjusted version of the previous. In this method, each estimate is always less than or equal to the true quotient, so that each new adjustment is always non-negative (unlike in non-restoring division methods).

The quotient is successively refined: after the N-th iteration, the quotient is accurate to approximately within 1 part in $2^{(14*N)}$ using the basic algorithm.

2.1 Notation

Denote the dividend register by X, the divisor register by Y, and the quotient register by Q. Denote a bit of a register using an index (e.g. Q_j) with higher indices indicating more significant bits. The LSB has index 0.

Without loss of generality, in this discussion, we consider that the binary representations of X, Y, Q, and other numbers are actually fixed-point fractional numbers between 0 and 1. The actual hardware operations work on binary numbers which could represent either integers or fixed-point numbers.

We describe the algorithm for positive numbers; handling negative numbers is straightforward using a sign-magnitude representation. All numbers are assumed to have the same width q; the algorithm can easily be modified for numbers with different width. The most significant bits are then denoted by $X_{(q-1)}$ and $Y_{(q-1)}$. (Note: later it will be explained that some of the registers must be extended with low-order bits to make registers of width greater than q. These extended bits have negative indices.)

2.2 Division by Approximating Q and Subtracting

Ordinary pencil-and-paper division can be generalized into the following method:

1. Initialize the initial quotient $Q = 0$ and a shift index $j = 0$.

2. Find an approximation Q_a to X/Y such that $Q_a \leq X/Y$.

3. $Q' = Q + Q_a * (1/2^j)$.

4. $X' = X - Q_a * Y$.

5. Shift X' left by k bits until the leading bit is 1. Set $j = j + k$.

6. $X = X'$, $Q = Q'$.

7. Repeat 2 through 6 until $j \geq q$.

8. The final X is the remainder.

If the approximation Q_a is quite accurate, then the number of bits k that are reduced in an iteration is large. The number of iterations is then small.

Unlike most division algorithms that calculate several bits of the quotient per iteration, this algorithm calculates an additive adjustment to the previous quotient estimate. The adjustments get exponentially smaller with each iteration. However, no bits of the quotient are guaranteed until the algorithm terminates; even the MSB of the quotient Q could change during the last iteration if a carry propagation occurs. For instance, if the intermediate result is for instance 10011111111..111, the actual result could be 1010000000..000001.

We now describe two versions of this method based on calculating Q_a using an approximate reciprocal multiplied by the leading bits of X. The second version also uses a novel approach to approximate the reciprocal.

2.3 A Basic Method

Suppose X and Y are left-shifted until their leading bits are both 1 (denote this operation as *pre-shifting or normalization*). The amount of shifting is remembered, so that the proper significance of the leading quotient bit is known. Given the first p bits of X, the value of X is known within an error of $1/(2^p)$. The minimal value of X would be $X_{(q-1)}..X_{(q-p)}000..0$ and the maximal value would be $X_{(q-1)}..X_{(q-p)}111..1$. The range of uncertainty is slightly smaller than the significance of $X_{(q-p)}$, which is $1/(2^p)$. Similarly, given the first m bits of Y, we know Y to within $1/(2^m)$.

Define X_h as the leading p bits of X extended with 0's to get a q-bit number:

$$X_h = X_{(q-1)}..X_{(q-p)}00000..0 \qquad (1)$$

Define Y_h as the leading m bits of Y extended with 1's to get a q-bit number:

$$Y_h = Y_{(q-1)}..Y_{(q-m)}11111..1 \qquad (2)$$

Let $\Delta X = X - X_h$ and $\Delta Y = Y - Y_h$. The deltas ΔX and ΔY are the adjustments needed to get the true X and Y from X_h and Y_h. Due to the definitions of X_h and Y_h, ΔX is always non-negative, and ΔY is always non-positive.

Suppose that we know the answer to $1/Y_h$. This is clearly less than or equal to the true reciprocal $1/Y$. Similarly, X_h/Y_h is always less than or equal to X/Y.

A division algorithm that uses a reduced number of iterations can then be conceptually summarized as follows:

1. Set the estimated quotient Q to 0 initially.

2. Let j and k denote the number of bits that X and Y have been left-shifted. Initially, both are set to the amounts of pre-shifting required to normalize X and Y.

3. Get an approximation of $1/Y_h$ from a look-up table called G_1. The index to the look-up table is the leading m bits of Y. Each table word is b_1 bits wide where $b_1 = m$ as discussed later in this paper.

4. Compute the new dividend as
   $X' = X - X_h * (1/Y_h) * Y$ (footnote [2]).

   Although this step appears to require two sequential multiplications to compute $X_h * (1/Y_h) * Y$, this is not really true. By doing the multiplication of $X_h * Y$ while the look-up of $(1/Y_h)$ is being performed, only one subsequent multiplication is needed in the first iteration. In the second and later iterations, $(1/Y_h) * Y$ should ideally be pre-computed as it is invariant across the iterations. For full speed, this requires 2 multipliers during the 1st iteration, one to compute $(1/Y_h) * (X_h * Y)$ and one to compute $(1/Y_h) * Y$.

5. Compute the new quotient
   $Q' = Q + (X_h/Y_h) * (1/2^{(j-k)})$
   $= Q + X_h * (1/Y_h) * (1/2^{(j-k)})$.

   In the above formulas for Q' and X', X_h can be composed of $p = m + 1$ leading digits of X followed by 0's.

6. Left shift X' by $m - 2$ bits to get rid of guaranteed leading 0's that occur (as discussed in the next subsection). Shift zeroes into the LSB of X' during this process.

   Calculate a revised $j' = j + m - 2$.

7. Set $j = j'$, $Q = Q'$, and $X = X'$.

8. Repeat steps 4 through 7 until $j \geq q$.

---

[2] Note: Although it might appear possible to simplify this to
$X' = X - X_h * (Y_h + \Delta Y)/Y_h = X_h + \Delta X - X_h * (1 + \Delta Y/Y_h)$
$= \Delta X - X_h * \Delta Y/Y_h$,
this does not actually work because the calculation of X' and Q' would be uncoordinated. Different round-off errors would occur to X' and Q' in each iteration.

9. The final Q is the quotient. Note that there are more bits of the quotient register than q. Let $Q_h = Q_q..Q_0$ and let $Q_l = Q_{-1}..Q_{-e}$ where e is the number of excess bits. The Q_h bits are correct to within one unit of the least significant bit. Some of the Q_l bits are not necessarily correct since another iteration of the algorithm would add into those bits.

10. The residual dividend X should be right-shifted by $j - q$ bits to get the remainder assuming the entire Q is the quotient. If Q is truncated to q bits, then the true remainder can be computed by then adding $Q_l * Y$ to X.

Since the algorithm reduces X' by $m - 2$ bits per iteration, the algorithm terminates after $\lceil q/(m - 2) \rceil$ iterations.

Next, we analyze this method to show why $m - 2$ bits per iteration are guaranteed in the worst-case.

2.4 How Many Bits per Iteration are Guaranteed Using the Basic Method?

In this section, an analysis is performed to show the following Theorem:

Theorem 1: Number of Bits Per Iteration Using the Basic Method

X' is reduced by at least $m-2$ bits per iteration if $m \geq 5$, the number p of significant bits in X_h is equal to $m + 1$, and the word width b_1 of Table G_1 is m.

Analysis

How do we compute the minimum number of bits that are reduced from X in each iteration?

The initial value of X is between $1/2$ and 1 since the leading bit is 1. By determining the largest possible value of X', we can determine the worst-case number of bits eliminated per iteration. If we can guarantee that $X' < 1/2^r$, then the most significant bit that could be non-zero has significance $1/2^{(r+1)}$. In this case, there are r guaranteed leading zeroes that can be eliminated per iteration.

Let us examine the formula for X':

$$X' = X - X_h * (1/Y_h) * Y \qquad (3)$$

In this case, two sources of inaccuracy affect the approximation $(1/Y_h) \approx 1/Y$. Let the error sources be defined by:

$$(1/Y_h) = 1/Y - R_b - R_c \qquad (4)$$

where R_b represents an error term due to using $1/Y_h$ instead of $1/Y$, and R_c represents the error in storing $1/Y_h$ to only a finite number of bits in the table G_1. The error R_b is always non-negative since $Y_h \geq Y$ meaning $1/Y_h \leq 1/Y$. The error R_c is always non-negative since $1/Y_h$ is stored into G_1 by always rounding down to the finite width b_1. Then

$$\begin{aligned} X' &= X - X_h * (1/Y_h) * Y \\ &= X_h + \Delta X - X_h * (1/Y - R_b - R_c) * Y \\ &= \Delta X + X_h * (R_b + R_c) * Y \qquad (5) \end{aligned}$$

If R_b and R_c can be accurately bounded, then this will give the maximal value of X'.

To compute the worst-case R_b, we can examine a Taylor series expansion of $1/Y$ about the point $Y = Y_h$:

$$\begin{aligned} B &= 1/Y \\ &= 1/Y_h - \Delta Y/Y_h{}^2 + (\Delta Y)^2/Y_h{}^3 \\ &\quad - (\Delta Y)^3/Y_h{}^4 + (\Delta Y)^4/Y_h{}^5 + ... \qquad (6) \end{aligned}$$

The $(w + 1)$-th term is denoted by $B_{w+1} = (-\Delta Y)^w/Y_h{}^{(w+1)}$. All terms of B are non-negative since ΔY is non-positive. The worst-case occurs when $Y_h \approx 1/2$ and $-\Delta Y = 1/2^m - 1/2^q$. In this case, the following bound holds:

$$B_{w+1} < 1/2^{(m*w-w-1)} \qquad (7)$$

In this case, the approximation $1/Y_h \approx 1/Y$ is equivalent to truncating the Taylor series after the first term. A bound on the sum of the remaining terms can serve as a bound on R_b:

$$R_b = \sum_{g=1}^{\infty} B_{g+1} < \sum_{g=1}^{\infty} 1/2^{(m*g-g-1)} = \frac{(1/2^{(m-2)})}{(1 - 1/2^{(m-1)})} \qquad (8)$$

For $m \gg 1$, this is just slightly greater than $1/2^{(m-2)}$. For $m \geq 5$, a non-stringent bound on R_b is:

$$R_b < 1.1 * (1/2^{(m-2)}) \qquad (9)$$

R_c comes from not using the exact value of $1/Y_h$ but instead using a finite-precision table G_1. Denote the output of the table by $(1/Y_h)_a$.

Let the error between the table lookup and the true value be ϵ_1:

$$(1/Y_h)_a = \epsilon_1 + 1/Y_h \qquad (10)$$

$$R_c = \epsilon_1 \qquad (11)$$

Suppose table G_1 is b_1 bits wide. Since $1/2 < Y_h < 1$, the maximal value of $1/Y_h$ is slightly less than 2. If words in the table G_1 can represent values up to but not including 2, the unit of the most significant binary digit in the table G_1 should have value 1. The unit of the least significant binary digit for table G_1 is then $1/2^{(b_1-1)}$. The error ϵ_1 is less than the unit of the least significant digit:

$$R_c = \epsilon_1 < 1/2^{(b_1-1)} \qquad (12)$$

Using this, we can determine the proper word width b_1. Suppose that b_1 is set to:

$$b_1 = m \qquad (13)$$

The error R_c is then bounded by:

$$R_c < 1/2^{(m-1)} \qquad (14)$$

Now we can substitute the bounds for R_b and R_c into equation 5 for X'. Since we set $Y_h \approx 1/2$ previously, Y should also equal 1/2. Since $X_h < 1$ and $\Delta X \leq 1/2^p - 1/2^q$, the worst-case X' is bounded by:

$$\begin{aligned} X' &= \Delta X + X_h * R_b * Y + X_h * R_c * Y \\ &< 1/2^p - 1/2^q + 1.1 * (1/2^{(m-2)}) * 1/2 \\ &\quad + (1/2^{(m-1)}) * 1/2 \end{aligned} \quad (15)$$

If $p = m + 1$, then this can be converted to:

$$X' < 1/2^{(m+1)} - 1/2^q + 1.6 * (1/2^{(m-1)}) \quad (16)$$

$$X' < 0.25/2^{(m-1)} - 1/2^q + 1.6/2^{(m-1)} \quad (17)$$

$$X' < 1/2^{(m-2)} \quad (18)$$

This gives an insight into a good value of p, the number of significant bits in X_h. If $p \geq m + 1$, then the worst-case value of $X' < 1/2^{(m-2)}$. In this case, the highest-order bit of X' that could possibly be 1 is the $(m-1)$-st bit, $X'_{(q-m+1)}$. Since p determines the size of the multipliers needed to evaluate X' and Q' and there is very little advantage to having $p > m+1$, the best value is $p = m + 1$.

Thus, at least $m - 2$ bits are skipped per iteration if $p = m + 1$ and $b_1 = m$.

The average number of bits per iteration is a couple of bits more than this worst-case value. However, to take advantage of this requires a leading 0's detect and variable-sized shift in step 6 of the above algorithm. This slows down the iteration time sufficiently that it is generally not worthwhile for $m \geq 10$ or so. In this case, we have omitted the analysis for the average number of bits.

2.5   An Advanced Method

Let us examine the Taylor series approximation equation for $1/Y$ about $Y = Y_h$:

$$\begin{aligned} 1/Y &= 1/Y_h - \Delta Y/Y_h{}^2 + (\Delta Y)^2/Y_h{}^3 \\ &\quad - (\Delta Y)^3/Y_h{}^4 + (\Delta Y)^4/Y_h{}^5 + ... \end{aligned} \quad (19)$$

All these terms are non-negative since $\Delta Y \leq 0$.

Our current method is to consider the terms beyond $1/Y_h$ to be an unavoidable error. Actually, however, it is possible to rapidly calculate one or more adjustment terms to attain additional accuracy in estimating $1/Y$.

The revised method would be as follows:

1. Set the estimated quotient Q to 0 initially.

2. Let j and k denote the number of bits that X and Y have been left-shifted. Initially, both are set to the amounts of pre-shifting required to normalize X and Y.

3. Get an approximation of $1/Y_h$ from look-up table G_1. The index to the look-up table is the leading m bits of Y. Each table word is b_1 bits long. (Note: later in the paper, the table size is optimized.)

   Simultaneously get approximations of $1/Y_h{}^2$, $1/Y_h{}^3$, etc. using look-up tables G_2, G_3, etc. with word widths b_2, b_3, etc. respectively. All tables are indexed using the first m bits of Y.

   In the next subsection, it will be shown that reasonable table widths b_i are given by:

   $$b_i = (m * t - t) + \lceil log_2 t \rceil - (m * i - m - i) \quad (20)$$

   This equation states that the tables of more significance require more bits of precision.

4. Compute an approximation B to $1/Y$ using the first t terms from the following series:

   $$\begin{aligned} B &= 1/Y_h - \Delta Y/Y_h{}^2 + (\Delta Y)^2/Y_h{}^3 \\ &\quad - (\Delta Y)^3/Y_h{}^4 + (\Delta Y)^4/Y_h{}^5 + ... \end{aligned} \quad (21)$$

   The number of terms t is at least 2; all terms after the second are optional. This can be done with a generation of partial products followed by a carry-save adder tree. (To accelerate the calculation of B, the least significant partial products in calculating powers of ΔY can be truncated for the higher-order terms since the higher-order terms are less significant (details in next subsection).)

5. Compute the new dividend as $X' = X - X_h * B * Y$

   Although this step appears to require two sequential multiplications to compute $X_h * B * Y$, in fact this is not true (see step 4 of basic algorithm).

6. Compute the new quotient $Q' = Q + X_h * B * (1/2^{(j-k)})$.

   It will be shown later in this paper that X_h in the formulas for Q' and X' can be composed of the leading $p = m * t - t + 2$ bits of X followed by 0's, where t is the number of terms used to calculate B. Also, B can be limited to $m * t - t + 4$ bits. This limits the size of the multiplies.

7. Left shift X' by $m * t - t - 1$ bits to get rid of guaranteed leading 0's that occur (as discussed in the next subsection). Shift zeroes into the LSB of X' during this process.

   Calculate a revised $j' = j + m * t - t - 1$.

8. Set $j = j'$, $Q = Q'$, and $X = X'$.

9. Repeat steps 5 through 8 until $j \geq q$.

10. The final Q is the quotient. Note that there are more bits of the quotient register than q. Let $Q_h = Q_q..Q_0$ and let $Q_l = Q_{-1}..Q_{-e}$ where e is the number of excess bits. The Q_h bits are correct to within one unit of the least significant bit. Some of the Q_l bits are not necessarily correct since another iteration of the algorithm would add into those bits.

11. The residual dividend X should be right-shifted by $j - q$ bits to get the remainder assuming the entire Q is the quotient. If Q is truncated to q bits, then the true remainder can be computed by then adding $Q_l * Y$ to X.

Since the advanced version reduces X' by $m*t - t - 1$ bits per iteration, the algorithm terminates after $\lceil q/(m*t - t - 1) \rceil$ iterations.

2.6 How Many Bits per Iteration are Guaranteed using the Advanced Method?

Theorem 2: Number of Bits Per Iteration Using the Advanced Method

X' is reduced by at least $m*t - t - 1$ bits per iteration if

- *The term Y_h includes $m \geq 5$ leading bits of Y.*
- *The number of Taylor-series terms from equation 21 used to construct B is t.*
- *The number p of significant bits in X_h is equal to $m*t - t + 2$.*
- *The word width b_i of Table G_i is $(m*t - t) + \lceil log_2 t \rceil - (m*i - m - i)$ for $i = 1, 2, ..t$.*
- *B is represented in $b_b = m*t - t + 4$ bits or more.*
- *The total arithmetic error in calculating B is less than $1/2^{(m*t-t+3)}$ (see below for explanation).*

Analysis

As in the basic method, m is the most important parameter, since an increase of 1 requires a doubling of the number of table words. Once m is chosen, the other parameters are set so that the total error from truncating other calculations is less than or equal to the error from including just the leading m bits of Y in Y_h. In this analysis, we show that this is true given the above parameter settings.

As before, we compute the worst-case number of bits that are reduced from X in each iteration. The initial value of X is between $1/2$ and 1 since the leading bit is 1. By determining the largest possible value of X', we can determine the worst-case number of bits eliminated per iteration.

Let us examine the formula for X':

$$X' = X - X_h * B * Y \tag{22}$$

where B is the first t terms from $B = 1/Y_h - \Delta Y/Y_h{}^2 + (\Delta Y)^2/Y_h{}^3 - (\Delta Y)^3/Y_h{}^4 + (\Delta Y)^4/Y_h{}^5 + ...$

In this case, three sources of inaccuracy affect the approximation $B \approx 1/Y$. Let the sources be defined by

$$B = 1/Y - R_b - R_c - R_d \tag{23}$$

where

- R_b represents an error term due to truncating the Taylor series after t terms. The error R_b is always non-negative since the truncated terms in the series are all non-negative.
- R_c represents the error in calculating B using lookup tables with finite width words. The error R_c is always non-negative since B is calculated using tables that are rounded down.

- R_d represents the error in truncating the arithmetic used to calculate B. The error R_d is non-negative since all the arithmetic is truncated, thus underestimating B.

Then

$$\begin{aligned} X' &= X - X_h * B * Y \\ &= X_h + \Delta X - X_h * (1/Y - R_b - R_c - R_d) * Y \end{aligned} \tag{24}$$

$$X' = \Delta X + X_h * (R_b + R_c + R_d) * Y \tag{25}$$

If R_b, R_c, and R_d can be accurately bounded, then this will give the maximal value of X'.

The $(w+1)$-th term of the Taylor-series for B is denoted by $B_{w+1} = (-\Delta Y)^w / Y_h{}^{(w+1)}$. All terms of B are non-negative since ΔY is non-positive. The worst-case occurs when $Y_h \approx 1/2$ and $-\Delta Y = 1/2^m - 1/2^q$. In this case, the following bound holds:

$$B_{w+1} < 1/2^{(m*w - w - 1)} \tag{26}$$

If the first t terms are used to construct B, then the remainder R_b is bounded by:

$$R_b = \sum_{g=t}^{\infty} B_{(g+1)} < \sum_{g=t}^{\infty} 1/2^{(m*g - g - 1)} = \frac{(1/2^{(m*t-t-1)})}{(1 - 1/2^{(m-1)})} \tag{27}$$

For $m \gg 1$, this is just slightly greater than $1/2^{(m*t-t-1)}$. For $m \geq 5$, a non-stringent bound on R_b is:

$$R_b < 1.1 * (1/2^{(m*t-t-1)}) \tag{28}$$

R_c comes from using finite-width lookup tables in calculating the value of the first t terms. R_d comes from truncating the arithmetic used to calculate B and truncating the value of B.

A cumulative error can be computed where the error in each table lookup is represented by ϵ_i and the truncation error in computing each multiplicative term is δ_i:

$$B = \sum_{i=1}^{t} [\delta_i + (\epsilon_i + 1/Y_h{}^i) * (-\Delta Y)^{(i-1)}] \tag{29}$$

$$B = \sum_{i=1}^{t} [\frac{(-\Delta Y)^{(i-1)}}{Y_h{}^i}] + \sum_{i=1}^{t} [(\epsilon_i) * (-\Delta Y)^{(i-1)}] + \sum_{i=1}^{t} \delta_i \tag{30}$$

Then R_c is equal to the sum of the terms involving ϵ_i:

$$R_c = \sum_{i=1}^{t} \epsilon_i * (-\Delta Y)^{(i-1)} \tag{31}$$

Let δ_0 be an additional term that represents truncating B to a certain number of bits after the summation. Then

$$R_d = \sum_{i=0}^{t} \delta_i \tag{32}$$

Let us focus first on R_c. Suppose table G_i is b_i bits wide. Since $1/2 < Y_h < 1$, the maximal value of $1/Y_h{}^i$ is slightly less than 2^i. If words in the table G_i can represent values up to but not including 2^i, the unit of the most significant binary digit in the table G_i should have value $2^{(i-1)}$. The unit of the least significant binary digit for table G_i is then $1/2^{(b_i-i)}$. Each ϵ_i is less than the unit of the least significant digit:

$$\epsilon_i < 1/2^{(b_i-i)} \qquad (33)$$

The worst-case value of $-\Delta Y$ is $-\Delta Y = 1/2^m - 1/2^q$. Substituting into equation 31 yields:

$$R_c < \sum_{i=1}^{t} 1/2^{(b_i-i)} * (1/2^{(m*i-m)}) \qquad (34)$$

$$R_c < \sum_{i=1}^{t} 1/2^{(b_i+m*i-m-i)} \qquad (35)$$

This equation helps to determine the proper word widths b_i. Suppose that the b_i are set to:

$$b_i = (m*t-t) + \lceil log_2 t\rceil - (m*i-m-i) \qquad (36)$$

As expected since the first tables have more significance, each table T_i is $m-1$ bits wider than the table $T_{(i+1)}$.

The total error R_c is then bounded by:

$$R_c < \sum_{i=1}^{t} 1/2^{((m*t-t)+\lceil log_2 t\rceil)} \qquad (37)$$

$$R_c < 1/2^{(m*t-t)} \qquad (38)$$

Now consider R_d. The δ_i represent maximal permissible truncation errors. This concept accelerates the calculation of B by allowing the arithmetic for calculating B to be reduced in size. For instance, suppose $t = 4$, $m = 16$, and $q = 53$. Then the last term of B is $B_4 = (-\Delta Y)^3 * (1/Y_h{}^4)$. Since ΔY has $q - m = 53 - 16 = 37$ bits, the full computation of $(-\Delta Y)^3$ would use large multipliers and result in a $37*3-2 = 109$ bit result. This is clearly much more than necessary because the unit of the most significant bit of B_4 will be $1/2^{(3*m-4)} = 1/2^{44}$. The allowable truncation errors δ_i can be used both to prune multiplier trees by discarding least significant partial products and to truncate results to smaller widths. The details are not presented here.

Since $1 \leq B < 2$, the MSB of B has unit 1. If B is truncated to b_b bits, then $\delta_0 <$ unit of LSB so $\delta_0 < 1/2^{(b_b-1)}$.

In this case, we wish to restrict R_d as follows:

$$R_d < 1/2^{(m*t-t+2)} \qquad (39)$$

For the purposes of this paper, this can be achieved by both restricting δ_0 to

$$\delta_0 < 1/2^{(m*t-t+3)} \qquad (40)$$

meaning that B can be truncated to $b_b = m*t-t+4$ bits, and restricting the remaining δ_i to

$$\sum_{i=1}^{t} \delta_i < 1/2^{(m*t-t+3)} \qquad (41)$$

Now we can substitute the bounds for R_b, R_c, and R_d into constraint 25 to determine the maximum value of X'. Since we set $Y_h \approx 1/2$ previously, Y should also equal $1/2$. Since $X_h < 1$ and $\Delta X \leq 1/2^p - 1/2^q$, the worst-case X' is bounded by:

$$X' = \Delta X + X_h * R_b * Y + X_h * R_c * Y + X_h * R_d * Y \qquad (42)$$

$$\begin{aligned}
X' \;<\; & 1/2^p - 1/2^q + 1.1 * (1/2^{(m*t-t-1)}) * 1/2 \\
& + (1/2^{(m*t-t)}) * 1/2 + (1/2^{(m*t-t+2)}) * 1/2
\end{aligned} \qquad (43)$$

If $p = m*t-t+2$, then this can be converted to:

$$X' < 1/2^{(m*t-t+2)} - 1/2^q + 1.6*(1/2^{(m*t-t)}) + 1/2^{(m*t-t+3)} \qquad (44)$$

$$\begin{aligned}
X' \;<\; & 0.25/2^{(m*t-t)} - 1/2^q + 1.6 * (1/2^{(m*t-t)}) \\
& + 0.125 * (1/2^{(m*t-t)})
\end{aligned} \qquad (45)$$

$$X' < 1/2^{(m*t-t-1)} \qquad (46)$$

If $p \geq m*t-t+2$, then the worst-case value of X' is bounded by the above inequality. In this case, the highest-order bit of X' that could possibly be 1 is the $(m*t-t)$-th bit, $X'_{(q-m*t+t)}$. In fact, using $p > m*t-t+2$ only improves the convergence by a fraction of a bit, so it is best to use $p = m*t-t+2$ to minimize the size of the multiplies.

In summary, at least $m*t-t-1$ bits are skipped per iteration if $p = m*t-t+2$, t is the number of terms used to construct B, and the other conditions in Theorem 2 hold. (Note that this works for the basic method where $t = 1$, $p = m+1$, and the predicted minimum number of bits is $m-2$.)

3 Comparison to Other High-Radix Methods Using Reciprocals

In this section, our method is compared to the Cyrix short-reciprocal method, the Newton-Raphson technique, the MacLaurin series technique, and an implementation of a modified Newton-Raphson technique in the IBM RISC System/6000.

We refer to approximations of the reciprocal in all methods using B. In some cases, such as our basic method, B is the direct result of a table look-up. In other cases, B requires some calculations.

3.1 Cyrix Short-Reciprocal Algorithm

The recently designed Cyrix arithmetic coprocessor briefly described in [4] and [11] employs a "short reciprocal" algorithm similar to the basic technique described in this paper to get 17 bits/iteration. We only discriminate between the two methods here. Unlike our method, the Cyrix method derives its reciprocal approximation using a combination of table lookup and Newton-Raphson. The iterative portion of the Cyrix method uses an intentional overestimate of the reciprocal rather than an underestimate. In addition, the Cyrix method is implemented using a redundant number system while ours is non-redundant. The Cyrix coprocessor can also do square root using a similar iterative scheme. A fuller description of their work may be forthcoming.

3.2 Comparison to Newton-Raphson

In the standard Newton-Raphson technique [13], an accurate reciprocal B is computed first by an iterative method. This reciprocal is accurate enough so that the quotient can be calculated directly by a final multiplication $Q = X * B$. The calculation of the reciprocal approximation in Newton-Raphson is necessarily iterative: the improved reciprocal is calculated using $B_{(n+1)} = B_n * (2 - Y * B_n)$ [13] [10]. Each iteration requires two multiplications in $B_n * Y * B_n$ that cannot be performed simultaneously. Each iteration doubles the number of bits of accuracy in B; to get 56-bit accuracy requires 2 iterations starting with a 14-bit accurate reciprocal. A separate calculation is required to produce a remainder, which has been shown to be non-negative in [13].

In contrast, our method calculates the reciprocal B using one or more terms from a Taylor-series that can be implemented in a non-iterative manner. Our method iterates by calculating a reduced dividend $X' = X - (X * B) * Y$ where $X * B$ is the approximate quotient based on the approximate reciprocal B. Each iteration reduces the dividend X' by a constant number of bits which depends on the accuracy of B. Since estimates of the quotient are calculated simultaneously in each iteration, no final multiplication is needed to get the quotient Q. A form of remainder is available directly in the register X assuming that the entire Q is the quotient. If the quotient is truncated, then the truncated bits multiplied by Y should be added to X. The method is designed so that the remainder is never negative.

3.3 Comparison to IBM RISC System/6000

The IBM RISC System/6000 implements floating-point division using an iterative algorithm employing its multiply and add hardware [12] [10]. The algorithm is a modified version of the Newton-Raphson method. One enhancement is used: after a reciprocal is calculated, the method performs one pass of a reduction similar to our method's iteration. Using our notation, it is as follows:

1. $Q = X * B$
2. $X' = X - Q * Y$
3. $Q' = Q + X' * B$

Reference [10] describes the algorithm and detailed analyses for guaranteed correct rounding.

In addition to the general differences between Newton-Raphson and our method, the above reduction is done using a full precision quotient, i.e. $X' = X - Q * Y$ instead of $X' = X - X_h * B * Y$ used in our method.

3.4 Comparison to MacLaurin Series

The MacLaurin series method [13] uses an approximation to the reciprocal:

$$\begin{aligned}
B &\approx 1/Y = 1/(Z + 1) \\
&= (1 - Z) * (1 + Z^2) * (1 + Z^4) \\
&\quad * (1 + Z^8) * (1 + Z^{16}) + ... \quad (47)
\end{aligned}$$

The number of bits of accuracy doubles with each additional factor. In the normal iterative implementation, the first few terms are approximated using a lookup table. Each iteration calculates an additional factor from the previous using one multiplication and an add (or a 2's complement in [13]). An additional multiplication is required to multiply the factor with the current approximation B. This multiplication can overlap the calculation of the next factor, so that each iteration adding one factor takes just 1 multiplication plus a 2's complement. As discussed in [13], the convergence per iteration is mathematically equivalent to the convergence of the Newton-Raphson method where $B_0 = 1$. The quotient and remainder must each be calculated following the generation of B. Unlike our method or Newton-Raphson, the implementation sizes are somewhat fixed. If one implementation includes the first k multiplicative terms in a lookup table of 2^v words, the next larger implementation includes the first $k + 1$ terms with table size about 2^{2*v}.

Like the Newton-Raphson method, the convergence of the MacLaurin series differs from our method. Also, our method automatically provides a quotient without extra calculations.

4 Brief Notes on Implementation

1. The look-up table(s) are indexed using the leading m bits of Y. Since the leading bit of Y is known to be 1 after normalization, it is not actually useful to use that bit as part of the index. Therefore, the actual table can be of size $2^{(m-1)}$ words instead of 2^m.

2. The look-up table words are always rounded down to fit into the number of bits in the word so that the estimated reciprocal is never more than the exact answer to $1/Y$.

3. In each iteration, the critical loops are the calculation of Q' and X'. These can be implemented using carry-save adder (Wallace) trees that perform a multiply-and-add or multiply-and-subtract function. The latter can be accomplished by using 2's complement numbers. The hardware should make a 1's complement of each row to be subtracted and add a fixed 1 to that row's LSB in the carry-save adder tree.

4. As shown in Appendix 1, the basic method requires three multipliers of size $(m+1) \times m$, $(m+1) \times q$, and $(m+1) \times (q+m)$. Also required are one subtraction of size $(q+2m-1)$ and one addition of size $\lfloor q/(m-2) \rfloor * (m-2) + 2m$ which can actually be merged with the carry-save adder trees for the multiplications.

5. For the advanced method, the minimum size of the additions and multiplications except those to calculate B are as follows:

   - Note that Y has q bits, X_h has $p = m*t - t + 2$ bits (rest are 0's), and B has $b_b = m*t - t + 4$ bits.

   - Three multipliers are needed of size $p \times b_b$, $b_b \times q$, and $b_b \times (q + b_b - 1)$. Also required are one subtraction of size $(q + p + b_b - 2)$ and one addition of size $\lfloor (q/(m*t - t - 1)) \rfloor * (m*t - t - 1) + p + b_b - 1$ which can actually be merged with the carry-save adder trees for the multiplications.

5   Comparative Speed Analysis

In this section, three schemes are analyzed for practical high-speed implementation to demonstrate why the above technique is fast. The other two techniques are ordinary 1 bit or 2 bit/iteration division and Newton-Raphson division.

The charter of our research project is to focus on very fast implementations of arithmetic functions using maximum parallelism and dense, state-of-the-art technologies [7]. For this analysis, we assume a fast hypothetical technology whose characteristics are similar to a 1991 VLSI ECL technology:

1. Simple gate delays are 250 ps.

2. In the worst-case, complex AND/OR gate structures take almost twice as long as simple gates in ECL and are assumed to be 500 ps.

3. The assumed delay for the registers required for iteration is 500 ps.

4. One-way communication time between a logic chip and a RAM chip is .75 ns.

5. Static RAM access takes 3 ns for the RAM and 1.5 ns for a round-trip communication with the division logic chip for RAMs up to 64K x n bits. A RAM access time of 3 ns was demonstrated in a 1988 paper on an experimental 64-Kbit RAM [15].

6. An addition of 64 bits takes 2 ns. This is based on another 1988 paper on a 32-bit ECL adder [2]. Smaller additions take slightly less time, and larger ones take slightly more time.

7. A 53x53 multiplication takes about 6 ns in ECL [3]. Smaller multiplications take less time; for instance, a 32x32 bit multiplication takes 5 ns.

8. A 53x53 multiply plus an add takes 6 ns. A multiply and add takes no more time than a multiply if the add function is incorporated into the carry-save adder tree.

9. Clock skew effects as discussed for example in [8] and [14] are not included in this analysis.

We analyze for critical path delay assuming a fully-parallel implementation, i.e. all operations that can be made independent are performed in parallel.

In this case, we calculate the extra delay to compute the remainder as a separate number. The remainder will be relative to the quotient truncated to q bits. In our method, the previously-discussed correction to the remainder is required to adjust for this truncation (see last step of the algorithm). In Newton-Raphson, the remainder must be computed after the quotient. Since $Q = X * B$ provides about $2 * q$ bits, a simultaneous computation of the remainder and quotient is not possible. The simultaneous equation for the remainder $X' = B * (X * Y)$ is relative to an untruncated quotient. A remainder for a truncated quotient requires either a correction like the one in our method or a sequential calculation of quotient and remainder. The time consumed is about the same.

The details of the speed analysis are presented in Appendix 2 for some example implementations.

Table 1 summarizes the division schemes and their projected speeds for 53-bit division using the given technology assumptions. Extra delays are cited separately for computing a true remainder needed for round-to-nearest calculations. The amount of table storage required is also given in kilobits. A coarse comparison of logic size can be made using the total number of partial product bits needed for the major multipliers.

Some observations are as follows. The Newton-Raphson technique with a 7-bit accurate lookup table uses fewer storage bits than our basic method with $m = 11$ but still requires large multipliers. The speeds of the two are about equal. The fastest implementations are our advanced method using $m = 11$ and $m = 15$ with 2 Taylor-series terms. Using more than 2 terms to approximate B is slightly faster but uses much larger tables than just using 2 terms. However, the number of product bits remains moderate since the iteration multipliers are not needed; only truncated multipliers to calculate B and plus one 53x53 multiplier for the quotient plus remainder are needed.

For comparison, the Cray-2, first delivered in 1986, performs a floating-point division in 152 ns using MSI ECL circuits [6].

6   Summary and Conclusions

By using large, accurate lookup tables for the reciprocal, division can be accomplished in just a few iterations using the basic method. The keys are to design the method to underestimate the quotient and to adjust the quotient by adding rather than appending low-order bits. Using adding, the design is freed from having to produce disjoint sets of quotient bits in each iteration. In addition, a more accurate reciprocal can be quickly calculated using a Taylor-series approximation. Theoretically, the Taylor-series can be calculated directly rather than iteratively, unlike Newton-Raphson. In practice, the higher powers of ΔY may be calculated with staged multiplications, although lookup tables for powers of ΔY remains an unexplored possibility. By using more accurate reciprocals, the advanced version can divide in 2 or even 1 iteration.

Method	Est. Delay	Remainder	Table Size	Product Bits
Canonical 1-bit	159 ns	0 ns	0	53
Radix-4 SRT with 4 levels cascaded/iter	76 ns	0 ns	0	212
Newton-Raphson using 7-bit Initial	35.5 ns	6 ns	896 bits	≈ 4614
Newton-Raphson using 14-bit Initial	30 ns	6 ns	224K bits	≈ 4536
Basic Method using $m = 11$	36.0 ns	3.5 ns	12K bits	1536
Basic Method using $m = 13$	32.0 ns	3.75 ns	56K bits	1848
Basic Method using $m = 16$	27.5 ns	4 ns	544K bits	2346
Advanced Method using $m = 11$ and $t = 2$	26.5 ns	4.5 ns	34K bits	3888
Advanced Method using $m = 15$ and $t = 2$	22 ns	5 ns	736K bits	5952
Advanced Method using $m = 15$ and $t = 4$	20.5 ns	6 ns	2432K bits	≈ 4056

Table 1: Table of Relative Speeds and Required Storage and Multiplier Sizes

Implementations of our method can be faster than Newton-Raphson in modern ECL technology. The advanced method using $m = 11$ and 2 Taylor-series terms is both faster and smaller than a fast Newton-Raphson implementation. The advanced version using 2 Taylor-series terms and $m = 15$ is another 15% faster than using $m = 11$ at the cost of larger tables and multipliers. In a hypothetical implementation in ECL, the advanced method can divide 53-bit numbers in an estimated 22 ns vs. 6 ns for a multiply, thus achieving a ratio between divide and multiply times of less than 4:1. The algorithm can produce an exact remainder, which makes the implementation of exact rounding specifications (e.g. IEEE floating-point) straightforward. Both Newton-Raphson and this technique are substantially faster than 1-bit or 2-bit/iteration schemes. The logic depth of each iteration is more than in 1-bit or 2-bit schemes, so the speedup is less than proportional to the ratio of bits per iteration.

7 Acknowledgements

Thanks to the reviewers, David Matula, and Tien-Chi Chen for providing many helpful comments and additional references.

Thanks to the entire Nano-Second Arithmetic Research Group at Stanford for many enlightening discussions and seminars. Nhon Quach and Eric Schwarz carefully reviewed drafts of this paper and made many helpful comments.

References

[1] D. Atkins. "Higher-Radix Division Using Estimates of the Divisor and Partial Remainders." Oct. 1968, IEEE Transactions on Computers, pp. 925-934.

[2] G. Bewick, P. Song, G. De Micheli, and M. Flynn. "Approaching a Nanosecond: A 32 bit Adder." 1988 International Conference on Computer Design, pp. 221-226.

[3] G. Bewick. Private communication in October 1990 about work in progress. Computer Systems Laboratory, Stanford University, Stanford, CA.

[4] T. Brightman. Slides from presentation on Cyrix co-processor. 1989, future Trends panel session of 9th Symposium on Computer Arithmetic.

[5] J. Fandrianto. "Algorithm for High Speed Shared Radix 8 Division and Radix 8 Square Root." Sept. 1989, 9th Symposium on Computer Arithmetic, pp. 68-75.

[6] M. Flynn. "Sub-nanosecond Arithmetic Proposal." 1989, unpublished report, Computer Systems Laboratory, Stanford University, Stanford, CA.

[7] M. Flynn et al. "Sub-nanosecond Arithmetic". May 1990, Technical Report CSL-TR-90-428, Computer Systems Laboratory, Stanford University, Stanford, CA.

[8] L. Glasser and D. Dobberpuhl. The Design and Analysis of VLSI Circuits, Chapter 6. 1985, Addison-Wesley, Reading, Massachusetts.

[9] E. Krishnamurthy. "On Range-Transformation Techniques for Division." February 1970, IEEE Transactions on Computers, pp. 157-160.

[10] P. Markstein. "Computation of Elementary Functions on the IBM RISC System/6000 Processor." January 1990, IBM Journal of Research and Development, vol. 34, no. 1, pp. 111-119.

[11] D. Matula. "Highly Parallel Divide and Square Root Algorithms for a New Generation Floating Point Processor." Oct. 1989, extended abstract from SCAN-89 Symposium on Computer Arithmetic and Self-Validating Numerical Methods.

[12] R. Montoye, E. Hokenek, and S. Runyon. "Design of the IBM RISC System/6000 floating-point execution unit." January 1990, IBM Journal of Research and Development, vol. 34, no. 1, pp. 59-70.

[13] S. Waser and M. Flynn. Introduction to Arithmetic for Digital Systems Designers, Chapter 5. 1982; Holt, Rinehart, and Winston; New York, New York.

[14] D. Wong, G. De Micheli, and M. Flynn. "Designing High-Performance Digital Circuits Using Wave Pipelining." Aug. 1989, VLSI '89 Conference, pp. 241-252.

[15] K. Yamaguchi, H. Nanbu, K. Kanetani, et al. "An Experimental Soft-error Immune 64-Kb 3ns ECL Bipolar RAM." 1988 Bipolar Circuits and Technology Meeting, pp. 26-27.

8 Appendix 1: The Size of Needed Multipliers and Dividers

For the basic method, the minimum size of the additions and multiplications are as follows:

- Note that Y has q bits, X_h has $p = m + 1$ bits (rest are 0's), and $1/Y_h$ has $b_1 = m$ bits.

- In $X' = X - X_h * (1/Y_h) * Y$:

  In the first iteration, $X_h * Y$ is calculated first using a $(m+1) \times q$ bit multiply giving a $q+m$ bit result. This is then multiplied by $(1/Y_h)$ using an $m \times (q + m)$ bit multiply to get a $(q + 2m - 1)$ bit result. The subtraction from X is done using a $(q + 2m - 1)$ bit subtract yielding a $(q + m + 1)$ bit result after the shift of $m - 2$ bits.

  The quantity $(1/Y_h) * Y$ is also calculated during the first iteration using a $m \times q$ bit multiply yielding a $(q + m - 1)$ bit result. In subsequent iterations, $X_h * ((1/Y_h) * Y)$ is computed using a $(m+1) \times (q+m-1)$ bit multiply yielding a $(q + 2m - 1)$ bit result. The subtraction is the same size $(q + 2m - 1)$ bits yielding a $(q + m + 1)$ bit result.

  With some input MUX'ing, it is possible to use two multipliers total of sizes $(m + 1) \times q$ and $(m + 1) \times (q + m)$.

- In $Q' = Q + X_h * (1/Y_h) * (1/2^{(j-k)})$, the actual multiplication is $X_h * (1/Y_h)$ which is $(m + 1) \times m$ bits yielding a $2m$ bit result. The add increases in size with each iteration. In the last iteration, it is about $\lfloor q/(m - 2) \rfloor * (m - 2) + 2m$ bits long.

- In summary, three multipliers are needed of size $(m+1) \times m$, $(m + 1) \times q$, and $(m + 1) \times (q + m)$. Some MUX'es are needed to select the inputs to the latter two. Also required are one subtraction of size $(q + 2m - 1)$ and one addition of size $\lfloor q/(m-2) \rfloor * (m-2) + 2m$ which can actually be merged with the carry-save adder trees for the multiplications.

The calculation of sizes for the advanced method is similar. The minimum size of the additions and multiplications except those to calculate B are as follows:

- Note that Y has q bits, X_h has $p = m * t - t + 2$ bits (rest are 0's), and B has $b_b = m * t - t + 4$ bits.

- In $X' = X - X_h * B * Y$:

  In the first iteration, $X_h * Y$ is calculated first using a $p \times q$ bit multiply giving a $q + p - 1$ bit result. This is then multiplied by B using an $b_b \times (q + p - 1)$ bit multiply to get a $(q + p + b_b - 2)$ bit result. The subtraction from X is done using a $(q + p + b_b - 2)$ bit subtract yielding a $(q + p + 3)$ bit result after the shift of $m * t - t - 1$ bits.

  The quantity $B * Y$ is also calculated during the first iteration using a $b_b \times q$ bit multiply yielding a $(q + b_b - 1)$ bit result. In subsequent iterations, $X_h * (B * Y)$ is computed using a $p \times (q + b_b - 1)$ bit multiply yielding a $(q + p + b_b - 2)$ bit result. The subtraction is the same size $(q + p + b_b - 2)$ bits yielding a $(q + p + 3)$ bit result.

With some input MUX'ing, it is possible to use two multipliers total of size $b_b \times q$ and $b_b \times (q + b_b - 1)$.

- In $Q' = Q + X_h * B * (1/2^{(j-k)})$, the actual multiplication is $X_h * B$ which is $p \times b_b$ bits yielding a $p + b_b - 1$ bit result. The add increases in size with each iteration. In the last iteration, it is about $\lfloor (q/(m * t - t - 1)) \rfloor * (m * t - t - 1) + p + b_b - 1$ bits long.

- In summary, three multipliers are needed of size $p \times b_b$, $b_b \times q$, and $b_b \times (q + b_b - 1)$. Some MUX'es are needed to select the inputs to the latter two. Also required are one subtraction of size $(q + p + b_b - 2)$ and one addition of size $\lfloor (q/(m * t - t - 1)) \rfloor * (m * t - t - 1) + p + b_b - 1$ which can actually be merged with the carry-save adder trees for the multiplications.

9 Appendix 2: Speed and Size Calculations

9.1 Speed of the Basic Technique

9.1.1 11 bits/iteration

Set $m = 13$.

The look-up table has size $2^{12} = 4K$ words. The table width is $p = m + 1 = 14$ bits, so the table size is 56K bits.

The number of iterations required is $\lceil 53/(m - 2) \rceil = \lceil 53/11 \rceil = 5$.

The initial RAM lookup takes 4.5 ns.

The time per iteration is limited by the loop to compute the remaining dividend X'. For X', the critical path can be implemented using a MUX, a $(m + 1) \times (q + m) = $ 14x66 multiply tree, and a $(q + 2m - 1) = $ 78-bit final add. The MUX takes 0.5 ns. The multiply tree takes about 2.25 ns to do the partial products generation and reduction. The addition takes 2.25 ns. Finally, iteration registers are assumed to take 0.5 ns.

The loop time is then $0.5 + 2.25 + 2.25 + .5 = 5.5$ ns.

The total time is then $4.5 + 5.5 * 5 = 32.0$ ns.

The final shift of X' can be done with wiring and takes no extra time. The remainder for the truncated quotient can be computed using a multiply of about 14x53 and an add which takes about an extra 3.75 ns.

The calculations for 14 bits/iteration using $m = 16$ and 9 bits/iteration using $m = 11$ are similar.

9.2 Speed of the Advanced Technique

9.2.1 27 bits/iteration using 2 Terms

Set $m = 15$.

Use 2 terms of the approximation to get B.

The look-up tables have size $2^{14} = 16K$ words. Table G_i has width $m * t - t + \lceil \log_2 t \rceil - (m * i - m - i) = 44 - 14 * i$ bits. Table G_1 has width 30 bits for a size of 480K bits. Table G_2 has width 16 bits for a size of 256K bits.

The number of iterations required is $\lceil 53/(m * t - t - 1) \rceil = \lceil 53/27 \rceil = 2$.

The initial RAM lookup takes 4.5 ns.

The computation of B takes 4.5 ns using some multiply and add hardware. The multiplication of $\Delta Y * (1/(Y_h)^2)$ is 16x38 bits. The addition is about 50 bits wide.

The time per iteration is limited by the loop to compute the remaining dividend X'. For X', the critical path can be implemented using a MUX, a $b_b \times (q + b_b - 1) = 32\text{x}84$ multiply tree, and a $(q + p + b_b - 2) = 113$-bit final add. The MUX takes 0.5 ns. The multiply tree takes about 3 ns to do the partial products generation and reduction. The addition takes 2.5 ns. Finally, iteration registers are assumed to take 0.5 ns.

The loop time is then $0.5 + 3 + 2.5 + 0.5 = 6.5$ ns.

The total time is then $4.5 + 4.5 + 2 * 6.5 = 22$ ns.

The final shift of X' can be done with fixed wiring and takes no extra time. The remainder for the truncated quotient can be computed in about an extra 5 ns using about a 32x53 multiply and add.

The calculation for 19 bits/iteration using $m = 11$ and $t = 2$ is similar.

9.2.2   55 bits/iteration using 4 Terms

Set $m = 15$.

Use 4 terms of the approximation to get B. The accuracy of B is about 55 bits.

The look-up tables have size $2^{14} = 16K$ words. Tables $G1$, $G2$, $G3$, and $G4$ have widths 59, 45, 31, and 17 bits respectively. This means that tables $G1$ through $G4$ have sizes of 944K, 720K, 496K, and 272K bits, respectively.

The number of iterations required is $\lceil 53/(m*t - t - 1) \rceil = \lceil 53/55 \rceil = 1$.

The initial RAM lookup takes 4.5 ns but is not on the critical path, as explained below.

The computation of B takes 14 ns using some multiply and add hardware. The slowest term is $(\Delta Y)^3 * (1/(Y_h)^4)$ which takes 3 multiplications, 2 to get $(\Delta Y)^3$ and 1 more to get the term. This takes 8 ns for the first two multiplications. Since 8 ns exceeds the lookup time for the corresponding multiplicative factor $1/Y_h{}^4$, the RAM lookup is not part of the critical path. The final multiplication is combined with the addition of all the terms and is estimated to take 6 ns.

The computation of Q' requires a full 53x53 multiplication which takes 6 ns. Finally, storage registers are assumed to take 0.5 ns. The time for this is then $6 + 0.5 = 6.5$ ns.

The total time is then $14 + 6.5 = 20.5$ ns.

In this case, the true remainder is most easily calculated using $X' = X - Q_h * B$ where Q_h is the truncated quotient. This requires a 53x53 multiply and an add which takes about 6 ns. (The correction method is no faster.)

Due to the increase in complexity for calculating B and the larger multiplication/additions needed in calculating Q' and X', this method is not much faster than using only 2 terms to approximate B even though the number of iterations is cut in half.

9.3   Speed of ordinary 1-bit iteration division

In each iteration of ordinary division, the divisor is subtracted from the dividend, the result is checked for negativity, the new dividend is selected, and the results are stored in the iteration registers.

The subtraction takes 2 ns. The negativity check is trivial since the high-order bit of the result can be directly used as the control to the MUX selecting the new dividend. The MUX requires 2 gate levels for a delay of .5 ns. The registers take .5 ns.

The total delay per iteration is $2.5 + .5 = 3$ ns. The number of iterations is 53 for a total delay of $53 * 3 = 159$ ns.

In this method, the remainder is available in the dividend register following the computation.

9.4   Speed of Newton-Raphson division using 14-bit Table Lookup

The initial table lookup of Newton-Raphson takes 4.5 ns. The table is about 16K words of 14 bits = 224K bits.

Each iteration of Newton-Raphson requires 1 multiply followed by a second multiply and add. The first iteration requires a 14x14 and a truncated 14x28 multiply; the second requires a 28x28 and a truncated 28x56 multiply.

In the 1st and 2nd iterations, multiplies take 4 and 5 ns respectively. Storage into the iteration registers takes 0.5 ns.

The first iteration takes $4 + 4 + 0.5 = 8.5$ ns. The second takes $5 + 5 + 0.5 = 10.5$ ns.

Following two iterations, a 56-bit accurate reciprocal is available in a register that we denote by B.

The iterations are followed by a 53x56 bit multiplication to get the quotient which takes 6 ns. A final register storage takes 0.5 ns.

The total delay is then about $4.5 + 8.5 + 10.5 + 6.5 = 30$ ns.

The remainder for the truncated quotient Q_h can be computed using $X' = X - Q_h * Y$. This requires a 53x53 bit multiply and subtract which takes 6 ns.

Table 1 cites multiplier sizes assuming that a dedicated multiplier is used for each iteration. Alternatively, one 53x56 multiplier would be sufficient but all multiplies would take 6 ns thus adding a total of 6 ns to the propagation delay.

Simple Radix 2 Division and Square Root with Skipping of Some Addition Steps

Paolo Montuschi[†]

Luigi Ciminiera[‡]

[†] Dip. di Automatica e Informatica,
Politecnico di Torino,
Corso Duca degli Abruzzi 24,
10129 Torino (Italy)

[‡] Dip. di Ingegneria Elettrica,
Università degli Studi di L'Aquila,
Poggio di Roio,
67040 L'Aquila (Italy)

Abstract

We present a new algorithm for shared radix 2 division and square root whose main characteristic is the ability to avoid any addition, when the digit 0 has been selected. Unlike other similar works, the solution presented uses a redundant representation of the partial remainder, while keeping the advantages of classical solutions. The paper shows how the next digit of the result can be selected even when the remainder is not updated, showing also the tradeoff arising. The average occurrences of 0 digit selections is also estimated in order to assess the benefits of the algorithm presented.

1 Introduction

Division and square root have always played primary roles in the computer arithmetic. Their importance has became more relevant with the recent introduction of the IEEE 754 floating point standard, as units complying with the IEEE 754 must include both division and square root in their instruction sets. Several architectures have been proposed for division, for square root and for shared division and square root in [11], [1] and more recently in [6], [4], [5], [10] and [2].

In this paper we present a new algorithm for shared radix 2 division and square root whose main characteristic is the possibility to avoid the addition required by *normal* steps when the digit 0 has been selected in the previous iteration. Actually, while *normal* iterations require that the partial remainder is updated by carrying out an addition/subtraction, *fast* iterations update the partial remainder simply with a left shift, with no addition/subtraction, and so can be executed faster.

The possibility to speed up the duration of the whole selection process by decreasing the number of additions, has been extensively explored in several works of the past literature. Robertson in [11] introduced the concept of redundancy with the precise aim of increasing the probability of selecting the digit 0, thus in effect reducing the number of additions necessary for updating the partial remainders. Wilson and Ledley in [14] claimed that with Robertson's method applied to division two digits per iteration were produced (on the average). The same paper [14] also cites a work by Smith and Weinberger [12], which concludes that for random sequences, the ratio of required cycles to quotient bits approaches 1/3 asymptotically. Freiman in [7] considers division and presents an analytical analysis based on Markov Chains, to determine the distributions of the remainders and of the divisors, for different ranges. In particular, in [7] it is shown that Robertson's algorithm produces 8/3 binary digits per iteration, a result which is confirmed by the simulation described in the same paper. Square root is considered in the detail by Metze in [8]. All the aforementioned methods refer to carry assimilated representations of the partial remainder, and then need long carry propagated additions, while a more attractive solution is certainly represented by partial remainders represented in redundant (e.g. carry save) form.

This paper studies again the problem of considering the selection of the digit 0, but from a different point of view. Our approach first considers the "classical" algorithm for shared radix 2 division and square root, with redundant representations of the partial remainders. Then, by introducing some modifications to the basic unit, an architecture is derived which could take the maximum advantage of the 0-selections, still maintaining the digit selection rules of the "classical" implementation. The simulation will provide tables reporting the average number of 0-selections in the case of different assimilations of the most significant bits of the partial remainders.

The paper is organized as follows: in section 2 we outline the proposed algorithm when applied to division, and we then provide the extension to square root in section 3. In section 4 our architecture for the proposed radix 2 division and square root algorithm is introduced. The evaluation is left to section 5, while the differences with previous works are discussed in section 6.

2 Division

The algorithm for computing the division x/d in base 2 [11], relies on the following formulae

$$w_i = 2(w_{i-1} - y_{i-1}d) = 2^{i+1}(x - Y_{i-1}d) \qquad (1)$$

Table 1: Definitions of symbols used

i	iteration step
w_{i-1}	shifted partial remainder value at step $i-1$, with $w_0 = 2x$
y_i	digit of the partially developed result which has been computed at the i-th iteration of (1); $y_i \in \{-1, 0, +1\}$
d	divisor
x	dividend
Y_i	value of the quotient after the i-th iteration, with $Y_{-1} = 0$

$$Y_i = Y_{i-1} + 2^{-i} y_i \qquad (2)$$

where the symbols used are defined in Table 1. For the purposes of this paper we define an iteration as the set of operations between two consecutive digit selections. Therefore, iteration i is the set of operations occurring between the moment when y_{i-1} is selected and the moment when y_i is selected. For division, we consider x and d as being normalized to $1/2 \le d, x < 1$. It turns out that the result $Y = x/d$ is normalized to $1/2 < Y < 2$. It is assumed for x and d to be available in their full precision and in carry assimilated form, at the beginning of the iterations, thus excluding from this analysis the on-line algorithms [13].

2.1   Previous work

Both the algorithm and the architecture for radix 2 division are well known from the existing literature [4] and Fig. 1 shows a sketch of the scheme of the "classical" architecture for radix 2 division. The shifted partial remainder w_{i-1} is stored in redundant form in the registers C and S. The carry lookahead adder CLA then operates the assimilation of the 4 most significant bits of w_{i-1} which is input to the digit selection table T. The selected digit y_{i-1} is then used to update the previous partial remainder by means of a carry save adder CSA and a shift of one position left (block SH). To handle conveniently the cases of positive and negative update of the partial remainder CSA can work in co-operation with a 1's complementor (COM). The on-the-fly-conversion [3] of the partially developed result Y_i is carried out by the sub-unit OTFC.

When the shifted partial remainder is represented in carry save form [4], the selection rules for the digit y_{i-1} are:

$$\begin{array}{llll}
select & y_{i-1} = +1 & if & 0 \le \widehat{w}_{i-1} \le 3/2 \\
select & y_{i-1} = 0 & if & \widehat{w}_{i-1} = -1/2 \\
select & y_{i-1} = -1 & if & -5/2 \le \widehat{w}_{i-1} \le -1
\end{array} \qquad (3)$$

where $\widehat{w}_{i-1}$ (expressed on 4 bits in two's complement) denotes the representation of w_{i-1} in carry assimilated form, truncated to the first fractional bit. It should be observed that the selection $y_{i-1} = +1$ (or $y_{i-1} = -1$)

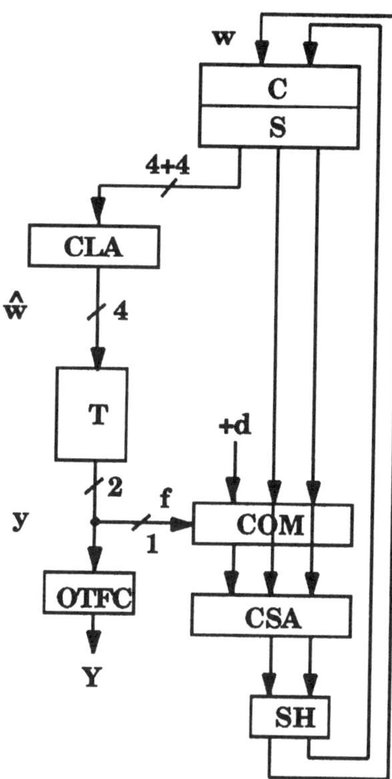

Figure 1: Conventional radix 2 division unit

implies the necessity to update the shifted partial remainder by subtracting (adding) the divisor and by performing a left shift of one position. Conversely, when the selection is $y_{i-1} = 0$, it is only necessary to shift one position left w_{i-1} in order to get the new w_i.

2.2   The algorithm

The proposed algorithm comes from two consequences which arise when the digit $y_{i-1} = 0$ has been selected: *i)* to get the new shifted partial remainder w_i it is not necessary to pass through any carry save adder, but only to perform a left shift, and *ii)* the value of $\widehat{w}_i$ can be easily deduced from $\widehat{w}_{i-1}$; in fact, the information provided in output of the carry lookahead adder at iteration $i-1$ is also valid for iteration i since it is not affected by any manipulations other than a simple shift. These concepts can be better presented by referring to the numerical values. Let us assume the selection $y_{i-1} = 0$. From the selection rules (3), we have $\widehat{w}_{i-1} = -1/2$. Let us now consider the left shift required by (1) to get the new w_i from the previous w_{i-1}. With reference to Fig. 2, we can observe that it is possible to obtain the new value of $\widehat{w}_i$ from the previous $\widehat{w}_{i-1}$ in a very simple way. The three bits of $\widehat{w}_i$ from the weight 2^2 to the weight 2^0 are respectively the same as the three bits from the weight 2^1 to the weight 2^{-1} of $\widehat{w}_{i-1}$. This is due to the fact that when a 0 has been selected it is necessary to perform only sin-

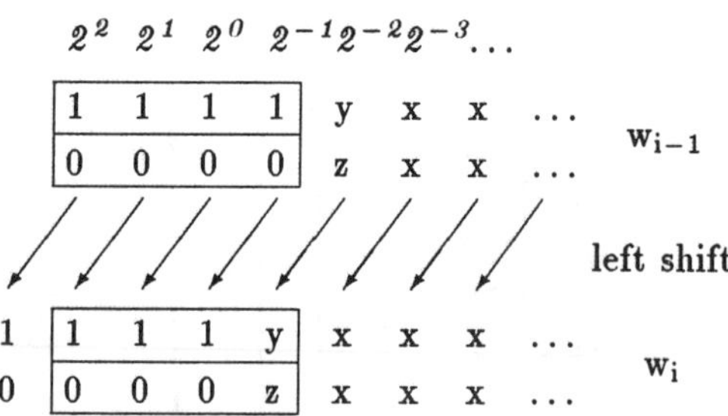

Figure 2: Updating of the partial remainder when the digit 0 has been selected

gle shift left, and in such a case the information given by the CLA is not corrupted. As can be seen in Fig. 2, the fourth bit (i.e. the least significant) which is necessary to determine completely the new $\widehat{w}_i$ used for the next digit selection, must come from the assimilation of the two bits of w_{i-1} with weight 2^{-2}, i.e. the two bits of w_i with weight 2^{-1}. Let us denote with a_i the assimilation of the two bits of weight 2^{-1} of w_i. Three possibilities exist:

1. The assimilation yields 0, which becomes the least significant bit of $\widehat{w}_i$. In fact, $\widehat{w}_{i-1} = (1111)_2$ implies $\widehat{w}_i = (1110)_2$, i.e. $\widehat{w}_i = 2\widehat{w}_{i-1} + 0 \cdot 2^{-1} = -1$. In such a case, according to the rules (3), the next digit selection must be $y_i = -1$.

2. The assimilation yields 1 and, according to (3), it must be $y_i = 0$.

3. The assimilation yields 2 and, according to (3), it must be $y_i = +1$.

Therefore, it should be noted that the digit selection at step i (when a 0 has been selected at step $i-1$), is very easy and it is only determined by the value of the assimilation of the two bits of w_{i-1} with weight 2^{-2}, because the information provided by output on the CLA is not corrupted by updates and can be reused in the selection procedure of the next iteration. Should a digit 0 be selected at step i too, the aforementioned process can still be applied for the selection procedure at step $i+1$, and so on. The proposed algorithm is formally expressed by the following selection rules:

- If $y_{i-1} \neq 0$

$$
\begin{aligned}
select \quad y_i = +1 \quad & if \quad 0 \leq \widehat{w}_i \leq 3/2 \\
select \quad y_i = 0 \quad & if \quad \widehat{w}_i = -1/2 \\
select \quad y_i = -1 \quad & if \quad -5/2 \leq \widehat{w}_i \leq -1
\end{aligned}
\tag{4}
$$

- If $y_{i-1} = 0$ then

$$
select \quad y_i = +1 \quad if \quad a_i = 2 \cdot 2^{-1}
$$

$$
\begin{aligned}
select \quad y_i = 0 \quad & if \quad a_i = 1 \cdot 2^{-1} \\
select \quad y_i = -1 \quad & if \quad a_i = 0 \cdot 2^{-1}
\end{aligned}
\tag{5}
$$

where a_i (expressed on 2 bits in binary form) denotes the result of the assimilation of the two bits of weight 2^{-1} of w_i.

3 Square root

The algorithm for square root, although different in the iteration step from the division recurrence (1), has been shown to present many similarities with division in the selection rules. This has allowed units to be designed where the same hardware is shared by division and square root [6]. The algorithm for computing the square root $\sqrt{x}$ in base 2 [10], relies on the following formulae

$$
z_i = 2[z_{i-1} - (2Y_{i-1} + y_i 2^{-i})]y_i = 2(z_0 - Y_i^2)2^{i+1}
\tag{6}
$$

$$
Y_i = Y_{i-1} + 2^{-i}y_i
\tag{7}
$$

where z_{i-1} is the shifted partial remainder value at step $i-1$, with $z_0 = 2x$, and the same notation introduced for division has been adopted. For square root we assume x as being normalized to $1/4 \leq x < 1$. It turns out that the result $Y = \sqrt{x}$ is normalized to $1/2 \leq Y < 1$. Also for square root, it is assumed for x to be available in its full precision and in carry assimilated form, at the beginning of the iterations. A key work in the study of binary square root is represented by [10], where an architecture which is very similar to the scheme of Fig. 1 is assumed as reference model, and where it is demonstrated that the selection rules are the same as for division (3), provided that the transformation of variables $\widehat{w}_{i-1} = \widehat{z}_{i-1}/2$ is taken into account, i.e. the fractional point is considered to be shifted by one position. It turns out that the same considerations expressed in section 2.2 are still valid, and also for square root our algorithm can be formally expressed in the same terms as the selection rules (4) and (5) with $\widehat{w}_{i-1} = \widehat{z}_{i-1}/2$.

4 Architecture for shared division and square root

4.1 Previous work

The selection rules for radix 2 division (3) need to be implemented by table T in Fig. 1. The quotient digit selection of y_i can be implemented in two ways [4], either by first using a carry lookahead adder (CLA) to assimilate the four most significant binary positions of w_i so as to produce $\widehat{w}_i$ and then using the four resulting bits as inputs to a combinatorial network for the selection, or by using the eight bits of the four most significant positions of the carry save representation of w_i as inputs to the combinatorial network. The main difference between the two choices is in the assimilation of the most significant part of w_i, which is performed by the carry lookahead adder CLA and by the selection table T itself, respectively. Therefore, the architecture sketched in Fig. 1 refers directly to an implementation related to the first choice, while it is necessary to remove the CLA in order to obtain the architecture related to the second choice (actually, with a different and a (slightly) more complex selection table T).

Let us consider the combinational functions to implement the selection procedure [4]. We denote with $s_{i,3}s_{i,2}s_{i,1}s_{i,0}$ and $c_{i,3}c_{i,2}c_{i,1}c_{i,0}$ the sum and carry bits of the four most significant binary positions of w_i respectively, and with $r_{i,3}r_{i,2}r_{i,1}r_{i,0}$ their assimilation (i.e. the four most significant bits of $\hat{w}_i$) and we assume that a quotient digit represented in sign and magnitude (q_s and q_m, respectively) is produced, where $+1$ is represented by $(q_s, q_m) = (0, 1)$, 0 is represented by $(q_s, q_m) = (1, 0)$, and -1 is represented by $(q_s, q_m) = (1, 1)$. When a CLA is used, a possible pair of selection functions would be

$$q_s = r_{i,3}$$
$$q_m = \overline{r_{i,2}r_{i,1}r_{i,0}} \qquad (8)$$

Conversely, when the assimilation is performed by the selection functions themselves, according to [4] we have

$$q_s = p_{i,3} \oplus (g_{i,2} + p_{i,2}g_{i,1} + p_{i,2}p_{i,1}g_{i,0})$$
$$q_m = \overline{p_{i,2}\overline{p_{i,1}}p_{i,0}} \qquad (9)$$

where

$$p_{i,j} = s_{i,j} \oplus c_{i,j} \quad and \quad g_{i,j} = s_{i,j}c_{i,j} \qquad (10)$$

It should be noted that for both types of implementations, either with or without CLA, it is still necessary to have the possibility of updating the partial remainder by adding both positive and negative terms (e.g. $+d$ and $-d$ in the case of division), a task which is assigned to the 1's complementor block COM in Fig. 1. Although other alternatives not needing COM exist when considering division, the block COM becomes necessary for square root, where the updating of the partial remainder must be performed according to the partially developed square root value, which in turn is not fixed to a given constant, but changes as the iterations evolve. The "correction" into 2's complement representation can be performed in a similar way to the one used in [15]. The block COM, can be implemented by a set of EX-OR gates, driven by an enable line, which below is referred to as f.

4.2 Proposed architecture

The reference model for our architecture is shown in Fig. 3, and is designed to implement the algorithm illustrated in section 2.2, by introducing some particular concepts and blocks which let the architecture to be particularly efficient. First of all, the partial remainder is stored in a redundant form which is different from the usual carry and sum form used in Fig. 1: register P contains the EXOR of the carry and sum bits with the same weight, which are output by the modified carry save adder MCSA, while register G stores the AND operation performed on the same bits. In other words, P holds the $p_{i,j}$ and G holds the $g_{i,j}$ bits as defined by relations (10). The contents of P and G are shifted

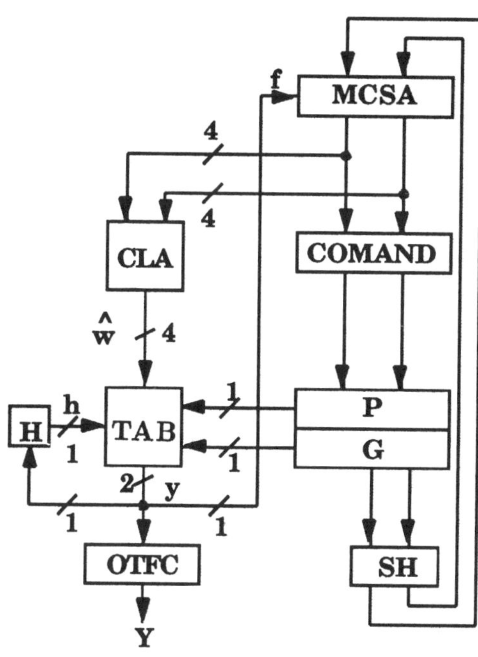

Figure 3: Architecture for fast division and square root

one position left by suitable wiring and then enter the block MCSA, i.e. a modified carry save adder. Also d (i.e. the divisor) and f (the enable signal for the 1's complementation) enter MCSA, and it is worth remembering that $f = 0$ has the implicit meaning to take d, while $f = 1$ has the meaning that the 1's complement of d is taken. The block MCSA plays an important role in our architecture. Let us consider the boolean functions for computing the "next" carry and sum of the partial remainder when a selection different from the digit 0 has been performed at the previous iteration:

$$s_{i+1,j} = c_{i,j} \oplus s_{i,j} \oplus (f \oplus d_j)$$
$$c_{i+1,j} = c_{i,j}(f \oplus d_j) + s_{i,j}(f \oplus d_j) + c_{i,j}s_{i,j}(11)$$

where d_j denotes the $j-th$ bit of the divisor d. Since the EXOR operator is associative, it is possible to rewrite expressions (11) as functions of $p_{i,j}$ and $g_{i,j}$ only.

$$s_{i+1,j} = p_{i,j} \oplus f \oplus d_j$$
$$c_{i+1,j} = g_{i,j} + f\bar{d}_jp_{i,j} + \bar{f}d_jp_{i,j} \qquad (12)$$

The block MCSA implements the functions of sum and carry given by (12), with a complexity comparable with that of the usual scheme of the carry save adder (e.g. CSA in Fig. 1) implementing relations (11) where, however, $(f \oplus d_j)$ represents a single variable (created by the block COM) and not an operation. The MCSA's output is the carry and sum representation of the new partial remainder, and corresponds to the point where the carry lookahead adder CLA takes the most significant bits of the carry save representation of the partial remainder, just as is done in the architecture of Fig. 1. This ensures the algorithm to be exactly the same for both schemes (however, when $y_{i-1} \neq 0$), although in Figures 1 and 3 different representations of the partial remainder are retrieved in the registers. The carry and sum representation coming out from MCSA is transformed in EXOR-AND representation by the block COMAND; the outputs are stored in P and G respectively, and the unit becomes ready for a new iteration. The role of TAB is to implement the selection rules of our algorithm defined in section 2.2, while the block H is a single bit register, destined to hold the information required by the algorithm on the previous digit selection. In particular, the following convention can be adopted: during iteration i, the block H stores 0 if $y_{i-1} = 0$ and 1 if $y_{i-1} \neq 0$. The boolean variable associated with the contents of the register H is denoted with h. The design of TAB plays a key role in the whole architecture, and must be split into two cases: when a CLA is, or is not, used. We again consider the digit to be selected as being represented in sign and magnitude, with the same notation of section 4.1. When a CLA is used to perform the assimilation, from (8) we have

$$
\begin{aligned}
q_s &= r_{i,3}h + \bar{h}g^-_{i,0} \\
q_m &= (\overline{r_{i,2}r_{i,1}r_{i,0}})h + \bar{h}p^-_{i,0}
\end{aligned}
\tag{13}
$$

Conversely, when a CLA is not used, the selection functions are again given by (9) since the assimilation is carried out by the table TAB itself. and it is not necessary for h to enter the table TAB. The block OTFC is used to perform the on-the-fly conversion into a non redundant form, by following the guidelines given by Ercegovac and Lang in [3]. Observe that, at the end of all the iterations, the remainder of the division operation, can be obtained directly from the output lines of the block MCSA in carry save form. Finally, a variable duration clock should be used in the architecture, to regulate the timings of *normal* and *fast* iterations. This circuit can be easily implemented, provided that the longer cycle is an integer multiple of the shorter. In this case, the selection of a non-zero digit disables the clock for P and G registers, for a number of cycles needed to obtain the required period; this can be implemented straightway by using a counter or a shift register used as a delay line for the clock enabling signal. As a small ratio between the two clocks is expected, the number of additional flip-flop required is only a small percentage of the $2n$ memory elements used to implement P and G registers, hence it will influence only to a limited extent the cost of the whole unit. The evaluation of the whole architecture will be discussed in section 5.

4.3 Remarks and extension to square root

At this point, the operating principle of the architecture of Fig. 3 turns out to be very simple. Our architecture behaves as a "classical" unit, by performing *normal* iterations, until a digit selection $y_{i-1} = 0$ occurs. In correspondence of this event, the $i-th$ iteration is carried out faster than *normal* iterations, since

it is not necessary to pass through any carry save updating. The initialization phase is the same as in the "classical" architectures, as when a new computation starts, the register P is loaded with the non redundant representation of the dividend (or radicand, for square root), while the register G is loaded with all 0's. The extension to square root is straightforward, since as we have seen in section 3 the algorithm is the same, and therefore the sharing of our unit between division and square root is implemented exactly in the same way as is done in the other architectures for both division and square root.

5 Evaluation

The complete evaluation of the proposed architecture involves two aspects: hardware requirements and performance evaluation. Concerning the hardware, only very limited additional resources are required by our architecture shown in Fig. 3 with respect to the "classical" unit in Fig. 1. In particular, they are: *i)* a "line" of AND gates to implement a part of the block COMAND, *ii)* a few flip-flops to implement the register H and the variable duration clock, *iii)* the amount of hardware necessary for the selector to permit to permit the loading through or bypass the CSA, and *iv)* (only when the CLA is used), a few logical gates so as to implement the slightly more complex selection functions of TAB with respect to the table T. The performance evaluation must take into account the two separate cases when a CLA is, or is not, used to assimilate the most significant bits of the partial remainder.

5.1 Architecture with carry lookahead adder

By looking at Fig. 3 we observe that the blocks COMAND and CLA operate in parallel. In fact, as we have seen in section 4.2, CLA is a 4 bit carry lookahead adder and COMAND is a line of EXORs in parallel to a line of AND gates, and hence it is reasonable to assume that, independently of the implementation technology, the delay of the CLA is larger than or, equal to, the delay of the block COMAND. Therefore, the execution time of a *normal* iteration of the proposed architecture in Fig. 3 is $T_{new,norm,CLA} = t_{MCSA} + t_{CLA} + t_{TAB} + t_{reg,PG}$ where, in general, t_X denotes the delay of the block X in the architecture in Fig. 3, and $t_{reg,PG}$ the delay for loading the registers P and G (including the delay of a selector to permit the loading through or bypass the CSA). On the other hand, the execution time of an iteration of the "classical" architecture is $T_{old,CLA} = t_{CSA,COM} + t_{CLA} + t_T + t_{reg,CS}$, where $t_{CSA,COM}$ is the total delay required for a carry save sum/subtraction and $t_{reg,CS}$ is the delay for loading the registers C and S. For *fast* iterations the execution time of our architecture becomes $T_{new,fast,CLA} = t_{TAB} + t_{reg,PG}$.

5.2 Architecture without carry lookahead adder

By removing from Fig. 3 the block CLA we observe that the critical path passes through MCSA, COMAND, the register load and TAB. The execution time of a *normal* iteration of the proposed architecture becomes $T_{new,norm,noCLA} = t_{MCSA} + t_{COMAND} + t_{TAB} + t_{reg,PG}$. Let us consider now the architecture in Fig. 1 when the CLA has been removed. From the expressions of the selection functions (9) we see that they require the knowledge of the terms $p_{i,j}$ and $g_{i,j}$, as they are defined in (10). This is equivalent to considering the functions for digit selection in the architecture in Fig. 1 as being implemented by the cooperation of a block COMAND (to compute the terms of (10)), and a table T to implement the functions (9). Therefore, the execution time of an iteration of the "classical" architecture in Fig. 1 is $T_{old,noCLA} = t_{CSA,COM} + t_{COMAND} + t_T + t_{reg,CS}$. It is worth remembering that as we have seen in section 4.2, both tables TAB and T implement the same relations (9). For *fast* iterations the execution time of our architecture becomes $T_{new,fast,noCLA} = t_{TAB} + t_{reg,PG}$.

5.3 Average execution time

From the previous computations, it emerges that technological factors determine whether *normal* iterations of the proposed architectures are longer or shorter than the iteration delays of the "classical unit", although in general it is expected for this difference to be small. Anyway, even if the *normal* iterations could have a longer duration, the average performances may be improved if a sufficient number of *fast* iterations is executed. The aim of this section is to determine the average execution time, that is to say how much the *fast* iterations contribute to reducing the average iteration time. Since a pure theoretical analysis is very complicated, we have preferred to direct our efforts towards the simulation in order to detect the number of occurrences of the *fast* iterations. We have first considered the square root, by performing an exaustive type of simulation, that is to say, we have considered all the possible radicands expressed on a given number of bits. For our purposes, the generic radicand x has been taken into account as expressed in carry assimilated form. Our simulation has explicitly considered the behavior of the proposed architectures (irrespective of whether the CLA is used). This means that, in correspondence to the selection of a 0, no sum operation has been performed. The results of the simulation for different lengths of the radicands and for different assimilations of the most significant bits of the remainder, are reported in Table 2. From Table 2 we see that, as the number of bits of the radicand increases, the percentage of 0 selections is approximately 28% when an assimilation of 4 bits of the result is considered by the digit selection process. As it will be explained in section 5.4, higher percentages are obtained with the increase of the number of bits of the remainder which are assimilated. Basically, we have used the same assumptions for simulating division as those adopted for square root. The results of the simulation for different lengths of the dividends and for different assimilations of the most significant bits of the remainder, are reported in Table 3. It should be observed that the rightmost column in Table 3 (i.e. corresponding to the whole partial remainder represented in carry assimilated form), reflects approximately the bound given by Freiman in [7] on the number of digits produced

per iteration by the Robertson's method. In fact, from [7], it is known that one iteration produces (on the average) 8/3 binary digits. This corresponds to having $(1 - 8/3) = 5/3$ selections of the digit zero every 8/3 binary digits produced. Therefore, with Robertson's method, according to Freiman's results, the percentage of 0-selections is 5/8=62.5%, which is approximately the same as the values shown in the CLA_N column in Table 3. From Table 3 we observe that, as the number of bits of the radicand increases, the percentages of selecting a 0 approach approximately 35% when the digit selection process considers an assimilation of 4 bits of the result. Higher percentages are obtained with the increase in the number of bits of the remainder which are assimilated.

5.4 General remarks

From the previous analysis it is clear the importance of the results of Tables 2 and 3. In fact, as we have seen, the percentages of 0-selections in Tables 2 and 3 offer a criterion for evaluating how different implementation technologies can favour the average execution time of the proposed algorithm and architecture, with respect to the delay of the "classical" architecture.

Several other conclusions can be deduced from an analysis of Tables 2 and 3, which involve the choice among a class of different architectural solutions for implementing our unit. The most important is that assimilating more bits of the partial remainder, than the strictly necessary (i.e. 4), has the effect of enlarging the region of the truncated partial remainder for which the digit 0 can be selected. This corresponds to passing from the region $[-1/2, -1/2]$ of a 4 bit assimilation, to $[-1/2, 0]$, $[-1/2, 1/4]$ and $[-1/2, 3/8]$ of 5, 6 and 7 bit assimilations, respectively. If we assume that the maximum size of these regions is being considered, then the final effect is that the percentage of selecting a 0 increases as the number of bits of the partial remainder which are assimilated increases, because of both the lower truncation error and of the larger region for a 0-selection. To take full advantage of the larger region for a 0 selection, the selection table TAB must consider all the bits of the assimilated part of the partial remainder, i.e. 5, 6 and 7 bits respectively in the case of 5, 6 and 7 bit assimilations. Although this increases the complexity of the implementation of the table TAB, it is necessary to consider all the assimilated bits of the partial remainder. In fact, as we have observed with simulation, the increase in the percentages of selecting 0's remains relatively small if the assimilated part of the partial remainder is still examined on "only" 4 bits of the "classical" architecture. For example, with a 7 bit assimilation and an inspection of 4 bits, the percentage of selecting a 0 is equal to 25.954%, that is about one half the corresponding value in Table 2, i.e. 51.724%, and does not substantially differ from the value 25.854% reported in the column of the 4 bit assimilation. This implies that the most significant role in the increase of the percentages of selecting 0's, is played by the enlargement of the regions for the 0 selection, and not by the assimilation of our extended subset of the partial remainder. For this reason, in our implementations we have considered that all the assimilated bits enter TAB.

Table 2: Square root: percentages of selection of 0

length of the radicand	length of the result	iterations performed	percentage of 0's				
			CLA_4	CLA_5	CLA_6	CLA_7	CLA_N
8 bits	5 bits	768	17.083	34.115	41.797	44.010	45.443
10 bits	6 bits	3840	20.521	36.667	44.271	46.458	48.073
12 bits	7 bits	18432	22.439	38.542	46.050	48.405	49.799
14 bits	8 bits	86016	23.865	39.983	47.363	49.805	51.267
16 bits	9 bits	393216	24.976	41.073	48.372	50.877	52.328
18 bits	10 bits	1769472	25.854	41.962	49.184	51.724	53.211
20 bits	11 bits	7864320	26.551	42.681	49.851	52.392	53.912
22 bits	12 bits	34603008	27.161	43.276	50.396	52.950	54.496
24 bits	13 bits	150994944	27.660	43.775	50.857	53.412	54.982
26 bits	14 bits	654311424	28.094	44.203	51.247	53.806	55.396

Table 3: Division: percentages of selection of 0

length of the dividend	length of the result	iterations performed	percentage of 0's				
			CLA_4	CLA_5	CLA_6	CLA_7	CLA_N
8 bits	5 bits	4096	34.766	52.539	60.059	62.500	63.965
10 bits	6 bits	40960	35.518	52.393	59.658	62.256	63.633
12 bits	7 bits	393216	35.508	52.182	59.171	61.812	63.198
14 bits	8 bits	3760016	35.434	52.053	58.869	61.466	62.914
16 bits	9 bits	33554432	35.364	51.958	58.664	61.198	62.681
18 bits	10 bits	301989888	35.296	51.891	58.526	61.011	62.501

The upper bound on the percentages of selecting a 0 is reached when the partial remainder is represented in assimilated form (see the columns CLA_N in Tables 2 and 3).

By examining the numerical values of Table 2 we observe that with an assimilation of 4 bits, percentages of 0 selections of about 28% are obtained. With the assimilation of 5 bits, the percentages increase to about 44%, while with 6 bits the increase is more limited (about 51%), then becoming almost insignificant for 7 bit assimilations (i.e. about 54%). Although it is beyond the scope of this paper to evaluate the proposed architecture in such a case since it would be highly technology dependent, it is clear that there exists a knee in the performance curve vs. the number of assimilated bits, such that the advantages of a larger assimilation are almost irrelevant compared with the price to be paid for, in terms of increased hardware complexity. In fact, a larger assimilation implies the use of a larger (and slower) CLA and of a larger (and slower) selection table TAB still considering all the bits of the assimilated part of the partial remainder [9]. Task of the designer is, therefore, to evaluate among all the most convenient implementation, given the technological constraints.

6 Differences with previous works

Several works have considered the possibility to speed up the duration of the selection process by decreasing the number of additions. Key works been provided by Metze in [8], by Freiman [7] and by Wilson in [14]. However, in all these works, partial remainders are represented in non-redundant form. In our algorithm the redundancy of the SRT-like algorithms is used to achieve two goals: firstly to allow selection rules based on limited precision comparisons (because of the carry save representation of the partial remainder), and secondly to obtain "good" probabilities of selecting a 0, and then to avoid the need of additions/subtractions for updating the partial remainder. These two goals are contradictory, and in the past the former has always been given preference with respect to the latter.

In our work we have not considered any priority between these two goals. We have not studied the derivation of selection rules for minimizing the number of non zero digits of the result, in the case of carry save representations, since it can be demonstrated that, in order to obtain a *minimal representation* in the sense used by Metze [8], for a radix 2 square root with carry save adder, the complexity of the digit selection tables is comparable to that required by radix 4 square root. Conversely, the attention has been focused on the study of the statistical properties of the "classical" radix 2 division and square root methods known in the literature, and in particular on the percentage of the 0-selections. We have then slightly modified the digit selection tables of the "classical" unit known in the literature, in order to take the full advantage of the existence of the 0-selections. Moreover, by properly tuning the portion used for the redundancy which is employed for limited length comparisons and digit selections, and the complementary portion used for the redundancy which has the role of increasing the percentage of the 0-selections, it is possible to achieve several different architectural solutions. Eventually, unlike other algorithms using the redundant representation of the partial remainder, we use a representation different from the "classical" non-assimilated carry sum form.

From these considerations, it is clear that both our algorithm and architecture differ radically from the previous proposed in the literature.

7 Conclusions

We have presented a new algorithm for shared radix 2 division and square root. The main characteristic of the proposed scheme is the subdivision of the iteration steps for the calculation of the result, into *fast* and *normal* iterations. For *fast* iteration we mean an iteration when the addition can be avoided, while for *normal* iterations we refer to all the other iterations.

Some implementation aspects have been discussed. In particular, it has been shown that a redundant representation of the remainder different from the non assimilated carry-sum form could improve the performance characteristics so that a *normal* cycle with duration close to that of a "classical" architecture can be achieved.

The proposed design methodology can be extended to architectures where larger assimilations of the partial remainder such as the "minimal" one (i.e. the 4 most significant bits) are taken into account, with the aim of increasing the average number of *fast* iterations occurring during a computation. It can be shown that, in such a case, an increased percentage of 0's selected during a computation leads to more complex selection functions. Among the proposed solutions, i.e. pure radix 2 and radix 2 with larger assimilations, it is the task of the designer to evaluate the most convenient implementation, depending on his specific technological constraints.

Acknowledgment

We thank Prof. Miloš Ercegovac for his helpful suggestions. Thanks are also due to the referees for their valuable comments.

References

[1] D. E. Atkins, "Higher-Radix Division Using Estimates of the Divisor and Partial Remainders," IEEE Trans. Comput., Vol.C-17, pp.925-934, October 1968.

[2] L. Ciminiera and P. Montuschi, "Higher Radix Square Rooting," IEEE Trans. Comput., Vol.39, No.10, October 1990, pp.1220-1231.

[3] M. D. Ercegovac and T. Lang, "On-the-Fly Conversion of Redundant into Conventional Representations," IEEE Trans. Comput., Vol.C-36, pp.895-897, July 1987.

[4] M. D. Ercegovac and T. Lang, "Division," Internal Report CS252, Computer Science Department, UCLA, 1988.

[5] M. D. Ercegovac and T. Lang, "Radix-4 Square Root Without Initial PLA,"IEEE Trans. Comput., Vol.C-39, pp.1016-1024, August 1990.

[6] J. Fandrianto, "Algorithm for High Speed Shared Radix 8 Division and Radix 8 Square-Root," Proc. 9th IEEE Symposium on Computer Arithmetic, Santa Monica, CA, pp.68-75, September 1989.

[7] C.V. Freiman, "Statistical Analysis of Certain Binary Division Algorithms," Proc. IRE, Vol.49, pp.91-103, January 1961.

[8] G. Metze, "Minimal Square Rooting," IEEE Trans. Electron. Comput., Vol.EC-14, pp.181-185, April 1965.

[9] P Montuschi and L. Ciminiera, "Fast Radix 2 Division and Square Root," Internal Report, Politecnico di Torino, Dipartimento di Automatica e Informatica, I.R. DAI/ARC 15-90.

[10] S. Majerski, "Square-Rooting Algorithms for High-Speed Digital Circuits," IEEE Trans. Comput., Vol.C-34, pp.724-733, August 1985.

[11] J. E. Robertson, "A New Class of Digital Division Methods," IRE Trans. Electron. Comput., Vol.EC-7, pp.218-222, September 1958.

[12] J.l. Smith and A. Weinberger, "Shortcut Multiplication for Binary Digital Computers," NBS Circular 591, Sec.1, pp.13-22.

[13] P. K.-G. Tu, "On-Line Arithmetic Algorithms for Efficient Implementation," PhD Dissertation, Computer Science Department, UCLA, CSD-900029, September 1990.

[14] J.B. Wilson and R.S. Ledley, "An Algorithm for Rapid Binary Division," IRE Transactions on Electronic Computers, Vol.EC-10, pp. 662-670, December 1961.

[15] J. H. P. Zurawski and J. B. Gosling, "Design of High-Speed Digital Divider Units," IEEE Trans. Comput., Vol.C-30, pp.691-699, September 1981.

A 160nS 54bit CMOS Division Implementation Using Self-Timing and Symmetrically Overlapped SRT Stages

Ted E. Williams
HaL Computer Systems
1315 Dell Avenue
Campbell, CA, 95008

Mark A. Horowitz
Center for Integrated Systems
Stanford University
Stanford, CA, 94305

Abstract

A full-custom VLSI chip demonstrates an arithmetic implementation for computing the mantissa of a 54bit (floating-point double-precision) division operation in 45nS to 160nS, depending on the data. The design uses self-timing to avoid the need to partition logic into clock cycles and the need for high-speed clocks. Self-timing allows the circuits to iterate with no overhead over the pure combinational logic delays. It also allows a greater efficiency symmetric overlapped execution of the SRT stages because of "dynamic" path ordering. The design has several other performance enhancements, and this paper tabulates their effect on the performance.

1   Introduction

Previous division implementations have generally tried to attain high-performance by increasing the complexity of the logic function performed in each "cycle." Higher radix arithmetic can utilize this complexity to reduce the number of clock cycles required in clocked designs [1], [2]. But as technology improves, it is increasingly more difficult to fully utilize all clock cycles. An alternate approach is to avoid clock cycle limitations altogether by using self-timed logic. This paper describes several implementation issues in the design of a self-timed CMOS divider, including the methodology used to eliminate latch delays between stages in an iterating ring, a modification to the quotient selection logic that narrows the remainder datapath, immediate done indication for repeating quotients, and dynamic overlapping of action between a stage and its neighbors *without* a pre-defined grouping into pairs. As in RISC processor design [3], each improvement can be judged by a version of Amdahl's law:

$$\frac{100}{T} = \frac{1}{f}\left(1 + \frac{100}{S}\right) - 1 \tag{1}$$

where S is the percentage speedup of one part, f is the fraction of the total delay attributable to that part, and T is the percentage improvement in the total performance.

Estimates for self-timed SRT division designs suggest a radix 2 approach obtains the best performance because its simplicity allows fast stages in a reasonable area. Measurements from fabricated and tested CMOS VLSI parts verify the ideas in this paper and demonstrate high-performance without special technology.

2   Self-Timed Methodology

Self-timed components avoid the need of distributing global clocks and the need of allowing for clock-skew in synchronous systems. Variances and data dependencies in delays can be used advantageously because each component can begin when its required operands actually arrive rather than always waiting for worst-case timing. The performance is also the best possible for the actual environmental conditions, without needing to de-rate specifications to allow additional margins for the conceived ranges of power supply voltage, die temperature, and fabrication spread.

Circuits can achieve self-timed operation by carefully matching delays between components or by encoding completion information within data signals. An example of the former method is a self-timed multiplier [6] chip, which uses a matched on-chip clock generator to provide a clock for the internal blocks. In contrast, our chip demonstrates the latter method by using local completion detectors and handshaking between fully asynchronous blocks and operates correctly for any values of gate delays. Completion-information is embedded in the data throughout the design by using a pair of wires for each bit. Called a "dual-monotonic pair," the wires transmit both a value and a timing signal by using the protocol in Table 1.

Precharged function blocks use merged n-channel pull-down networks to choose which of the wires in each pair to set high when input data arrives. Completion detector NOR gates examine the dual-monotonic pairs to generate local done signals used to control the precharge signals. If

Wire A^T	Wire A^F	Signal A
0	0	Reset = Not Ready
0	1	Evaluated FALSE
1	0	Evaluated TRUE
1	1	Not used = Never occurs

Table 1: Encodings on a Dual-monotonic Wire Pair

the circuit is embedded in a synchronous system, the chip Done signal can be used to stretch clock cycles as in [12], or to indicate on which clock cycle the system may take the outputs from the self-timed chip.

Because the iterative steps of SRT division need a repetitive structure, the core of our chip is a ring of five stages that iterates completely under self-timed control. The ring achieves minimal latency operation because it has directly concatenated precharged logic blocks in a looped domino chain, as shown in Figure 1, without any explicit latches. Each stage "falls like a domino" because it evaluates as soon as its predecessor provides valid data, without waiting for control or clocking. The looped chain is thus like a ring-oscillator that computes. This is possible because the precharge (reset) signals for each block are controlled separately so each block can be used as an implicit latch without adding any additional transistors. The self-timed control precharges each block after data has passed it, and removes the precharge signal (thereby enabling its evaluation), before data has looped around to its inputs again. The critical path goes solely through the combinational data elements. Thus, the data flows continually at the same rate it would flow through an "unwrapped" combinational array implementing the same functions. While previous asynchronous approaches [5], [9] have suffered delays due to handshaking control, this method of self-timing adds zero control overhead [11] to the latency of the raw function computation.

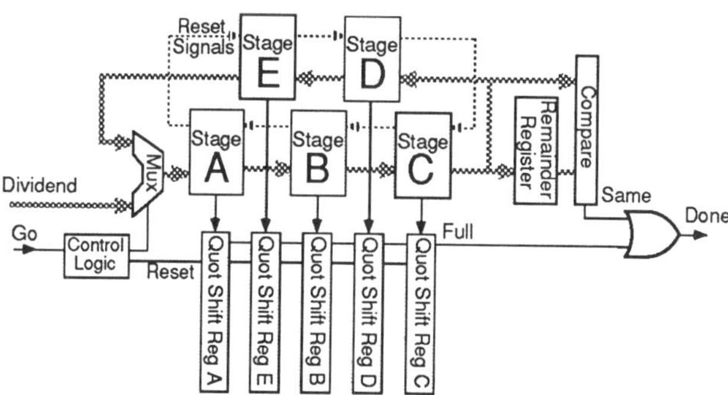

Figure 1: Block diagram for the division circuit using a ring of domino stages which iterates using self-timing. Dotted lines show control signals; shaded lines are dual-monotonic datapaths.

3   Quotient Digit Selection

The blocks of each stage in the iterative ring implement one step of a modified radix 2 SRT division algorithm:

$$P_{i+1} = 2P_i - Dq_i \tag{2}$$

where D is the divisor, P_i is the partial remainder at stage i, and q_i is the quotient digit selected by that stage. The probabilistic distribution of quotient digits is not uniform due to the asymmetry of the two's complement number system. We wrote a computer program to collect statistics by enumerating all possible data patterns in the most significant bit positions.   Because the SRT algorithm deals only with the most significant bits, the collected statistics became asymptotic for input data patterns exceeding about 9 bits in significance. The asymptotic values are therefore the correct statistics for uniformly distributed inputs of all larger mantissa widths. Table 2 shows the asymptotic distribution of the values of the partial remainder approximation and the quotient digits selected for radix 2 division.

The asymmetric distribution of quotient digits can be used to speed the more frequently used circuit paths since the self-timed implementation can take advantage of the improvement. For example, our circuit used different sized transistors in the replicated short adders (replicated because of the overlapped execution feature to be discussed in Section 5) preceding the next quotient digit selection logic. We designed the $q_i = -1$ adder arm with larger transistors because the critical path goes through it most frequently. We did not use larger transistors in the less frequently chosen $q_i = +1$ arm because the additional loading on the wires also going to the $q_i = -1$ arm would have actually decreased the total performance. The effect of this transistor optimization decreases the expected value of the delay of driving these adder inputs by about 30%, resulting in a 4% improvement in the total divider performance.

When the most negative or two most positive possible values for the partial remainder approximation occur, the selected quotient digit does not change the sign of the next remainder. So, in these cases the next quotient digit can be chosen in advance to be the same as the current quotient

Remainder	- 4	- 3	- 2	-1	0	+1	+2	+3
Probability	3.2%	11%	28%	35%	18%	4.6%	.40%	.05%
Quot Digit	-1				0	+1		
Probability	42%				35%	23%		

Table 2: Statistics for Radix 2 SRT division.

digit. Table 2 shows the most negative approximated remainder value occurs with a 3% probability and the two most positive values occur with less than a 1% probability. Based on these statistics, we chose to implement a force-ahead for only the negative case with the quotient selection logic equations:

$$q_i = +1 \quad \text{if} \quad \hat{P}_i \geq 0 \quad \text{and} \quad F_{i-1} = 0 \quad (3)$$

$$q_i = 0 \quad \text{if} \quad \hat{P}_i = -1 \quad \text{and} \quad F_{i-1} = 0 \quad (4)$$

$$q_i = -1 \quad \text{if} \quad \hat{P}_i \leq -2 \quad \text{or} \quad F_{i-1} = 1 \quad (5)$$

$$F_i = 1 \text{ if } \hat{P}_i = -4 \text{ or } \left(F_{i-1}=1 \text{ and } \left[\hat{P}_{i-1}\right]_{\text{msb}} = 0\right) \quad (6)$$

where $\hat{P}_i$ is the approximated partial remainder at stage i, and F_i is the flag set to force the next quotient digit. Allowing the quotient digit selection logic to leap ahead one digit in 3% of all selections improves the total performance by about 1% in itself, but it also enables the additional optimization described in the next paragraph by reducing the required size of the remainder datapath and the adder widths.

The possible shifted remainder values in any maximally redundant radix r SRT division step (just before reduction by a quotient digit) range from $-rD$ to $+rD$. When the remainder is represented in carry-save form as the sum of two numbers each in two's complement form, it is possible for each of these numbers to be in the range $-2rD$ to $+2rD$ and previous implementations have required the datapath be wide enough to hold the bits necessary to represent those numbers. However, the approximation of the remainder, formed by a short CPA combining the sum and carry bits, must still have its actual value in the range $-rD-2U$ to $+rD$ where U is the maximum value of the unpropagated bits in either the sum or the carry terms. Since the unpropagated bits are always positive, U is always positive. If the minimum representable number were only $-rD$, then values in the range $-rD-2U$ to $-rD$ would be aliased as positive numbers in the range $+rD-2U$ to $+rD$ because of the sign bit "falling off." The incorrect interpretation of the approximation would, of course, cause the wrong quotient digit to be chosen. However, for radix 2, this can only happen in the cases in which the remainder is at the extreme of its possible valid range where it was possible to predict two quotient digits in advance. The quotient selection logic in equations (3)-(5) never misinterprets an aliased remainder as a positive number even when the sign bit falls off. Equation (6) sets the force-next-digit flag when the most negative quotient digit occurs or if the flag was set before and the remainder is already aliased. The partial remainder need only be able to represent numbers in the range $-2D$ to $+2D$, thus narrowing the datapath and shortening by one bit the adders used to form the irredundant approximation.

The trimming of the datapath is only applicable in maximally redundant SRT algorithms because the minimum approximated remainder $-rD\frac{\rho}{r-1}-2U$ of lower redundancy choices (such as $\rho=2$, $q_i \in \{-2,-1,0,1,2\}$ in radix 4) can be represented in the same number of bits as $-rD\frac{\rho}{r-1}$. For radix 2, the only choice of redundancy, $\rho=1$, is maximally redundant, and therefore the trimming can be applied to make the adders need only 3 bits instead of the 4 bits used in previous implementations [4]. Since the logic to force both the current and next quotient digits has about 40% less delay than the delay of an extra bit in the adders and this accounts for a sixth of the total delay through a stage, the net total performance is improved by 5%.

4   Quotient Accumulation and Early Done Detection

As the quotient selection logic in the five stages of the self-timed ring output quotient bits, they are collected by five separate asynchronous shift registers composed of the cells shown in Figure 2. Between each valid quotient digit sent into a shift register, a reset spacer is sent to separate the digits. The ring loops a maximum of 11 times to fill the five shift registers with a total of 55 bits for a double-precision result. On each iteration, the remainder comparison on the right side of Figure 1 determines if the partial remainder has remained unchanged during the last iteration:

$$P_{i+5} = P_i \quad (7)$$

If the remainder repeats, then subsequent remainders and quotient digits will also repeat:

$$\begin{aligned} P_{i+10} &= P_{i+5} = P_i, \quad (8) \\ q_{i+5} &= q_i, \\ q_{i+6} &= q_{i+1}, \\ q_{i+7} &= q_{i+2}, \\ q_{i+8} &= q_{i+3}, \\ q_{i+9} &= q_{i+4} \end{aligned}$$

Since there is no need to compute the repeating digits again, the iterations terminate and the division done signal is generated early. Even when the iterations terminate early, the full quotient is immediately available from the shift registers because the asynchronous design using C-elements correctly ripples the quotient digits to their final positions as they arrive rather than waiting for a fixed number of clocks as would a synchronous shift register. The repeated quotient digits are also immediately available because they fill all of the positions behind each new quotient digit as it ripples through a shift register. Only

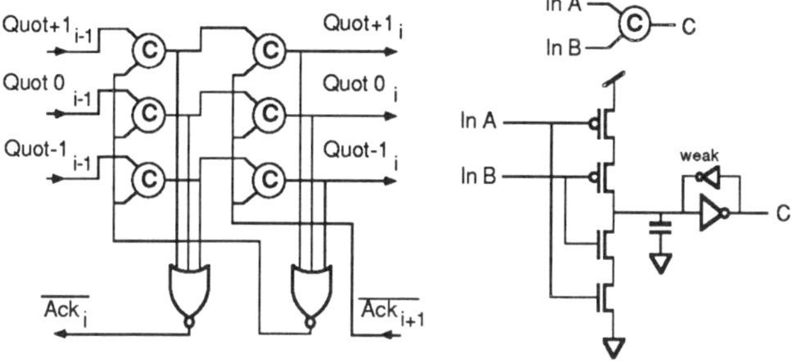

Figure 2: A cell of the asynchronous shift registers for capturing quotient digits on a triple-monotonic wire set. Each quotient digit arrives when one of the three input wires is set high, and is followed by a "spacer," where all three are again low. A static C-element is defined at right.

after the remainder comparison determines more iterations will be needed are reset spacers sent into the shift registers to wipe out the repeated digits and prepare the shift registers to accept the next digits. This interlocking of the reset spacers does not add any delay to the overall computation because it occurs in parallel with the evaluation of other quotient digits.

The effect on performance of detecting repeating quotients and finishing early is dependent upon the distribution of input operands. Data from some algorithmic applications may be likely to have more round numbers, and data from an external input, like a sensor, may be uniformly distributed only within a limited precision. For example, the early done detection will speedup 12% of the cases in a uniform distribution of 8bit input operands, for a total performance improvement of 9%. A typical division instruction operand mix might get half that speedup.

The final quotient can be rounded correctly even when the iterations terminate early. In both the early done and the normal case requiring all of the iterations, the remainder at the stage where the iterations stopped can be sent through a carry-look-ahead adder (CLA) to determine its sign. If the remainder is negative, the quotient must be decremented at the least-significant bit position to which the remainder corresponds. This operation and the conversion of the redundant quotient into a standard binary form can be performed by a carry-select-adder with multiple carry chains operating in parallel. The different rounding possibilities and the remainder sign select the correct CLA output. Thus, after the iterations terminate, only a single CLA delay is required to resolve both the final remainder and rounded quotient.

5 Dynamic Overlapping of Stages

Since the carry-save adders (CSA) used for the computation of partial remainders provide only a redundant result, short 3bit carry-propagate adders (CPA) are needed to form an irredundant remainder approximation for input to the quotient digit selection logic. While the traditional SRT algorithm is purely sequential, the implementation here, shown in Figure 3, overlaps the execution of adjacent stages by replicating the CPAs for each possible quotient digit so they can begin operation before the actual quotient digit arrives and chooses the correct branch. Two of the three CPAs are also preceded by CSAs to combine the remainder respectively with the divisor and the negation of the divisor. Actually, the logic for the sum terms in these two CSAs is shared because the dual-monotonic data convention already provides both the true and complement of each bit. The carry terms cannot be shared. While the arm for a zero quotient digit still requires a CPA to combine the sum and carry terms of the redundant representation, the zero arm requires no preceding CSA.

The self-timed control for each stage is also shown in Figure 3 and uses C-elements to combine the completion-detector signals to produce the signals that reset each block as soon as its outputs have been consumed. Since these C-elements are never in the overall critical path they introduce zero overhead.

Figure 4 shows the concatenation of the data elements for any two adjoining stages. The overlapping of execution allows the average delay through a stage to be the average rather than the sum of the propagation delays through the remainder and quotient digit selection paths. The

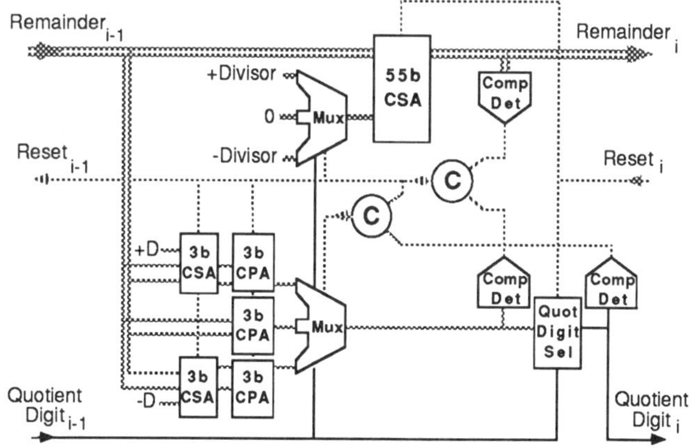

Figure 3: Internal structure of each stage in the ring implementing an SRT division step with overlapped execution and self-timed reset (precharge) control.

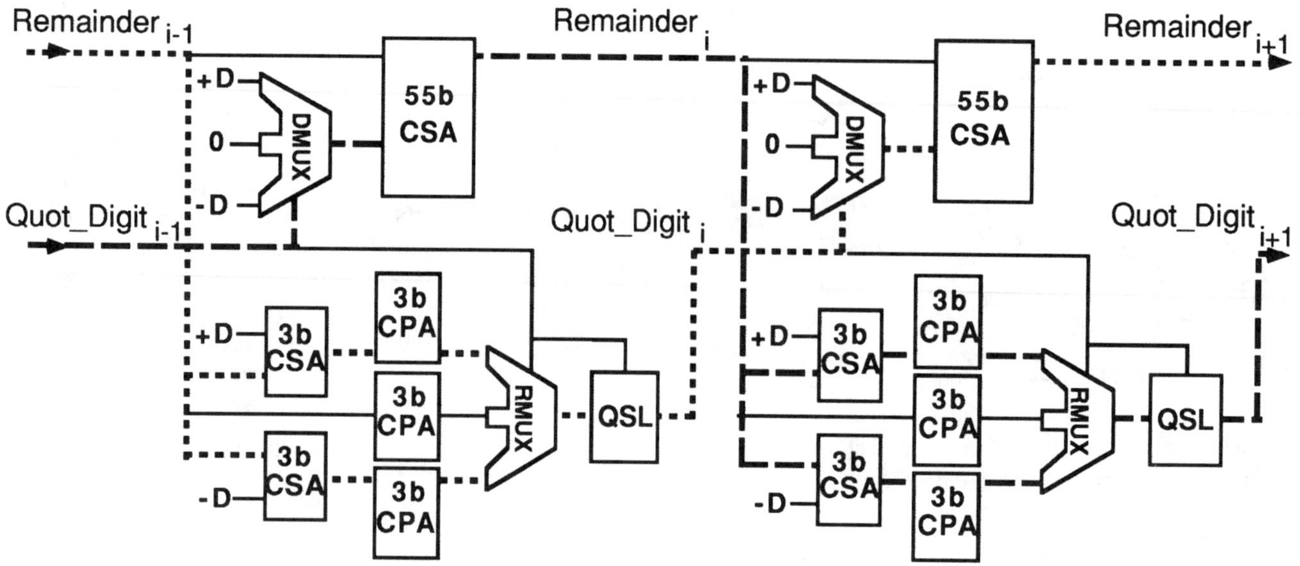

Figure 4: Dataflow through a pair of stages in the present overlapped execution scheme.

overlapping of these paths can be abstracted to the arrangement of overlapping blocks shown in Figure 5.

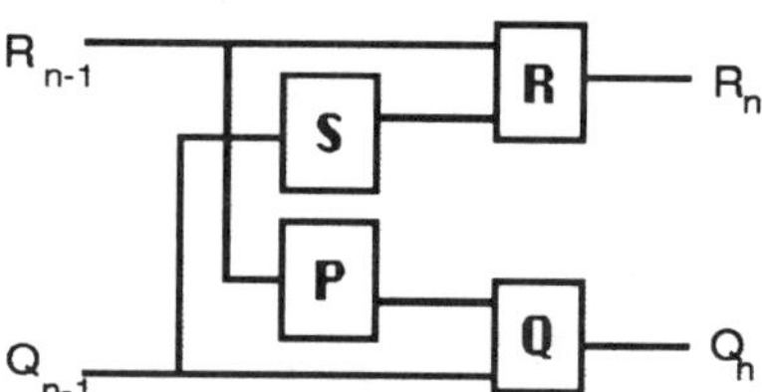

Figure 5: Model for overlapped execution dataflow in each stage

When these blocks are self-timed and therefore operate as soon as their required operands arrive, the average delay per stage[1] in a chain of identical stages of the overlapped arrangement is

$$\frac{1}{2}\left\{ P{+}Q{+}R{+}S\ +\max\left[0,\ abs(R{-}Q){-}(P{+}S)\right]\right\} \qquad (9)$$

The last term is usually negative and drops out, giving a performance increase due to the factor of $\frac{1}{2}$ in front. In the overlapping of the stages for SRT division, the delay of block P in the quotient selection path is the largest of the delays because it contains the CPAs. The overlapping

reduces by one-half the effect of the delay in block P on the total delay. When the added mux delays are taken into account, the overlapping of the stages in a radix 2 design increases performance by 35% over a standard sequential arrangement of the same blocks.

The structure of our overlapping scheme results in a data wavefront that leapfrogs down the succession of stages. If the critical path goes through the quotient selection path in one stage, it will likely go through the partial remainder path in the next stage, and vice-versa. However, data-dependent variances in delays make it possible for the overall minimal critical path to go through the same path in two adjacent stages. Delay variances arise because of the varying number of bits propagated in the carry chains, the occurrence of some zero quotient digits, and the cases in which a negative quotient digit can be selected in advance when a single stage can determine two quotient digits. The self-timing of the datapath ensures data always flows through the minimal critical path. Previous overlapped execution schemes such as the one from [7] have pre-grouped stages into pairs. Whereas our approach makes all the stages symmetric, the scheme in Figure 6 does not replicate the first CPA. This lack of symmetry

---

[1] An exact analysis of the delays of Figure 5 shows that the time the nth R block finishes in a chain of identical stages is

$$R_n = \tfrac{n}{2}(P{+}Q{+}R{+}S) + \max\left[S{+}\tfrac{n}{2}(R{-}S{-}P{-}Q),\ \left(\tfrac{n}{2}-1\right)(Q{-}P{-}R{-}S),\ \tfrac{e}{2}(Q{-}P{-}R{-}S),\ R{-}P{-}Q{+}\tfrac{e}{2}(Q{+}P{-}R{+}S)\right]$$

when the chain's inputs start at $R_0 = Q_0 = 0$ and where $e \equiv \left\{ \begin{array}{l} 1 \text{ if } n \text{ odd} \\ 0 \text{ if } n \text{ even} \end{array} \right\}$.

Symmetrically, the nth Q block finishes at time

$$Q_n = \tfrac{n}{2}(P{+}Q{+}R{+}S) + \max\left[P{+}\tfrac{n}{2}(Q{-}P{-}R{-}S),\ \left(\tfrac{n}{2}-1\right)(R{-}S{-}P{-}Q),\ \tfrac{e}{2}(R{-}S{-}P{-}Q),\ Q{-}R{-}S{+}\tfrac{e}{2}(R{+}S{+}P{-}Q)\right].$$

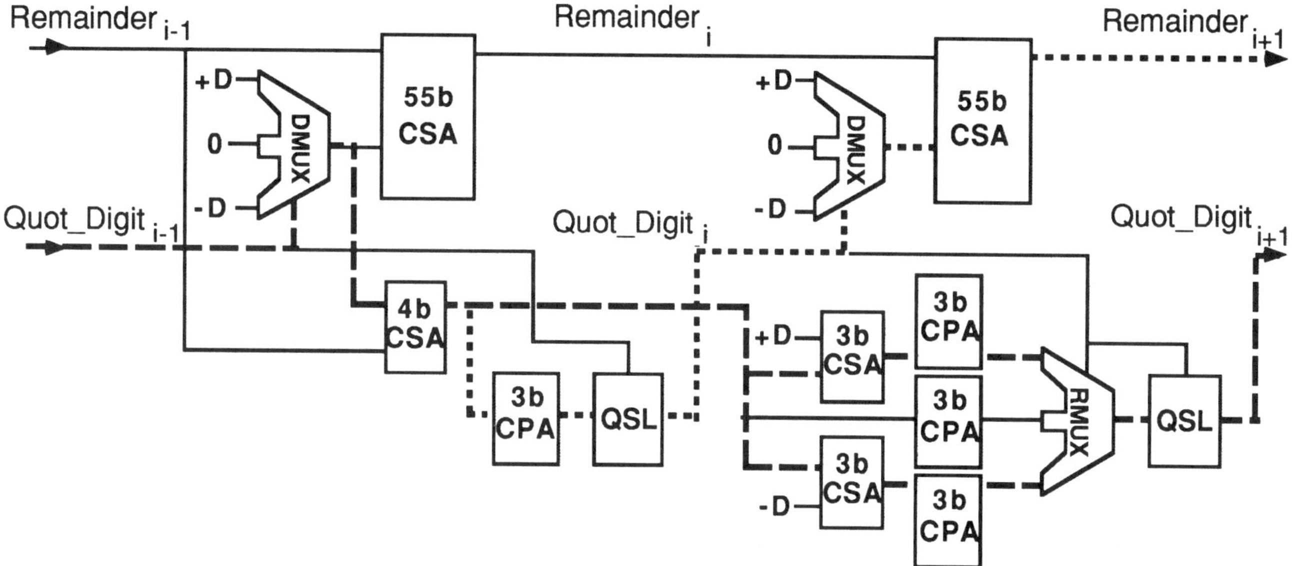

Figure 6: Other implementations used asymmetric dataflows that enforced a specific grouping of the stages into pairs.

makes the critical path go through the same blocks every time. Such a grouping loses about 5% in performance because it enforces extra waiting in some cases and does not achieve the additional path minimization possible with self-timed overlapped execution, which allows a "dynamic" adjustment of the execution order instead of a static grouping of the stages into pairs.

The latch-free methodology introduced in Section 2 can attain the full benefit of overlapped execution. Because the remainder and quotient digits passed between the stages are skewed, any imposed latch with a common clock would add the delay of that skew, in addition to the propagation delay through the latch and added margin to tolerate skew between the clock and data. The propagation delay through a latch would be about 15% of the other delays through a pair of stages; the required clock skew margin 5%; and the skew between remainder and quotient digits is about 10%. So, since adding latches would increase the delay through a pair of stages by a total of 30%, using self-timing to remove the latches results in a 23% performance improvement.

6 Choice of Radix

In an integrated circuit implementation, there are always tradeoffs between speed, power, and chip area. Self-timing provides a useful means of comparing performance without constraints from fixed clock cycles, or delays from latches or input/output considerations. In particular, VLSI implementations of different radix approaches for SRT division can tradeoff complexity amongst the various arithmetic components [8]. This section states fair relative comparisons by using the delay of a gate with unity fan-in and fan-out as the basic speed unit. Real gates have delays many times this basic unit because of stacked transistors (fan-in) and loading (fan-out). The delays stated for blocks include the delays due to buffers on inputs and due to the loading of outputs.

Estimates for SRT radix 2 and radix 4 design choices, with and without overlapped execution, are based on circuit simulations and have been updated and calibrated with measurements from the fabricated chips. Table 3 summarizes the comparisons of these implementation choices. All of the parameters are for an implementation with a minimal-latency self-timed ring. The radix 2 designs require five stages in the ring, while the radix 4 designs only require four stages since the propagation delay of each individual stage is longer [10]. None of the figures include the final 55bit CLA required for rounding and converting the redundant representations back to standard binary. The right two columns contain numbers specific to the CMOS fabrication technology available.

Table 3 shows that radix 2 is slightly better than radix 4 when neither have overlapped execution. This is because the additional complexity of the radix 4 quotient selection does not quite justify the use of radix 4 when clocking does not need to be considered. However, if a design were clocked, the difficulty of supplying a clock at twice the frequency might make radix 4 preferable. Overlapped execution in either radix 2 or radix 4 gives a significant performance increase, about 30% for radix 2 and 35% for radix 4. The key advantage of the self-timed overlapped execution style here is that the average critical path per

Radix & Style (OverExec = Overlapped Execution)	Average Critical Path per Pair of Stages In Unity fan-in, Unity fan-out gate delays	Unity fan in/out Gate Delays per Quot Bit	Latency for 54 bits with 250pS Unit gate delays	Silicon Area in 1.2µ Technology
Radix 2	2 (CPA3+QSL3+DMUX3+CSA55) 2 (5.7 + 3.8 + 2.8 + 4.5) = 33.6	16.8	225 nS	7 mm^2
Radix 2, OverExec	CSA3+CPA3+RMUX3+QSL3+DMUX3+CSA55 3.9 + 4.9 + 3.5 + 3.8 + 2.8 + 4.7 = 23.6	11.8	160 nS	10 mm^2
Radix 4	2 (CPA7+QSL5+DMUX5+CSA56) 2 (11 + 16 + 3.5 + 4.5) = 70.0	17.5	235 nS	12 mm^2
Radix 4, OverExec	CSA7+CPA7+RMUX5+QSL5+DMUX5+CSA56 3.9 + 10 + 5.0+ 16 + 3.5 + 6.2 = 44.6	11.2	150 nS	18 mm^2

Table 3: Tradeoffs in Speed and Area for different implementation approaches.

stage has a factor of $\frac{1}{2}$ times the delay from the CPAs and quotient selection logic. Since, for higher radices, these components occupy bigger proportions of the total delay, the effect of overlapped execution is even more significant for radix 4. To summarize, with overlapped execution, radix 4 is faster than radix 2, but the area cost is much higher because of the replication of the carry-propagate adders. Not only are five adders required instead of only three, but they are also larger. Still higher radices, such as radix 8, would accentuate these tradeoff effects. Overlapped execution would have an even greater percentage reduction in delay, but at a formidable cost in area.

7   Test Results

The tradeoffs discussed in the previous section led to the choice of implementing radix 2 with overlapped execution. We used MAGIC for the full-custom layout of the self-timed divider. The 45K transistor design was fabricated in 1.2µ CMOS technology through MOSIS. The ring's five stages are columns which are mirrored appropriately to weave the datapath and achieve equal path lengths. By careful cell design and over-cell routing, the area cost of using two wires for each bit in dual-monotonic pairs added only about 20% to the total area. The die photo shown in Figure 7 shows an active area of 9.7mm^2, which contains test registers surrounding a core iterating ring in the central 6.8mm^2.

The chips generate the correct data outputs over a wide range of operating conditions. The actual operating conditions determine the actual performance, and Figure 8 shows measured speeds for various voltages and temperatures. For operation at 5V and 35°C ambient temperature, quotient bits are produced internally every 2.8nS. The measured total latency for a 54bit quotient is 160nS for worst-case data, and 45nS for best-case data requiring only two ring iterations. This large data-dependency in timing shows the effect of the early-done detection.

Since exponent logic can operate in parallel, a complete floating-point division operation could be formed by adding the measured delays for the mantissa operation to

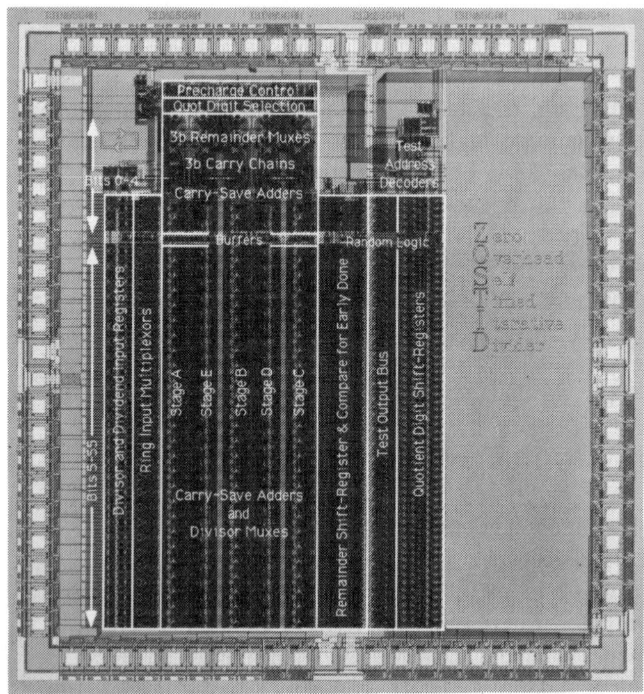

Figure 7: Micrograph of the zero-overhead self-timed 54bit divider in 1.2µ CMOS technology.

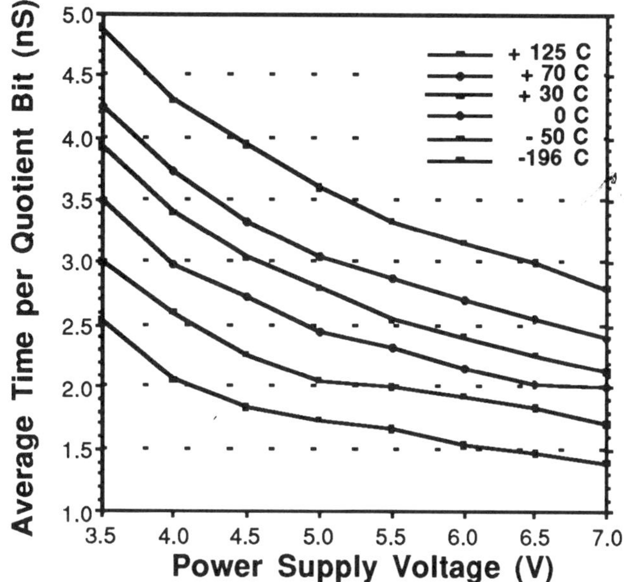

Figure 8: Measured performance per quotient digit at various voltages and temperatures

only the time required for the additional 55bit CLA to round and resolve out of a redundant representation. In the same technology, this delay would likely be 4nS to 8nS.

8 Summary

The set of enhancements in this paper and the performance gained from each of them are summarized in Table 4. The total effect of all the performance enhancements provides a factor of two increase in performance due to architectural improvements over a "straightforward" SRT approach. Moreover, the table does not quantify some advantages such as the benefit that self-timing does not require the cost of global clock distribution.

The self-timed design methods in this paper enable an iterative circuit to compute at the speed of a combinational

Enhancement	Speedup
Resizing CSA inputs based on statistics	4%
Forcing q_{i+1} = -1 when possible	1%
Shortening Remainder Approx CPA by 1 bit	5%
Early Done Detection of Repeating Quotients	4%
Overlapped Stage Execution	35%
Dynamic Overlapping	5%
Latch-Free Self-Timing	23%
Total Effect of Performance Enhancements	100%

Table 4: Summary of gains from performance enhancements

array while only requiring a fraction of its silicon area. Since the design produces a done indication, the outputs can be used as soon as they are available without waiting for worst case margins over the ranges of possible data values, temperature, voltage, and fabrication spread. Self-timing allows an overlapping of the execution of the SRT stages to attain a dynamically adjusting data-dependent minimal critical path. Measurements on fabricated parts verified the architectural techniques achieve high-performance even with ordinary CMOS technology.

References:

[1] D. Atkins, "Higher-Radix Division Using Estimates of the Divisor and Partial Remainder," *IEEE Tran. on Computers*, vol. 17, no. 10, pp. 925-934, Oct. 1968.

[2] J. Fandrianto, "Algorithm for High Speed Shared Radix 8 Division and Radix 8 Square Root," Proc. 9th Symp. Comp. Arith., pp. 68-75, Sept. 1989.

[3] J. Hennessy, D. Patterson, *Computer Architecture: A Quantitative Approach*, Palo Alto: Morgan Kaufmann, pp. 5-12, 1990.

[4] W. McAllister , D. Zuras, "An nMOS 64b Floating-Point Chip Set," ISSCC Digest of Technical Papers, pp. 34-35, Feb. 1986.

[5] T. Meng, R. Brodersen, D. Messerschmitt, "Automatic Synthesis of Asynchronous Circuits from High-Level Specifications," *IEEE Tran. on Computer-Aided Design*, vol. 8, no. 8, pp. 1185-1205, Nov. 1989.

[6] M. Santoro, M. Horowitz, "SPIM: A Pipelined 64x64-bit Iterative Multiplier," *IEEE Journal of Solid-State Circuits*, vol. 24, no. 2, pp. 487-493, Apr. 1989.

[7] G. Taylor, "Radix 16 SRT Dividers with Overlapped Quotient Selection Stages," Proc. 7th Symp. Comp. Arith., pp. 64-71, June 1985.

[8] T. Williams, M. Horowitz, "SRT Division Diagrams and Their Usage in Designing Custom Integrated Circuits for Division," Stanford Tech Report CSL-TR-87-326, Nov. 1986.

[9] T. Williams, M. Horowitz, et. al., "A Self-Timed Chip for Division," Proc. Stanford Conference on Advanced Research in VLSI, pp. 75-95, Mar. 1987.

[10] T. Williams, "Latency and Throughput Tradeoffs in Self-Timed Asynchronous Pipelines and Rings," Stanford Tech Report CSL-TR-90-431, Aug. 1990.

[11] T. Williams, M. Horowitz, "A Zero-Overhead Self-Timed 160nS 54b CMOS Divider," ISSCC Digest of Technical Papers, pp. 98-99, Feb. 1991.

[12] G. Wolrich, E. McLellan, et.al., "A High Performance Floating-Point Coprocessor," *IEEE Journal of Solid-State Circuits*, vol. 19, no. 5, pp. 690-696, Oct. 1984.

GCD and Standard Functions

Chair:
S.M. Sedjelmaci
Université d'Oran Es-Senia

A Redundant Binary Euclidean GCD Algorithm

Shrikant N. Parikh

Programming Systems
IBM
Westlake, Texas   76299

David W. Matula

Dept. of Comp.Sci. and Eng.
SMU
Dallas, Texas   75275

Abstract

An efficient implementation of the Euclidean gcd algorithm employing the redundant binary number system is described. The time complexity is $O(n)$ utilizing $O(n)$ 4-2 signed 1-bit adders to determine the gcd of two n-bit integers. The process is similar to that employed in SRT division. The efficiency of the algorithm is competitive, to within a small factor, with floating point division in terms of the number of shift and add/subtract operations. The novelty of our algorithm is based on properties derived from our scheme of normalization of signed bit fractions. Our implementation is noted to be well suited to systolic hardware design.

1   Introduction

A redundant binary version of the Euclidean algorithm for the greatest common divisor (gcd) is developed. Both our methodology and the performance achieved is similar to that of the SRT division process [1,4,7,12,14,17,18, 20,21].

The traditional form of the original Euclidean algorithm [5] employs recursive integer division and is based on the fact that $\gcd(p - kq, q) = \gcd(p, q)$ for any k. The standard binary form of the Euclidean algorithm employs binary shift and subtract, and thus relies specifically on the fact that $\gcd(p - 2^i q, q) = \gcd(p, q)$. Using carry look ahead adders, the gcd of two n-bit numbers can be found by an implementation of the binary form of the Euclidean algorithm in time $O(n \log n)$ with $O(n)$ hardware. The binary form relies on strict adherence to the monitoring of shifts and argument normalization to determine when to swap the "remainder" $p' = p - 2^i q$ and the "divisor" q, corresponding to what would be the next recursive division in the traditional algorithm.

As noted in [11], the direct attempt to introduce redundant binary constant time subtraction into the binary form of the Euclidean algorithm yields non trivial side effects in compensating for the lack of an appropriate notion of normalization of signed bit numbers. Our first contribution in this paper is the introduction of a convenient and useful notion of normalization of signed bit fractions. The resulting normalized signed bit fraction will have magnitude in the range $(\frac{1}{4}, 1)$, and may be confirmed to be normalized by simply checking at most three leading signed bits.

Our principal contribution is then the specification of a redundant binary form of the Euclidean algorithm (RBEA). The algorithm treats the n-bit integers p, q as $n + 1$ signed bit fractions $b_0.b_1 b_2 \cdots b_n$ which are each normalized to the range $(-1, -\frac{1}{4})$ or $(\frac{1}{4}, 1)$ by a modified shift operation, with b_0 available for a subsequent possible overflow condition. A brief computation on a few leading bits of p, q (equivalently, table lookup) allows us to determine a particular one of the three terms $p - q$, $p - 2q$, or $2p - q$ which must have magnitude less than $\frac{1}{2}$. This term replaces p and is appropriately left shifted at least once until normalized according to our signed bit normalization criteria.

Note that the selection of one of the terms $p - 2q$, $2p - q$, or $p - q$ to replace p in the presence of redundant binary representation of p and q is analogous to the combined SRT division steps of i) digit selection (table lookup), ii) divisor multiple formation, and iii) partial remainder update. As in SRT division, we are similarly assured that the new value may be appropriately shifted at least once. As a consequence of our redundant binary normalization criteria and the term selection process, we then are able to prove that Algorithm RBEA determines the gcd of two n-bit integers employing at most $2n$ shift/subtract operations. With this algorithm we thus need only $O(n)$ time with $O(n)$ 4-2 signed 1-bit adders.

In section 2 we state for reference the Euclidean gcd algorithm in both its traditional and binary forms.

In section 3 we introduce criteria for the normalization of signed bit fractions. We specify our redundant binary Euclidean algorithm in section 4, and then prove it may be implemented in $O(n)$ time with $O(n)$ 4-2 signed 1-bit adders.

Regarding extensions we first note that the alternative right shift binary gcd algorithm has been shown [3] to be implementable in an elegant systolic design. Algorithm RBEA appears similarly viable for systolic design. In addition, algorithm RBEA may readily be extended [11] to also obtain a solution α, β of

$$\alpha p - \beta q = \gcd(p, q),$$

according to the generalized Euclidean gcd result. Thus our methodology provides the foundation for an efficient systolic design yielding the more comprehensive results of the expanded Euclidean gcd algorithm, results which can not be obtained from the right shift binary gcd algorithm.

2 Euclidean Algorithm

A summary of the Euclidean algorithm and its binary version is given in this section based on notation and discussion from [11].

Algorithm EA (Euclidean Algorithm)

For any (p, q), $q \neq 0$, this algorithm computes the gcd of p, q.

1. Let $b_{-2} = p$; $b_{-1} = q$.
2. For $i = 0, 1 \ldots$, while $b_{i-1} \neq 0$ determine a_i as the quotient and b_i as the remainder (of the same sign as b_{i-2}) of the division of b_{i-2} by b_{i-1}, so $b_i = -b_{i-1} \times a_i + b_{i-2}$.

The algorithm terminates when $b_i = 0$. At this point b_{i-1} is the gcd of p and q.

The binary version [11] replaces each division by a sequence of shift-and-subtracts climaxed by a swap. The swap corresponds to the interchange of role of divisor and remainder for initiating the next division step of the traditional Euclidean algorithm.

Algorithm BEA (Binary Euclidean Algorithm)

Given initializations p,q of the registers P,Q such that $(p,q) \neq (0,0)$, this algorithm terminates with the gcd value right justified in the P register. U_P and U_Q are auxiliary registers initialized to 1 which identify the unit-position of each argument.

```
While P and Q not normalized do
    leftshift P, Q, U_P, U_Q;
While Q ≠ 0 do begin
    While Q not normalized do
        leftshift Q and U_Q;
    Loop
        While P not normalized do
            Exit Loop if U_Q= U_P;
            Leftshift P and U_P;
        End;
        If sign(P) = sign(Q) then
            P := P−Q
            else P := P+Q;
    EndLoop;
    Swap(P,Q); Swap(U_P,U_Q);
End;
While U_P≠ 1 do rightshift P and U_P.
```

In order to introduce redundant binary representation to expedite the addition/subtraction operations in Algorithm BEA, an appropriate notion of normalization of redundant binary operands must be developed.

3 Redundant Binary Representation and Normalization

Redundant binary representation employs the signed bits $\{\bar{1},0,1\}$. We here introduce a notion of normalization of signed bit fractions that guarantees such a number has magnitude in the range $(\frac{1}{4},1)$, where verification of normalization is determined by the leading three signed bits.

The signed bit string with radix point after b_0, $p = b_0.b_1\bar{b}_2\ldots b_k, b_i = \bar{1},0,1$ for all i, of length $k + 1$ is termed a signed bit fraction if and only if

$$-1 < \sum_{i=0}^{k} b_i 2^{-i} < 1.$$

A signed bit fraction is in complement form whenever $b_0 \neq 0$. b_0 is termed the complement bit, and when it is nonzero, its sign must be opposite to that of the sign of the necessarily nonzero value of the remaining portion of the number. The complement bit may thus be interpreted as a magnitude complementing bit in the sense that the absolute value of $b_0.b_1b_2\ldots b_k$ is $1 - |x|$, where

$$x = \sum_{i=1}^{k} b_i 2^{-i} \neq 0.$$

The signed bit fraction $b_0.b_1b_2\ldots b_k$ is termed <u>normalized</u> whenever exactly one of b_0 and b_1 is nonzero, where also $b_2 \neq -b_1$, when $b_0 = 0$.

Thus we may define both standard and complement normalized forms.

<u>Standard form</u>: $b_0 = 0,\ b_1 \neq 0$ and $b_2 \neq -b_1$.
<u>Complement form</u>: $b_0 \neq 0,\ b_1 = 0$.

A signed bit fraction that is not normalized is termed <u>unnormalized</u>.

Observation 1: A normalized signed bit fraction always is in the range $(\frac{1}{4},1)$ or $(-1,-\frac{1}{4})$. An unnormalized signed bit fraction with $b_0 = 0$ must be in the range $(-\frac{1}{2},\frac{1}{2})$.

Given the unnormalized signed bit fraction $b_0.b_1b_2\ldots b_k$, the complement bit may be forced to zero by the following <u>decomp</u> function.

$$\text{decomp}(p) = \begin{cases} 0.b_0b_2b_3\ldots b_k, & \text{if } b_0 = -b_1 \neq 0, \\ p, & \text{if } b_0 = 0. \end{cases}$$

Note we cannot have $b_0 = b_1 \neq 0$ since $-1 < p < 1$ for signed bit fraction. Note further that the decomp function does not change the value of p. The result may or may not be normalized.

The operation termed simplifying ("absorbing") shift (<u>simshift</u>) is defined for every noncomplemented unnormalized signed bit number $p = 0.b_1b_2\ldots b_k$ to yield a signed k bit fraction given by

$$\text{simshift}\,(p) = \begin{cases} 0.b_1b_3b_4\ldots b_k, & \text{if } b_1 \neq 0 \wedge b_1 = -b_2 \\ 0.b_2b_3b_4\ldots b_k, & \text{if } b_1 = 0. \end{cases}$$

Observation 2:

1. simshift(p) has twice the value of p,

2. simshift never creates a complemented signed bit fraction,

3. simshift(p) has length one bit less than p.

Normalization, a unary operation, defined for all nonzero signed bit fractions can be stated as follows:

Normalization Algorithm:

If p not normalized do p = decomp(p);
While p not normalized do p = simshift(p);

Lemma: Given a signed bit fraction $p \neq 0$ of length $k + 1$,

$$p = \sum_{i=0}^{k} b_i 2^{-i} \neq 0,\ b_i = -1, 0, 1 \text{ for all } i,$$

then the normalization of p will be achieved after at most $k - 1$ simshift operations, and this bound is best possible.

To implement the Euclidean algorithm for two signed nonzero numbers, the step of selectively determining their difference when similarly signed or their sum when oppositely signed is aided by introducing the following <u>diff</u> operation:

$$a \text{ diff } b = \begin{cases} a - b \text{ for } ab \geq 0, \\ a + b \text{ for } ab < 0. \end{cases}$$

The diff operation always returns a magnitude equal to the difference of the original two magnitudes.

Observation 3: For the normalized signed bit numbers a, b, the range of a diff b will be $(-\frac{3}{4}, \frac{3}{4})$.

It is useful to contrast standard binary normalization with redundant binary normalization:

Standard binary representation: The range of a normalized binary fraction is $[\frac{1}{2},1)$, $(-1, -\frac{1}{2}]$. The difference of two positive normalized binary fractions is never normalized, and when nonzero can always be normalized by one or more left shift operations.

Redundant binary representation: The range of a normalized signed bit fraction is $(\frac{1}{4}, 1)$, $(-1, -\frac{1}{4})$. The result of the diff operation of two normalized signed bit fractions when nonzero can always be normalized by zero or more simshift operations. Note, the difference of two normalized signed bit fractions might be normalized.

4   Digit Selection

The digit selection process refers to a selection of one of the terms p diff q, p diff $2q$, or $2p$ diff q, as the new value of p. Specifically, after p and q are

```
    1/4       1/2       3/4        1        5/4
    |-------|-------|-------|-------|
Case 1:                   |<-------->|
p>=2q         ^q         ^2q          ^p

Case 2:                |<--->|
1>=2q>p       ^q     ^p    ^2q

Case 3:                         |<---------->|
2q>1>p>q>1/2            ^q     ^p            ^2q
```

Figure 1: Examples of diff operations

normalized, the result of the p diff q operation is previewed (by look ahead) to check if it would be normalized or not. The occurrence of a normalized result would guarantee the resulting magnitude to be greater than $\frac{1}{4}$, in which case, depending on the signs of p, q, and p diff q, one of the terms $(2p$ diff $q)$ or $(p$ diff $2q)$ is then chosen as the new value of p. With this selection, the value of p can be guaranteed to be less than $\frac{1}{2}$. This is diagrammatically shown above. Without loss of generality consider the case where p and q are positive, and $1 > p > q > \frac{1}{4}$. In each of three cases, given $p - q > \frac{1}{4}$, the position of p, q and the difference between $2q$ and p are illustrated with $|2q - p| < \frac{1}{2}$ in all cases (see Figure 1).

Lemma: Given $|p$ diff $q| > \frac{1}{4}$, where p, q are normalized signed bit fractions, then

$$|p \text{ diff } 2q| < \tfrac{1}{2} \text{ if } |p| > |q|,$$
$$|2p \text{ diff } q| < \tfrac{1}{2} \text{ if } |p| < |q|.$$

Lemma: Given $\frac{1}{4} < |p|, |q| < 1$,

$$\text{Min}(|p \text{ diff } q|, |p \text{ diff } 2q|, |2p \text{ diff } q|) < \frac{1}{2}.$$

Corollary:

After the selection of a diff operation term such that the magnitude of the result is less than $\frac{1}{2}$, it will always be possible to normalize the result by a process including at least one simshift operation.

Note that if a complement condition initially exists, it will be possible to first perform the decomp function. Note further that the imposition of at least one simshift can result in the b_0 bit becoming non zero. We are assured this result corresponds to a complement form signed bit fraction as the initial result was

guaranteed to be of absolute value less than $\frac{1}{2}$ before the simshift operation.

This digit selection is based on the result of the operation $(p$ diff $q)$, specifically on the state of the result – whether it was normalized or not normalized. This in turn suggests that it is necessary and sufficient to know at most the first five positions of p and q (positions 0,1,2,3,4) if an alternative table lookup were used, since the carry propagation in redundant binary addition is limited to two places. Digit selection can be obtained either from a table or by implementing in hardware, simple combinational logic to perform a diff operation on the first five bits, and checking for normalization of the result.

Algorithm RBEA (Redundant Binary Euclidean Algorithm)

Given initializations p,q of the registers P,Q such that $(p,q) \neq (0,0)$, this algorithm terminates with the gcd value right justified in the P register. U_P and U_Q are auxiliary registers initialized to 1 which identify the unit-position of each argument.

```
While P and Q not normalized do
    Simshift P, Q and leftshift U_P, U_Q;
LoopA
    While Q not normalized do
        Simshift Q and leftshift U_Q;
        Exit LoopA if U_Q overflows;
    LoopB
        While P not normalized do
            Exit LoopB if U_P=U_Q;
            Simshift P and leftshift U_P;
            If (P diff Q) is normalized then
                If sign(P diff Q) ≠ sign(P) then
                    If U_P≠U_Q then
                        leftshift U_P; P:=2P diff Q;
                    Else exit LoopB;
                Else P:=P diff 2Q;
            Else P:=P diff Q;
            P:=decomp(P);
            Simshift P;
            If U_P≠U_Q then leftshift U_P;
            Else leftshift U_P and exit LoopB;
        End LoopB;
        Swap(P,Q); Swap(U_P,U_Q);
    End LoopA;
While U_P≠1 do rightshift P, U_P.
```

Theorem 1: Given the signed bit integers p and q, the number of diff operations in the execution of algorithm RBEA is at most the sum of the lengths of p and q.

Proof: In each major cycle ending with a swap operation, each diff operation is followed by one or more simshift operations on register P. The number of diffs is bounded above by the number of simshift operations. A simshift operation reduces the length of one of the arguments by one, and the algorithm terminates when $q = 0$. Since the number of simshifts is bounded by the total of the lengths of p and q, the number of diff operations is at most the sum of the lengths of p and q.

$\square$

Since n-bit redundant binary addition is a constant time operation with O(n) 4-2 signed 1-bit adders, the gcd of two n-bit numbers can be determined by Algorithm RBEA in O(n) time with O(n) bit level processors.

Observation 4 The absolute value of the chosen term $(p \text{ diff } q)$, $(p \text{ diff } 2q)$, $(2p \text{ diff } q)$ in Algorithm RBEA can always be simshifted. This fact can be used to combine the simshift and diff operation thereby avoiding the cost of a simshift immediately following a diff operation.

Algorithm RBEA terminates with a number of diff operations (previously shown to be equivalent to add/sub operations) not exceeding $j + k$, where j and k are the lengths of the two operands. This is a guaranteed upper bound. This upper bound is achieved by a guaranteed progress of at least one bit after each diff operation. This phenomenon is made possible by the digit selection process. The algorithm in the form shown needs swap operations but these can be eliminated by using two symmetrical registers in hardware. In practice, simulation of the algorithm with various length arguments has shown that the number of diff operations per bit of input ('progress') approaches 0.41.

Acknowledgements

We wish to thank Søren P. Johansen for helpful comments and improvements in the statement of Algorithm RBEA.

References

[1] D.E.Atkins, "The Theory and Implementation of SRT Division," *Report No. 230*, Dept. of Computer Science, University of Illinois, June, 1967.

[2] A.Avizienis, "Signed-digit Number Representation for Fast Parallel Arithmetic," *IRE Transaction in Electronic Computing*, Vol. 10, pp. 389-400, 1961.

[3] R.P.Brent, H.T.Kung, "A Systolic Algorithm for GCD Computation," *Proc. 7th IEEE Symp. on Comp. Arith.*, pp 118-125, 1985

[4] M.D.Ercegovac, "A Higher-Radix Division with Simple Selection of Quotient Digits," *Proceedings of Sixth IEEE Symposium on Computer Arithmetic*, pp 94-98, June, 1983.

[5] Euclid, *Proposition 1 I& 2 of Elements*, Book 7, appx.300BC.

[6] M.J.Foster and H.T.Kung, "The Design of Special Purpose VLSI Chips," *IEEE Computer*, Vol.13, pp.26-40, Jan. 1980.

[7] C.C.Freiman, "Statistical Analysis of Certain Binary Division Algorithm," *Proc. IRE, Vol.49*, pp 91-103, 1961.

[8] G.H.Hardy and E.M.Wright, *An Introduction to the Theory of Numbers*, 4th ed., Oxford, England: Clareton Press, 1959.

[9] Y. Harata et al., "High-Speed Multiplier LSI Using a Redundant Binary Adder Tree," *International Conference on Computer Design*, 1984.

[10] D.E.Knuth , *The Art of Computer Programming: Vol II,Seminumerical Algorithms*, (Section 4.5), Addison-Wesley Publishing Company, 1981.

[11] P.Kornerup and D.W.Matula, "Finite Precision Rational Arithmetic: An Arithmetic Unit," *IEEE Transactions on Computers*, Vol C-32, No.4, pp. 378-388, April 1983.

[12] S. Kuninobu et at., "Design of High-Speed Multiplier and Divider Using Redundant Binary Representation," *Proc. of the 8th ISCA*, pp 80-86, May 1987.

[13] O.L. MacSorley, "High-speed Arithmetic in Binary Computer," *Proc. IRE*, Vol. 49, no. 1, pp. 67-91, Jan. 1961.

[14] G.Metze, "A Class of Binary Divisions Yielding Minimally Represented Quotients," *IRE Transactions on Electronic Computers*, Vol. EC-10, December 1962.

[15] S. Parikh, *An Architecture For a Rational Arithmetic Unit,* Ph.D. dissertation, Southern Methodist University, December 1988.

[16] G.W.Reitwiesner, "Binary Arithmetic", *Advances in Computers*, Vol. 1, F.L.Alt, Ed. New York: Academic, 1960.

[17] J.E.Robertson, "A New Class of Digital Division Methods," *IRE Trans. El. Comp.*, Vol EC-7, no.3, pp.218-222, Sept. 1958.

[18] N.R.Scott, *Computer Number Systems and Arithmetic,* Prentice-Hall Inc., NJ, 1985, Section 7.11.

[19] B. Shirazi et al., "RBCD: A Redundant Binary-coded Decimal Adder," *IEE Proceedings - Computers and Digital Techniques*, Vol. 136, No.2, pp.156-160, 1989.

[20] G.S.Taylor, "Radix 16 SRT Dividers with Overlapped Quotient Selection Stages," *Proceedings of 7th Symposium on Computer Arithmetic*, 1985.

[21] K.D.Tocher, "Techniques of Multiplication and Division for Automatic Binary Computers," *Quart. J. Mech. Appl. Math.*, Vol. 11, pt.3, pp. 364-384, 1958.

[22] N. Takagi et al., "High-Speed VLSI Multiplication Algorithm With a Redundant Binary Addition Tree," *IEEE Trans. on Computers*, Vol. C-34, No.9, pp. 789-796, September 1985.

OCAPI: Architecture of a VLSI Coprocessor for the GCD and the Extended GCD of Large Numbers

Alain GUYOT

Laboratoire TIM3/IMAG
46 Avenue Félix Viallet F38031 Grenoble FRANCE

Abstract

In this paper, various algorithms for finding the greatest common divisor (GCD for short) and extended GCD of very large integers are explored. In particular the trade-off between computation time and area is examined. Two of the algorithms, from which the method to derive variants is straightforward, are detailed. Then the architecture of a VLSI processor dedicated to GCD as well as multiply, divide, square root, ... of very large numbers (> 600 decimal digits), using an internal radix 2 redundant representation and supporting multiple precision, is devised.

Index terms : GCD, extended GCD, redundant number system, most significant digit first algorithm, very large integers

1. Introduction

Computation of the GCD of two integers is used in integer arithmetic (e.g. normal form) as well as rational arithmetic [1] either to get the canonic form or perform rounding. Extended GCD is necessary for multiple precision GCD [2] and to work over $\mathbb{Z}/p\mathbb{Z}$. Computer algebra programs (reduce, macsyma, ...) reportedly spend 85% of their time in GCD [3] for some applications because of the frequency and the comparatively high cost of this operation. So, while all the algorithms studied in this paper have a complexity of $O(n)$, it is important to find a good one.

2. Background

The "binary algorithm" proposed by Stein [4] relies on shift, subtraction, exchange and comparison; all operations except comparison suitable for large integers. For convenience, right shift n position(s) is noted "$/2^n$" and left shift "$*2^n$".

This work was supported by the GCIS

function GCD (A,B) ; { assume A odd }
begin
  while B $\neq$ 0 **do**
  if B mod 2 = 0 **then** $\begin{pmatrix} A \\ B \end{pmatrix} := \begin{pmatrix} A \\ B/2 \end{pmatrix}$ { B is even }
  else if A $\geq$ B **then** $\begin{pmatrix} A \\ B \end{pmatrix} := \begin{pmatrix} B \\ (A-B)/2 \end{pmatrix}$ {A and B are odd}
  else $\begin{pmatrix} A \\ B \end{pmatrix} := \begin{pmatrix} A \\ (B-A)/2 \end{pmatrix}$;
  GCD := A ;
end { GCD } .

This algorithm gives the GCD because 1) every transformation preserves the GCD(A,B), and 2) GCD(A,0) = A.

It is not very far from Euclid's algorithm [5] when binary non restoring division is developed:

function GCD (A,B); {assume A or B odd, A $\geq$ B, $2^n \leq A < 2^{n+1}$}
begin
  while B $\neq$ 0 **do**
  if B $< 2^n$ **then** $\begin{pmatrix} A \\ B \end{pmatrix} := \begin{pmatrix} A \\ B*2 \end{pmatrix}$
  else if A $\geq$ B **then** $\begin{pmatrix} A \\ B \end{pmatrix} := \begin{pmatrix} B \\ (A-B)*2 \end{pmatrix}$ {$2^n \leq B < A < 2^{n+1}$}
  else $\begin{pmatrix} A \\ B \end{pmatrix} := \begin{pmatrix} A \\ (B-A)*2 \end{pmatrix}$;
  while A mod 2 = 0 **do** A := A / 2; GCD := A ;
end { GCD } .

Basically, Stein's algorithm tests and forces to zero the LSB of B while Euclid's does the same with the MSB.

Brent and Kung [6] have built a systolic algorithm based on Stein's one that, instead of comparing actual numbers, compares an estimate of their number of bits, and uses serial binary addition/subtraction. Due to the estimates, addition becomes necessary to ensure stability [6]. Kornerup and Matula [1] use signed binary digits for a carry propagation free addition/subtraction and only the sign, as in the SRT division. Purdy [7] proposed use the same notation [1] and the same evaluator as Brent's [8]. Yun and Zhang [9] later improved their algorithm to make it about twice as fast and to include as well the extended GCD.

Those algorithms rely on the affectation $B := (B - q*A)*2^p$, and it is easy to build new algorithms by changing the ranges of q and p. Since we want one hardwired addition/subtraction per cycle but we can afford several shifts we will limit $|q|$ to powers of 2. In this paper, depending on the realization, the transistor cost of another shift varies from 2 (parallel) to 10 (serial) transistors, the cost of an adder is from 20 (2's complement) to 40 (redundant) transistors per digit.

3 . LSB first approach

Let us give in pseudo-Pascal, as a first example, an algorithm that divides the greatest of the two numbers by 4 at each iteration by forcing to two of its LSBs 0 whenever necessary, where $q \in \{-1,0,1,2\}$ and $p = -2$. This algorithm simplifies when conventional binary notation is used since then $a_0+b_0=0$ never happens and absolute value is no longer needed as an operation and of course the adder is simpler in term of gates as we will see later.

function GCD $(A, B, \delta_a, \delta_b)$;

{assume A odd, $\delta_a = \lceil \log_2 A \rceil$, $\delta_b = \lceil \log_2 B \rceil$, # of bits of A and B}

begin
while $\delta_b > 0$ **do**   { $B \neq 0$ }
begin

    if $b_0 = 0$ **then**
    begin
    if $b_1 = 0$ **then** $B := (B+0)/4$ { q = 0 }
    else $B := (B+ 2*A)/4$ { q = 2 }
    end
    else if $\delta_a \geq \delta_b$ **then**
    begin swap (δ_a, δ_b) ; { exch }
    if ($a_0+b_0=0$ **and** $a_1+b_1=0$) **or**
    ($a_0+b_0 \neq 0$ **and** $|a_1+b_1| = 1$) **then** { $q = \pm 1$ }

$$\binom{A}{B} := \binom{B}{(A+B)/4} \quad \textbf{else} \quad \binom{A}{B} := \binom{B}{(B-A)/4}$$

    end else
    if ($a_0+b_0=0$ **and** $a_1+b_1=0$) **or**
    ($a_0+b_0 \neq 0$ **and** $|a_1+b_1| = 1$) **then**

$$\binom{A}{B} := \binom{A}{(A+B)/4} \quad \textbf{else} \quad \binom{A}{B} := \binom{A}{(B-A)/4} ;$$

    $\delta_b := \delta_b - 2$;
  end ;
   GCD := A ;
end.

4 . MSB first approach

The second example forces the MSB of B to 0 and tries to predict if the 2nd MSB will also turn to be 0. To test the MSB, redundant notation must be used whether the addition/subtraction is done in serial or in parallel. Let us note $\bar{1}$, 0 or 1 the values of a signed binary digit. Due to the many ways of writing integers in redundant notation, the numbers of digits δ_a and δ_b of A and B alone are useless for comparison. For example $A = 1 \bar{1} \bar{1} \bar{1} \bar{1}$ is in fact smaller than $B=10$ despite $\delta_a=5$ and $\delta_b=2$. So besides δ_a and δ_b we use also the first k digits μ_a and μ_b of A and B. A normalization forces the 1st digit of μ_a and μ_b to be non-zero, and all the k digits to have the same sign (0 has both signs), μ_a and $\mu_b \in [-2^k+1..-2^{k-1}] \cup [2^{k-1}..2^k-1]$. They are stored as the common sign and the absolute value of the digits.

<table>
<tr><td>$2^{\delta-1}$ $2^{\delta-k}$</td><td></td><td>2^0</td></tr>
<tr><td>S</td><td>μ</td><td>$(\delta - k)$ digits $\longrightarrow$</td></tr>
<tr><td>normalized part</td><td></td><td>redundant part</td></tr>
</table>

Of course the larger k is, the more digits in μ_a and μ_b , the more accurate the predictors and the quicker the convergence of the GCD algorithm, in number of cycles. But since there is a carry propagate to normalize μ, each cycle would also last longer. So in the present paper we propagate the carry on 3 bits because it is the minimum allowing us to perform another operation of interest: division [10,11], and we try also propagation on 4 bits. When μ_a and μ_b are normalized A and B are seminormalized, that is we always have:

$$(|\mu_a|-1)2^{\delta_a-k} < |A| < (|\mu_a|+1)2^{\delta_a-k} \,\&\, (|\mu_b|-1)2^{\delta_b-k} < |B| < (|\mu_b|+1)2^{\delta_b-k} .$$

Since A and B are only seminormalized, for the following GCD algorithm, values $|\mu_b| = 2^{k-1}$ and $|\mu_a| = 2^k-1$ deserves special attention. Let us take as an example k = 3, B = $100\bar{1}0$ ($\delta_b=5$, $\mu_b=4$) and A= $1\bar{1}10$ ($\delta_a=4$, $\mu_a=7$). If $B:=B-A*2^{\delta_b-\delta_a}$ is performed, the result will be $B=\bar{1}00\bar{1}0$, that also is seminormalized and negatif, so the next operation will be $B:=B+A*2^{\delta_b-\delta_a}$ that gives back the previous value, and the algorithm <u>never ends</u>. By rewriting B when necessary before swapping A and B, the algorithm prevents μ_a from taking the values 7 or -7. A similar loop forever is reported in [1] and treated by forcing a shift in the loop. The table on next page summarizes the new range of μ_b from the value μ_a and the previous value of μ_b after B:=B-A, in bold the values that should not be taken, and in italics the column $\mu_a=7$ that is avoided. A symmetrical table can be drawn for B:=B+A.

$\mu_b \setminus \mu_a$	4	5	6	7
1	-1,+1	-2,0	-3,-1	*-4,-2*
5	0,+2	-1,+1	-2,0	*-3,-1*
6	+1,+3	0,+2	-1,+1	-2,0
7	+2,+4	+1,+3	0,+2	*-1,+1*

The following pseudo-Pascal program is the behavioral description of a chip for the GCD with $q \in \{-1,0,1\}$ and $p \in \{0,1,2\}$. The program manipulates four integers: δ_a and δ_b with $\log_2(\log_2 ($ max. value of A or B)) bits, μ_a and μ_b with k bits (unsigned) plus one bit sign.

Let n be δ_b-k-1, b_n be the most significant signed digit of B and $b_{n..0}$ denote the string $b_n, b_{n-1}, ..., b_0$.

```
function GCD (A, B, k, δa ,δb) ;
{ δa=⌈log2A⌉ δb=⌈log2B⌉, number of bits of A and B }
begin
while δb > 0 do          {B ≠ 0 }
begin
    if |μb| < 2^(k-1) then
    begin
    if ((δa>δb) or  ((δa = δb) and |μa| > 2^(k-1) )))
    and ( bn * μb) = 2^(k-1)-1 then
            begin μb := μb + bn ; bn := - bn  end
    else if |μb| < 2^(k-2) then
    begin μb:=4*μb+2*bn+bn-1; bn..2:=bn-2..0; δb:= δb -2 end
    else begin  μb:=μb+μb+bn ; bn..1:=bn-1..0; δb := δb -1 end
    end else begin
    if (δa > δb) or ( (δa = δb) and (|μa| > |μb| )) then
            begin swap ( δa , δb ) ; swap (A , B) end ;
    if | μa + μb| < 2^(k-2) then
            begin B := (B + A) * 4 ; δb := δb - 2 end
    else if | μa - μb | < 2^(k-2) then
            begin B:= (B - A) * 4; δb := δb - 2 end
    else if | μa + μb | < 2^(k-1) then
            begin B:= (B + A) * 2; δb := δb - 1 end
    else   begin B:= (B - A) * 2 ; δb := δb - 1 end ;
    end;
end ;
  GCD := A ;
end.
```

5. Generalization

One addition/subtraction can force to 0 at least either the 2 least significant positions or the most significant position of B, and sometimes more positions. To take advantage of the 0s in more positions, that is to shift them out, is more costly in term of transistors. On the contrary forcing just one LSB position makes the circuit slower but the operation part and the control simpler. In a systolic approach, with a distributed control, the control part becomes preponderant [6]. Thus the time/area of different

approaches has been plotted, where the time is approximated by the number of cycles and the area by the number of transistors. Distributivity of addition/subtraction and shift as well as serialization of less frequent operations were used to keep the number of transistors low.

Approaches with examination of the LS digits

≠	q (quotient)	p (position)	bits or digits examined	
L1	-1 , +1	-1	b_1	
L2	-1, 0 , +1	-1	a_1, b_1	note 1
L3	-1, 0 , +1	-1 , -2	a_0, a_1, b_0, b_1	note 2
L4	-1,0,+1,+2	-2	a_0, a_1, b_0, b_1	note 3
L5	-2,-1,0,+1,+2,+4	-2 , -3	a_0, a_1, b_0, b_1, b_2	

note 1: This is the Brent-Kung [6] variant
note 2: This is the Yun-Zhang [9] variant
note 3: This is the first example detailed in this paper

Approaches with examination of the MS digits

≠	q (quotient)	p (position)	k	
M1	-1 , 0 , +1	1	3	
M2	-1 , 0 , +1	1 , 2	3	note 4
M3	-1 , 0 , +1	1	4	
M4	-1 , 0 , +1	1 , 2	4	

note 4: This is the second example detailed in this paper

Some of the approaches were realized in serial, in parallel or both:

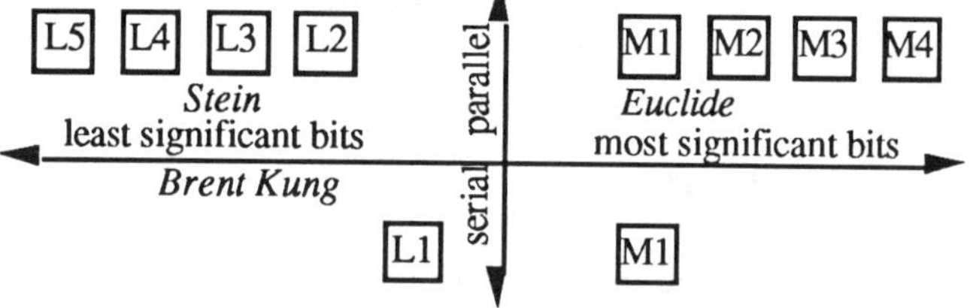

Pseudo Pascal description of the variants, as well as the corresponding hardware are detailed in [12] available through the IMAG librarian.

6. Extended GCD

Now we want the time/area trade off of a circuit that computes a pair of integers R and S that follows the Bezout's equation, $A*R - B*S = H$, H denotes the GCD(A,B). This is done by executing in the reverse order a sequence of operations corresponding to the ones that led to H (1st column of the tables of next page) while preserving Bezout's identity (3th column). For that phase, the initial values of (A,B,R,S) are (H,0,1,1). Unfortunately, for the LSB approach, that leads to R and S that may not be integer, so for this approach we first preserve $A*R - B*S = H*2^n$ (4th column) and then simplify R and S while keeping them integers.

Let us give the table for the LSB example

compute H $=GCD(A,B)$	restore A and B	Preserves $A*R-B*S=H$	Preserves $A*R-B*S=H*2^n$
$B := B / 4$	$B := B * 4$	$S := S / 4$	$R := R * 4$
$B := B/4+A/2$	$B:=B*4-A*2$	$R := R - S/2$; $S := S / 4$	$R := R*4-S*2$
$\begin{pmatrix} A \\ B \end{pmatrix} := \begin{pmatrix} B \\ A \end{pmatrix}$	$\begin{pmatrix} A \\ B \end{pmatrix} := \begin{pmatrix} B \\ A \end{pmatrix}$	$\begin{pmatrix} R \\ S \end{pmatrix} := \begin{pmatrix} -S \\ -R \end{pmatrix}$	$\begin{pmatrix} R \\ S \end{pmatrix} := \begin{pmatrix} -S \\ -R \end{pmatrix}$
$B := (B+A)/4$	$B:=B * 4 - A$	$R := R - S/2$; $S := S / 4$	$R := R * 4 - S$
$B := (B-A)/4$	$B:=B*4 + A$	$R := R - S/2$; $S := S / 4$	$R := R * 4 + S$

and the table for the MSB example

compute H $= GCD(A,B)$	restore A and B	preserves $A*R - B*S = H$ when restoring A and B	
$B := B * 2$	$B := B / 2$	$S := S * 2$	
$B := B * 4$	$B := B / 4$	$S := S * 4$	
$\begin{pmatrix} A \\ B \end{pmatrix} := \begin{pmatrix} B \\ A \end{pmatrix}$	$\begin{pmatrix} A \\ B \end{pmatrix} := \begin{pmatrix} B \\ A \end{pmatrix}$	$\begin{pmatrix} R \\ S \end{pmatrix} := \begin{pmatrix} -S \\ -R \end{pmatrix}$	
$B := (B+A)*2$	$B:=B/2 - A$	$R := R - S * 2$;	$S := S * 2$
$B := (B-A)*2$	$B:=B/2 + A$	$R := R + S * 2$;	$S := S * 2$
$B := (B+A)*4$	$B:=B/4 - A$	$R := R - S * 4$;	$S := S * 4$
$B := (B-A)*4$	$B:=B/4 + A$	$R := R + S * 4$;	$S := S * 4$

Simplification of R and S is straightforward, but adds cycles and transistors, since it is necessary to keep A and B in registers, to the extended GCD computation in the LSB approach. Since A,B and H are odd, R and S exibit the same parity.

while $n > 0$ **do**
begin $n := n-2$;
    if $r_0 = 0$ **then begin if** $r_1 = 0$
    then $\begin{pmatrix} R \\ S \end{pmatrix} := \begin{pmatrix} R/4 \\ S/4 \end{pmatrix}$ **else** $\begin{pmatrix} R \\ S \end{pmatrix} := \begin{pmatrix} R/4+B/2 \\ S/4+A/2 \end{pmatrix}$ **end**
    else begin if $(a_0+r_0 = 0$ **and** $a_1+r_1 = 0)$ **or**
    $(a_0+r_0 \neq 0$ **and** $|a_1+r_1| = 1)$
    then $\begin{pmatrix} R \\ S \end{pmatrix} := \begin{pmatrix} (R+B)/4 \\ (S+A)/4 \end{pmatrix}$ **else** $\begin{pmatrix} R \\ S \end{pmatrix} := \begin{pmatrix} (R-B)/4 \\ (S-A)/4 \end{pmatrix}$ **end** ;
end .

If we have only one adder, the two additions/subtraction that are presented here in parallel must be done serially.

7. Measures

The first plot below shows the performance-cost ratio of the different approaches and variants for the GCD computation. The cost was obtained by carefully designing an operation part for each algorithm [12], and the performance by running a simulation of each of the operation parts with the same set of 200 samples of randomly generated 2,000-bit integers whose first and last

bit was forced to be non-zero. The trade off of each variant has been locally optimized. The time to load in the operands was not counted.

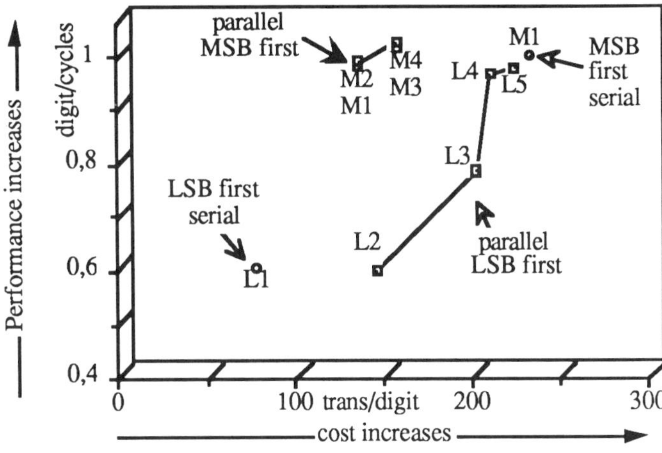

Algorithm L5 that removes an average of 2.5 digits from the least significant position of B at each cycle happens to be slower than algorithm M1 that removes 1 digit at the most significant position when it is not stalling to rewrite B. The size, or number of significant digits, of the result of an addition/subtraction is not deductible from the size of the operands, and errors tend to accumulate. After several trials, we settled on no size adjustment, the size of the result being the size of the biggest operand.

Having computed the GCD in n cycles, the number of additional cycles required for the extended GCD is n minus the cycles when the algorithm was stalling for the MSB approaches, and n plus the number of cycles to simplify the result in the LSB one.

The LSB first serial approach is very cheap for the GCD, and the GCD only. Otherwise the MSB approaches are far more interesting for the extended GCD, as shown on the following plotting the cost of extended GCD.

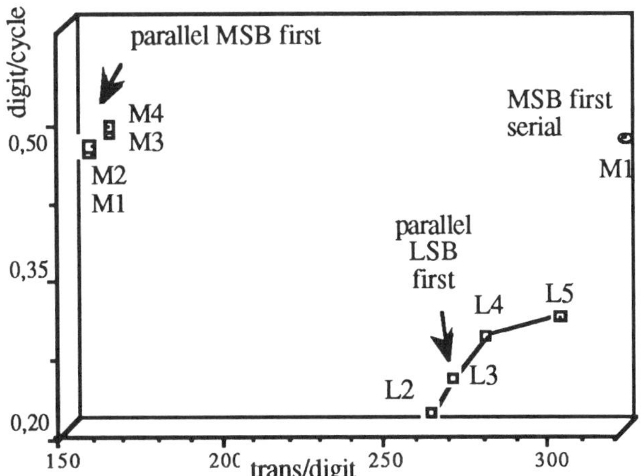

8. Coprocessor for high precision

This VLSI circuit is designed to be placed close to a microprocessor and plugged into the memory bus, in order to significantly reduce the time required for high precision software computation. Due to the limited word size of the memory and pad limitation of the circuit, long numbers have to be transmitted serially in time O(n).

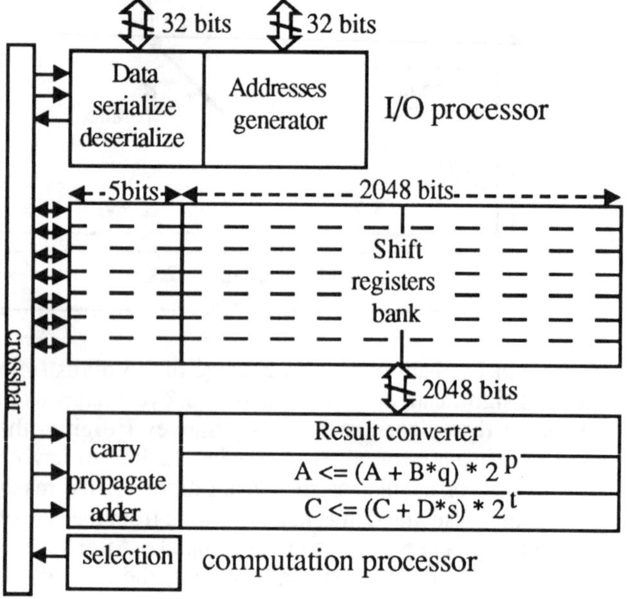

The goals of the architecture are:

1- To overlap computation and communication, so on line computation is desirable, but also pipelining of load in, execution and store back [13].

2- To minimize external bus occupancy by keeping intermediate results or local variables in an on-chip register bank

3- To benefit from a very large operator to avoid multiple precision operations. Internal redundant notation keeps the carry propagation path short [14,15]. The operator must take advantage of the parallel transmission of internal variables with the register bank and of the serial transmission of external variables.

4- To offer add, subtract, multiply, divide, GCD, extended GCD, square root [16,17] and also provide support for rational numbers [1] and matrices.

5- To efficiently use the silicon. So the datapath and the operator are configurable to perform several operations in parallel on short operands and/or on operands transmitted in parallel.

The I/O processor acts as a DMA. It spies on the buses for the address and the value of the first word of the operand, and then requests the bus and generate the addresses sequence for the next words [13]. Whenever available it uses the burst mode to speed up the exchange. It can interleave two loads and one store.

Each register of the bank is accessible in parallel or in serial mode. The most significant bit occupies the rightmost position. One parallel and six serial accesses can take place simultaneously. A crossbar allows to serially load or store registers while computing on others in the pipeline mode, or to bypass the bank in on line mode. The crossbar includes a serial adder. For the extended GCD, three of the registers are used as a stack to record the sequence of operations.

Two adder/subtractor-shifters, each with two registers for redundant numbers, compose the computation processor. For multiply, divide [14] or square root [16,], they are used together in the on line mode or independently in the parallel/serial mode to perform two different operations. On the figure, A,B,C,D represent registers for large numbers, $q,s \in \{-1,0,1\}$ and $r,t \in \{0,1,2\}$. Registers B and D can be loaded serially, from the left to the right position[14]. A pair of register converts on the fly serial signed binary result into conventional representation (-1 or 0 in the most significant position, 0 or 1 elsewhere) [18] and into a representation with no 0 (only -1 and 1). This late representation is useful for matrix triangularization by Givens' rotation method since it does not change the scale factor [19].

9. Conclusion

By simulation we have found that, in the GCD computation, algorithms that decimate the most significant positions are faster than the ones that decimate the least significant positions and are even faster for the extended GCD, while at the same time demanding a lower transistor count.

This has led to the architecture of a VLSI coprocessor for high precision arithmetic that support the GCD and the extended GCD on top of standard operations. The overall complexity is expected to be about 700 K transisors with a density of 5 Kt/mm^2 in 1.5μ and a very high regularity. It will be linked to a software package [20,21] which is now used to tune the circuit.

Acknowledgements

We are indebted to the referees specially for suggestions of new applications on rational arithmetic.
Many thanks to Dr. J.L. Roch for his introduction to computer algebra, to Dr. J.M. Muller for his introduction to on line operations, to S. Chebbi who carried out all the simulations with an equal temper and to R. Bouraoui who has designed a prototype chip for the extended GCD.

References

[1] *P. Kornerup & D. W. Matula* "Finite Precision Rational Arithmetic: An Arithmetic Unit" IEEE Trans. on Computers, Vol.C34-4 1983

[2] *D.H. Lehmer* "Euclid's Algorithm for Large Numbers" AMMMM 45, pp. 227,233, 1938

[3] *J. Davenport & Y. Robert* "VLSI and Computer Algebra: The GCD Example" Dynamical Systems and Cellular Automata Academic Press, London 1985

[4] *J. Stein* "Computing GCD without division"
   J. Comp. Phys. Vol. 1, pp 397-405 1967

[5] *D.W. Knuth* "Seminumerical Algorithms" (The Art of Computer Programming) pp.317-336,Addison-Wesley, Reading, Massachusetts 1981

[6] *R.P. Brent & H.T. Kung* "Systolic VLSI Array for linear-time GCD computation" proc. VLSI'83, Elsevier Science Publishing, North Holland-IFIP 1983

[7] *G. B. Purdy* "A carry-free algorithm for finding the greatest common divisor of two integers" Comp. & Maths. with Appls, Vol.9 pp311-316, 1983

[8] *R.P. Brent & H.T. Kung* "A systolic algorithm for integer GCD computation" Proc. 7th Symposium on Computer Arithmetic 1985

[9] *D.Y. Yun & C.H. Zhang* "A fast carry-free algorithm and hardware design for extended integer GCD computation" Proc. ACM Symposium on Symbolic Algebraic Computing pp 82-84 1986

[10] *V.C Hamacher & J. William* "A linear time divider array" Canadian Electr. Engineering Journal, Vol6, N° 4, 1981

[11] *M. Ercegovac & K. Trivedi* "On line algorithms for division and multiplication" IEEE Trans. on Computers, Vol.C26-7 pp 681-687 1977

[12] *R. Bouraoui, A. Guyot & J.L. Roch* " The best way to build a circuit for the extended GCD of large integers" IMAG research report 1990

[12] *R. Bouraoui, A. Guyot & J.L. Roch* " The best way to build a circuit for the extended GCD of large integers" IMAG research report 1990

[13] *M. Cosnard, A. Guyot, B. Hochet, J.M. Muller, H. Ouaouicha, Ph. Paul & E. Zysman* "The FELIN arithmetic coprocessor" Proc. 8th Symposium on Computer Arithmetic, 1987

[14] *A. Guyot, Y. Herreros & J.M. Muller* "JANUS, an on-line multiplier/divider for manipulating large numbers" Proc. 9th Symposium on Computer Arithmetic 1989

[15] *G. Metze & J.E. Robertson* "Elimination of carry propagation in digital computer" proc. IFIP conference, 1959 reprinted in IEEE Computer Arithmetic Vol 1, E. E. Swartzanlander ed.

[16] *V. Oklobdzija & M. Ercegovac* "An on-line square root algorithm" IEEE Trans. on Computers, Vol.C31-1 1982

[17] *A. Guyot & Y. Kusumaputri* "OCAPI: A circuit for on line radix 2 high precision arithmetic" submitted to ESSCIRC91, Milano, September 1991

[18] *M. Ercegovac & T. Lang* "On-the-fly conversion of Redundant into Conventional Representation" IEEE Trans. on Computers, Vol.C36-7 1987

[19] *M. Ercegovac & T. Lang* "Redundant and On-Line CORDIC: Matrix Triangularisation and SVD" IEEE Trans. on Computers, Vol.C39-6 1990

[20] *J.L. Roch* "L'architecture du Système PAC et son Arithmétique Rationnelle" Ph.D. diss., IMAG/INPG 1989

[21] *J.L. Roch* "The PAC System, general presentation" Proc. CAP90. E. Kaltofeln ed. North Holland. 1990

Table-Lookup Algorithms for Elementary Functions and Their Error Analysis

Ping Tak Peter Tang

Mathematics and Computer Science Division

Argonne National Laboratory

9700 South Cass Ave.

Argonne, IL 60439-4801

Abstract

Table-lookup algorithms for calculating elementary functions offer superior speed and accuracy when compared with more traditional algorithms. We show that, with careful design, it is feasible to implement table-lookup algorithms in hardware. Furthermore, we present a uniform approach to carry out a tight error analysis for such implementations.

1 Introduction

Since the adoption of IEEE Standard 754 for floating-point arithmetic [7], there has been a revival of interest in implementing elementary functions to near-perfect accuracy (see [12], [5], and [8], for example). A common thread of the recent work in this area is table-lookup algorithms. From a mathematical and algorithmic point of view, table-lookup algorithms are straightforward generalizations of more traditional algorithms such as those in [2] and [4]. According to W. Kahan, J. C. P. Miller had already pointed out in late 1950s that such approaches would be desirable as soon as memory became inexpensive. More recently, work by Gal ([3]) and the new runtime library by IBM ([1] and [6]) are notable in promoting table-lookup algorithms. Although these recent efforts use rather large tables and implement the functions in software, we illustrate here that, with careful design, the table sizes can be made so small that these table-lookup algorithms become easily realizable in hardware. Despite the size reduction, the speed and accuracy benefits brought about by table-lookup algorithms are preserved. Furthermore, we show that these table-lookup algorithms lend themselves to a uniform error analysis that yields tight error bounds. Typically, a theoretical bound of 0.57 unit in the last place (ulp) can be obtained for an implementation whose maximum error observed over several million arguments is 0.55 ulp.

The rest of the paper is organized as follows. Section 2 presents the general idea behind table-lookup algorithms. Section 3 illustrates the idea by three specific examples. Section 4 presents a uniform approach for a tight error analysis and an illustrative example. Section 5 discusses the feasibility of hardware implementations of table-lookup algorithms. Section 6 makes some concluding remarks on the advantages of table-lookup algorithms over CORDIC ([9]) and ordinary (without table-lookup) polynomial algorithms.

2 Table-Lookup Algorithms

Let f be the function to be implemented and I be the domain of interest. A typical table-lookup algorithm contains a set of "breakpoints" $c_k, k = 1, 2, \ldots, N$ in I and a table of N approximations T_k to $f(c_k)$. For any input argument $x \in I$, the algorithm calculates $f(x)$ in three steps.

Reduction: For this given x, the algorithm selects an appropriate breakpoint c_k. (In general, c_k is the breakpoint closest to x.) It then applies a "reduction transformation"

$$r = \mathcal{R}(x, c_k).$$

A typical case is $\mathcal{R}(x, c_k) = x - c_k$.

Approximation: The algorithm now calculates $f(r)$ using some approximation formula

$$p(r) \approx f(r).$$

Very often, p is a polynomial.

Reconstruction: Based on the reduction transformation $\mathcal{R}$, the values $f(c_k)$ and $f(r)$, and the nature of f, the algorithm calculates $f(x)$ by a reconstruction formula $\mathcal{S}$:

$$
\begin{aligned}
f(x) &= \mathcal{S}(f(c_k), f(r)) \\
&\approx \mathcal{S}(f(c_k), p(r)) \\
&\approx \mathcal{S}(T_k, p(r)).
\end{aligned}
$$

Although a traditional polynomial or rational-function-based algorithm (see [2] and [4]) can also be expressed in these three steps, there are two properties peculiar to a table-lookup algorithm. First, the reduction process in a table-lookup algorithm is much more flexible since, unlike a traditional algorithm, the choice of breakpoints is basically independent of the function f in question. In most situations, the breakpoints are chosen so that the reduced argument r can be computed efficiently on the particular machine in question. Second, in a table-lookup algorithm, the magnitude of the reduced argument r can be made as small as one wishes, limited only by the table size one can accommodate. We now illustrate these ideas by three realistic examples.

3   Algorithms for 2^x, $\log x$, and $\sin x$

The functions 2^x, $\log x$, and $\sin x$ are among the most commonly used elementary functions. We have purposely chosen different bases for the exponential and logarithm functions (base 2 and e, respectively) to illustrate the flexibility of table-lookup algorithms.

The algorithms here calculate the functions on "primary" domains. Transformations of arguments to such domains are standard and well known (see [2] and [4]).

3.1 2^x on [-1,1]

Reduction: Find the breakpoint $c_k = k/32, k = 0, 1, \ldots, 31$ such that

$$|x - (m + c_k)| \leq 1/64,$$

where $m = -1, 0,$ or 1. Calculate r by $r = [x - (m + c_k)] \cdot (\log 2)$. Note that $|r| \leq \log 2/64$.

Approximation: Approximate $e^r - 1$ by a polynomial $p(r)$,

$$p(r) = r + p_1 r^2 + \cdots + p_n r^{n+1}.$$

The coefficients $p_1, p_2, \ldots, p_n$ can be determined by applying the Remes algorithm [14] to the function $e^r - 1$ on the interval $[-\log 2/64, \log 2/64]$.

Reconstruction: Reconstruct 2^x by the relationships

$$
\begin{aligned}
2^x &= 2^{m+c_k} \cdot e^r \\
&= 2^m \{2^{c_k} + 2^{c_k}(e^r - 1)\} \\
&\approx 2^m \{2^{c_k} + 2^{c_k} \cdot p(r)\} \\
&\approx 2^m \{T_k + T_k \cdot p(r)\}.
\end{aligned}
$$

$T_k \approx 2^{c_k}$, $k = 0, 1, \ldots, 31$, are the tabulated values.

3.2 $\log x$ on [1,2]

Reduction: If $x < e^{1/16}$, compute $\log(x)$ by a simple polynomial approximation and exit. This polynomial is designed to approximate $\log(x)$ on $[1, e^{1/16}]$ (see [13] for details). Otherwise, find the breakpoint $c_k = 1 + k/64, k = 0, 1, \ldots, 64$ such that

$$|x - c_k| \leq 1/128.$$

Calculate r by $r = 2(x - c_k)/(x + c_k)$. Note that $|r| \leq 1/128$.

Approximation: Approximate $\log(x/c_k)$ by an odd polynomial $p(r)$:

$$p(r) = r + p_1 r^3 + p_2 r^5 + \cdots + p_n r^{2n+1}.$$

The coefficients $p_1, p_2, \ldots, p_n$ can be determined by applying the Remes algorithm to the function $\log([1 + r/2]/[1 - r/2])$ on the interval $[0, 1/128]$ using odd polynomials. Note that

$$
\begin{aligned}
\log\left(\frac{x}{c_k}\right) &= \log\left(\frac{1 + \frac{r}{2}}{1 - \frac{r}{2}}\right) \\
&= 2\left(\left(\frac{r}{2}\right) + \frac{1}{3}\left(\frac{r}{2}\right)^3 + \frac{1}{5}\left(\frac{r}{2}\right)^5 + \cdots\right).
\end{aligned}
$$

Reconstruction: Reconstruct $\log x$ by the relationships

$$
\begin{aligned}
\log(x) &= \log(c_k) + \log(x/c_k) \\
&\approx \log(c_k) + p(r) \\
&\approx T_k + p(r).
\end{aligned}
$$

$T_k \approx \log(c_k)$, $k = 0, 1, \ldots, 64$, are the tabulated values.

3.3 $\sin x$ on $[0, \pi/4]$

Reduction: If $|x| < 1/16$, calculate $\sin x$ by a simple polynomial approximation. Otherwise, find the breakpoint c_{jk} of the form

$$c_{jk} = 2^{-j}(1 + k/8) \qquad j = 1, 2, 3, 4; \quad k = 0, 1, \ldots, 7$$

that is closest to x. Calculate r by $r = x - c_{jk}$. Note that $|r| \leq 1/32$.

Approximation: Approximate $\sin r - r$ and $\cos r - 1$ by polynomials p and q, respectively:

$$
\begin{aligned}
p(r) &= p_1 r^3 + p_2 r^5 + \cdots + p_n r^{2n+1}, \\
q(r) &= q_1 r^2 + q_2 r^4 + \cdots + q_m r^{2m}.
\end{aligned}
$$

The coefficients $p_1, p_2, \ldots, p_n; q_1, q_2, \ldots, q_m$ can be determined by applying the Remes algorithm to the functions $\sin r - r$ and $\cos r - 1$ on the interval $[0, 1/32]$ using odd and even polynomials, respectively.

Reconstruction: Reconstruct $\sin(x)$ by the relationships

$$
\begin{aligned}
\sin(x) &= \sin(c_{jk} + r) \\
&= \sin(c_{jk})\cos r + \cos(c_{jk})\sin r \\
&\approx \sin(c_{jk}) + r\cos(c_{jk}) \\
&\quad + (\sin(c_{jk})q(r) + \cos(c_{jk})p(r)) \\
&\approx S_{jk} + rC_{jk} + (S_{jk}q(r) + C_{jk}p(r)).
\end{aligned}
$$

Table 1: A Realistic Set of Algorithms for IEEE Double Precision

Function	No. of Table Entries	No. of Coefficients
2^x	32	$n = 5$
$\log x$	64	$n = 2$
$\sin x$	64	$n = 3, m = 3$

To conclude this section, we tabulate in Table 1 the table size and coefficient requirements for a realistic set of algorithms tailored to IEEE double precision. Clearly, this is one particular choice out of the many possible combinations of breakpoints and approximating polynomials. Take 2^x, for example: one can reduce the table size to 16 by increasing n to 6 (hence the price of an add and multiply). Optimal choice is an engineering issue whose answer depends on the particular project in question.

4   Error Analysis

Recall the three steps of calculating f at x:

Reduction: $r = \mathcal{R}(x, c_k)$.

Approximation: $p(r) \approx f(r)$

Reconstruction: $f(x) = \mathcal{S}(f(c_k), f(r)) \approx \mathcal{S}(T_k, p(r))$

Because of inexact computations, we obtain $\hat{r}$ instead of r, $\hat{p}$ instead of p, and $\hat{\mathcal{S}}$ instead of $\mathcal{S}$. Hence, the computed result is

$$\hat{\mathcal{S}}(T_k, \hat{p}(\hat{r})).$$

The goal of the error analysis is to estimate accurately the difference

$$|\mathcal{S}(f(c_k), f(r)) - \hat{\mathcal{S}}(T_k, \hat{p}(\hat{r}))|.$$

We apply the triangular inequality. Thus,

$$
\begin{aligned}
&|\mathcal{S}(f(c_k), f(r)) - \hat{\mathcal{S}}(T_k, \hat{p}(\hat{r}))| \\
\leq\ & |\mathcal{S}(f(c_k), f(r)) - \mathcal{S}(f(c_k), f(\hat{r}))| + \\
& |\mathcal{S}(f(c_k), f(\hat{r})) - \mathcal{S}(f(c_k), p(\hat{r}))| + \\
& |\mathcal{S}(f(c_k), p(\hat{r})) - \hat{\mathcal{S}}(T_k, \hat{p}(\hat{r}))| \\
\leq\ & E_1 + E_2 + E_3.
\end{aligned}
$$

In most situations,

$$
\begin{aligned}
E_1 &\leq \text{constant} \cdot |f(r) - f(\hat{r})| \\
&\approx \text{constant} \cdot |f'(r)| \cdot |r - \hat{r}|
\end{aligned}
$$

and

$$E_2 \leq \text{constant} \cdot \max_t |f(t) - p(t)|.$$

E_1 can be easily estimated because the reduction process $\mathcal{R}$ is usually so simple that $|r - \hat{r}|$ can be estimated tightly. E_2 can also be easily estimated since the numerical value

$$\max_t |f(t) - p(t)|$$

is obtained when the polynomial is sought, usually by the use of the Remes algorithm. The maximum is taken over the domain of approximation.

Let us make a short digression for those interested in our implementation of the Remes algorithm. The version of Remes algorithm we use is the one-point exchange algorithm (see [10] for example) as opposed to a multiple-exchange algorithm ([14]). Moreover, the exchange scheme is based on the simplex algorithm which is a generalization of that based on sign alternations (see [11] for details). These specifics allow the algorithm to handle non-Haar systems as well as approximation of discontinuous functions. Moreover, the domain does not need to be discretized, because

$$E_p = \max_t |f(t) - p(t)|$$

is calculated by searching the zeros of $f'(t) - p'(t)$ by numerical root finder. Consequently, the numerical value of E_p we obtain is accurate to many digits. This accuracy is definitely sufficient for the purpose of error analysis, because most of the time we need only one or two digits of E_p (see below).

We return to the discussion on errors. The rounding error

$$E_3 = |\mathcal{S}(f(c_k), p(\hat{r})) - \hat{\mathcal{S}}(T_k, \hat{p}(\hat{r}))|$$

in calculating the polynomial and reconstruction is usually the most difficult to estimate tightly. With the use of table-lookup algorithms, however, the analysis is greatly simplified. The reason is that the magnitude of the reduced argument r (or $\hat{r}$) is typically so small that rounding errors associated with r^k, $k \geq 2$, are practically zero. The simplicity of the analysis in [12] and [13] illustrates the situation.

To illustrate the ideas here, we carry out the analysis of 2^x for a typical IEEE double-precision implementation. Let ϵ denote 1 ulp of 1, i.e., $\epsilon = 2^{-52}$. Clearly, the subtraction $s := x - (m + c_k)$ is exact. Hence the errors in the reduction step are those caused by the multiplication by $\log 2$ and the fact that the value "$\log 2$" used is only a 53-bit approximation:

$$
\begin{aligned}
r &= s \log 2, \quad \text{and} \\
\hat{r} &= s(\log 2 + \delta_1) + \delta_2,
\end{aligned}
$$

where δ_1 is the error in the approximation to $\log 2$, and δ_2 is the error caused by the finite-precision product. Now,

$$\log 2 = .69314718\ldots \in 2^{-1} \cdot [1, 2).$$

Therefore, rounding (as opposed to truncating) $\log 2$ to working precision (53 bits in our case) gives an error no bigger than $\frac{1}{2}\frac{1}{2}\epsilon = 2^{-2}\epsilon$. Hence $|\delta_1| \leq 2^{-2}\epsilon$. Next,

$$|s \cdot (\log 2 + \delta_1)| \leq \log 2/64 + \delta_1/64.$$

Since $\log 2/64 + \delta_1/64 \in 2^{-7} \cdot [1, 2)$, rounding the product $s(\log 2/64 + \delta_1)$ to working precision gives an error no bigger than $2^{-8}\epsilon$. Hence, $|\delta_2| \leq 2^{-8}\epsilon$. Therefore,

$$|r - \hat{r}| \leq |s\delta_1| + |\delta_2| \leq 2^{-7}\epsilon.$$

Hence,

$$|E_1| \leq |\frac{d}{dt}e^t||r - \hat{r}| \leq 1.02 \times 2^{-7}\epsilon.$$

Next, the best approximating polynomial p obtained by a Remes algorithm gives

$$|(e^t - 1) - p(t)| \leq 2^{-63} = 2^{-11}\epsilon, \quad |t| \leq \log 2/64.$$

Thus,

$$|E_2| \leq 2^{-11}\epsilon.$$

Finally, we estimate the errors in computing $p(\hat{r})$ and the final reconstruction. Since 2^{c_k} is not representable in 53 bits, we use 2 variables T_1 and T_2 where T_1 is 2^{c_k} rounded to 53 bits and T_2 is a correction term that makes $T_1 + T_2 = 2^{c_k}$ for all practical purposes. (Note that a T_2 having 6 significant bits will be sufficient.) The reconstruction is

$$2^m(T_1 + (T_1 * p + T_2)).$$

Scaling by 2^m is exact. The last add contributes no more than 0.5 ulp of error. Estimating the error in $T_1 * p + T_2$ is simple: Since $|\hat{r}^2| < 2^{-12}$, the only significant error in calculating p is the last add. Thus the computed result is

$$(2^{c_k} + \delta_1)(p + \delta_2) + T_2 + \delta_3$$
$$= 2^{c_k}p + T_2 + \delta_1 p + \delta_2 2^{c_k} + \delta_3,$$

where $|p| \leq 2^{-6}$, $|\delta_1| \leq 2^{-1}\epsilon$, $|\delta_2| \leq 2^{-7}\epsilon$, and $|\delta_3| \leq 2^{-6}\epsilon$.

The rounding error is bounded by

$$
\begin{aligned}
|E_3| &\leq \frac{1}{2}\,\text{ulp} + (2^{-7} + 2^{-6} + 2^{-6})\epsilon \cdot 2^m \\
&\leq \frac{1}{2}\,\text{ulp} + (2^{-7} + 2^{-5})\,\text{ulp} \\
&\leq 0.54\,\text{ulp}.
\end{aligned}
$$

Consequently,

$$|E_1| + |E_2| + |E_3| \leq 0.556\,\text{ulp}.$$

Note that in the case $c_k = 0$ (in particular, $|x| \leq 1/64$), the bound given here is pesimistic because no error will be committed in $T_1 * p + T_2$, for $T_1 = 1$ and $T_2 = 0$.

5 Feasibility of Hardware Implementations

Table-lookup algorithms have been implemented successfully in software ([6] and [12], for example). With today's VLSI technology and the algorithms' flexibility, hardware implementations also seem extremely feasible. In what follows, we reexamine the algorithm for 2^x in Section 3.1 from the point of view of hardware implementation.

Table Size: The table required here has 32 double-precision values. Assuming 8 bytes of storage per entry (although one can device more compact storage schemes), this table is only 2 kbits. Such a table is easily accommodated by today's standard.

Reduction and Table Lookup: Consider the reduction for $|x| \geq 1/16$ where x is an IEEE double-precision number. Thus

$$x = (-1)^s \cdot 2^k \cdot (1.b_1 b_2 \ldots b_{52})_{\text{base2}},$$

where $s \in \{0, 1\}, k \in \{-1, -2\}$. Now, mathematically,

$$m + c_k = \text{round-to-integer}(32x)/32,$$

and both m and c_k are determined by the seven bits of information s, $\text{lsb}(k)$, and $\{b_1, b_2, \ldots, b_5\}$. Clearly, simple multiplexers can be constructed to deliver both $m + c_k$ and T_k.

Approximation and Reconstruction: Most hardware performs IEEE sums and products via an internal data format that has a longer mantissa and a wider exponent field than the working precision in question (say double precision or double-extended precision) in question. The longer mantissa typically contains the guard, round, and sticky bits that ensure correct IEEE rounding; and the internal exponent field typically contains two extra bits designed to accommodate products of working-precision values of extreme magnitudes, say, two smallest denormalized numbers. Rounding and exception handling are usually performed separately during conversion of internal formats back to the working precision.

It is not hard to see that the approximation and reconstruction steps in Section 3.1 can be carried out solely in the internal format. First, provided the internal exponent field has two extra bits, the recurrence

$$
\begin{aligned}
p(r) &= r + p_1 r^2 + \cdots + p_5 r^6 \\
&= r + r \cdot r \cdot (p_1 + r \cdot (p_2 + \cdots + r \cdot p_5) \cdots)
\end{aligned}
$$

and the reconstruction

$$2^m(T_k + T_k \cdot p(r))$$

where $|r| \leq \log 2/64$ will not overflow or underflow. (Note that the coefficients $p_j \approx 1/j!$ are moderate in magnitude.) Second, note that there is no need to round any of the intermediate results to working precision. In fact, keeping the extra guard and round bits would enhance the accuracy. Performing these two steps without the need of rounding or error checking will clearly save a good amount of time compared to full IEEE operations everywhere.

6 Concluding Remarks

Table-lookup algorithms offer several advantages over traditional polynomial/rational-funtion algorithms and CORDIC algorithms. In comparison, a table-lookup algorithm is generally

1. faster because it requires less work in the approximation steps,

2. more accurate because the rounding error made in the approximation step is tiny, and

3. amenable to tight error analysis.

CORDIC algorithms have been attractive from a hardware point of view because they require only shifts and adds (see [9] for example). The price one has to pay for this simplicity is a relatively large number of iterations of pseudo-division (reduction) and pseudo-multiplication (reconstruction). Advances in VLSI technology have now made it possible to realize basic operations such as primitive floating-point adds and multiplies (that is, without rounding or exception handling) on the order of a few clocks on modern hardware. In such hardware environments, table-lookup algorithms are extremely feasible alternatives to CORDIC.

Acknowledgement

This work was supported by the Applied Mathematical Sciences subprogram of the Office of Energy Research, U. S. Department of Energy, under Contract W-31-109-Eng-38.

References

[1] R. C. Agarwal, J. W. Cooley, F. G. Gustavson, J. B. Shearer, G. Slishman, and B. Tuckerman, New scalar and vector elementary functions for the IBM System/370, *IBM Journal of Research and Development*, 30, no. 2, March 1986, pp. 126–144.

[2] W. Cody and W. Waite, *Software Manual for the Elementary Functions*, Prentice-Hall, Englewood Cliffs, N.J., 1980.

[3] S. Gal, Computing elementary functions: A new approach for achieving high accuracy and good performance, in *Accurate Scientific Computations*, Lecture Notes in Computer Science, Vol. 235, Springer, New York, 1985, pp. 1–16.

[4] J. F. Hart et al., *Computer Approximations*, John Wiley and Sons, New York, 1968.

[5] D. Hough, Elementary functions based upon IEEE arithmetic, *Mini/Micro West Conference Record*, Electronic Conventions, Inc., Los Angeles, 1983.

[6] IBM Elementary Math Library, *Programming RPQ P81005, Program number 5799-BTB, Program Reference and Operations Manual*, SH20-2230-1, August 1984.

[7] IEEE standard for binary floating-point arithmetic, *ANSI/IEEE Standard 754-1985*, Institute of Electrical and Electronic Engineers, New York, 1985.

[8] P. W. Markstein, Computation of elementary functions on the IBM RISC System/6000 processor, *IBM Journal of Research and Development*, 34, no. 1, January 1990, pp. 111–119.

[9] R. Nave, Implementation of transcendental functions on a numeric processor, *Microprocessing and Microprogramming*, 11, 1983, pp. 221–225.

[10] M. J. D. Powell, *Approximation Theory and Methods*, Cambridge University Press, Cambridge, 1981.

[11] P. T. P. Tang, A fast algorithm for linear complex Chebyshev approximations, *Mathematics of Computation*, 51, no. 184, October 1988, pp. 721–739.

[12] P. T. P. Tang, Table-driven implementation of the exponential function in IEEE floating-point arithmetic, *ACM Transactions on Mathematical Software*, 15, no. 2, June 1989, pp. 144–157.

[13] P. T. P Tang, Table-driven implementation of the logarithm function in IEEE floating-point arithmetic, *ACM Transactions on Mathematical Software*, 16, no. 2, December 1990, pp. 378-400.

[14] L. Veidinger, On the numerical determination of the best approximation in the Chebyshev sense, *Numerische Mathematik*, 2, 1960, pp. 99–105.

Accurate and Monotone Approximations of Some Transcendental Functions

Warren E. Ferguson, Jr.

Tom Brightman

Cyrix Corporation, P.O. Box 850118, Richardson, Texas 75081-0118

Abstract

A technique for computing monotonicity preserving approximations $F_a(x)$ of a function $F(x)$ is presented. This technique involves computing an extra precise approximation of $F(x)$ that is rounded to produce the value of $F_a(x)$. For example, only a few extra bits of precision are used to make the accurate transcendental functions found on the $Cyrix^{TM}$ $FasMath^{TM}$ line of 80387 compatible math coprocessors monotonic.

1   Introduction

The IEEE Standard for Binary Floating-Point Arithmetic [1] is a landmark in the field of computer arithmetic. This standard establishes the formats used to store floating-point numbers and defines the results that must be returned by the fundamental mathematical operations of addition, subtraction, multiplication, division, square root, and remainder.

This standard, however, omits any discussion of what should be returned as the value of transcendental functions like the sine or the logarithm. One reason for this omission is that no simple algorithms that yield correctly the rounded value of these functions are known. It is therefore natural to focus on easily computable approximations of these functions that preserve their important properties.

This paper shows that monotonic approximations of the sine, cosine, tangent, arctangent, exponential, and logarithm, on restricted domains, can be derived by computing extra precise approximations. For example, an approximation of the sine function on $[-\pi/4, \pi/4]$ that is accurate to 66 bits of precision will necessarily be monotonic when rounded to 64 bits of precision. We have used this technique to establish the monotonicity of a Cyrix FasMath coprocessor's polynomial based approximations of transcendental functions [2]. Since this technique does not depend on how the approximation is determined, then it also can be applied to approximations derived by other means, e.g., CORDIC based approximations.

2   Machine Numbers

The IEEE Standard [1] defines four floating-point number formats: single basic, single extended, double basic, and double extended. Within a given format the representable numbers can be described as

$$x = (-1)^s\, 2^E\, (b_0 \bullet b_1 b_2 \cdots b_{p-1})$$

where $s = 0$ or 1, $E_{min} \leq E \leq E_{max}$, $b_i = 0$ or 1.

We consider exclusively the situation where arguments, and the approximate values of transcendental functions, are representable as double extended format numbers (machine numbers). Math coprocessors compatible with the 80387 math coprocessor use a double extended format with $E_{max} = +16383$, $E_{min} = -16382$, $p = 64$.

Of particular interest are the following two classes of *machine numbers*: *denormalized numbers* (including zero) and *normalized numbers*. Consecutive nonnegative denormalized numbers $m < \bar{m}$ have the form

$$m = (I)\, 2^{-16445} \quad \text{and} \quad \bar{m} = m + 2^{-16445}$$

where I is an integer that satisfies $0 \leq I \leq 2^{63}$; note that $-16445 = -16382 - 63$.

Normalized numbers can be classified by the *binade* to which they belong; in the binade E consecutive positive normalized numbers $m < \overline{m}$ have the form

$$m = (1+f)2^E \quad \text{and} \quad \overline{m} = m + 2^{E-63}$$

where $2^{63}f$ and E are integers satisfying $0 \le 2^{63}f < 2^{63}$ and $-16382 \le E \le +16383$.

In the following discussion the symbols $\underline{m}$, m, and $\overline{m}$ will always be used to denote consecutive machine numbers ordered so that $\underline{m} < m < \overline{m}$.

3 Transcendental Functions

Cyrix FasMath coprocessors use polynomial based methods [3,4,5] to approximate the five basic transcendental functions: $2^x - 1$ on $[-1/2, 1/2]$, $\text{Log}_2(1+x)$ on $[1/\sqrt{2}-1, \sqrt{2}-1]$, $\text{Sin}(x)$ and $\text{Tan}(x)$ on $[-\pi/4, \pi/4]$, and $\text{ATan}(x)$ on $[0,1]$.

Argument reduction techniques are used to expand the domain of these basic transcendental functions. For example, FasMaths can return the approximate value of $2^x - 1$ for arguments in $[-1,1]$. Given such an argument it first uses a polynomial based method to determine the approximate value of $2^{x/2} - 1$ and then recovers the approximate value of $2^x - 1$ via the identity

$$2^x - 1 = (2^{x/2} - 1)(\{2^{x/2} - 1\} + 2) \ .$$

FasMaths can determine the approximate value of $\text{Log}_2(x)$ for all positive values of x. To do so it first determines the value of p for which $2^{-p}x$ belongs to $[1/\sqrt{2}, \sqrt{2}]$. Using this value of p polynomial based methods are used to determine the approximate value of $\text{Log}_2(1 + \{2^{-p}x - 1\}) = \text{Log}_2(2^{-p}x)$. Finally the identity

$$\text{Log}_2(x) = p + \text{Log}_2(2^{-p}x)$$

is used to recover the approximate value of $\text{Log}_2(x)$.

FasMaths can determine the approximate value of $\text{Sin}(x)$, $\text{Cos}(x)$, and $\text{Tan}(x)$ for arguments x satisfying $|x| \le 2^{63}$. To do so they first use the symmetric partial remainder instruction to determine the last three bits of the quotient q and the exact remainder r, with $|r| \le \pi/4$, when x is divided by an approximation to $\pi/2$ with 68 bits of precision. Once the approximate value of $\text{Sin}(r)$ or

$\text{Tan}(r)$ is determined by polynomial based methods standard trigonometric identities are used to recover the values of $\text{Sin}(x)$, $\text{Cos}(x)$, and $\text{Tan}(x)$. These trigonometric identities require the last three bits of q to determine the appropriate transformation. FasMaths also use the identity

$$\text{Cos}(r) = \sqrt{1 - \text{Sin}^2(r)}$$

to recover the approximate value of $\text{Cos}(r)$ for arguments $|r| \le \pi/4$ from the approximate value of $\text{Sin}(r)$. A FasMath's approximation of $\text{Sin}(r)$, $\text{Cos}(r)$, and $\text{Tan}(r)$ preserve the identities

$$\begin{aligned} \text{Sin}(x) &= -\text{Sin}(-x), \\ \text{Cos}(x) &= \text{Cos}(-x), \text{ and} \\ \text{Tan}(x) &= -\text{Tan}(-x). \end{aligned}$$

The use of the symmetric partial remainder instruction to reduce input arguments also means that FasMaths return rounded-period approximations of trigonometric functions.

Finally, FasMaths can determine the approximate value of $\text{ATan}(x)$ for all values of x. To do so it uses the identities

$$\begin{aligned} \text{ATan}(x) &= -\text{ATan}(-x), \text{ and} \\ \text{ATan}(x) &= \pi/2 - \text{ATan}(1/x) \end{aligned}$$

to reduce a general argument to one between 0 and 1. Once again, an approximate value of $\pi/2$ with 68 bits of precision is used when the last identity is applied.

4 Measures of Errors

Two of the most frequently cited measures of error associated with an approximation F_a of a nonzero number F are the error *err* and relative error *relerr* defined by

$$\text{err}(F_a) \equiv F_a - F \ , \text{ and } \quad \text{relerr}(F_a) \equiv \frac{F_a - F}{F} \ .$$

Note that $F_a = (1 + \text{relerr}(F_a))\, F$.

We say that F_a is a *p-bit approximation* of F if the error amounts to no more than one-half of one bit in the p-th significant bit of the normalized binary representation of F.

To understand the implication of this definition let us suppose that F is a nonnegative normalized number in the binade E,

$$F = (1 + f)\, 2^E \quad \text{where} \quad 0 \le f < 1.$$

If

$$F_a = (1 + f + \epsilon)\, 2^E, \text{ with}$$

$$|\epsilon| \le \tfrac{1}{2}\, 2^{-P+1} = 2^{-P},$$

then by definition F_a is a p-bit approximation of F. Note that the error in F_a is represented as an error ϵ in its *significand* f. If F_a is a p-bit approximation of F, then it follows that

$$|\,\mathrm{err}(F_a)\,| = |\,\epsilon\,|\, 2^E, \text{ and}$$

$$|\,\mathrm{relerr}(F_a)\,| = \frac{|\,\epsilon\,|}{1 + f} \le 2^{-P}.$$

Since $1 \le 1 + f < 2$, then we infer that p-bit approximations that are not (p+1)-bit approximations have relative errors whose magnitudes lie between 2^{-P-1} and 2^{-P}; the larger the significand the smaller the associated relative error. We summarize these observations as follows:

Fact: Let $F = \pm(1 + f)\, 2^E$ belong to the binade E. If the magnitude of the relative error $\mathrm{relerr}(F_a)$ in an approximation F_a of F is no larger than $2^{-P}/(1 + f)$, then F_a is a p-bit approximation of F.

This variation by a factor of 2 in the upper bound on the magnitude of the relative error of p-bit approximations, called precision wobble [5], is one reason why base 2 arithmetic is preferred over base 16 arithmetic. Base 16 (hexadecimal) arithmetic has a precision wobble 8 times larger than that of base 2 (binary) arithmetic.

5 Accurate Approximations

Consider an approximation $F_a(x)$ of a function $F(x)$. If $F(x)$ belongs to the binade $E(x)$, then we can represent $F(x)$ as

$$F(x) = \pm\{1 + f(x)\}\, 2^{E(x)},$$

where $1 \le 1 + f(x) < 2$. If we express the error in the approximation of $F(x)$ by $F_a(x)$ as an error $\epsilon(x)$ associated with the significand of $F(x)$,

$$F_a(x) = \pm\{1 + f(x) + \epsilon(x)\}\, 2^{E(x)},$$

then we find that

$$|\,\mathrm{relerr}(F_a(x))\,| = \frac{|\,\epsilon(x)\,|}{1 + f(x)}.$$

If $F_a(x)$ is a p-bit approximation of $F(x)$, then $|\,\epsilon(x)\,| \le 2^{-P}$ and so

$$|\,\mathrm{relerr}(F_a(x))\,| \le \frac{2^{-P}}{1 + f(x)}.$$

Note that the upper bound on the magnitude of the relative error of p-bit approximations wobbles not when the argument x crosses a binade's boundary, but rather when the value $F(x)$ crosses a binade's boundary.

For each of the five basic transcendental functions, the upper-most dashed lines in Figures 2 depict the upper bound on the relative error that must be satisfied by any 64-bit precise approximation.

6 Monotone Approximations

Informally, the monotonic behavior of an approximation can be established by making use of three facts: the monotonic behavior exhibited by the function, the discrete nature of the machine numbers, and the accuracy of the approximation.

For example, consider the sine function on $[0, \pi/4]$. On this domain $\mathrm{Sin}(x)$ is a monotone increasing function whose slope is never less than $\mathrm{Cos}(\pi/4) \simeq 0.7$. If $m < \bar{m}$ are two consecutive machine numbers that belong to the binade E, then $\mathrm{Sin}(m)$ and $\mathrm{Sin}(\bar{m})$ will differ by at least $0.7*(\bar{m} - m) \simeq (0.7)\, 2^{E-63}$. If the error in the approximation to $\mathrm{Sin}(x)$ is less than half of this difference throughout the binade E, then the approximation will also behave monotonically.

This monotonicity argument implies that the approximation is determined using a higher precision arithmetic than that required in the final result. This is relatively easy to achieve for FasMaths because they implement the full IEEE 754 Standard, and such an implementation requires that the five basic operations of addition, subtraction, multiplication, division and square root be computed internally to a higher precision. A mathematically precise version of this monotonicity argument is presented in the following theorem.

Theorem: Let $F(x)$ be a monotonic function defined on the interval $[a,b]$. Let $F_a(x)$ be an approximation of $F(x)$ whose associated relative error admits the bound

$$| \text{relerr}(F_a(x)) | \leq R$$

for some constant $R < 1$. If for every pair $m < \bar{m}$ of consecutive machine numbers in $[a,b]$

$$R < R(m,\bar{m}) \equiv \frac{| F(\bar{m}) - F(m) |}{| F(\bar{m}) | + | F(m) |},$$

then $F_a(x)$ exhibits on the set of machine numbers in $[a,b]$ the same monotonic behavior exhibited by $F(x)$ on $[a,b]$.

Proof: Let $m < \bar{m}$ be any pair of consecutive machine numbers in $[a,b]$. From the assumptions we find that

$$| F_a(m) - F(m) | + | F_a(\bar{m}) - F(\bar{m}) |$$

$$= | \text{relerr}(F_a(m))F(m) | + | \text{relerr}(F_a(\bar{m})) F(\bar{m}) |$$

$$= R \{ | F(m) | + | F(\bar{m}) | \}$$

$$\leq R(m,\bar{m}) \{ | F(m) | + | F(\bar{m}) | \}$$

$$\leq | F(\bar{m}) - F(m) |.$$

As illustrated in Figure 1, this inequality shows that the error bar centered at $F(m)$ with half-width $R\,|\,F(m)\,|$ containing $F_a(m)$, and the error bar centered at $F(\bar{m})$ with half-width $R\,|\,F(\bar{m})\,|$ containing $F_a(\bar{m})$, cannot overlap. Therefore the function F on $[a,b]$, and the approximation F_a on the machine numbers in $[a,b]$, must exhibit the same monotonic behavior. $\square$

Of course the user never sees the value of the underlying higher precision approximation, but only its value truncated to double extended format. To insure that the value presented to the user also behaves monotonically it is important that this truncation to double extended format preserves monotonicity. It is simple to verify that if the truncation trunc() used is consistently truncation by rounding or truncation by chopping, then monotonicity is preserved because $x < y$ implies that $\text{trunc}(x) \leq \text{trunc}(y)$. Therefore when the assumptions of the previous theorem hold, then the truncated approximation is weakly monotonic whenever the function is strictly monotonic.

Before applying this theorem to the five basic transcendental functions it is convenient to simplify the expression for $R(m,\bar{m})$ as follows. First note that each of these transcendental functions is monotone increasing with a single zero at a machine number m_0 in the domain. The monotonic behavior of the approximation at this zero is ensured by the fact that $R < 1$, for this fact implies that the approximation is zero only when the function is zero and that the approximation has the same sign as the function. Since m_0 is a machine number, then the function has the same sign at consecutive machine numbers $m < \bar{m}$. Therefore, the denominator in the expression for $R(m,\bar{m})$ can be simplified to

$$| F(\bar{m}) | + | F(m) | = | F(\bar{m}) + F(m) |.$$

Note that

$$F(\bar{m}) + F(m) =$$
$$= \begin{cases} 2F(m) \{1 + \dfrac{F(\bar{m}) - F(m)}{2\,F(m)}\}; & F(\bar{m}) > F(m) > 0 \\[2ex] 2F(\bar{m}) \{1 + \dfrac{F(m) - F(\bar{m})}{2\,F(\bar{m})}\}; & F(m) < F(\bar{m}) < 0 \end{cases}$$

where the fraction inside the curly braces is a positive quantity. Using these facts the expression for $R(m,\bar{m})$ can be simplified to yield

$$R(m,\bar{m}) = \left| \frac{F(\bar{m}) - F(m)}{F(\bar{m}) + F(m)} \right| = \frac{S(m,\bar{m})}{1 + S(m,\bar{m})}$$

where

$$S(m,\bar{m}) \equiv \frac{| F(\bar{m}) - F(m)|}{2 \min\{ | F(m) |, | F(\bar{m}) | \}}.$$

If we interpret $S(m,\bar{m})$ as ∞ when either m or $\bar{m}$ is equal to m_0, and $R(m,\bar{m})$ as 1 when $S(m,\bar{m})$ is ∞, then this simplified expression for $R(m,\bar{m})$ remains valid when either m or $\bar{m}$ is equal to m_0.

In Table 1 we present the expression for S for each of the five basic transcendental functions. Note that the expressions associated with $2^x - 1$ and $\text{Log}_2(1 + x)$ have been split into two cases, one case for consecutive negative machine numbers $\underline{m} < m < 0$ and another case for consecutive positive machine numbers $0 < m < \bar{m}$. These expressions for S have been written so that they clearly show their dependence on the average and difference of the consecutive machine numbers. Table 2 displays highly accurate approximations of S that were obtained from the expressions in Table 1.

These approximations of S were obtained by replacing functions depending on the difference of two consecutive machine numbers by the first term of the relevant MacLaurin expansion. The high accuracy of these approximations is a result of the fact that the difference between consecutive machine numbers is never greater than 2^{-63} on the domains under consideration.

For each of the five basic transcendental functions, the dotted line of Figures 2 depict upper bounds on the relative error that must be satisfied by any monotonic approximation whose arguments are obtained from double extended format machine numbers. These curves were obtained from the relationship between R and S and the expressions for S displayed in Table 2.

Note that the upper bounds on the relative error of approximations wobble when the value of the function crosses a binade boundary, while the upper bounds on the relative error of monotonic approximations wobble when the argument of the function crosses a binade boundary.

The bottom-most dash-dotted curves Figure 2 depict rigorous bounds on the relative error in the approximations used by FasMaths. In all cases the upper bound on this error lies below the bounds required of accurate and monotonic double extended format approximations. In particular, note that the relative errors lie below the curve characterizing approximations that have full double extended format (64-bit) accuracy. This observation demonstrates mathematically that the approximate values of the basic transcendental functions computed by a FasMath are either the correctly rounded value or one of its two immediate double extended format neighbors.

The appendix illustrates the type of analysis used to derive the worst case bound on the relative error of monotonic approximations to Sin(x).

7 Conclusion

The techniques presented in this paper has been used to prove mathematically that the approximations delivered by FasMath coprocessors posses the desirable properties of accuracy and monotonicity [2]. While FasMaths use polynomial based approximations, the techniques presented in this paper are generally applicable to any method for which one can compute upper bounds on the relative error of the approximation. In particular, it could be used to justify the monotonic behavior of approximations derived by CORDIC based methods.

One interesting application of the relative error curves depicted in Figures 2 can be described as follows. The gap between the total error curve and the bound on accurate approximations suggests that the larger the gap, the more frequently a FasMath should return a correctly rounded answer, see also [6]. By using Brent's MP package, owners of FasMaths can verify this claim. Remember, however, that by the nature of its derivation the curves depicting the relative error in a FasMath's approximations are usually a significant overestimate of the actual relative error.

Accuracy and monotonicity results for transcendental functions can also be found in [7]. Recently Dr. W. H. Kahan [8] has informed us that he has presented at UC Berkeley techniques similar to those presented in this paper.

Appendix

Since R() is an increasing function of S(), then R() achieves its least value when S() achieves its least value. For the sine function the bound on the relative error of monotonic approximations is given as

$$S(m,\bar{m}) \;=\; \frac{\bar{m}-m}{2\,\mathrm{Tan}(m)}\,.$$

Consider the case when $m < \bar{m}$ are consecutive machine numbers in the binade E. The numerator of S is constant at value 2^{E-63} while the denominator of S is an increasing function of m. Therefore within the binade E the least value of S is assumed at its right endpoint.

The relevant domain for the sine function is $[0,\pi/4]$. The machine numbers in this domain consist of the whole of the binades with $-16382 \le E \le -2$ and that portion of the binade with $E = -1$ consisting of machine numbers less than $\pi/4$. We therefore conclude that

$$\min S(m,\bar{m}) =$$

$$= 2^{-65} \min\left\{ \min_{\substack{x = 2^{E+1} \\ -16382 \le E \le -2}} \left(\frac{x}{\mathrm{Tan}(x)} \right), \frac{1}{\mathrm{Tan}(\pi/4)} \right\}.$$

Since the function x/Tan(x) is a decreasing function of x for $x \leq 1/2$, then the least value of x/Tan(x) is achieved at the largest value of x. Consequently

$$\min S(m,\overline{m}) = 2^{-65} \min \left\{ \frac{1/2}{Tan(1/2)} , \frac{1}{Tan(\pi/4)} \right\}$$
$$= \frac{2^{-66}}{Tan(1/2)} \simeq (1.83)\, 2^{-66}$$

and so to a high degree of approximation

$$\min R(m,\overline{m}) \simeq \min S(m,\overline{m}) \simeq (1.83)\, 2^{-66}.$$

It is interesting to note that the informal argument would have predicted that the most stringent bound on $R(m,\overline{m})$ would occur when $m \simeq \pi/4$ rather than when $m = 1/2$.

References

[1] "IEEE Standard for Binary Floating-Point Arithmetic", ANSI/IEEE Std 754-1985, The Institute of Electrical and Electronics Engineers, Inc., New York, New York, 10017.

[2] *FasMath 83D87 Accuracy Report*, Cyrix Corporation, Richardson, Texas, May 1989.

[3] *Computer Approximations*, John F. Hart et. al., Robert E. Krieger Publishing Company, Malabar, Florida, 1978.

[4] *Software Manual for the Elementary Functions*, William J. Cody, Jr. and William Waite, Prentice-Hall, Inc., Englewood Cliffs, New Jersey, 1980.

[5] *Computer Evaluation of Mathematical Functions*, C. T. Fike, Prentice-Hall, Inc., Englewood Cliffs, New Jersey, 1968.

[6] "Evaluating Elementary Functions in a Numerical Coprocessor Based on Rational Approximations", Israel Koren and Ofra Zinaty, IEEE Trans. Computers, Vol. 39, No. 8, August 1990, pp. 1030-1037.

[7] "High Accuracy Standard Functions for Real and Complex Intervals", K. Braune and W. Krämer, in *Computerarithmetic: Scientific Computation and Programming Languages*, E. Kaucher, U. Kulisch, Ch. Ullrich (ed.), Teubner, Stuttgart, pp. 81-114, 1987.

[8] Personal communication, W. H. Kahan, February, 1990.

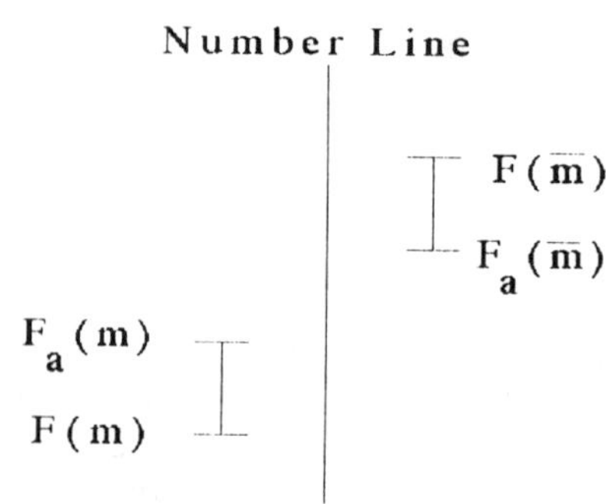

Figure 1: Relationship between F and F_a.

Function	Interval	S
$\text{Sin}(x)$	$[0, \pi/4]$	$\text{Cos}(\frac{m+\bar{m}}{2}) \dfrac{\text{Sin}(\frac{\bar{m}-m}{2})}{\text{Sin}(m)}$
$\text{Tan}(x)$	$[0, \pi/4]$	$\dfrac{1}{\text{Cos}(\bar{m})} \dfrac{\text{Sin}(\bar{m}-m)}{2\,\text{Sin}(m)}$
$\text{ATan}(x)$	$[0, 1]$	$\dfrac{\text{ATan}(\frac{\bar{m}-m}{1+m\bar{m}})}{2\,\text{ATan}(m)}$
$2^x - 1$	$[-1, 0]$	$2^{-\frac{m-\underline{m}}{2}} \dfrac{\text{Sinh}(\frac{m-\underline{m}}{2}\,\text{Log}_e(2))}{2^{-m}-1}$
$2^x - 1$	$[0, 1]$	$2^{\frac{\bar{m}-m}{2}} \dfrac{\text{Sinh}(\frac{\bar{m}-m}{2}\,\text{Log}_e(2))}{1-2^{-m}}$
$\text{Log}_2(1+x)$	$[\frac{1}{\sqrt{2}}-1, 0]$	$\dfrac{\text{Log}_e(1-\frac{m-\underline{m}}{1+m})}{2\,\text{Log}_e(1+m)}$
$\text{Log}_2(1+x)$	$[0, \sqrt{2}-1]$	$\dfrac{\text{Log}_e(1+\frac{\bar{m}-m}{1+m})}{2\,\text{Log}_e(1+m)}$

Table 1: Expressions for S.

Function	Interval	S
$\text{Sin}(x)$	$[0, \pi/4]$	$\dfrac{\bar{m}-m}{2\,\text{Tan}(m)}$
$\text{Tan}(x)$	$[0, \pi/4]$	$\dfrac{\bar{m}-m}{\text{Sin}(2m)}$
$\text{ATan}(x)$	$[0, 1]$	$\dfrac{\bar{m}-m}{2\,(1+m^2)\,\text{ATan}(m)}$
$2^x - 1$	$[-1, 0]$	$\dfrac{m-\underline{m}}{2\,\text{Log}_2(e)\,(2^{-m}-1)}$
$2^x - 1$	$[0, 1]$	$\dfrac{\bar{m}-m}{2\,\text{Log}_2(e)\,(1-2^{-m})}$
$\text{Log}_2(1+x)$	$[\frac{1}{\sqrt{2}}-1, 0]$	$\dfrac{m-\underline{m}}{2\,(1+m)\,\text{Log}_e(\frac{1}{1+m})}$
$\text{Log}_2(1+x)$	$[0, \sqrt{2}-1]$	$\dfrac{\bar{m}-m}{2\,(1+m)\,\text{Log}_e(1+m)}$

Table 2: Accurate approximations of S.

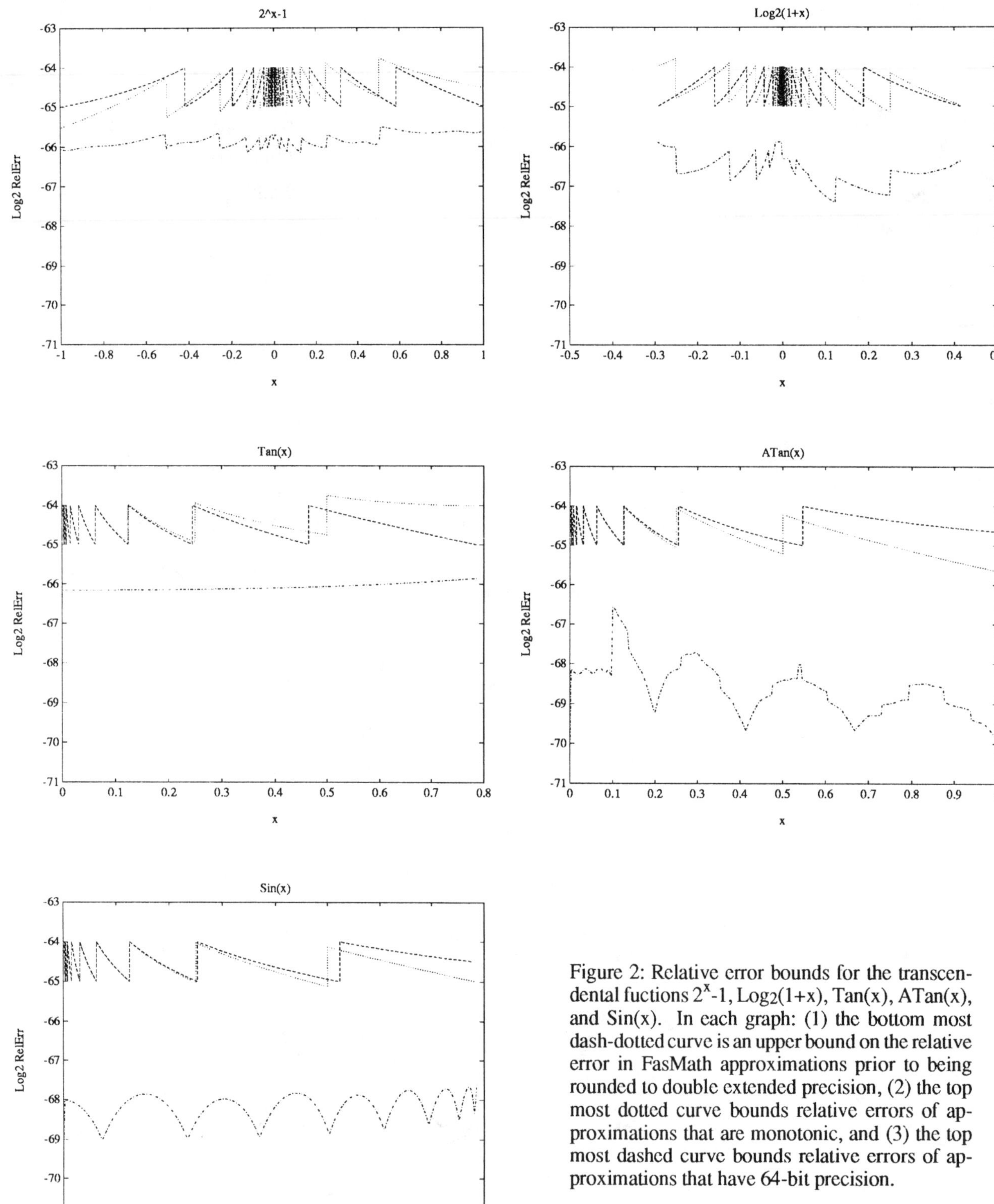

Figure 2: Relative error bounds for the transcendental fuctions 2^x-1, Log2(1+x), Tan(x), ATan(x), and Sin(x). In each graph: (1) the bottom most dash-dotted curve is an upper bound on the relative error in FasMath approximations prior to being rounded to double extended precision, (2) the top most dotted curve bounds relative errors of approximations that are monotonic, and (3) the top most dashed curve bounds relative errors of approximations that have 64-bit precision.

On-Line and Signal Processing

Chair:
Henk J. Sips
Delft University of Technology

Application of On-Line Arithmetic Algorithms to the SVD Computation: Preliminary Results*

Paul K.-G. Tu

Advanced Workstation Division
IBM Corporation
Austin, TX 78758

Miloš D. Ercegovac

Computer Science Department
University of California, Los Angeles
Los Angeles, CA   90024

Abstract

A scheme for the singular value decomposition (SVD) problem, based on on-line arithmetic, is discussed. The design, using radix-2 floating-point on-line operations, implemented in the LSI[1] HCMOS gate array technology, is compared with a compatible conventional arithmetic implementation. The preliminary results indicate that the proposed on-line approach achieves a speedup of 2.4-3.2 with respect to the conventional solutions, with 1.3 - 5.5 more gates and more than 6 times fewer interconnections.

1   Introduction

In this paper we explore the effectiveness of implementing numerical computations with on-line arithmetic algorithms. Of interest are the characteristics of computations that make them suitable for on-line implementation and the cost and performance features of such on-line computations. Since on-line algorithms operate in a digit-pipelined fashion, it is essential that the critical path of its application contain long sequences of arithmetic operations. The singular value decomposition (SVD)[13] is one such application. Other approaches in implementing SVD are described in [2, 4, 6, 10, 8, 7].

In order to conduct comparison with conventional approaches, we give estimates of the complexity and performance parameters of the on-line approach based on the LSI gate array designs given in [15, 16, 17]. The on-line algorithms have been implemented according to the LSI Logic gate array design guidelines using Viewlogic Workview CAD tools[18]. Complexity figures are calculated directly from the designs, and delays are measured from library components using the data given in [15]. Although the discussed on-line SVD system has not been completely implemented, the presented gate counts and delays provide, in our opinion, useful estimate about the feasibility and performance of on-line approach. The results presented here are preliminary and limited. We did not study in depth the effects of cancellation of leading digits in floating-point additions which present challenging design problems. We do, however, discuss the problems and outline some solutions.

*Supported in part by the NSF grant No. MIP-8813340.

[1]LSI is a trademark of LSI Logic Corporation.

We have not compared the cost/performance of the on-line SVD with other special-purpose designs [6, 2] because of incompatible technology or lack of design details.

In the following, we first give an overview of the SVD algorithm. Then the on-line scheme for the SVD computation is presented. It is assumed that standardized single-operation on-line units are used for the implementation. The on-line scheme is compared with the conventional approach in Section 4.

2   The SVD algorithm

The singular value decomposition (SVD) is one of the basic matrix operations required for many important modern signal processing computations[13]. The SVD of a matrix $\mathbf{A}$ is defined as the factorization

$$\mathbf{A} = \mathbf{U}\mathbf{\Sigma}\mathbf{V}^T$$

where $\mathbf{U}^T\mathbf{U} = \mathbf{I}$, $\mathbf{V}^T\mathbf{V} = \mathbf{I}$, and $\mathbf{\Sigma}$ is a diagonal matrix with non-negative diagonal elements. For many real-time signal processing applications, numerically stable SVD algorithms that are suitable for implementation on parallel computer architectures are desirable.

We adopt the algorithm by Brent, Luk and van Loan[12] which computes the SVD of an $n \times n$ matrix $\mathbf{A}$ on a mesh-connected processor array. It has been shown in [11] that this algorithm converges when n is odd, and does not always converge when n is even. In our discussion on-line arithmetic is used as the low level realization of arithmetic operations of a numerical computation, and it is assumed that for the on-line computation to generate the correct result, the high level algorithm must be numerically correct when realized by conventional parallel arithmetic algorithms. Our emphasis is on the implementation efficiency and performance, not the convergence of the computation. For the sake of clarity in our illustration, we assume that $\mathbf{A}$ is real and n is even. The same conclusions apply when n is odd.

In this scheme, the array consists of $n/2 \times n/2$ processors, where each processor operates on a 2×2 submatrix of $\mathbf{A}$ (Fig. 1). A series of 2×2 SVDs is performed on submatrices along the main diagonal of $\mathbf{A}$, indicated by the shaded areas in Fig. 1. Each 2×2 SVD is realized

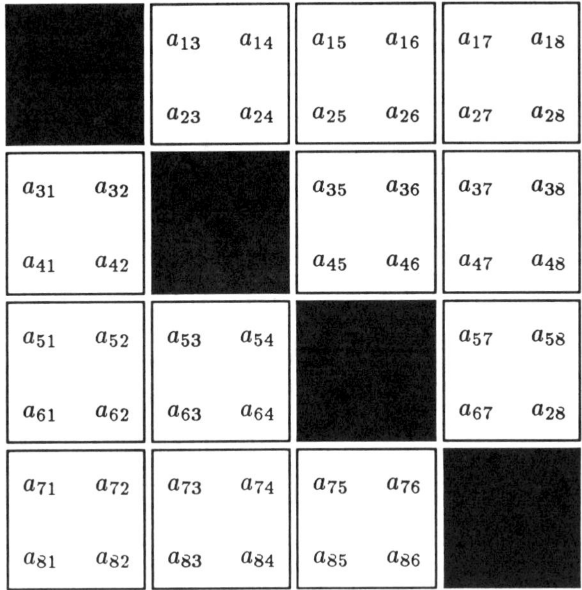

Figure 1: Mapping a matrix onto a processor array ($n = 8$)

by a 2-sided rotation that diagonalizes the submatrix. The rotation angles are computed by Algorithm FHSVD [12].

Algorithm FHSVD

$$\mu_1 = a_{i+1,i+1} - a_{i,i} \quad \nu_1 = a_{i+1,i} + a_{i,i+1}$$

if $|\nu_1| \leq \epsilon |\mu_1|$ **then**

$$\chi_1 = 1 \quad \sigma_1 = 0$$

else

$$\rho_1 = \frac{\mu_1}{\nu_1} \quad \tau_1 = \frac{\text{sign}(\rho_1)}{|\rho_1| + \sqrt{1 + \rho_1^2}}$$

$$\chi_1 = \frac{1}{\sqrt{1 + \tau_1^2}} \quad \sigma_1 = \chi_1 \tau_1$$

$$\mu_2 = a_{i+1,i+1} + a_{i,i} \quad \nu_2 = a_{i+1,i} - a_{i,i+1}$$

if $|\nu_2| \leq \epsilon |\mu_2|$ **then**

$$\chi_2 = 1 \quad \sigma_2 = 0$$

else

$$\rho_2 = \frac{\mu_2}{\nu_2} \quad \tau_2 = \frac{\text{sign}(\rho_2)}{|\rho_2| + \sqrt{1 + \rho_2^2}}$$

$$\chi_2 = \frac{1}{\sqrt{1 + \tau_2^2}} \quad \sigma_2 = \chi_2 \tau_2$$

$$c_i^L = \chi_1 \chi_2 + \sigma_1 \sigma_2 \quad s_i^L = \sigma_1 \chi_2 - \chi_1 \sigma_2$$

$$c_i^R = \chi_1 \chi_2 - \sigma_1 \sigma_2 \quad s_i^R = \sigma_1 \chi_2 + \chi_1 \sigma_2$$

end.

Then the rotation is computed as

$$\begin{bmatrix} a'_{i,i} & 0 \\ 0 & a'_{i+1,i+1} \end{bmatrix} =$$

$$\begin{bmatrix} c_i^L & s_i^L \\ -s_i^L & c_i^L \end{bmatrix}^T \begin{bmatrix} a_{i,i} & a_{i,i+1} \\ a_{i+1,i} & a_{i+1,i+1} \end{bmatrix} \begin{bmatrix} c_i^R & s_i^R \\ -s_i^R & c_i^R \end{bmatrix}$$

where c_i^L, s_i^L, c_i^R and s_i^R are the cosine and sine values of the left and right rotation angles, respectively.

During each iteration, $n/2$ 2×2 SVDs are performed in parallel. Values of the left rotation angles are passed to processors of the corresponding rows while values of the right rotation angles are passed along the corresponding columns, where 2×2 rotations are performed on the submatrix in each processor,

$$\begin{bmatrix} a'_{i,j} & a'_{i,j+1} \\ a'_{i+1,j} & a'_{i+1,j+1} \end{bmatrix} =$$

$$\begin{bmatrix} c_i^L & s_i^L \\ -s_i^L & c_i^L \end{bmatrix}^T \begin{bmatrix} a_{i,j} & a_{i,j+1} \\ a_{i+1,j} & a_{i+1,j+1} \end{bmatrix} \begin{bmatrix} c_j^R & s_j^R \\ -s_j^R & c_j^R \end{bmatrix}$$

The 2-sided 2×2 rotation is computed as

$$a'_{i,j} =$$
$$c_j^R (c_i^L a_{i,j} - s_i^L a_{i+1,j}) - s_j^R (c_i^L a_{i,j+1} - s_i^L a_{i+1,j+1})$$

$$a'_{i,j+1} =$$
$$s_j^R (c_i^L a_{i,j} - s_i^L a_{i+1,j}) + c_j^R (c_i^L a_{i,j+1} - s_i^L a_{i+1,j+1})$$

$$a'_{i+1,j} =$$
$$c_j^R (s_i^L a_{i,j} + c_i^L a_{i+1,j}) - s_j^R (s_i^L a_{i,j+1} + c_i^L a_{i+1,j+1})$$

$$a'_{i+1,j+1} =$$
$$s_j^R (s_i^L a_{i,j} + c_i^L a_{i+1,j}) + c_j^R (s_i^L a_{i,j+1} + c_i^L a_{i+1,j+1})$$

After each iteration, rows and columns of the matrix are exchanged between adjacent processors according to the *parallel ordering* which is described in [1]. Under this ordering, for each i and j, such that $1 \leq i < j \leq n$, columns i and j of **A** are paired and rotated in the same column of processors exactly once in every $n-1$ consecutive iterations which are referred to as a *sweep*. Likewise, for each $1 \leq i < j \leq n$, rows i and j of **A** are paired and rotated in the same row of processors exactly once in every sweep. To simplify control, a fixed number of S sweeps is performed to diagonalize **A**. It is shown in [12] that $S < 9$ for $n \leq 50$.

Assuming that there is no data broadcasting among processors, Figure 2 shows the processor array by Brent, Luk and van Loan. The dashed arrow lines represent the transmission of the rotation angles and the solid arrow lines represent the transmission of matrix elements. Processors on the main diagonal compute the rotation angles and the main diagonal elements of the matrix, while those not on the main diagonal perform 2×2 rotations.

The performance bottleneck of the SVD is clearly the calculation of the rotation angle, which contains long sequences of sequential operations and cannot be effectively sped up by conventional techniques.

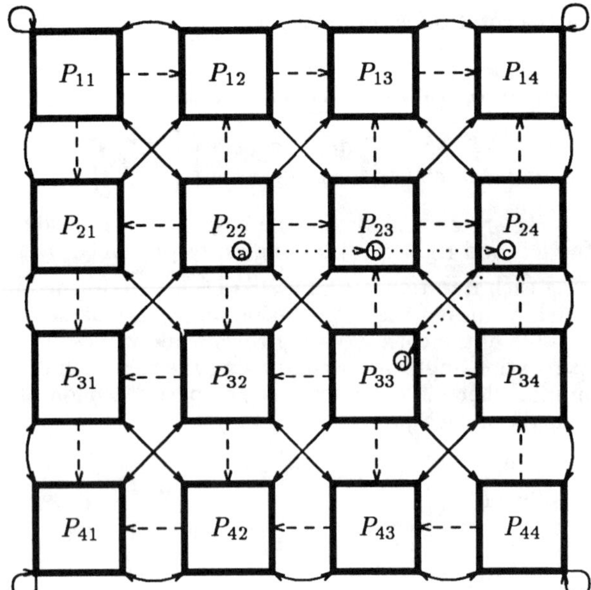

Figure 2: Systolic array for the SVD by Brent, Luk and van Loan ($n = 8$)

3 On-line scheme for computing the SVD

We now describe the SVD scheme based on on-line arithmetic[3, 5]. In this scheme each arithmetic operation in the computation is realized by a single-operation on-line unit. All on-line units have the same interfaces, so successive units can be interconnected directly with no intermediate data conversion. We assume that the inter-processor transmission of the angles and the matrix elements are on-line. On-line algorithms for floating-point operations used are given in the appendix. Their design appears in [15] and [17].

For each on-line operation, it is assumed that the input operands are quasi-normalized, which is defined as $|x| \in [\frac{1}{4}, 1)$. The output is also quasi-normalized. This is achieved by incorporating post-normalization step into the on-line algorithms for addition, multiplication and division operations. The on-line result of square root operation is always quasi-normalized. Details of post-normalization steps for different operations are given in the Appendix. The most-significant bit of the mantissa is always non-zero thus making the sign determination simple. This also holds in the case of cancellation of the most significant digits, since no output is delivered before the result is quasi-normalized. While the on-line delays of multiplication, division and square root given in Table 3 are bounded by 7, 7 and 4, respectively, the on-line delay of addition can be as large as 56 due to cancellation of the most significant digits. This is a serious problem that has not been included in the perfromance analysis in this paper. However, the design includes necessary on-line buffers to synchronize inputs arriving out of phase due to cancellation.

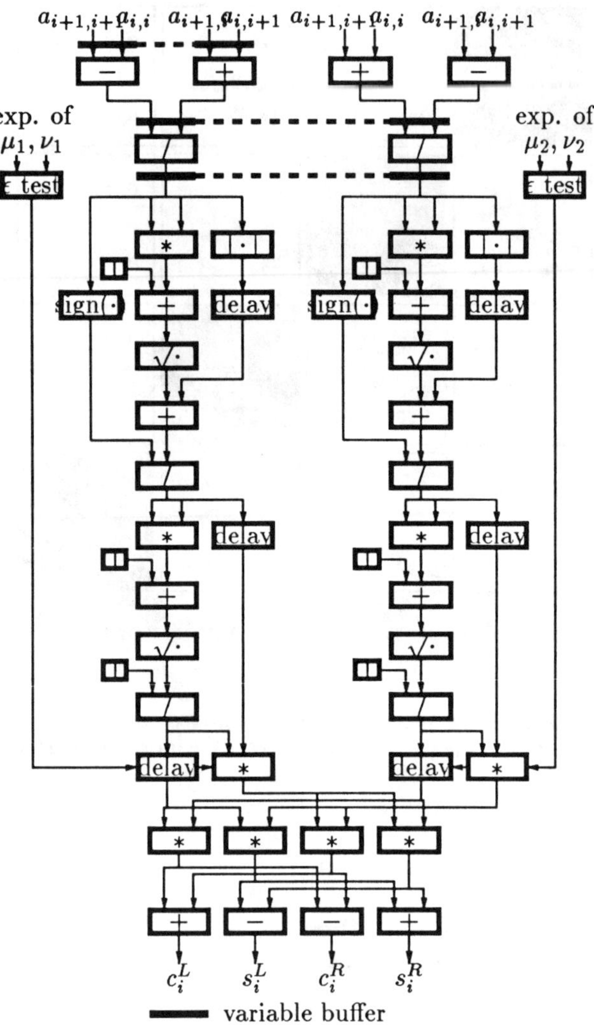

Figure 3: On-line scheme for algorithm FHSVD

3.1 Complexity of the on-line SVD

Figure 3 shows the on-line network for the FHSVD algorithm. The ϵ tests in the FHSVD algorithm are performed concurrently with the computation of ρ_i, τ_i, χ_i and σ_i. If a test succeeds, the corresponding variables are forced to 1 and 0, respectively. Two types of delay buffers are used in the implementation. Fixed delay buffers (FDB) are used to synchronize the arrival of operands for some of the on-line units over the paths with a fixed difference in on-line delays. The out-of-phase inputs, caused by cancellation, are synchronized through variable delay buffers (VDB). These buffers use a design similar to the mantissa alignment [15] and have an estimated cost of 1,150 gates/buffer. The sign of a number can be obtained directly from the sign bit of the first mantissa digit. Constants can be stored in registers and transmitted to the on-line units when needed. The control of on-line units is simple and it is not included in this study.

The on-line network for the rotation computation of

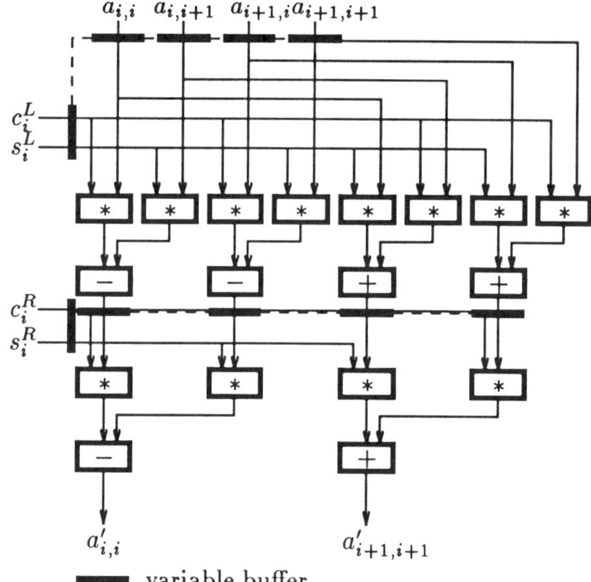

Figure 4: On-line scheme for rotation of a diagonal processor

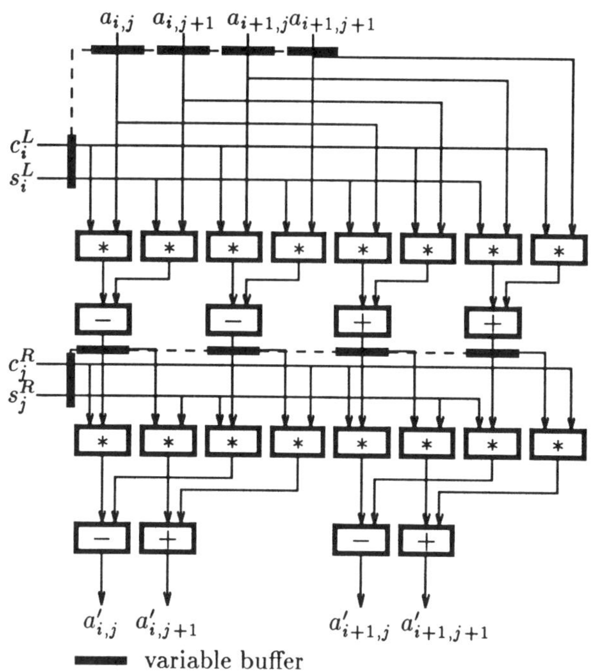

Figure 5: On-line scheme for rotation of an off-diagonal processor

main diagonal processors is shown in Figure 4. We assume that six inputs to the upper and lower level multipliers are synchronized using 6 variable delay buffers each. Figure 5 shows the on-line network for the rotation computation of off-diagonal processors. The number of each type of arithmetic units and estimates of their gate count for each processor are given in Table 1 for radix-2 computation, based on the designs described in [15]. The cost of postnormalization is included. Each input and output data requires 10 pins, 8 for the exponent and 2 for the mantissa. Let G_{diag} and G_{off} denote the gate count for each main diagonal and off diagonal processor, respectively. The total cost of an $n/2 \times n/2$ processor array, which computes the SVD of an $n \times n$ matrix, is

$$G_{\frac{n}{2} \times \frac{n}{2}} = G_{diag} \times \frac{n}{2} + G_{off} \times \left(\left(\frac{n}{2} \right)^2 - \frac{n}{2} \right) \quad (1)$$

3.2 Performance of the on-line SVD

The SVD computation is composed of the rotation angle calculation and the rotation calculation. The operations in the critical paths of these calculations are listed in Table 2. Let T_{cyc} denote the *iteration cycle time*, defined as the number of clock cycles between starts of consecutive iterations, and T_{ite} the *complete iteration time*, which is the number of clock cycles for a complete iteration. Assume that it takes one clock cycle for data to pass through a processor, and the exchange of matrix elements between processors takes no extra time. Then the *complete sweep time*, defined as the number of clock cycles to compute a sweep, which consists of $n - 1$ iterations, is

$$T_{swe} = (n - 2) \cdot T_{cyc} + T_{ite} \quad (2)$$

Arith. unit	Gate count	Data pins	Main diag. proc.				Off diag. proc.	
			ang. units	rot. units	total units	cost	rot. units	cost
Add	1337	30	14	6	20	26700	8	10700
Mult	3941	30	10	12	22	86700	16	63100
Div	4261	30	6	0	6	25600	0	0
Sqrt	4261	20	4	0	4	17000	0	0
VD Buffer	1150		5	12	17	19950	12	13800
FD Buffer	100		6	0	6	600	0	0
ϵ Tests	200		2	0	2	400	0	0
Total cost			~ 176950				~ 87600	

Table 1: Processor cost of on-line SVD

Computation	$+/-$	$\times$	$\div$	$\sqrt{}$
Angle calc.	5	4	3	2
Rotation	2	2		

Table 2: Critical path operations for the SVD

Operation	minimum step-time	δ_{min}	δ_{max}
Add*	$14\ t_g$	5 cycles	5 cycles
Mult	$15\ t_g$	5 cycles	7 cycles
Div	$17\ t_g$	6 cycles	7 cycles
Sqrt	$17\ t_g$	4 cycles	4 cycles

* no cancellation of leading digits

Table 3: On-line delay and step-time for radix-2 on-line units

The total number of clock cycles for the SVD consisting of S sweeps is

$$T_{svd}=(S \cdot (n-1) - 1) \cdot T_{cyc} + T_{ite} \qquad (3)$$

When post-normalization is performed in an on-line operation, it costs an extra cycle to the on-line delay since the second digit of the result is examined before determining whether the first digit can be output. When the result contains leading zeros, the zeros are discarded and more digits are examined. In the optimal case, there is no leading zero generated in the mantissa, while in the worst case (assuming no leading digit cancellation for addition), up to 1 (addition), 2 (division) or 3 (multiplication) leading zeros may be present. Based on the designs presented in [15], the required minimum step-time and minimum and maximum on-line delays of the radix-2 implementation of the basic operations are listed in Table 3. The step-time is measured in units of t_g, which is roughly the delay of a nand gate with no more than 4 inputs. In the following discussion, we define *cycles* as the maximum of the minimum required step-time for all individual operations. We also assume δ_{max} for each on-line unit. The delays of the components used in the designs are listed in [15]. The on-line delays for the rotation angle calculation and the rotation calculation, denoted as Δ_{ang} and Δ_{rot}, respectively, are

$$\Delta_{ang} = 82\ cycles, \Delta_{rot} = 24\ cycles$$

The iteration cycle time of the on-line SVD is limited by two factors. 1. The availability of input. In the SVD computation, the rotation angle calculation of each iteration requires the result of the rotation from the previous iteration. The time between the start of the rotation angle calculation and the arrival of data for the next iteration, denoted as Δ_{ite}, includes the on-line delays of the rotation angle calculation, of transmitting the computed angles through one processor, and of the rotation calculation on an off-diagonal processor, as is indicated in Figure 2 by the dotted path $a \rightarrow b \rightarrow c \rightarrow d$. For radix-2 computation, we have

$$\Delta_{ite}=\Delta_{ang} + \delta_{pas} + \Delta_{rot}$$
$$=82 + 1 + 24 = 107\ cycles$$

where δ_{pas} is the number of cycles for data to pass through a processor and is assumed to be 1.

2. The availability of the arithmetic units. For each on-line unit, the next iteration can not start until it has finished the computation of the previous iteration. Since in on-line computations all units work in pipelined fashion, the cycle time between iterations must be no less than the largest busy cycle time of all on-line units in the critical path of the computation. Assuming $m+1$ output digits are computed, the busy cycle of an on-line unit is

$$t_{busy}=\delta + m$$

where δ is the on-line delay of the on-line unit. For radix-2 computation, division has the largest busy cycle with $\delta = 7$. For double precision computation of 8 exponent bits and 56 mantissa bits, we have

$$t_{busy}=63\ cycles$$

Both factors mentioned above present lower bounds on T_{cyc}. Assuming that the matrix size $n \geq 4$, we have

$$T_{ite}=\Delta_{ang} + (\frac{n}{2} - 2) \cdot \delta_{pas} + \Delta_{rot} + m$$

then for radix-2 double precision floating-point computation, we have

$$T_{cyc}=\Delta_{ite} = 107\ cycles$$
$$T_{ite}=160 + \frac{n}{2}\ cycles$$
$$T_{swe}=(107 + \frac{1}{2}) \cdot n - 54\ cycles$$
$$T_{svd}=S \cdot (n-1) \cdot 107 + \frac{n}{2} + 53\ cycles$$

The hardware utilization is measured in terms of $unit \cdot cycle$, which is defined as an arithmetic unit resource for one clock cycle during the computation. The total amount of resource available for the SVD computation is

$$\text{total resource}=(\text{total number of units}) \times T_{svd}$$

and the amount of resources utilized during the computation is the summation of busy cycles of all units

$$T_{busy} \equiv \sum_{\text{all units}} t_{busy}$$

The utilization factor of the computation is defined as the ratio of T_{busy} and the total resource,

$$U \equiv \frac{T_{busy}}{\text{total resource}}$$

For the on-line SVD computation, each main diagonal processor has 20 addition units, 22 multiplication units, 6 division units and 4 square root units. Each off-diagonal processor has 8 addition units and 16 multiplication units. Assuming the worst case on-line delay (δ_{max}) for each on-line operation, and no leading digit calcellation in addition, the busy cycle for each addition, multiplication, division and square root unit is 61,

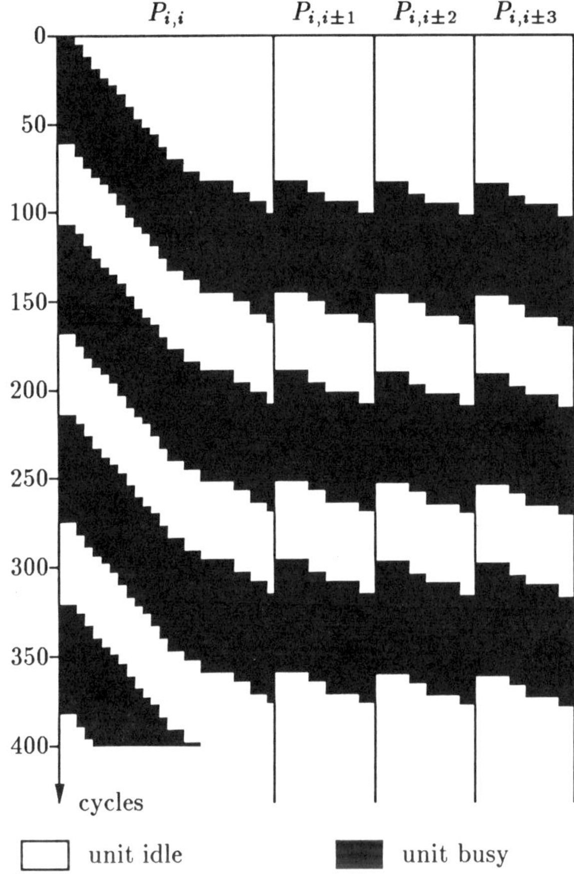

Figure 6: Timing diagram for on-line SVD ($n = 8$)

63, 63 and 60, respectively. The utilization factor for the on-line SVD is

$$U_{OL} = \frac{\left(3224 \cdot \dfrac{n}{2} + 1496 \cdot \left(\left(\dfrac{n}{2}\right)^2 - \dfrac{n}{2}\right)\right) \cdot (n-1) \cdot S}{\left(52 \cdot \dfrac{n}{2} + 24 \cdot \left(\left(\dfrac{n}{2}\right)^2 - \dfrac{n}{2}\right)\right) \cdot T_{svd}}$$

It can be shown that for $S \geq 10$ and $n \geq 8$, we have $U_{OL} > 0.57$. Figure 6 is a timing diagram that shows the busy and idle periods of each of the main diagonal and off-diagonal processors.

4 Comparisons

We now compare the on-line implementation with conventional special purpose designs for the SVD computation. The delays are measured in terms of *gate delays*, denoted as t_g, which is approximately the delay time of a nand gate with no more than 4 inputs. The effect of fanout is ignored in the comparison since it depends on low level design details, and we assume that it affects designs of both approaches similarly. Costs of the designs are measured in terms of LSI *equivalent gates*, which is defined in [9]. The data format of input and

Arith.	Main diag. proc.		Off diag. proc.	
unit	# inputs	num.	# inputs	num.
Add	5	8	2	8
Mult	3	12	2	16
	2	4		
Div	3	4		
Sqrt	2	4		
Total	2	8	2	24
	3	16		
	5	8		

Table 4: Multiplexers for arithmetic units (Case 1)

output of each arithmetic operation is double precision floating-point with 8-bit exponent and 56-bit mantissa.

The designs of the on-line and conventional implementations are based on components available from the LSI HCMOS gate array libraries[9]. The cost and performance figures of the components for both approaches are based on the discussion in [15]. Conventional floating-point algorithms used in the study correspond to typical designs described in the literature or implemented on commercial chips. The floating-point adder uses a barrel shifter and a 2-level carry-lookahead mantissa adder, performing a double-precision floating-point addition in two clock cycles. The floating-point multiplier uses radix-4 recoded multiplier, a tree of carry-save adders, and a 2-level carry-lookahead adder to produce a double- precision product in six clock cycles. The floating-point radix-4 divider corresponds to Taylor's scheme[14] augmented with the on-the-fly quotient converter. It uses redundant remainders. Double-precision quotient is obtained in 30 cycles. Similar characteristics are assumed for the floating-point square-root unit.

From Table 3, the step-time of the on-line implementation is $16\,t_g$. The step-time of the conventional approach is assumed to be that of the 56-bit parallel adder, which is $21\,t_g$. The difference of step-time is taken into consideration when comparing the schemes.

Case 1: Maximum parallelism

In this case we assume for the conventional approach that there are as many arithmetic units available as needed by the computation whenever data is ready. For the processors on the main diagonal of the array, there are 4 adders, 8 multipliers, 2 dividers and 2 square root units. During each iteration, each adder is used 5 times, each multiplier 3 times, each divider 3 times and each square root unit 2 times. For the off-diagonal processors, there are 4 adders and 8 multipliers. Each adder and multiplier is used 2 times per iteration. Table 4 lists the multiplexers that are needed for each processor. The total cost of each main diagonal processor is

$$\begin{aligned}
G_{diag} = &\; 4 \cdot g_{add} + 8 \cdot g_{mult} + 2 \cdot g_{div} + 2 \cdot g_{sqrt} \\
&+ 8 \cdot g_{mux2} + 16 \cdot g_{mux3} + 8 \cdot g_{mux5} \\
\approx &\; 149,400
\end{aligned}$$

Arith. unit	Main diag. proc.		Off diag. proc.	
	# inputs	num.	# inputs	num.
Add	5	8	4	4
Mult	3	12	4	8
	2	4		
Div	3	4		
Sqrt	2	4		
Total	2	8	4	12
	3	16		
	5	8		

Table 5: Multiplexers for arithmetic units (Case 2)

and the cost of each off diagonal processor is

$$G_{off}=4 \cdot g_{add} + 8 \cdot g_{mult} + 24 \cdot g_{mux2}$$
$$\approx 120,600$$

The number of cycles for each arithmetic operation are

$$t_{add}=2 \; cycles$$
$$t_{mult}=6 \; cycles$$
$$t_{div}=30 \; cycles$$
$$t_{sqrt}=29 \; cycles$$

From the critical path operations given in Table 2, the number of cycles for the rotation angle calculation and the rotation calculation are

$$T_{ang}=5 \cdot t_{add} + 4 \cdot t_{mult} + 3 \cdot t_{div} + 2 \cdot t_{sqrt} = 182$$
$$T_{rot}=2 \cdot t_{add} + 2 \cdot t_{mult} = 16$$

Then we have the iteration cycle time and the complete iteration time for $n \geq 4$,

$$T_{cyc}=T_{ang} + \delta_{pas} + T_{rot} = 199 \; cycles$$
$$T_{ite}=T_{ang} + (\frac{n}{2} - 2) \cdot \delta_{pas} + T_{rot}$$
$$=196 + \frac{n}{2} \; cycles$$

Case 2: Reduced complexity

Assume that each conventional processor on the main diagonal of the array still has 4 adders, 8 multipliers, 2 dividers and 2 square root units, but let the off diagonal processors have 2 adders and 4 multipliers each. Each off diagonal processor performs the rotation computation twice per iteration, each time computing the rotations of 2 of the matrix elements associated with the processor. If a matrix element is going to be transmitted to a main diagonal processor, it is always computed in the first run. Each adder and multiplier of an off diagonal processor is used 4 times per iteration. Table 5 shows the number of multiplexers used in each processor. The cost for each off diagonal processor is

$$G_{off}=2 \cdot g_{add} + 4 \cdot g_{mult} + 12 \cdot g_{mux4}$$
$$\approx 62800$$

Because of the long computation delay of the rotation angle calculation performed by the main diagonal processors, T_{cyc} remains the same as in case 1. The complete iteration time becomes

$$T_{ite}=T_{ang} + (\frac{n}{2} - 2) \cdot \delta_{pas} + 2 \cdot T_{rot}$$
$$=212 + \frac{n}{2} \; cycles$$

The utilization factor for this scheme is

$$U_{C2}=\frac{\left(468 \cdot \frac{n}{2} + 112 \cdot \left(\left(\frac{n}{2}\right)^2 - \frac{n}{2}\right)\right) \cdot (n-1) \cdot S}{\left(16 \cdot \frac{n}{2} + 6 \cdot \left(\left(\frac{n}{2}\right)^2 - \frac{n}{2}\right)\right) \cdot T_{svd}}$$

and for $n = 8$, $S = 10$ we get $U_{C2} = 0.124$.

Case 3: Maximum utilization

Assume that each main diagonal processor has one each of adder, multiplier, divider and square root unit, and each off-diagonal processor has one adder and one multiplier. For each iteration, each diagonal processor performs 20 additions, 22 multiplications, 6 divisions, 4 square root operations, and each off diagonal processor performs 8 additions and 16 multiplications. Each processor needs about 12 registers for buffering input, output and intermediate results. The complexity of the processors are

$$G_{diag}=g_{add} + g_{mult} + g_{div} + g_{sqrt} \approx 26800$$
$$G_{off}=g_{add} + g_{mult} \approx 16800$$

Assume that the order of performing the arithmetic operations is such that those operations in the critical path always proceed when ready. The performance figures are

$$T_{ang}=235 \; cycles$$
$$\Delta_{ang}=221 \; cycles$$
$$T_{rot}=\begin{cases} 74 \; cycles & \text{diagonal processor} \\ 98 \; cycles & \text{off-diagonal processor} \end{cases}$$
$$\Delta_{rot}=38 \; cycles$$
$$T_{cyc}=269 \; cycles$$
$$T_{ite}=\Delta_{ang} + (\frac{n}{2} - 2)\delta_{pass} + T_{rot} = 317 + \frac{n}{2} \; cycles$$
$$T_{swe}=(n-2)T_{cyc} + T_{ite} = 269.5n - 221 \; cycles$$
$$T_{svd}=(S(n-1) - 1)T_{cyc} + T_{ite}$$
$$=269S(n-1) + \frac{n}{2} - 48 \; cycles$$

The utilization factor is

$$U_{C3}=\frac{\left(468 \times \frac{n}{2} + 112 \times \left(\left(\frac{n}{2}\right)^2 - \frac{n}{2}\right)\right) \times (n-1) \times S}{\left(4 \times \frac{n}{2} + 2 \times \left(\left(\frac{n}{2}\right)^2 - \frac{n}{2}\right)\right) \times T_{svd}}$$

Parameter	on-line (radix-2)	conventional		
		Case 1	Case 2	Case 3
T_{cyc}	1	2.45	2.45	3.28
T_{ite}	1	1.60	1.73	2.56
T_{swe}	1	2.27	2.30	3.15
T_{svd}	1	2.42	2.43	3.21
G_{diag}	1	0.84	0.84	0.15
G_{off}	1	1.38	0.77	0.19
Total cost	1	1.16	0.77	0.18

Table 6: Performance and cost ratio against on-line scheme for $n = 8$ and $S = 10$

Operation	On-line	Conventional
Add	1337	4773
Multiply	3941	12063
Divide	4261	4993
Square root	4261	4993

Table 7: Cost comparison of on-line and conventional arithmetic unit design

For $n = 8$ and $S = 10$ we have $U_{C3} = 0.3$.

Table 6 shows the performance and cost ratio for the conventional and on-line SVD with $n = 8$ and $S = 10$, assuming no cancellation of leading digits. The figures also reflect the ratio between the step-times of the schemes.

The comparison shows that for $n = 8$, the on-line scheme achieves a speedup of about 2.4 over Case 2 of the conventional implementation and about 3.2 over Case 3, and the gate count ranging from factor 1.3 to 5.5 of the conventional approach. For each processor, the on-line scheme requires 80 data connections, while the conventional scheme needs 512 data connections.

5   Concluding remarks

This paper reports a preliminary results on the implementation cost and performance of on-line arithmetic with respect to conventional arithmetic for a particular technology (gate arrays) and a particular application (SVD). The system itself has not yet been fully implemented. Our estimates are based on the gate-level designs of individual on-line units and provide a good estimate of the system gate-count. The results indicate that for double precision floating-point computations, the designs of the on-line arithmetic units have less complexity than the conventional designs. Table 7 compares the gate counts of the designs. For the SVD computation, the on-line scheme has a much larger number of arithmetic units and a greater gate count than the conventional scheme. We note that because of the overlapping of the on-line operations, the hardware utilization of the on-line scheme is much higher than the conventional designs. Moreover, on-line approach reduces for each processor the communication bandwidth from 512 to 80 data connections which is of significance in system packaging.

The results presented here are limited in several respects. We did not analyze performance in the case of cancellation of leading digits. This is a serious problem especially for on-line arithmetic because of performance degradation. Cancellation can increase the on-line delay up to the working precision, i.e., 56. The numerical behavior and its relation to cost and performance of an SVD on-line arithmetic system need to be addressed. In our study of the system cost we included the necessary variable delay buffers to synchronize operands which get out of phase due cancellation. To reduce the buffering cost, an alternative is to have control with blocking mechanisms. We note that when operands to a floating-point on-line adder are generated by an on-line multiplier, the least significant half of the product is available in parallel form and, at a cost of two shift registers, it can be used to extend the number of significant digits to the adder. Another issue is the interface with a host system and mechanisms for significance monitoring in on-line algorithms.

References

[1] R. P. Brent and F. T. Luk. The solution of singular-value and symmetric eigenvalue problems on multiprocessor arrays. *SIAM Journal on Scientific and Statistical Computing*, 6(1):69–84, January 1985.

[2] J. R. Cavallaro and F. T. Luk. CORDIC arithmetic for an SVD processor. *Proceedings of the 8th Symposium on Computer Arithmetic*, pages 113–120, 1987.

[3] M. D. Ercegovac. On-line arithmetic: an overview. In *Proceedings of the SPIE, Real Time Signal Processing VII*, volume 495, pages 86–93, San Diego, CA, August 1984.

[4] M. D. Ercegovac and T. Lang. On-line schemes for computing rotation angles for SVDs. In *Proceedings of the SPIE, Advanced Algorithms and Architectures for Signal Processing II*, volume 826, pages 160–169, August 1987.

[5] M. D. Ercegovac and T. Lang. On-line arithmetic: A design methodology and applications in digital signal processing. In R. W. Brodersen and H. Moscovitz, editors, *VLSI Signal Processing, III*, chapter 24, pages 252–263. IEEE Press, 1988.

[6] M. D. Ercegovac and T. Lang. Redundant and on-line CORDIC: Application to matrix triangularization and svd. *IEEE Transactions on Computers*, 39(6):725–740, June 1990.

[7] J.-A. Lee. *Redundant CORDIC: Theory and its application to matrix computations*. PhD thesis, University of California, Los Angeles, 1990.

[8] J.-A. Lee and T. Lang. On-line CORDIC for generalized singular value decomposition. *Proceedings of the SPIE, High Speed Computing II*, pages 235–247, 1989.

[9] LSI Logic Corporation, 1551 McCarthy Blvd, Milpitas CA 95035. *Databook and Design Manual for HCMOS Macrocells and Macrofunctions*, October 1986.

[10] F. T. Luk. Architectures for computing eigenvalues and SVDs. *Proceedings of the SPIE, Highly Parallel Signal Processing Architectures*, 614:24–33, 1986.

[11] H. Park. *On the equivalence and convergence of parallel Jacobi SVD algorithms*. PhD thesis, Cornell University, August 1987.

[12] F. T. L. Richard P. Brent and C. V. Loan. Computation of the singular value decomposition using mesh-connected processors. *Journal of VLSI and Computer Systems*, 1(3):242–270, 1985.

[13] J. M. Speiser and H. J. Whitehouse. A review of signal processing with systolic arrays. In *Proceedings of the SPIE, vol. 431, Real Time Signal Processing VI*, pages 2–6, August 1983.

[14] G. S. Taylor. Radix 16 SRT dividers with overlapped quotient selection stages. In *Proceedings of the 7th Symposium on Computer Arithmetic*, pages 64–71, 1985.

[15] P. K.-G. Tu. *On-line arithmetic algorithms for efficient implementations*. PhD thesis, University of California, Los Angeles, 1990.

[16] P. K.-G. Tu and M. D. Ercegovac. Design of on-line division unit. In *Proceedings of the 9th Symposium on Computer Arithmetic*, pages 42–49, 1989.

[17] P. K.-G. Tu and M. D. Ercegovac. Gate array implementation of on-line algorithms for floating-point operations. In *Proceedings of the Twenty-Fourth Asilomar Conference on Signals, Systems and Computers*, 1990.

[18] VIEWlogic Systems, Inc., 313 Boston Post Road West, Marlboro, Massachusetts 01752. *WORKVIEW Manual Set, Release 3.0*, June 1988.

Appendix: Floating-Point Algorithms

A   Radix-2 on-line addition

Step 1. [Initialization and alignment]

$e_d \leftarrow e_x - e_y$

$e_z \leftarrow \max(e_x, e_y) + 1$

$A[-2] = 0, \quad z_0 = 0$

for $j = 1, \cdots, |e_d|$ do
  if $e_d \geq 0$ then
    $x'_j = x_j, \quad y'_j = 0$
  else
    $x'_j = 0, \quad y'_j = y_j$

for $j = |e_d| + 1, \cdots, m$ do
  if $e_d \geq 0$ then
    $x'_j = x_j, \quad y'_j = y_{j-|e_d|}$
  else
    $x'_j = x_{j-|e_d|}, \quad y'_j = y_j$

for $j = -1, \cdots, 0$ do
  $A[j] \leftarrow 2A[j-1] + (x'_{j+2} + y'_{j+2})2^{-2}$

Step 2. [result generation]

for $j = 1, \cdots, m+1$ do
  $A[j] \leftarrow 2(A[j-1] - z_{j-1})$
    $+(x'_{j+2} + y'_{j+2})2^{-2}$
  /* $A[j]$ has 3 bits of precision */

$$z_j \leftarrow \begin{cases} -1 & \text{if} \quad A[j] < -2^{-1} \\ 0 & \text{if} \quad -2^{-1} \leq A[j] < 2^{-1} \\ 1 & \text{if} \quad A[j] \geq 2^{-1} \end{cases}$$

$z_{out} \leftarrow \text{postnorm}_{add}(z_j, z_{j+1}), z_{m+2} = 0$

end.

The post-normalization for addition works as follows:

1. Examine the first two digits (z_1, z_2) of the result.

2. If the values of the digits being examined, (z'_1, z'_2), are (1,0), (1,1), (-1,0), or (-1,-1), then the result is quasi-normalized. These digits and the rest of the result are output in sequence.

3. Otherwise, we combine the two digits as

$$z'_1 = 2 \cdot z'_1 + z'_2 \tag{4}$$

the result exponent is decremented, and we get the next result digit $z'_2 \leftarrow z_{next}$.

4. Steps 2 and 3 are repeated until either m result digits have been examined without satisfying the condition of step 2, in which case the output is zero, or the condition in step 2 is satisfied, in which case z'_1, z'_2 and the remaining result digits are output.

In the optimal case where the condition of step 2 is satified for the first 2 digits of the result, the total on-line delay is $\delta = 5$, where 1 cycle is nedded to calculate the difference of the exponents, 3 cycles are needed to generate the first digit of the result, and 1 cycle is needed to get the second result digit.

In the worst case when all digits are exausted and the output is zero, the total on-line delay is the total number of cycles needed to obtain all digits of the result, which is $\delta_{cancel} \leq m + 3$, where m is the working precision.

B   Radix-2 on-line multiplication

Step 1. [Initialization]

$e_z \leftarrow e_x + e_y$

$X[-3] = 0, Y[-3] = 0, A[-3] = 0, z_0 = 0$

for $j = -2, \cdots, 0$ do
  $X[j] \leftarrow X[j-1] + x_{j+3}2^{-j-3}$
  $A[j] \leftarrow 2A[j-1] + (x_{j+3}Y[j-1] + y_{j+3}X[j])2^{-3}$
  $Y[j] \leftarrow Y[j-1] + y_{j+3}2^{-j-3}$

Step 2. [product generation]

for $j = 1, \cdots, m+3$ do
  $X[j] \leftarrow X[j-1] + x_{j+3}2^{-j-3}$
  $A[j] \leftarrow 2(A[j-1] - z_{j-1})$
    $+(x_{j+3}Y[j-1] + y_{j+3}X[j])2^{-3}$

$$Y[j] \leftarrow Y[j-1] + y_{j+3}2^{-j-3}$$
/* $\widehat{A[j]}$ has 3 bits */
$$z_j \leftarrow \begin{cases} -1 & \text{if } \widehat{A[j]} < -2^{-1} \\ 0 & \text{if } -2^{-1} \le \widehat{A[j]} < 2^{-1} \\ 1 & \text{if } \widehat{A[j]} \ge 2^{-1} \end{cases}$$
$$z_{out} \leftarrow \text{postnorm}_{mul}(z_j, z_{j+1}), \quad z_{m+4} = 0$$
end.

The post-normalization for multiplication works as the following.

1. Examine the first two digits (z_1, z_2) of the result.

2. If the values of the digits being examined, (z_1', z_2'), are (1,0), (1,1), (-1,0), or (-1,-1), then the result is quasi-normalized. The digits being examined and the rest of the result are output in sequence.

3. Otherwise, we combine the two digits as

$$z_1' = 2 \cdot z_1' + z_2' \tag{5}$$

   the result exponent is decremented, and we get the next result digit $z_2' \leftarrow z_{next}$.

4. Steps 2 and 3 are repeated until either the condition in step 2 is satisfied, when z_1', z_2' and the remaining result digits are output, or z_4 has been examined, in which case the output digits are $z_1', z_5, \cdots$ and the rest of the result digits.

In the optimal case where the condition of step 2 is satisfied for the first 2 digits of the result, the total on-line delay is $\delta_{min} = 5$, where 4 cycles are needed to generate the first digit of the result, and 1 cycle is needed to get the second result digit.

In the worst case there can be at most 3 leading zeros in the result. The total online delay is the total number of cycles needed to obtain the first four digits of the result, which is $\delta_{max} = 7$.

C Radix-2 on-line division

Step 1. [Initialization and shifting]
$$e_q \leftarrow e_n - e_d + 1$$
$$A[0] \leftarrow \sum_{i=1}^{3} n_i 2^{-i-1}$$
$$D[0] \leftarrow \sum_{i=1}^{4} d_i 2^{-i}$$
if $D[0] < 2^{-1}$ **then**
$$D[0] \leftarrow 2 \cdot D[0] + d_{next}2^{-4};$$
$$e_q \leftarrow e_q + 1$$
$$Q[0] \leftarrow 0; \qquad q_0 \leftarrow 0$$
Step 2. [quotient generation]
for $j = 1, \cdots, m+2$ **do**
$$D[j] \leftarrow D[j-1] + d_{next}2^{-j-4}$$
$$A[j] \leftarrow 2(A[j-1] - q_{j-1}D[j])$$
$$+n_{next}2^{-4} - d_{next}Q[j-2]2^{-4}$$

/* $\widehat{A[j]}$ has 6 bits */
$$q_j = \begin{cases} -1 & \text{if } \widehat{A[j]} < -\frac{1}{4} \\ 0 & \text{if } -\frac{1}{4} \le \widehat{A[j]} < \frac{3}{16} \\ 1 & \text{if } \widehat{A[j]} \ge \frac{3}{16} \end{cases}$$
$$Q[j] \leftarrow Q[j-1] + q_j 2^{-j}$$
$$q_{out} \leftarrow \text{postnorm}_{div}(q_j, q_{j+1}), \quad q_{m+3} = 0$$
end.

The post-normalization for division works as follows.

1. Examine the first two digits (q_1, q_2) of the result.

2. If the values of the digits being examined, (q_1', q_2'), are (1,0), (1,1), (-1,0), or (-1,-1), then the result is quasi-normalized. The digits being examined and the rest of the result are output in sequence.

3. Otherwise, we combine the two digits as

$$q_1' = 2 \cdot q_1' + q_2' \tag{6}$$

   the result exponent is decremented, and we get the next result digit $q_2' \leftarrow q_{next}$.

4. Steps 2 and 3 are repeated until either the condition in step 2 is satisfied, when q_1', q_2' and the remaining result digits are output, or q_3 has been examined, in which case the output digits are $q_1', q_4, \cdots$ and the rest of the result digits.

In the optimal case where the condition of step 2 is satisfied for the first 2 digits of the result, the total on-line delay is 6, where 5 cycles are needed to generate the first digit of the result, and 1 cycle is needed to get the second result digit.

In the worst case there can be at most 2 leading zeros in the generated result. The total online delay is the total number of cycles needed to obtain the first three digits of the result, which is $\delta_{max} = 7$.

D Radix-2 on-line square root

Step 1. [Initialization and shifting]
if e_x *odd* **then**
$$A[0] \leftarrow \sum_{i=1}^{2} x_i 2^{-i-1}$$
else
$$A[0] \leftarrow x_1 2^{-3}$$
$$e_y \leftarrow \lfloor e_x/2 \rfloor + 1$$
Step 2. [output generation]
for $j = 1, \cdots, m$ **do**
$$A[j] \leftarrow 2 \cdot (A[j-1] - y_{j-1}(Y[j-2]$$
$$+Y[j-1])) + x_{next}2^{-4}$$
/* $\widehat{A[j]}$ has 7 bits */
/* S_{sqrt} is the selection function defined in [15]. */
$$y_j \leftarrow S_{sqrt}(\widehat{A[j]})$$
$$Y[j] \leftarrow Y[j-1] + y_j 2^{-j}$$
end.

The CORDIC Householder Algorithm

Shen-Fu Hsiao   Jean-Marc Delosme

Department of Electrical Engineering
Yale University
New Haven, CT 06520

Abstract

A novel n-dimensional (n-D) CORDIC algorithm, for
Euclidean and pseudo-Euclidean rotations, is proposed.
The new algorithm is closely related to Householder
transformations. It is shown to converge faster than
CORDIC algorithms developed earlier for $n = 3$ and 4.
Processor architectures for the algorithm are presented.
The area and time performance of n-D CORDIC processors are evaluated. For a comparable time performance, the processors require significantly less area
than parallel Householder processors. Furthermore, arrays of n-D Euclidean CORDIC processors are shown
to speed up the QR decomposition of rectangular matrices by a factor of $n-1$ with respect to a 2-D CORDIC
processor array.

1 Introduction

1.1 2-D CORDIC Algorithms

The COordinate Rotation DIgital Computer (CORDIC),
introduced by Volder [13] and extended later by
Walther [15], is an algorithm operating on 2-
dimensional (2-D) vectors to perform rotations using
simple arithmetic primitives. The algorithm has several variants and can be used to calculate a variety of
elementary functions. While the traditional approach
to vector rotation calls for square-root extraction, division, and multiplication, the CORDIC algorithm requires only additions and shifts. The algorithm approx-
imates a 2-D rotation R_2 (Euclidean or hyperbolic) by
the product $\prod_{i=1}^{p} R_{2,i}$ of p elementary rotations

$$R_{2,i} = (1 + \sigma t_i^2)^{-1/2} U_{2,i}, \; U_{2,i} = \begin{pmatrix} 1 & \delta_i t_i \\ -\sigma \delta_i t_i & 1 \end{pmatrix},$$

$$(1)$$

where $U_{2,i}$ is called the unscaled part of $R_{2,i}$,

$$\sigma = \begin{cases} 1 & \text{for a Euclidean rotation} \\ -1 & \text{for a hyperbolic (or pseudo-Euclidean)} \\ & \text{rotation,} \end{cases}$$

and t_i is a non-positive power of 2. When $t_{i+1} = t_i$, two
consecutive rotations have the same magnitude; this is
called a 'repetition'.

When evaluating the angle of a vector $[x_1 \; x_2]^T$ with
the first axis, a sequence of unscaled rotations $U_{2,i}$ is

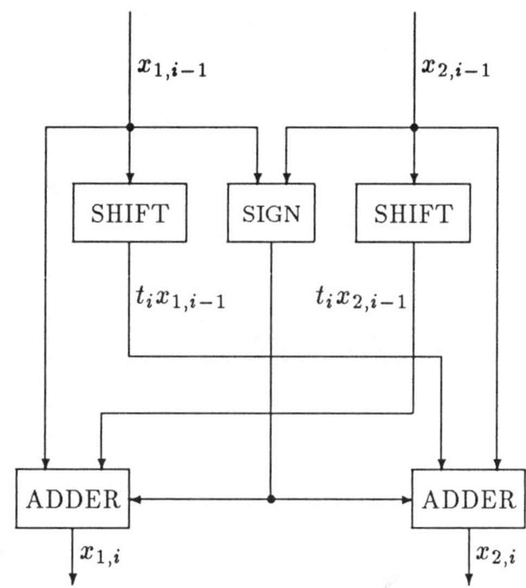

Figure 1: The architecture of a 2-D CORDIC processor.

applied to bring the vector along the direction $[1 \; 0]^T$
(evaluation process). The signs δ_i are selected according to $\delta_i = sign(x_{1,i-1} \cdot x_{2,i-1})$, where $x_{j,i-1}$ denotes the jth component of the vector at the begin-
ning of iteration i and, initially, $x_{j,0} = x_j$. The sequence $\delta_i, 1 \leq i \leq p$, thus obtained may be used
to rotate other vectors by the same amount (application process). Like the multiplication by the unscaled rotations $U_{2,i}$, multiplication by the scaling factor $\prod_{i=1}^{p}(1 + \sigma t_i^2)^{-1/2}$ is implemented by additions and
shifts, see *e.g.* [9]. While no repetition is needed
for the 2-D Euclidean CORDIC algorithm, it is necessary in the hyperbolic case to have a repetition for
$t_i = 2^{-4}, 2^{-13}, 2^{-40}, \dots, 2^{-k}, 2^{-(3k+1)}, \dots$ [15]. Fig. 1
illustrates the 2-D CORDIC processor architecture.

1.2 3-D and 4-D CORDIC Algorithms

One trivial way of rotating an n-dimensional vector
is to apply $(n - 1)$ 2-D CORDIC rotations on $\lfloor n/2 \rfloor$
2-D CORDIC processors in the time of $\lceil \log_2 n \rceil$ 2-D
CORDIC rotations. For example, in order to align
a 4-D vector with the $[1 \; 0 \; 0 \; 0]^T$ direction, two 2-D

CORDIC rotations are applied in parallel to zero out the second and fourth components, followed by another 2-D CORDIC rotation to zero out the third component. Delosme *et al.* [6][7] have proposed faster 3-D and 4-D CORDIC algorithms which require a number of iterations only slightly larger than for a single 2-D CORDIC rotation. These algorithms are derived from the rational representation theorem of Cayley. According to that theorem, any n-D Euclidean rotation R_n with no eigenvalue equal to -1, can be expressed as

$$R_n = (I - T_n)(I + T_n)^{-1}$$

where T_n is an $n \times n$ skew-symmetric matrix.

The matrix $T_{3,i}$ associated to a 3-D elementary rotation matrix was selected in [6] as

$$T_{3,i} = \begin{pmatrix} 0 & -\gamma_i t_i & \delta_i t_i \\ \gamma_i t_i & 0 & -\gamma_i t_i \\ -\delta_i t_i & \gamma_i t_i & 0 \end{pmatrix}, \ \gamma_i, \delta_i \in \{1, -1\}, \tag{2}$$

where t_i is a negative power of 2. The corresponding elementary rotation matrix is

$$R_{3,i} = (I - T_{3,i})(I + T_{3,i})^{-1} = (1 + 3t_i^2)^{-1} \times$$
$$\begin{pmatrix} 1 - t_i^2 & 2\gamma_i t_i + 2t_i^2 & -2\delta_i t_i + 2t_i^2 \\ -2\gamma_i t_i + 2t_i^2 & 1 - t_i^2 & 2\gamma_i t_i + 2t_i^2 \\ 2\delta_i t_i + 2t_i^2 & -2\gamma_i t_i + 2t_i^2 & 1 - t_i^2 \end{pmatrix}. \tag{3}$$

Its rotation axis is $[\gamma_i\ \delta_i\ \gamma_i]^T$, that is, either $[1\ 1\ 1]^T$ or $[1\ \text{-}1\ 1]^T$. When evaluating a rotation bringing a vector $[x_1\ x_2\ x_3]^T$ along the first axis $[1\ 0\ 0]^T$, the signs γ_i and δ_i are selected according to the control law

$$\gamma_i = sign(x_{1,i-1} \cdot x_{2,i-1}),$$
$$\delta_i = -sign(x_{1,i-1} \cdot x_{3,i-1}), \tag{4}$$

where $x_{j,i-1}$ denotes the jth component of the vector at the beginning of iteration i.

In the 4-D case, the 4×4 skew-symmetric matrix $T_{4,i}$ defined as

$$T_{4,i} = \begin{pmatrix} 0 & \gamma_i t_i & \delta_i t_i & \epsilon_i t_i \\ -\gamma_i t_i & 0 & -\epsilon_i t_i & \delta_i t_i \\ -\delta_i t_i & \epsilon_i t_i & 0 & -\gamma_i t_i \\ -\epsilon_i t_i & -\delta_i t_i & \gamma_i t_i & 0 \end{pmatrix} \tag{5}$$

was selected in [6] [7]. Because the rotation matrix

$$\hat{R}_{4,i} = (I - T_{4,i})(I + T_{4,i})^{-1}$$

is the square of the matrix

$$R_{4,i} = \frac{1}{\sqrt{1 + 3t_i^2}} \begin{pmatrix} 1 & -t_i\gamma_i & -t_i\delta_i & -t_i\epsilon_i \\ t_i\gamma_i & 1 & t_i\epsilon_i & -t_i\delta_i \\ t_i\delta_i & -t_i\epsilon_i & 1 & t_i\gamma_i \\ t_i\epsilon_i & t_i\delta_i & -t_i\gamma_i & 1 \end{pmatrix}, \tag{6}$$

which has a simpler unnormalized part than $\hat{R}_{4,i}$, the matrices defined by equation (6) were selected as 4-D CORDIC elementary rotation matrices in [6] [7].

To prove the convergence of an n-D CORDIC algorithm over a given range, one must consider all the vectors in that range, which is a cone with vertex the origin in n-D space, and show that the sequence of cones which consist of all the vectors after each iteration converges toward the first axis. We wrote a program to compute these cones, determine—in case repetitions are needed—for which magnitudes t_i rotations should be repeated, and prove the convergence of the 3-D and 4-D CORDIC algorithms [7]. In the 3-D case, at the first iteration, the whole 3-D space C_0 is partitioned into the the 8 canonical orthants $C_0^1, C_0^2, \cdots, C_0^8$, and each of them is rotated by $R_{3,1}$ with the control signs selected according to equation (4). The union of these rotated orthants forms a cone C_1 with vertex the origin. At the second iteration, C_1 is partitioned into 8 sub-cones $C_1^1, C_1^2, \cdots, C_1^8$ corresponding to the 8 possible assignments for the control signs. Each sub-cone is rotated by $R_{3,2}$ with the appropriate control signs. The union of the rotated sub-cones forms a cone C_2. The procedure is repeated on C_2, and so forth for every iteration. The convergence behavior can be characterized by the rays of the cones C_i which are the farthest away from the first axis in some norm. The convergence of the 4-D CORDIC algorithm is analyzed similarly except that the cones C_i are partitioned into 16, instead of 8, sub-cones at each iteration. From these computations, we found out that it is indeed necessary to perform several repetitions to obtain convergent algorithms for R_3 and R_4. In [7], the repetitions were selected to achieve the fastest convergence rate.

These earlier studies did not provide a systematic method for constructing the n-D CORDIC elementary rotation matrices for arbitrary n. In Section 2, we propose a solution to that problem, applicable to Euclidean and pseudo-Euclidean CORDIC spaces of arbitrary dimension. The solution employs elementary rotations which are essentially Householder reflections. In Section 3, two CORDIC processor architectures are proposed. Their performance compares favorably with that of a parallel implementation of the Householder transformation. Section 4 develops an example indicative of the increase in throughput achievable in application-specific array processors when n-D CORDIC processors, with $n > 2$, are used in lieu of 2-D CORDIC processors.

2   CORDIC Householder Algorithm

2.1   The 3-D Euclidean Case

As mentioned above, the 3-D elementary rotation $R_{3,i}$ in equation (3) has either $[1\ 1\ 1]^T$ or $[1\ \text{-}1\ 1]^T$ as rotation axis. If instead the rotation axis is chosen to be either $[0\ 1\ 1]^T$ or $[0\ 1\ \text{-}1]^T$, the matrix $T_{3,i}$ is replaced by

$$\bar{T}_{3,i} = \begin{pmatrix} 0 & -t_i s_{1,i} & -t_i s_{2,i} \\ t_i s_{1,i} & 0 & 0 \\ t_i s_{2,i} & 0 & 0 \end{pmatrix}, \tag{7}$$

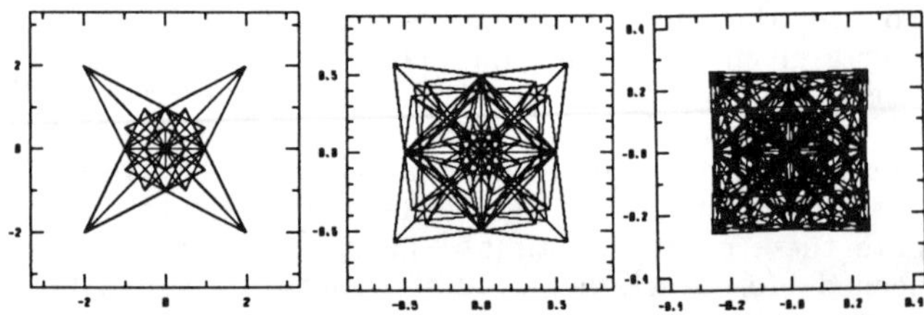

Figure 2: Domains after the first three iterations of $\bar{R}_3$.

and the elementary rotation matrices become

$$\bar{R}_{3,i} = (1 + 2t_i^2)^{-1} \times
\begin{pmatrix}
1 - 2t_i^2 & 2t_i s_{1,i} & 2t_i s_{2,i} \\
-2t_i s_{1,i} & 1 & -2t_i^2 s_{1,i} s_{2,i} \\
-2t_i s_{2,i} & -2t_i^2 s_{2,i} s_{1,i} & 1
\end{pmatrix}, \qquad (8)$$

with signs selected during the evaluation process according to

$$s_{1,i} = sign(x_{1,i-1} \cdot x_{2,i-1}), \quad s_{2,i} = sign(x_{1,i-1} \cdot x_{3,i-1}). \qquad (9)$$

The matrix $\bar{R}_{3,i}$ is simpler than $R_{3,i}$. Moreover, as is discussed next, no repetition is necessary with this new family of elementary rotations.

Using the method outlined in Section 1, we can determine whether any repetition is needed and prove the convergence of the new 3-D algorithm by determining the cone C_i of all the rotated vectors after each iteration i. The cones C_i have the origin for vertex and surround the first axis. The intersections of the cones C_1, C_2 and C_3 with the plane $x_1 = 1$ are shown in Fig. 2. The cones are invariant under a permutation of x_2 and x_3. The new 3-D CORDIC algorithm, $\bar{R}_3$, converges slightly faster than R_3 since it does not require any repetition while R_3 calls for the repetition of the iterations with $t_i = 2^{-4}, 2^{-7}, 2^{-12}, 2^{-22}, \cdots$ [7]. Furthermore, the domain obtained by intersecting the cone C_i with the plane $x_1 = 1$ is bounded by a square box with half-side $2t_i/(1 - 2t_i^2)$.

2.2   The General n-D Case

The pseudo-norm $\|x\|_{\Sigma_n}$ of an n-D vector x in pseudo-Euclidean space is defined as $\|x\|_{\Sigma_n}^2 = x^T \Sigma_n x$, where Σ_n is a signature matrix of the form

$$\Sigma_n =
\begin{pmatrix}
1 & 0 & 0 & \cdot & 0 \\
0 & \sigma_1 & 0 & \cdot & 0 \\
0 & 0 & \sigma_2 & \cdot & 0 \\
\cdot & \cdot & \cdot & \cdot & \cdot \\
0 & 0 & 0 & \cdot & \sigma_{n-1}
\end{pmatrix}, \quad \sigma_i \in \{-1, 1\}. \qquad (10)$$

Without loss of generality, we will assume the values of σ_i ordered according to:

$$\sigma_i = \begin{cases} 1 & 1 \leq i \leq k-1 \\ -1 & k \leq i \leq n-1 \end{cases}. \qquad (11)$$

If $k = n$, $\Sigma_n = I_n$ and $\|x\|_{\Sigma_n}$ is the Euclidean norm of the vector x. We shall view the Euclidean norm

(and rotations) as a particular case of pseudo-Euclidean norm (and rotations).

For an $n \times n$ matrix H_n to represent a pseudo-Euclidean rotation, it should preserve the pseudo-norm of an arbitrary vector x, $i.e.$,

$$(H_n x)^T \Sigma_n (H_n x) = x^T \Sigma_n x.$$

Since the above equation is true for any vector, H_n must satisfy

$$H_n^T \Sigma_n H_n = \Sigma_n. \qquad (12)$$

$i.e.$, it must be orthogonal with respect to Σ_n.

If $\bar{T}_{3,i}$ in equation (7) is generalized to an n-dimensional space with signature matrix Σ_n, $i.e.$,

$$\bar{T}_{n,i} = S_{n,i} \Sigma_n
\begin{pmatrix}
0 & -t_i & -t_i & \cdot\cdot & -t_i \\
t_i & 0 & 0 & \cdots & 0 \\
t_i & 0 & 0 & \cdots & 0 \\
\vdots & \vdots & \vdots & \vdots & \vdots \\
t_i & 0 & 0 & \cdots & 0
\end{pmatrix} S_{n,i}, \qquad (13)$$

where the signs in the matrix $S_{n,i}$

$$S_{n,i} =
\begin{pmatrix}
1 & 0 & \cdots & 0 \\
0 & s_{1,i} & \cdots & 0 \\
\vdots & \vdots & \vdots & \vdots \\
0 & 0 & \cdots & s_{n-1,i}
\end{pmatrix}, \qquad (14)$$

are selected during the evaluation process according to

$$s_{j,i} = \sigma_i \cdot sign(x_{1,i-1} \cdot x_{j+1,i-1}), \qquad (15)$$

the matrix $H_{n,i} = (I - \bar{T}_{n,i})(I + \bar{T}_{n,i})^{-1}$ is equal to

$$H_{n,i} = (1 + ct_i^2)^{-1} \times$$

$$S_{n,i}
\begin{pmatrix}
1 - ct_i^2 & 2t_i & \cdots & 2t_i \\
-2t_i & 1 + (c-2)t_i^2 & \cdots & -2t_i^2 \\
\vdots & \vdots & \cdots & \vdots \\
-2t_i & -2t_i^2 & \cdots & 1 + (c-2)t_i^2 \\
2t_i & 2t_i^2 & \cdots & 2t_i^2 \\
2t_i & 2t_i^2 & \cdots & 2t_i^2 \\
\vdots & \vdots & \cdots & \vdots \\
2t_i & 2t_i^2 & \cdots & 2t_i^2
\end{pmatrix}$$

$$\begin{matrix} 1 & 2 & \cdots & k \end{matrix}$$

$$\begin{pmatrix}
2t_i & \cdots & 2t_i \\
-2t_i^2 & \cdots & -2t_i^2 \\
\vdots & \vdots & \vdots \\
-2t_i^2 & \cdots & -2t_i^2 \\
1 + (c+2)t_i^2 & \cdots & 2t_i^2 \\
2t_i^2 & \cdots & 2t_i^2 \\
\vdots & \vdots & \vdots \\
2t_i^2 & \cdots & 1 + (c+2)t_i^2
\end{pmatrix} S_{n,i},$$

$$\begin{matrix} k+1 & \cdots & n \end{matrix}$$

$$\qquad (16)$$

where $c = 2k - n - 1$. $H_{n,i}$ does satisfy equation (12) and hence can be used as an elementary rotation matrix. Equation (16) defines the elementary rotation matrix for a general n-D CORDIC algorithm, applicable to both Euclidean and pseudo-Euclidean spaces. When $\Sigma_n = I_n$, $H_{n,i}$ is an elementary Euclidean rotation matrix and the more specific notation $\bar{R}_{n,i}$ is used instead of $H_{n,i}$. If the signature matrix for an *odd* dimensional pseudo-Euclidean space is such that $k = (n+1)/2$, *i.e.*, $c = 0$, *no scaling* is required to normalize the rotated vectors. Elementary CORDIC rotation matrices for $n = 2, 3$ and 4 are exhibited below:

- $n = 2$, $\sigma_1 = 1$:

$$\bar{R}_{2,i} = \frac{1}{1 + t_i^2} S_{2,i} \begin{pmatrix} 1 - t_i^2 & 2t_i \\ -2t_i & 1 - t_i^2 \end{pmatrix} S_{2,i}, \qquad (17)$$

- $n = 2$, $\sigma_1 = -1$:

$$H_{2,i} = \frac{1}{1 - t_i^2} S_{2,i} \begin{pmatrix} 1 + t_i^2 & 2t_i \\ 2t_i & 1 + t_i^2 \end{pmatrix} S_{2,i}, \qquad (18)$$

- $n = 3, \sigma_1 = \sigma_2 = 1$: $H_{3,i} = \bar{R}_{3,i}$ in equation (8),

- $n = 3$, $\sigma_1 = 1, \sigma_2 = -1$:

$$H_{3,i} = S_{3,i} \begin{pmatrix} 1 & 2t_i & 2t_i \\ -2t_i & 1 - 2t_i^2 & -2t_i^2 \\ 2t_i & 2t_i^2 & 1 + 2t_i^2 \end{pmatrix} S_{3,i}, \qquad (19)$$

- $n = 4$, $\sigma_1 = \sigma_2 = \sigma_3 = 1$:

$$\bar{R}_{4,i} = (1 + 3t_i^2)^{-1} \times$$
$$S_{4,i} \begin{pmatrix} 1 - 3t_i^2 & 2t_i & 2t_i & 2t_i \\ -2t_i & 1 + t_i^2 & -2t_i^2 & -2t_i^2 \\ -2t_i & -2t_i^2 & 1 + t_i^2 & -2t_i^2 \\ -2t_i & -2t_i^2 & -2t_i^2 & 1 + t_i^2 \end{pmatrix} S_{4,i},$$
$$(20)$$

- $n = 4$, $\sigma_1 = 1, \sigma_2 = \sigma_3 = -1$:

$$H_{4,i} = (1 - t_i^2)^{-1} \times$$
$$S_{4,i} \begin{pmatrix} 1 + t_i^2 & 2t_i & 2t_i & 2t_i \\ -2t_i & 1 - 3t_i^2 & -2t_i^2 & -2t_i^2 \\ 2t_i & 2t_i^2 & 1 + t_i^2 & 2t_i^2 \\ 2t_i & 2t_i^2 & 2t_i^2 & 1 + t_i^2 \end{pmatrix} S_{4,i}.$$
$$(21)$$

The elementary rotations $\bar{R}_{2,i}$ and $H_{2,i}$ are the square of the elementary rotations $R_{2,i}$ in equation (1) with σ set to 1 and -1, respectively. We call the algorithm of equation (1) a 'square-root' 2-D CORDIC algorithm and the algorithms of equations (17) and (18) 'rational' CORDIC algorithms. The rational 2-D Euclidean CORDIC algorithm has been used in [5] [8] to speed up the Singular Value Decomposition (SVD) computations and in [12] to obtain a redundant arithmetic algorithm with constant scaling. For $n = 4$, while the algorithm with elementary rotations $R_{4,i}$ given by equation (6) is a square-root CORDIC algorithm, the new algorithm with elementary rotations $\bar{R}_{4,i}$ is a rational CORDIC algorithm whose unnormalized elementary rotations are only slightly more complex.

The convergence of the new 4-D CORDIC algorithm can be analyzed and proved using the procedure outlined in Section 1. The cones C_i have for vertex the origin and surround the first axis, thus they are fully characterized by their intersection with the hyperplane $x_1 = 1$. Since they are symmetric with respect to x_2, x_3, x_4, these 3-D polytopes can further be projected orthogonally onto the hyperplane $x_2 = 0$ to study the convergence behavior. Fig. 3 represents this projection for the cones C_1, C_2 and C_3.

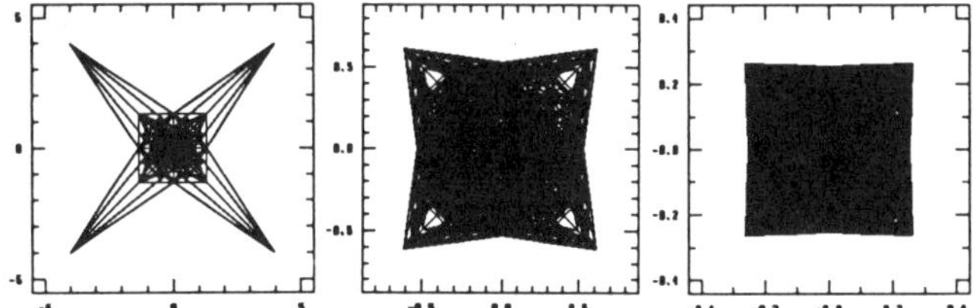

Figure 3: Domains after the first three iterations of $\bar{R}_4$.

Algorithm $\bar{R}_4$ does not require any repetition and converges faster than R_4, which requires repetition of the iterations with $t_i = 2^{-3}, 2^{-5}, 2^{-9}, 2^{-16}, \ldots$ [7]. The domain obtained by intersecting C_i with the hyperplane $x_1 = 1$ is bounded by a 3-D cube with half-side $2t_i/(1 - 3t_i^2)$.

2.3 Structure of the Elementary Rotations

The n-D elementary rotation matrix $H_{n,i}$ in equation (16) is the product of a reflection with respect to the hyperplane $x_1 = 0$ and a 'pseudo-Householder' reflection. Indeed, $H_{n,i}$ can be rewritten as

$$H_{n,i} = E_n \left(I_n - \frac{2\Sigma_n u u^T}{u^T \Sigma_n u} \right), \qquad (22)$$

where

$$E_n = \begin{pmatrix} -1 & 0 & 0 & \cdots & 0 \\ 0 & 1 & 0 & \cdots & 0 \\ \cdot & \cdot & \cdot & & \cdot \\ \cdot & \cdot & \cdot & & \cdot \\ 0 & 0 & 0 & \cdots & 1 \end{pmatrix} \text{ and } u = \begin{pmatrix} 1 \\ t_i s_{1,i} \\ \cdot \\ \cdot \\ t_i s_{n-1,i} \end{pmatrix}.$$

The matrix $I_n - 2\Sigma_n u(u^T \Sigma_n u)^{-1} u^T$ is a reflection with respect to the hyperplane normal to the vector u. It is orthogonal with respect to Σ_n and has determinant -1. It is a generalization of the Householder reflection to pseudo-Euclidean spaces. The role of the multiplication by the reflection E_n is to transform the pseudo-Householder reflection into a proper rotation, with determinant 1. Thus, our new elementary rotation matrix $H_{n,i}$ is essentially a Householder reflection in pseudo-Euclidean space. This is the origin for the name given to the new CORDIC algorithm.

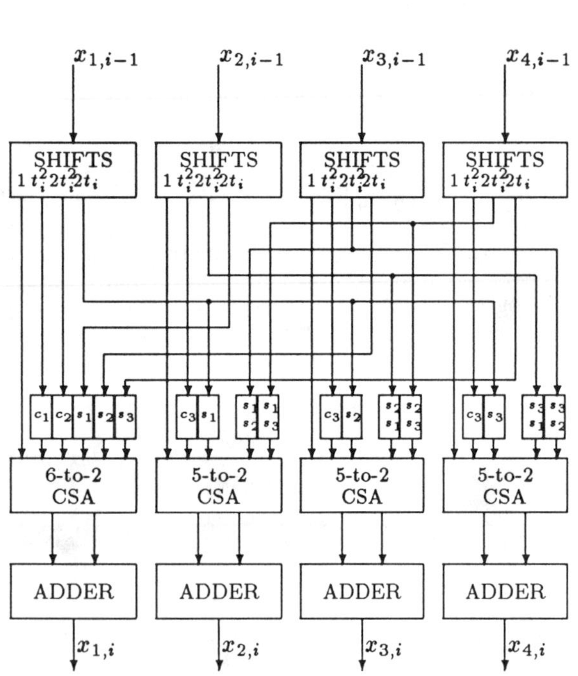

Figure 4: A possible architecture for a 4-D CORDIC processor.

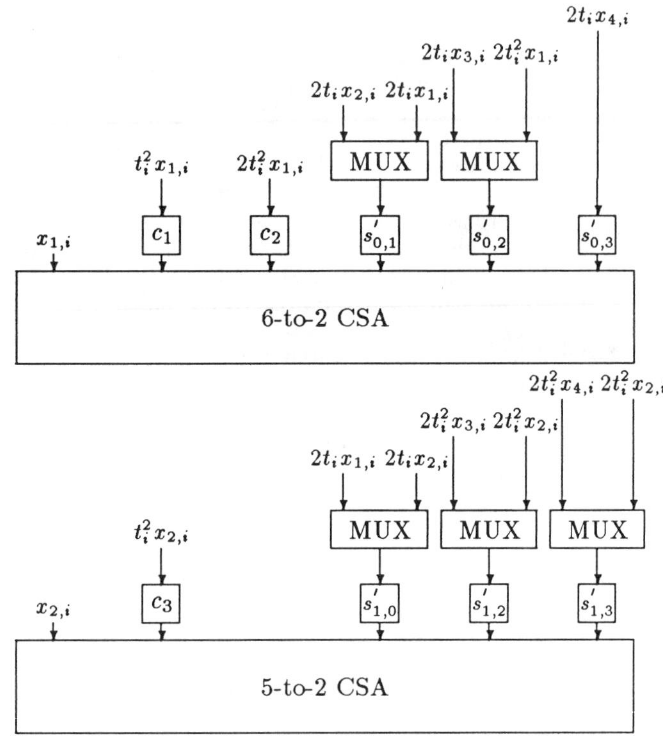

Figure 5: Modifications to the 4-D CORDIC processor architecture of Fig. 4 in order to perform scaling iterations in the processor

3 Processor Architectures

3.1 A 4-D Processor Architecture

A possible architecture to implement the 4-D CORDIC algorithm in equation (20) is illustrated in Fig. 4. By adding proper control inputs, c_1, c_2 and c_3, to the sign operators, this 4-D CORDIC processor can also execute the 2-D CORDIC algorithm in equation (17) and the 3-D CORDIC algorithm in equation (8). More precisely, in the (unscaled) rotation mode, the set of coefficients $(c_1, c_2, c_3) = (-1, 0, -1)$ selects the 2-D CORDIC function if $x_{3,i-1}$ and $x_{4,i-1}$ are preset to zero; $(c_1, c_2, c_3) = (0, -1, 0)$ selects the 3-D CORDIC function if $x_{4,i-1}$ is preset to zero; and $(c_1, c_2, c_3) = (-1, -1, 1)$ selects the 4-D CORDIC function.

Multiplication by a p-bit accurate approximation of the scaling factor $S_4 = \prod_{i=1}^{p}(1 + 3t_i^2)^{-1}$ can be performed by shifts and additions using q additional iterations, where q is small compared to p. However, the processor in Fig. 4 must be slightly modified to also perform the scaling operations. Several 2-to-1 multiplexers should be inserted at the input of the Carry-Save-Adders (CSA). Fig. 5 shows the modifications of the processor to enable scaling of the first and second components. The modifications for the third and fourth components are similar to those for the second component; the sign controls $(s'_{1,0}, s'_{1,2}, s'_{1,3})$ are replaced by $(s'_{2,0}, s'_{2,1}, s'_{2,3})$ and $(s'_{3,0}, s'_{3,1}, s'_{3,2})$, respectively.

The scaling factor S_4 is approximated by $\tilde{S}_4 = \prod_{i=1}^{q}(1 + a_i t_i + b_i t_i^2)$ where $a_i \in \{0, \pm 2\}$ and $b_i \in$ $\{0, \pm 1, \pm 2, \pm 3, \pm 4, \pm 5\}$. In the rotation mode, the multiplexers select data from the left-hand side inputs. The sign controls $s'_{k,l,i}, 0 \le k \ne l \le 3$, at iteration i are selected such that $s'_{k,l,i} = s_{k,i} s_{l,i}$ where $s_{0,i} = 1$. The iteration index i is dropped for clarity in Fig. 5. In the scaling mode, the multiplexers select the data from their right-hand side input. The sign controls are such that

$$c_1 = c_3, \qquad c_2 = s'_{1,3} = s'_{2,3} = s'_{3,2},$$
$$s'_{0,1} = s'_{1,0} = s'_{2,0} = s'_{3,0}, \qquad s'_{0,2} = s'_{1,2} = s'_{2,1} = s'_{3,1}.$$

We found that $q = 5$ extra iterations are required to perform the scaling operations in the 4-D CORDIC algorithm for 32 bits of accuracy.

A custom chip implementing the rational 2-D CORDIC rotation $\bar{R}_{2,i}$ in equation (17) has been designed at Yale and fabricated through MOSIS with a 2μ CMOS process. It operates on 32-bit fixed-point words with 5 extra guard bits. The $2t_i$ and t_i^2 shifters are realized by trees of 2-to-1 multiplexers, providing higher performance than barrel shifters. The area of two shifters by t_i and t_i^2, A_{sh2}, is about 1.4 times the area of a single shifter, A_{sh}, due to the possibility of overlapping layouts. The adders in this chip are realized by a variant of the Abraham-Gajski tree structure [1]. The depth of the trees to realize both shifters and adders is $\lceil \log_2 b \rceil$ where b is the word length, including guard bits. The area of such an adder is about 1.6 times A_{sh}. However, it is possible to design a faster adder with smaller area using multiple output domino logic and carry-look-ahead [10]. The area of this smaller adder, A_{add}, is about 0.6 times A_{sh}. The area of a 3-to-2 Carry-Save-Adder array, A_{csa}, is about 0.13 times A_{sh}. The area of a 2-to-1 word multiplexer, A_{mux}, is about 0.08 times A_{sh}. To sum up,

$$A_{sh2} = 1.4A_{sh} \quad A_{add} = 0.6A_{sh}$$
$$A_{csa} = 0.13A_{sh} \quad A_{mux} = 0.08A_{sh}.$$

Note that the sensitivity of these factors—1.4, 0.6, 0.13 and 0.08—to the bit accuracy is low since A_{sh}, A_{sh2} and A_{add} increase with bit accuracy slightly faster than linearly, while A_{csa} and A_{mux} are proportional to the bit accuracy.

The total area for the 4-D CORDIC processor, A_{4D}, is well approximated by the sum of the areas of the adders, shifters, CSA arrays, multiplexers and interconnection wires. The area of the wires, $A_{4D-wires}$, is about 2 times A_{sh} in this 4-D processor. Therefore

$$\begin{aligned} A_{4D} &\simeq 4A_{sh2} + 4A_{add} + A_{6-2CSA} \\ &\quad + 3A_{5-2CSA} + 11A_{mux} + A_{4D-wire} \\ &\simeq 5.6A_{sh} + 2.4A_{sh} + 0.52A_{sh} \\ &\quad + 1.17A_{sh} + 0.88A_{sh} + 2A_{sh} \\ &= 12.57A_{sh} \end{aligned}$$

Since the square-root 2-D CORDIC implementation in Fig. 1 requires

$$\begin{aligned} A_{2D} &\simeq A_{sh2} + 2A_{add} + 2A_{mux} + A_{2D-wire} \\ &\simeq 1.4A_{sh} + 1.2A_{sh} + 0.16A_{sh} + 0.58A_{sh} \\ &\simeq 3.34A_{sh}, \end{aligned}$$

the area of a 4-D CORDIC processor is about 3.8 times the area of a 2-D CORDIC processor.

The delay of the 6-to-2 and 5-to-2 CSA, T_{CSA}, is about three gate delays, $3T_{gate}$, where T_{gate} denotes the average delay per gate. The delay of a 2-to-1 multiplexer, T_{mux}, is T_{gate}. The delay of a shifter, $T_{shifter}$, is about $c_1\lceil \log_2 b \rceil$ gates delay for b-bit internal wordlength, where the factor c_1 accounts for the delay of the intermediate buffer stages driving the wires ($c_1 \simeq 1.3$). The delay of an adder is about $c_2\lceil \log_2 b \rceil$ gate delays, where c_2 accounts for the large carry-look-ahead gates ($c_2 \simeq 1.3$). The delay to drive the interconnection wires, T_{wires}, is about $2T_{gate}$. Therefore, for 32-bit accuracy, the delay of one 4-D CORDIC iteration is

$$\begin{aligned} T_{4D} &\simeq T_{shifter} + T_{mux} + T_{CSA} + T_{adder} + T_{wires} \\ &\simeq 6.5T_{gate} + T_{gate} + 3T_{gate} + 6.5T_{gate} + 2T_{gate} \\ &= 19T_{gate} \end{aligned}$$

while the delay for one 2-D CORDIC iteration is

$$\begin{aligned} T_{2D} &\simeq T_{shifter} + T_{mux} + T_{adder} + T_{wires} \\ &\simeq 6.5T_{gate} + T_{gate} + 6.5T_{gate} + 2T_{gate} \\ &= 16T_{gate}. \end{aligned}$$

To evaluate or apply a single 4-D rotation the proposed processor takes 60% of the time of two 2-D CORDIC processors at the expense of an area increase by a factor of 1.9. The gain in speed is thus obtained with a slight increase in the area-time product.

3.2 n-D Processor Architectures

Turning now to general n-D CORDIC algorithms, we found computationally that the first iteration (with $t_1 = 2^{-2}$) should be repeated twice for $n = 5, 6, 7$ and 8 to ensure convergence. The architecture in Fig. 4 can be easily generalized to implement an n-D CORDIC processor. For instance, a 5-D CORDIC processor requires an area $A_{5D} \simeq 1.3A_{4D}$ and a computation time $T_{5D} \simeq 1.1T_{4D}$. Therefore, a 5-D CORDIC processor takes about 40 % of the time ($= 3T_{2D}$) of two 2-D CORDIC processors to perform a 5-D rotation, with essentially no increase in the area-time product.

An alternative architecture for an n-D CORDIC processor, based on the the representation in terms of pseudo-Householder reflection given in equation (22), is shown in Fig. 6. Because of an additional shifter delay, this 'CORDIC Householder architecture' requires a longer computation time for each CORDIC iteration than the plain 'CORDIC architecture' exemplified in Fig. 4. On the other hand, it requires less area for high dimensional rotations ($n > 8$).

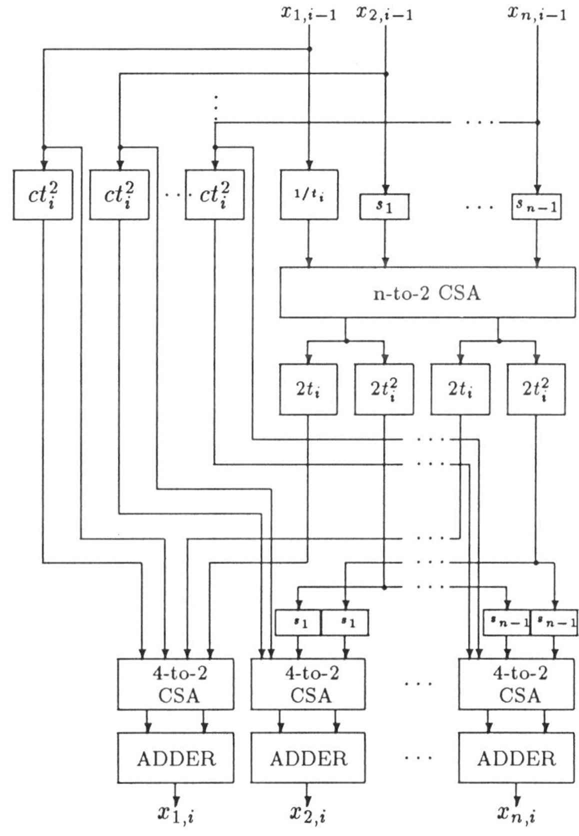

Figure 6: Architecture of an n-D pseudo-Euclidean CORDIC processor ($c = 2k - n - 1$).

3.3 Area and Time Comparisons

Two architectures have been presented: 'CORDIC architecture', which is the generalization of Fig. 4, and 'CORDIC Householder architecture' of Fig. 6. Since the pseudo-Householder transformation is considered to be the most efficient way to implement n-D vector rotations, we shall compare the two CORDIC architectures to a highly parallel implementation of the pseudo-Householder transformation. To any n-D vector $x = [x_1 \ x_2 \ \cdots \ x_n]^T$ in a space with signature matrix Σ_n may be associated a pseudo-Householder reflection bringing x along the axis $e_1 = [1 \ 0 \ \cdots 0]^T$:

$$H = I_n - 2\Sigma_n u(u^T \Sigma_n u)^{-1} u^T,$$

where $u = x + sign(x_1)\|x\|_{\Sigma_n} e_1$. The pseudo-Householder transformation calls for division, square-root and multiplications. Division and square-root extraction can be implemented by the Newton-Raphson method using multiplier and ROMs [14]. The multiplications are performed using n multipliers in parallel. For comparison purposes, the area and delay of a multiplier will be expressed in terms of the units A_{sh} and T_{gate}, respectively. The 32×32 multiplier presented in [11], which uses the Booth algorithm and a Wallace tree, will be used for our comparison. Compared to the 32×32 adder in [10], its area and delay are

$$A_{mpy} \simeq 35A_{add} \simeq 21A_{sh},$$
$$T_{mpy} \simeq 5A_{add} \simeq 5c_2\lceil\log_2 b\rceil T_{gate}.$$

In Fig. 7, $A_C, A_{CH}, A_H, T_C, T_{CH}$ and T_H denote the respective areas and delays of 'CORDIC', 'CORDIC Householder' and 'Householder' implementations for the Euclidean rotation of an n-D vector. Fig. 7(a) compares the areas of these implementations, Fig. 7(b) their delays, and Fig. 7(c) their area-time products. The two implementations of the new CORDIC algorithm are clearly preferable in terms of area-time product to the parallel implementation of the pseudo-Householder transformation.

4 Application to the QR Decomposition on a Processor Array

Givens rotations are widely used in modern signal processing algorithms and the square-root 2-D CORDIC algorithm is often employed to implement these rotations. Two well-known examples are the QR Decomposition (QRD) and Singular Value Decomposition (SVD)[2][3]. We shall examine here the QRD of a rectangular matrix to show how higher dimensional CORDIC algorithms can speed up array implementations of matrix computations.

The n-D CORDIC algorithm discussed in Section 2 enables $n-1$ elements to be zeroed out at each rotation instead of a single element for a Givens rotation. To triangularize an $l \times m$ matrix A with elements $a_{i,j}$, an n-D rotation is applied at time step 1 to zero out $a_{2,1}, \cdots, a_{n,1}$ in the first column of A. Simultaneously, this rotation is applied to the other elements of rows 1 to n. At time step 2, another n-D rotation is applied to rows 1 and $n+1$ to $2n-1$ to zero out $a_{n+1,1}, \cdots, a_{2n-1,1}$, while an $(n-1)$-D rotation is applied to rows 2 to n to zero out $a_{3,2}, a_{4,2}, \cdots, a_{n,2}$. Proceeding with this greedy approach, an $l \times m$ matrix is triangularized in

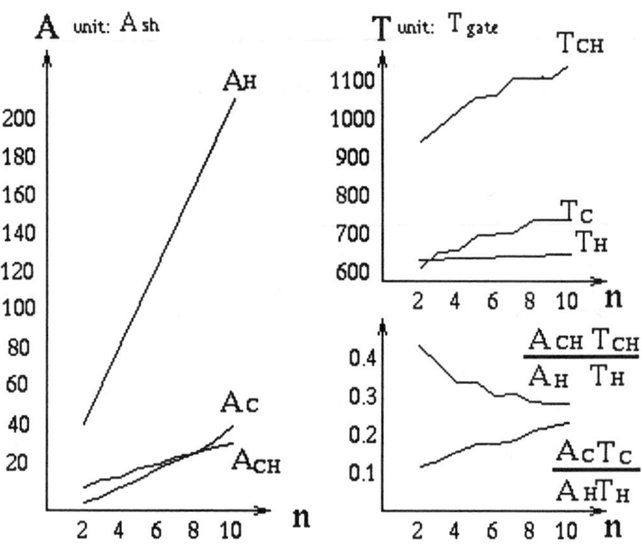

Figure 7: Area, time and area-time product as function of dimension n ($b=32$).

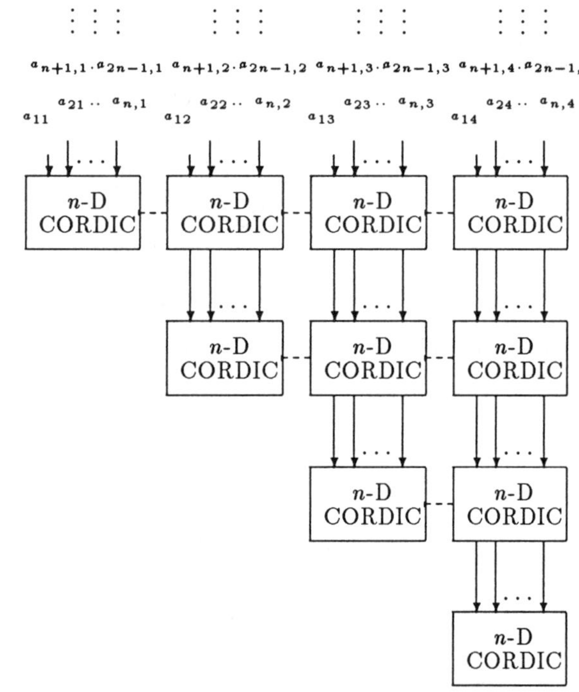

Figure 8: Array of n-D CORDIC processors for decomposing an $l \times 4$ matrix.

QRD		2-D CORDIC	4-D CORDIC
time		$16(l + m - 2)\times$ $(\bar{p} + q_1)$	$19(\lceil\frac{l-1}{3}\rceil + m - 1)\times$ $(\bar{p} + 1 + q_2)$
period		$16(l - 1)(\bar{p} + q_1)$	$19(\lceil\frac{l-1}{3}\rceil)(\bar{p} + 1 + q_2)$
area		$1.7m(m + 1)$	$6.4m(m + 1)$

Table 1: Comparison, for the QRD of $l \times m$ matrices, of the computation times for one problem, and of the periods and processor areas. The time unit is T_{gate} and the area unit is A_{sh}.

$\lceil(l - 1)/(n - 1)\rceil + (m - 1)$ time steps on a triangular array of $m(m + 1)/2$ n-D CORDIC processors as shown in Fig. 8. Each time step allows for a whole n-D CORDIC rotation. If several $l \times m$ matrices are pipelined into the array, the pipeline period to compute one QRD is $\lceil(l - 1)/(n - 1)\rceil$ time steps. Note that the ability of an n-D CORDIC processor to also perform lower dimensional rotations is crucial in this context.

Table 1 compares the computation time, the pipeline period and the processor area required for the QRD of $l \times m$ matrices using the square-root 2-D CORDIC algorithm and the new 4-D CORDIC algorithm; $\bar{p}$ is the bit accuracy, and q_1 and q_2 are the number of additional iterations required for scaling operations in the 2-D and 4-D algorithms, respectively. The 4-D CORDIC algorithm provides a speed up of 2.5 at the expense of an area increase by a factor of 3.8. The time performance is thus significantly improved with only a 50% increase in the area-time product.

5 Conclusion

A general n-D CORDIC algorithm, applicable to rotations in Euclidean and pseudo-Euclidean spaces, has been introduced. In the 3-D and 4-D Euclidean cases, the algorithm does not require any repetition and thus improves on previously developed CORDIC algorithms. The elementary rotations in the new algorithm are essentially pseudo-Householder reflections. Parallel implementations of the algorithm require, for $n \leq 10$, much less area than a parallel implementation of the Householder transformation to achieve roughly the same computation time and, in addition, they are simpler to design. Finally, the example of the QR decomposition illustrates the use of the new n-D algorithms in application-specific processor arrays in order to speed up matrix computations.

6 Acknowledgement

This work was supported by the Defense Advanced Project Research Agency (DARPA) under contract N00014-88-K-0573.

References

[1] J. Abraham and D. Gajski, *Easily Testable, High-speed Realization of Register-Transfer-Level Operations*, Proc. 10th Fault-Tolrant Computing Symp., pp. 339-344, Oct. 1980.

[2] H.M. Ahmed, J.-M. Delosme and M. Morf, *Highly Concurrent Computing Structures for Matrix Arithmetic and Signal Processing*, IEEE Computer, Vol. 15, No. 1, pp. 65-82, Jan. 1982.

[3] J.R. Cavallaro and F.T. Luk, *CORDIC Arithmetic for an SVD processor*, Journal of Parallel and Distributed Computing, pp. 271-290, June 1988.

[4] J.-M. Delosme, *VLSI Implementation of Rotations in Pseudo-Euclidean Spaces*, Proc. 1983 IEEE Conf. on ASSP, pp. 927-930, Apr. 1983.

[5] J.-M. Delosme, *A Processor for Two-dimensional Symmetric Eigenvalue and Singular Value Arrays*, Proc. 21st Asilomar Conf. on Circ., Syst., and Comp., pp. 217-221, Nov. 1987.

[6] J.-M. Delosme, *CORDIC Algorithms: Theory and Extensions*, Advanced Algorithms and Architectures for Signal Processing IV, Proc. SPIE 1152, pp. 131-145, Aug. 1989.

[7] J.-M. Delosme and S.-F. Hsiao, *CORDIC Algorithms in Four Dimensions*, Advanced Signal Processing Algorithms, Architectures and Implementations, Proc. SPIE 1348, pp. 349-360, July 1990.

[8] J.-M. Delosme, *Bit-level Systolic Algorithm for the Symmetric Eigenvalue Problem*, Proc. Int. Conf. on Application Specific Array Processors, pp. 770-781, Princeton, Sept. 1990.

[9] G. Haviland and A. Tuszynski, *A CORDIC Arithmetic Processor Chip*, IEEE Trans. on Computers, Vol. C-29, No. 2, pp. 68-79, Feb. 1980.

[10] 1.S. Hwang and A.L. Fisher, *Ultrafast Compact 32-bit CMOS Adders in Multiple-Output Domino Logic*, IEEE J. Solid State Circuits, Vol. SC-24, pp. 358-369, Apr. 1989.

[11] M. Nagamatsu et al., *A 15-ns 32 x 32-b CMOS Multiplier with an Improved Parallel Structure*, IEEE J. Solid State Circuits, Vol. SC-25, pp. 494-496, Apr. 1990.

[12] N. Takagi, T. Asada and S. Yajima, *A Hardware Algorithm for Computing Sine and Cosine Using Redundant Binary Representation*, Trans. IECE Japan, Vol. J69-D, No. 6, pp. 841-847, June 86 (in Japanese). English translation available in *Systems and Computers in Japan*, Vol. 18, No. 8, pp. 1-9, Aug. 1987.

[13] J.E. Volder, *The CORDIC Trigonometric Computing Technique*, IRE Trans. on Electronic Computers, Vol. EC-8, No. 3, pp. 330-334, Sept. 1959.

[14] S. Waser and M.J. Flynn, *Introduction to Arithmetic for Digital Systems Designs*, CBS College Publishing, 1982.

[15] J.S. Walther, *A Unified Algorithm for Elementary Functions*, AFIPS Conf. Proc., Vol. 38, pp. 379-385, 1971.

SVD by Constant-Factor-Redundant-CORDIC *

Jeong-A Lee

Department of Electrical Engineering
University of Houston
Houston, TX 77204-4793

Tomás Lang

Computer Architecture Department
Universitat Politecnica de Catalunya
Barcelona, Spain

Abstract

We develop a Constant-Factor-Redundant-CORDIC (CFR-CORDIC) scheme where the scale factor is forced to be constant while computing angles for SVD. Based on the scheme, we present a fixed-point implementation of SVD with the following additional features: (1) the final scaling operation is done by shifting, (2) the number of iterations in CORDIC rotation unit is reduced by about 25% by expressing the direction of the rotation in radix-2 and radix-4, and (3) the conventional number representation of rotated output is obtained on-the-fly, not from a carry-propagate adder. We compare this scheme with previously proposed ones and show that it provides similar execution time as redundant CORDIC with variable scaling factor with significant saving in area.

1 Introduction

Singular value decomposition(SVD) is an important tool in a variety of applications of digital signal processing [1, 2]. Since the decomposition consists of computation-intensive tasks, parallel algorithms and architectures have been developed to get an adequate throughput rate (processing time) [3, 4].

Among the parallel algorithms and architectures for SVD, we consider the one presented by Brent, Luk, and Van Loan [3]. It consists of a mesh-connected processor array, where diagonal processors compute left and right angles to diagonalize a 2x2 matrix. These angles are passed to off-diagonal processors in the column and row directions to perform the two-sided plane rotations.

The angle calculation and the rotation can be performed using standard arithmetic modules such as adders, multipliers, dividers, and square root units [4]. However, in this case the angle calculations require a long sequence of operations resulting in a large number of modules and a long execution time. An attractive alternative presented by Cavallaro and Luk [5] is to use the CORDIC algorithm which is effective for both the angle calculations and the rotations. The execution time of this approach was reduced by the redundant-CORDIC scheme presented by Ercegovac and Lang [11], which has the drawback that the scaling factor is not constant and has to be calculated for each angle.

In this paper we develop an algorithm and an implementation for SVD using Constant-Factor-Redundant-CORDIC, which is based on the scheme developed by Takagi, Asada, and Yajima for the computation of sine and cosine [12].

We review previous work on CORDIC for SVD in Section 2. Following that, we develop a Constant-Factor-Redundant-CORDIC scheme, evaluate the performance of this new scheme with respect to time and space, and compare it with previously proposed schemes in Section 3. The summary and conclusion are presented in Section 4.

2 Previous CORDIC work for SVD

We briefly review the conventional(non-redundant) and redundant CORDIC schemes to compute angles for SVD and perform the corresponding rotations.

2.1 Non-redundant CORDIC

Cavallaro and Luk presented a CORDIC processor for SVD [5] using conventional(non-redundant) arithmetic based on the original CORDIC algorithm [6, 7]. The two modes of operation are as follows:

Angle calculation(vectoring mode): The angle $\theta = \tan^{-1}(\frac{Y_a[0]}{X_a[0]}) = Z_a[n]$ is computed from the following recurrence equations with $Z_a[0] = 0$.

$$X_a[i+1] = X_a[i] + \sigma_i 2^{-i} Y_a[i]$$
$$Y_a[i+1] = Y_a[i] - \sigma_i 2^{-i} X_a[i]$$
$$Z_a[i+1] = Z_a[i] + \sigma_i tan^{-1} 2^{-i} \qquad (1)$$

where, the direction of the rotation σ_i is obtained as

$$\sigma_i = \begin{cases} 1 & \text{if } Y_a[i] \geq 0 \\ -1 & \text{if } Y_a[i] < 0 \end{cases} \qquad (2)$$

To compute θ with n-bit precision, n CORDIC iterations need to be executed. The step time of one CORDIC iteration is determined by σ computation,

---

*This research was done while the authors were at the Computer Science Department, UCLA, and was supported in part by the NSF Grant No. MIP-8813340 *Composite Operations Using On-Line Arithmetic for Application-Specific Parallel Architectures: Algorithms, Design, and Experimental Studies.*

variable shifting(for 2^{-i} term), and addition. Notice that to use non-redundant selection function(to know the sign of $Y[i]$), a carry-propagate adder is needed.

For SVD, θ_{left} and θ_{right} have to be calculated. For the calculation of these angles, Cavallaro and Luk adopted the direct two angle method; this requires the computation of the intermediate angles θ_{sum} and θ_{diff} such that

$$\theta_{sum} = \theta_{left} + \theta_{right} = \tan^{-1}\frac{y+x}{z-w}$$

$$\theta_{diff} = \theta_{right} - \theta_{left} = \tan^{-1}\frac{y-x}{z+w}$$

where x, y, w, z are the elements of the 2 x 2 matrix.

After that, θ_{left} and θ_{right} are computed. The computation of these angles is performed using an adder/subtractor and two CORDIC modules as described in Figure 1.

Rotation(rotating-mode): To perform a plane rotation on input vector $(X_r[0], Y_r[0])$ with an angle ($\theta = Z[0]$), the following recurrence equations are used.

$$X_r[i+1] = X_r[i] + \sigma_i 2^{-i} Y_r[i]$$
$$Y_r[i+1] = Y_r[i] - \sigma_i 2^{-i} X_r[i]$$
$$Z[i+1] = Z[i] - \sigma_i \tan^{-1} 2^{-i} \tag{3}$$

where, σ_i is computed from $Z[i]$

$$\sigma_i = \begin{cases} 1 & \text{if } Z[i] \geq 0 \\ -1 & \text{if } Z[i] < 0 \end{cases} \tag{4}$$

Notice that the recurrence equations of X and Y are the same for both angle and rotation modes, which only require shifters and adders for implementation. To perform the two-sided rotation(with the σ-encoded values of θ_{left} and θ_{right}), two CORDIC operations are performed as shown in Figure 1.

Scaling: CORDIC changes the length of the vector while rotating. To maintain the same length of the input vector $(X_r[0], Y_r[0])$, the following CORDIC scaling operation is necessary, where (x^R, y^R) is the rotated vector.

$$\begin{bmatrix} x^R \\ y^R \end{bmatrix} = \frac{1}{K} \begin{bmatrix} X_r[n] \\ Y_r[n] \end{bmatrix} \quad \text{where } K = \prod_{i=0}^{n-1} \sqrt{1 + \sigma_i^2 2^{-2i}} \tag{5}$$

Note that K is *constant* for non-redundant CORDIC since $\sigma_i^2 = 1$ for all i.

Since two angles (left and right angles) are associated with CORDIC for SVD, the corresponding scale factor becomes K^2. Instead of performing the scaling operation by multiplication by $1/K^2$, Cavallaro and Luk proposed a method similar to the ones proposed in [8, 9, 10], which includes additional number of iterations to make the scaling factor K^2 equal to 2 and perform the scaling by shifting. For an efficient implementation, the total number of these additional iterations need to be minimized; they showed that there exists a solution with 25% more iterations to force K^2 to be 2.

Evaluation: The timing diagram is shown in Figure 2. This SVD processor takes $3.25T_C$ cycles where T_C corresponds to the time to complete either an angle calculation or a plane rotation. The three CORDIC operations (one to compute the two angles θ_{sum} and θ_{diff} simultaneously, and two to perform the two-sided rotations) take $3T_C$ and the final scaling operation for the two-sided rotation takes $0.25T_C$ cycles.

2.2 Redundant CORDIC

Angle computation and Rotation: Ercegovac and Lang [11] have shown that the performance of the processor can be improved by using the redundant arithmetic(carry-free addition) and by determining the direction of the rotation based on an estimate instead of an exact value. This requires the modification of the recurrence and selection function to determine the direction of rotation, $\hat{\sigma}$.

In addition, $\hat{\sigma}_i^{left}$ and $\hat{\sigma}_i^{right}$ (corresponding to θ_{left} and θ_{right}) are obtained on-the-fly using $\hat{\sigma}_i^{sum}$ and $\hat{\sigma}_i^{diff}$, which eliminates Z recurrence since θ_{sum} and θ_{diff} do not need to be computed. This computation is done by the following on-line algorithm,

$$W[i] = Z[i] + \hat{\sigma}_{i+p+1} 2^i \tan^{-1}(2^{-(i+p+1)}) \tag{6}$$

$$Z[i+1] = 2(W[i] - \hat{\gamma}_i 2^i \tan^{-1}(2^{-i})) \tag{7}$$

with an initial condition

$$Z[0] = \sum_{j=0}^{p} \hat{\sigma}_j \tan^{-1}(2^{-j}) \tag{8}$$

where $\hat{\sigma}$ belongs to the set $\{ \pm 1, \pm\frac{1}{2}, 0 \}$ and $\hat{\gamma}$ to the set $\{ \pm 1, 0 \}$. The corresponding selection function, which determines $\hat{\gamma}_i$ using an estimate of $W[i]$ and on-line delay $p = 2$ is

$$\hat{\gamma}_i = \begin{cases} 1 & \text{if } \hat{W}[i] \geq \frac{1}{2} \\ 0 & \text{if } -\frac{1}{4} \leq \hat{W}[i] \leq \frac{1}{4} \\ -1 & \text{if } \hat{W}[i] \leq -\frac{1}{2} \end{cases} \tag{9}$$

where $\hat{W}[i]$ is computed using two fractional bits of the redundant representation of $W[i]$.

The computations of these steps are shown in Figure 3. Notice that since $\hat{\gamma}$ is produced from $\hat{\sigma}$, Z recurrence is not needed for CORDIC rotation units, either.

Scaling: The same scaling operation as in non-redundant CORDIC has to be performed. However, in redundant CORDIC, the scale factor K is *not a constant* as $\hat{\gamma}_i \in \{-1, 0, 1\}$. Therefore, K must be computed for each $\hat{\gamma}$-encoded angle, which has two steps: (1) K^2 computation using $P[j+1] = P[j] + |\hat{\gamma}_j| 2^{-2j} P[j]$ with initial condition $P[0] = 1$ and (2) $K = \sqrt{P}$. K computation is overlapped with angle computation to minimize the loss of timing. However, the division operation is needed for the final scaling operation.

Evaluation: The timing diagram is shown in Figure 4. This scheme produced a factor of 4 speed-up compared with the non-redundant scheme [11]. The speed-up is due to the following factors: redundant arithmetic, use of an estimate to determine the direction of the rotation, generation of encoded angles of θ_{left} and θ_{right} ($\hat{\gamma}_i$) on-the-fly from encoded angles of θ_{sum} and θ_{diff}. However, it introduced additional cost since the scale factor is a variable.

3 CFR-CORDIC for SVD

In this section, we develop a new redundant scheme, called Constant-Factor-Redundant-CORDIC(CFR-CORDIC) for SVD to reduce the implementation cost of redundant CORDIC.

3.1 Development of CFR-CORDIC

Angle computation: To compute the direction of the rotations, $\hat{\sigma}_i$(more specifically $\hat{\sigma}_i^{sum}$ and $\hat{\sigma}_i^{diff}$), the modified recurrence equations and selection function for signed-digit representation [11] are used.

$$X[i+1] = X[i] + \hat{\sigma}_i 2^{-2i} W[i]$$
$$W[i+1] = 2(W[i] - \hat{\sigma}_i X[i]) \tag{10}$$
$$\text{where } W[i] = 2^i Y[i] \tag{11}$$

$$\hat{\sigma}_i = \begin{cases} 1 & \text{if } \hat{W}[i] \geq \frac{1}{2} \\ 0 & \text{if } \hat{W}[i] = 0 \\ -1 & \text{if } \hat{W}[i] \leq -\frac{1}{2} \end{cases} \tag{12}$$

where $\hat{W}[i]$ is an estimate computed using one fractional bit of the redundant representation of $W[i]$.

The $\hat{\sigma}$-encoded angles of θ_{left} and θ_{right} are obtained again on-the-fly while $\hat{\sigma}$-encoded angles of θ_{sum} and θ_{diff} are produced. That is,

$$\hat{\sigma}_i^{left} \tan^{-1} 2^{-i} = \frac{\hat{\sigma}_i^{sum} \tan^{-1} 2^{-i} - \hat{\sigma}_i^{diff} \tan^{-1} 2^{-i}}{2}$$

$$\hat{\sigma}_i^{right} \tan^{-1} 2^{-i} = \frac{\hat{\sigma}_i^{sum} \tan^{-1} 2^{-i} + \hat{\sigma}_i^{diff} \tan^{-1} 2^{-i}}{2}$$

As discussed earlier, the result of direct computation of $\hat{\sigma}_i^{left}$ and $\hat{\sigma}_i^{right}$ from $\hat{\sigma}_i^{sum}$ and $\hat{\sigma}_i^{diff}$ produce the set $\{\ \pm 1, \pm\frac{1}{2},\ 0\ \}$. (Hereafter we will again skip the superscript to simplify the notation.)

To have a constant scale factor, we now need to find a set of $\hat{\gamma}_j$, which is in $\{-1, 1\}$, satisfying

$$\sum_{j=0}^{n-1} \hat{\sigma}_j \tan^{-1}(2^{-j}) = \sum_{j=0}^{n-1} \hat{\gamma}_j \tan^{-1}(2^{-j})$$

As in [11] we define the residual Z as:

$$Z[i] = 2^i \left(\sum_{j=0}^{i+p} \hat{\sigma}_j \tan^{-1}(2^{-j}) - \sum_{j=0}^{i-1} \hat{\gamma}_j \tan^{-1}(2^{-j}) \right)$$

where p is the on-line delay. The resulting recurrence equation is

$$Z[i+1] = 2(Z[i] + \hat{\sigma}_{i+p+1} 2^i \tan^{-1}(2^{-(i+p+1)}) - \hat{\gamma}_i 2^i \tan^{-1}(2^{-i}))$$

with an initial condition

$$Z[0] = \sum_{j=0}^{p} \hat{\sigma}_j \tan^{-1}(2^{-j}) \tag{13}$$

To simplify the selection function, the recurrence equation is decomposed into two parts by introducing an intermediate variable W, as shown before [11],

$$W[i] = Z[i] + \hat{\sigma}_{i+p+1} 2^i \tan^{-1}(2^{-(i+p+1)}) \tag{14}$$
$$Z[i+1] = 2(W[i] - \hat{\gamma}_i 2^i \tan^{-1}(2^{-i})) \tag{15}$$

To have a constant scale factor, the direction of the rotation $\hat{\gamma}$, now need to be in $\{-1, 1\}$. However, since $\hat{\gamma}$ is determined from an estimate and from a partial set of $\hat{\sigma}$, the convergence can not be assured.

Takagi et al. [12] proposed *a correcting iteration scheme* to solve the similar problem, which occurs when redundant CORDIC is used to compute cosine and sine function. For these computations, the direction of the rotation $\hat{\sigma}_i$ is obtained from the estimate of $Z[i]$ and is forced to be in $\{-1, 1\}$ to have a constant scale factor. To assure convergence, some iterations of CORDIC, called *correcting iterations* , are repeated at fixed intervals, where the frequency of repetition depends on the precision of the estimate. Since this scheme is proposed for cosine and sine computation, only the rotation mode of CORDIC is discussed.

We have shown the extension of this scheme to compute angle [13], especially for matrix triangularization [14], which requires to deal with the inter-dependency of the recurrences of X and Y.

In the case of CFR-CORDIC for SVD, the problem is aggravated by the fact that not only an estimate of $Z[i]$ is used but its computation is on-line. We use the concept of correcting iterations and on-line delays to solve this problem. In addition, we improve the scheme by minimizing the number of extra correcting iterations by dividing the iterations of CORDIC(for $i = 0$ to $i = n-1$) into two groups: one group where the direction of the rotation is in $\{-1, 1\}$ for $i = 0$ to $i = \frac{n}{2}$ and the other in $\{-1, 0, 1\}$ for $i = \frac{n+1}{2}$ to $i = n - 1$. With these two groups, we reduce the number of correcting iterations by 50 % since correcting iteration is not needed for the second half of the iterations and we still obtain a constant scale factor K since the value of K in n-bit precision does not depend on the value of direction of the rotation for $\frac{n+1}{2} \leq i \leq (n-1)$.

The corresponding selection functions to make the constant scale factor are :

$$(1)\ 0 \leq i < \frac{n}{2}: \quad \hat{\gamma}_i = \begin{cases} 1 & \text{if } \hat{W}[i] \geq 0 \\ -1 & \text{if } \hat{W}[i] < 0 \end{cases} \tag{16}$$

where the estimate $\hat{W}[i]$ is computed using t fractional bits of the redundant representation of $W[i]$. As discussed later, the value of t and on-line delay p shown in Equation 14 depends on the frequency of correcting iterations.

$$(2) \quad \frac{n}{2} \leq i \leq n-1: \quad \hat{\gamma}_i = \begin{cases} 1 & \text{if } \hat{W}[i] \geq \frac{1}{2} \\ 0 & \text{if } -\frac{1}{4} \leq \hat{W}[i] \leq \frac{1}{4} \\ -1 & \text{if } \hat{W}[i] \leq -\frac{1}{2} \end{cases} \tag{17}$$

Notice that the selection function of the second half iterations is the same one as shown in Equation 9.

This selection function forces the scale factors for the two-sided rotations to be constant but does not satisfy the convergence requirement. In fact, the convergence requirement was satisfied in the previous redundant scheme by setting the $\hat{\gamma}$ to be 0 for the region $U_{-1} < W[i] < L_1$ (more accurately, by forcing $L_0 \leq U_{-1}$ and $L_1 \leq U_0$) where $L_1 = 2^i \tan^{-1}(2^{-(i+p+1)})$ and $U_{-1} = -L_1$. This interval $[L_k, U_k]$ of $W[i]$ for the redundant scheme of Ercegovac and Lang was determined such that $Z[i+1]$ remained bounded when choosing $\hat{\gamma}_i = k$.

As we already pointed out, $\hat{\gamma}$ must not be zero to have a constant scale factor. Furthermore, the region to select 1 for $\hat{\gamma}$ is bigger than that to select -1 due to the use of the estimate. More specifically, for the region $-2^{-t} < W[i] \leq 0$, $\hat{\gamma}$ is selected to be 1. The worst case(where the amount outside of the bound for $W[i+1]$ is maximum) occurs when $W[i] = -2^{-t} + \epsilon$. The resulting $Z[i+1]$ and $W[i+1]$ are

$$Z[i+1] = -2 - 2 * 2^{-t} \quad \text{since } \hat{\gamma}_i = 1$$

$$W[i+1] = -2 - 2 * 2^{-t} - 2^{-(p+1)}$$

$$\text{assuming } \sigma_{i+p+2} = -1 \text{ for the worst case}$$

In the following iterations, the amount outside of the bound increases and $W[i+m]$ becomes:

$$W[i+m] = -2 - 2^m 2^{-t} - 2^{m-1} 2^{-(p+1)}$$
$$-2^{m-2} 2^{-(p+1)} - \ldots - 2^{-(p+1)}$$

To recover the bound, we use here a correcting iteration defined as:

$$Z[i+m]^c = W[i+m] - \hat{\gamma}_{i+m}^c 2^{i+m+1} \tan^{-1} 2^{-(i+m)}$$

$$W[i+m]^c = Z[i+m]^c$$

where $\hat{\gamma}_{i+m}^c$ is determined by the same selection function shown in Equation 16. After this, $Z[i+m+1]$ is obtained by the original equation 15 using $W[i+m]^c$ instead of $W[i+m]$. Now to satisfy the convergence requirement, $W[i+m]^c \geq L_{-1}$ must be forced.[1] From this condition, we get

$$2^{-t} + 2^{-(p+1)} < 2^{1-m} \tag{18}$$

---

[1] $L_{-1} = -2*2^{i+m} \tan^{-1} 2^{-(i+m)} + 2^{i+m} \tan^{-1}(2^{-(i+m+p+1)})$,as determined in [11].

By solving this, we can find out the relationship among t(the number of fractional bits), p(the online-delay), and m(the frequency indicator for a correcting iteration). We would like to maximize the value m so as not to perform a correcting iteration too often. At the same time, we want to minimize the value t for an efficient implementation of the selection function. A solution of $m = p+1$ and $t = p+2$ satisfies these restrictions. For an example, with on-line delay $p = 3$, a correcting iteration must occur with an interval of 4 or less and 5 fractional bits are used for the estimate. The computation steps are shown in Figure 5.

Rotation: As shown in Figure 5, since $\hat{\gamma}$ is produced from $\hat{\sigma}$, Z recurrence is not needed again for CORDIC rotation units.

The rotation processor can also incorporate two schemes we developed to speed up its processing. The first one is to reduce the number of iterations in CORDIC rotation unit by about 25% by expressing the direction of the rotation in radix-4 $\hat{\gamma}$ for the second half of the iterations: this method is based on the fact that for $i \geq \frac{n}{2}$, $\tan^{-1} 2^{-i} = 2^{-i}$ in n-bit precision. The second one is to convert the redundant number representation of the rotated output into the conventional number representation on-the-fly, not using a carry-propagate adder. The detailed development of these two schemes can be found in [13]. The incorporation of these schemes into an application system, specifically for matrix triangularization, was discussed in [14].

Scaling: When the final scaling operation is needed, it can be simply a shifting by forcing the scale factor K to be 2 (similar to the work for non-redundant CORDIC mentioned earlier), which requires to repeat iterations of CORDIC and scaling iteration of the following form:

$$X[i] = X[i] + \beta_j X[i] 2^{-j}$$

$$Y[i] = Y[i] + \beta_j Y[i] 2^{-j} , \ \beta_j \in \{-1, 1\}$$

The scaling iteration manipulates the extension or reduction of the vector norm without changing $\tan^{-1} \frac{X[i]}{Y[i]}$. Consequently, the scaling iteration does not affect the output of CORDIC angle unit. Moreover, when the precision of the output is fixed, β_j can be pre-computed.

For an efficient implementation, we need to minimize the number of additional iterations of CORDIC and scaling iterations. As discussed in [13], the problem of searching for the minimum number of additional iterations for CFR-CORDIC is more complicated(time-consuming) than the one in non-redundant CORDIC since the CFR-CORDIC requires *correcting iterations* to assure convergence. We developed an efficient searching method, called decomposed search, to reduce the searching time from T to $\sqrt{T}$ [13].

3.2 Performance analysis

Evaluation: Consider the case of $n = 16$ (16-bit precision data) and 5 fractional bits in the estimate. From the condition 18, we know that the interval of a correcting iteration is 4 or less, $m = 4$. Under these conditions,

Type	Total-time	$n = 16$
Non-redundant	$16.25n$	260
Redundant CORDIC	$5n + 10$	90
CFR-CORDIC	$4n + \alpha$	84

Table 1: Time comparisons (basic cycles)

Type	Total-time	$n = 16$
Non-redundant	$13.75n$	220
Redundant CORDIC	$4n + 10$	74

Table 2: Time with radix-4 scheme(basic cycles)

Type	Area estimate
Non-redundant CORDIC	3
Redundant CORDIC	6.4
CFR-CORDIC	3.4

Table 3: Area comparisons (units)

the minimum number of additional iterations is searched to force the scale factor to be 2 using *decomposed search* , which results in 6 additional iterations: repetitions of the 4th, the 8th and the 9th CORDIC iterations and scaling iterations of $\{+2, +10, -5 \}$.

The corresponding overall system timing is shown in Figure 6. As can be seen in the diagram, for $n = 16$ and $m = 4$, the timing of both processors matches well, resulting in an overall time of 42 iterations of CORDIC.

Comparisons: To compare the various schemes we need to define a basic cycle. As in [11], we use as this basic cycle the time taken by an iteration similar to radix-2 SRT division, that is, consisting of a 3-1 multiplexer, a carry-save adder, a simple selection function, and the loading of a register. Using this basic cycle, we determine in each scheme the number of basic cycles to complete one iteration of CORDIC and the total time to complete two-sided plane rotations for SVD, namely, angle computations(θ_{left} and θ_{right}) and two-sided plane rotations.

For non-redundant scheme, we estimate that one iteration takes 5 basic cycles since variable shifting and carry-propagate addition need to be performed. Consequently, T_C takes $5n$ basic cycles and the total time takes $16.25n$ basic cycles. For redundant CORDIC with variable scaling, one iteration takes 2 basic cycles as discussed in [11], resulting in an overall time of $5n + 10$ basic cycles shown in the timing diagram of Figure 4. For CFR-CORDIC, assuming that the number of fractional bits for the estimate is chosen not to increase the step time (that is, the selection will be done within the time of variable shifting), one iteration again takes 2 basic cycles. In this case, the total number of iterations depends on n(precision of data) and m(frequency of correcting iterations). Consequently, the overall time becomes $4n + \alpha$, where $4n$ corresponds to the two-sided rotations and α to the correcting iterations and on-line delays. For 16-bit precision and an estimate of 5 fractional bits, the CFR-CORDIC scheme takes 42 iterations, resulting in 84 basic cycles. Table 1 summarizes the time comparisons, which shows that the CFR-CORDIC scheme is the fastest scheme for $n = 16$.

The performance of non-redundant and redundant CORDIC can also be improved by applying our scheme with radix-4 direction of the rotation(for the second half of the iterations) since it reduces the total number of iterations as shown in Table 2. In this case, the scheme with variable scaling is faster.

To estimate the area requirement in each scheme, let's define the basic unit to be the area consumed by one conventional CORDIC unit to perform one iteration of angle computation with X, Y, and Z recurrence equations. We also assume that the CORDIC rotation unit used for the left angle is reused for the right angle.

Then, the non-redundant scheme takes about 3 units since it requires 3 CORDIC units. For redundant CORDIC, all 3 CORDIC units do not have Z recurrence equations and carry-free additions are employed, resulting in 2.6 units for CORDIC units. However, additional areas are necessary to compute $\hat{\gamma}$ on-the-fly(0.8 unit used twice for left and right in sequence) and to compute K_{left}(1 unit) and K_{right}(1 unit), on-line computation of $K_{left} * K_{right}$ and division for final scaling (1 unit). The resulting area consumption for redundant CORDIC becomes 6.4 units. In the case of CFR-CORDIC, we again require 2.6 units for CORDIC modules similar to the redundant case. Additional area is now for $\hat{\gamma}$ computations, which is estimated to be 0.8. Then, the total area consumption becomes 3.4. These area estimates are summarized in Table 3.

4   Conclusion

We briefly reviewed the non-redundant and redundant CORDIC schemes for SVD and presented the Constant-Factor-Redundant-CORDIC(CFR-CORDIC) scheme where the scale factor is forced to be constant while computing angles for plane rotations.

It is important to have a constant scale factor for the SVD CORDIC processor since multiplying a constant to the matrix does not change the output of CORDIC angle unit. In fact, when we skip the final scaling operation for all the plane rotations, the final singular values are multiplied by a constant term K^r, where K is the constant factor of a plane rotation and r is total number of plane rotations for SVD. Consequently, with the floating-point implementation, we can completely eliminate the final scaling operation for SVD applications where only the ratio of singular values need to be found. When we need to find singular values themselves, we can skip the final scaling operation for each plane rotation and perform once with K^r at the end. Notice that with the fixed-point implementation, the scaling operation is required in each two-sided rotation to avoid an overflow/precision error.

In addition, we presented schemes to make the scaling factor equal to 2 by including additional scaling iterations and CORDIC iterations, to reduce the number of iterations in the rotation by using radix-4 direction of the rotation, and to convert on-the-fly from the redundant output to a conventional implementation. We compared this scheme with previously proposed ones and showed that it provides similar execution time as redundant CORDIC with variable scaling factor with significant saving in area.

References

[1] G. H. Golub and C. F. Van Loan. *Matrix Computations*. The Johns Hopkins University Press, 1983.

[2] K. Bromley and J.M. Speiser. Signal Processing Algorithms, Architectures, and Applications. *Tutorial 31, SPIE 27th Annual Internat. Tech. Symp.*, 1983.

[3] R.P. Brent, F.T. Luk, and C.F. Van Loan. Computation of the Singular Value Decomposition using Mesh-Connected Processors. *Journal of VLSI and Computer Systems*, 1(3):242–270, 1985.

[4] F.T. Luk. Architectures for Computing Eigenvalues and SVDs. *SPIE Highly Parallel Signal Processing Architectures*, 614:24–33, 1986.

[5] J.R. Cavallaro and F.T. Luk. CORDIC Arithmetic for an SVD Processor. *Proceeding of 8th Symposium on Computer Arithmetic*, :113–120, 1987.

[6] J.E. Volder. The CORDIC Trigonometric Computing Technique. *IRE Trans. Electronic Computers*, EC-8:330–334, September 1959.

[7] J.S. Walther. A Unified Algorithm for Elementary Functions. *AFIPS Spring Joint Computer Conference*, :379–385, 1971.

[8] G. L. Haviland and A. A. Tuszynski. A CORDIC Arithmetic Processor Chip. *IEEE Transactions on Computers*, C-29(2):68–79, February 1980.

[9] H.M. Ahmed. *Signal Processing Algorithms and Architectures*. Ph.D. Dissertation, Department of Electrical Engineering, Stanford University, 1982.

[10] J.M. Delosme. VLSI Implementation of Rotations in Pseudo-Euclidean Spaces. In *Proceedings of IEEE Int. Conf. Acoustics, Speech, and Signal Processing 2*, pages 927–930, 1983.

[11] M. Ercegovac and T. Lang. Redundant and On-line CORDIC: Application to Matrix Triangularization and SVD. *IEEE Transactions on Computers*, C-39(6):725–740, June 1990.

[12] N. Takagi, T. Asada, and S. Yajima. *Redundant CORDIC methods with a Constant Scale Factor for Sine and Cosine Computation*. Submitted to IEEE Trans. on Computers, 1989.

[13] J. Lee. *Redundant CORDIC:Theory and its Application to Matrix Computations*. Ph.D. Dissertation, Computer Science Department, University of California, 1990.

[14] J. Lee and T. Lang. Matrix Triangularization by Fixed-Point Redundant CORDIC with a Constant Scale Factor. *Proc. SPIE Conference on Advanced Signal Processing Algorithms, Architectures, and Implementations*, July 1990.

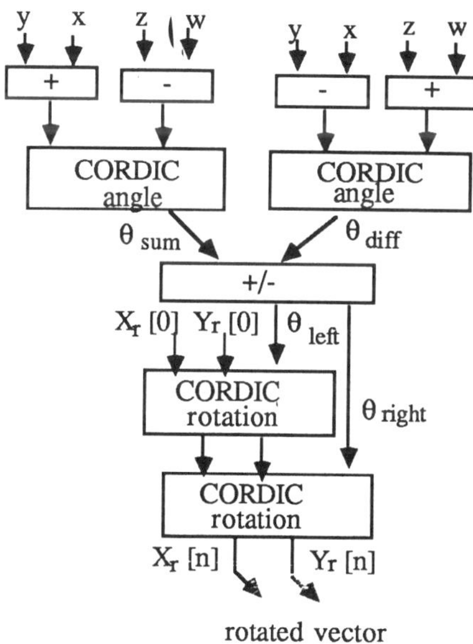

Figure 1: Non-redundant CORDIC scheme

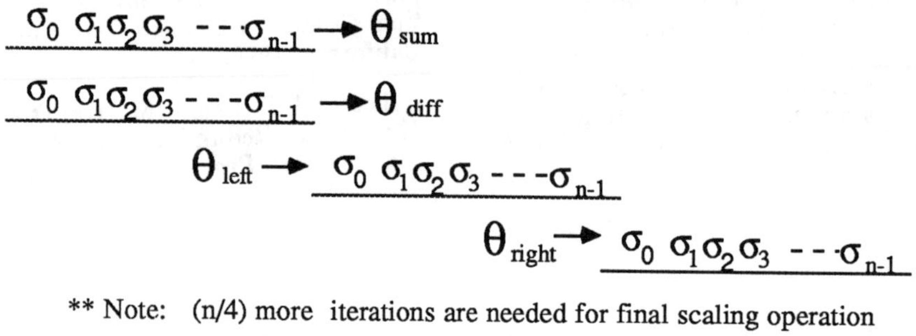

$$\underline{\sigma_0 \;\; \sigma_1 \sigma_2 \sigma_3 \;\; -- \;\; \sigma_{n-1}} \longrightarrow \theta_{sum}$$

$$\underline{\sigma_0 \;\; \sigma_1 \sigma_2 \sigma_3 \;\; ---\; \sigma_{n-1}} \longrightarrow \theta_{diff}$$

$$\theta_{left} \longrightarrow \underline{\sigma_0 \;\; \sigma_1 \sigma_2 \sigma_3 \;\; ---\; \sigma_{n-1}}$$

$$\theta_{right} \longrightarrow \underline{\sigma_0 \;\; \sigma_1 \sigma_2 \sigma_3 \;\; -- \;\; \sigma_{n-1}}$$

** Note: (n/4) more iterations are needed for final scaling operation

Figure 2: Timing of non-redundant scheme

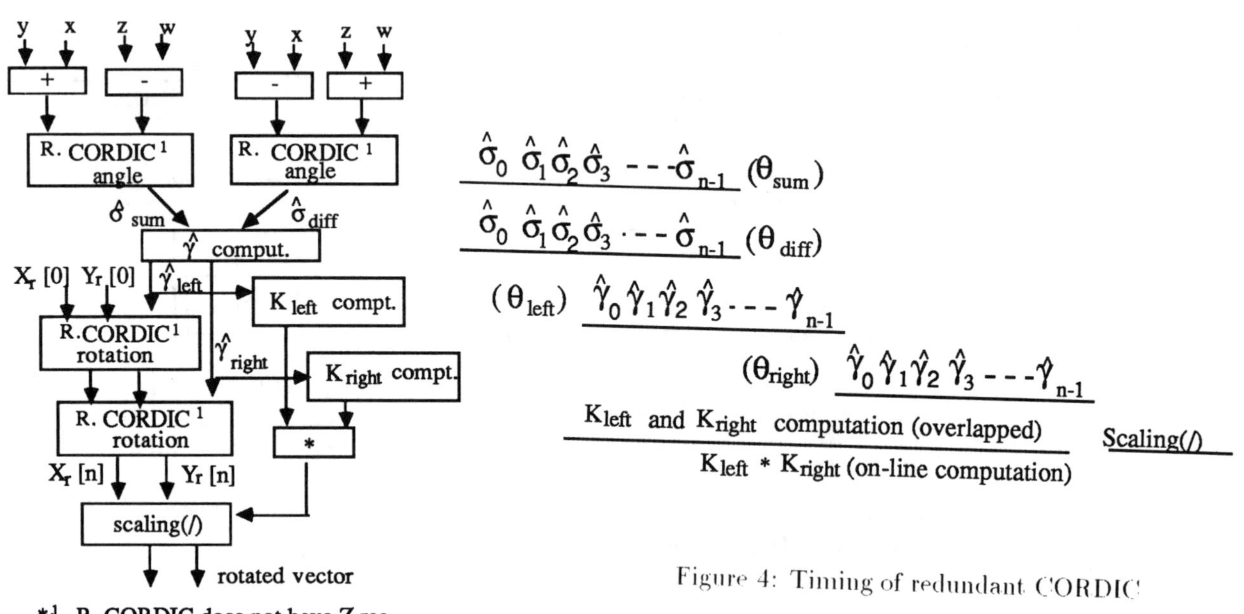

Figure 3: Redundant CORDIC(var. scale factor)

$$\underline{\hat{\sigma}_0 \;\; \hat{\sigma}_1 \hat{\sigma}_2 \hat{\sigma}_3 \;\; --\; \hat{\sigma}_{n-1}} \; (\theta_{sum})$$

$$\underline{\hat{\sigma}_0 \;\; \hat{\sigma}_1 \hat{\sigma}_2 \hat{\sigma}_3 \;\; \cdot --\; \hat{\sigma}_{n-1}} \; (\theta_{diff})$$

$$(\theta_{left}) \; \underline{\hat{\gamma}_0 \; \hat{\gamma}_1 \hat{\gamma}_2 \; \hat{\gamma}_3 \; ---\; \hat{\gamma}_{n-1}}$$

$$(\theta_{right}) \; \underline{\hat{\gamma}_0 \; \hat{\gamma}_1 \hat{\gamma}_2 \; \hat{\gamma}_3 \; ---\; \hat{\gamma}_{n-1}}$$

$$\underline{K_{left} \;\; \text{and} \;\; K_{right} \;\; \text{computation (overlapped)}} \qquad \underline{\text{Scaling}(/)}$$

$$K_{left} * K_{right} \; \text{(on-line computation)}$$

Figure 4: Timing of redundant CORDIC

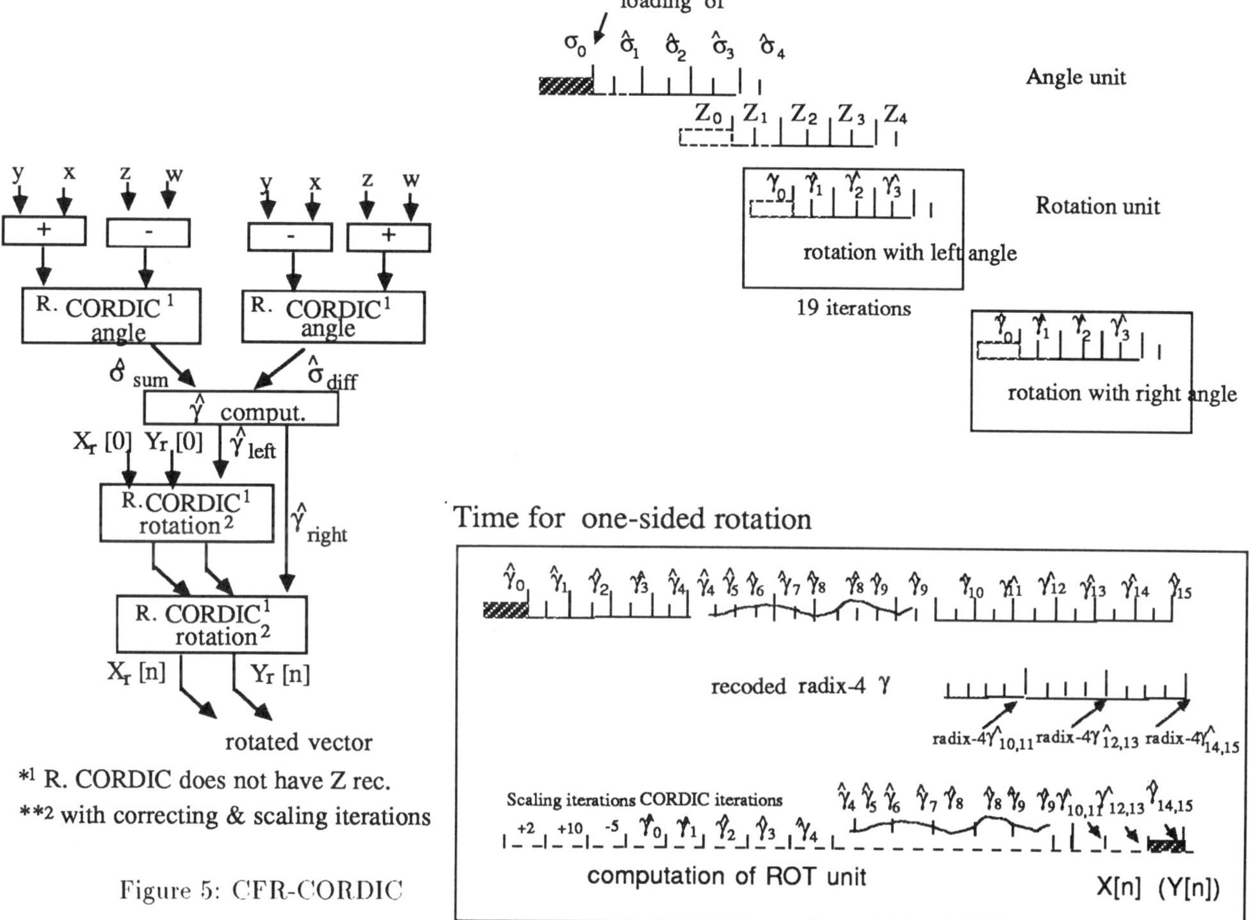

Figure 5: CFR-CORDIC

Figure 6: CFR-CORDIC timing for SVD

Design and Implementation of a Floating-point Quasi-Systolic General Purpose CORDIC Rotator for High-rate Parallel Data and Signal Processing

Alfons A.J. de Lange and Ed F. Deprettere
Department of Electrical Engineering
Delft University of Technology
2628 CD Delft, The Netherlands

Abstract

We describe the design and implementation of an algorithm and a processor which can be used to accelerate computations in which large amounts of rotations (circular as well as hyperbolic) are involved. The processor is a low-cost high-throughput vlsi implementation of the algorithm. With 10^7 rotations per second, many real-time and interaction-time applications in scientific computation become feasible.

1. Introduction

A central problem in image processing (HDTV, video phone), document processing (digital copiers), imaging (medical, acoustic radar), dynamic system modeling, simulation and visualization (computational sciences), computer graphics (photo realism, animation) and other time critical computational intensive applications [Kun88]) is : how can the algorithm be so (re-)designed that it can meet the time/space requirements. A solution to this problem is ideally obtained by deriving a parallel algorithm in which the basic functions have been carrefully converted into opimally designded software and/or hardware operations. In applications from time-critical domains such as computer graphics, image processing etc. fast hardware functions are usally paramount. In such cases, the performance of multi-processor or processor array implementations will heavily depend on the performance of its basic software functions (in programmable processors) or hardware functions (in application specific hardware). In case the basic functions are simple, say of type multiplication, the corresponding software/hardware algorithms will not degrade performance when they are also kept simple. However, when basic functions become more complex, it will not in general be acceptable to realize them in terms of a composition of the simple functions since their overall efficiency, in terms of time and/or space, will be much lower than for newly designed operations.

For example, in many signal processing and matrix computation applications, i.e. speech processing, radar, computer graphics and antenna array processing [Jai89, Vee88, Dep89]. Algorithms in these domains *can most advantageously* be expressed in terms of basic plane rotations, yielding much more stable and often also more structured algorithms than the conventional algorithms. For a typical example, see [Jai89]. However, the advantage may quickly turn into a disadvantage when the basic plane rotation itself is not implemented in an optimal way. The latter will usually mean non-iteratively and unconditionally (if possible), minimal number of evaluation steps, inherent accuracy etc.

To pursue the plane rotation example, it is well known that circular implemented using Taylor or Chebychev expansions, which requires a number of multiplication and addition/subtraction steps. See also [Ahm85, Sch83, Nav83] and[Mul85]. Another solution that is more appealing for applications where rotation evaluation time must be a constant is the bit-recursive shift and add *Coordinate Digital Computer* (CORDIC) approach. See [Vol59]. In application specific algorithm design, Volder's algorithm is easily implemented in both software and hardware. However, in case one would like to go a bit more general, in the sense that one wants a bit-recursive shift-and-add implementation of Walter's unified plane rotation

$$R_m(\alpha) = \begin{bmatrix} \cos(m^{\frac{1}{2}}\alpha) & -m^{\frac{1}{2}}\sin(m^{\frac{1}{2}}\alpha) \\ m^{-\frac{1}{2}}\sin(m^{\frac{1}{2}}\alpha) & \cos(m^{\frac{1}{2}}\alpha) \end{bmatrix}$$

which is circular, linear or hyperbolic in case m=1, m=0 and m=-1 respectively, the task is much more involved. The reason is that the hyperbolic rotation can not be implemented in the same (minimal bit-recursive) way as the circular rotation, whence it is not at all obvious wheter a unified algorithm for the unified function exists and if so, what the optimal implementation will be. See e.g.[Coc72, Hav70], and [Hwa79].

Several implementations of the CORDIC algorithm in hardware are described in [Hav70, Coc72]. AT&T, Motorola and Intel have released products using CORDIC arithmetic. These are respectively the WE32206, the 68881 and the 80x87. HP built a CORDIC floating point processor box for their 2100 series of minicomputers, however it was never sold. The CORDIC hardware in these products is meant to compute transcendental functions like sin, cos, arcsin, arccos, sinh, cosh, arcsinh, arccosh, exp, log and sqrt. Currently, CORDIC hardware is only used in desk-calculators. This is because it is generally felt that a CORDIC co-processor adds much hardware and communication links to the multiplier-adder unit of a standard computer. It is cheaper to add some simple sequencing logic or micro-programs to a multiplier-adder unit, which allows polynomial approximation of transcendental functions. In short, the application of a CORDIC processor for the computation of transcendental functions is not very efficient. This is not true for the application of fast CORDICs in massive-parallel algorithms where they can significantly accelerate system performance. An example of a more useful CORDIC processor that was designed recently is the TMC2330 of TRW LSI Products. It is a pipelined implementation that is used to accelerate circular coordinate transformations in computer graphics. See[Wil89]. Another useful CORDIC processor is the word-level pipelined core processor applied in the Plessey (Plessey Research Caswell Ltd, UK) PDPSP 163XX series products. However, no vectorization mode has been built in this core processor, which makes it useless in most of the previously mentioned applications. We even believe that a CORDIC processor is only useful if *both vectorization and rotation are possible and can be executed simultaneously*. With the latter we moreover mean *also working on the same angle*.

In this paper, we present a CORDIC algorithm and processor which allow a substantial acceleration of many known and novel algorithms in the application domains of computational algebra and physics, in particular time critical applications which must be implemented in parallel machines or arrays of processors.

In the *next section* a modified version of the CORDIC algorithm as originally proposed by Volder [Vol59] and Walther [Wal71] is presented. This modification enables a low-cost simultaneous implementation of the various rotation functions in several coordinate systems. Tthis section also describes the derivation of optimal parameters for the proposed CORDIC algorithm. *Section 3* briefly discusses several CORDIC (hardware) architectures with different speed and hardware complexity properties. High-speed architectures are carefully considered, because a CORDIC processor implementation must be applicable in real-time signal processing computations. *Section 4* describes a word-level floating point extension to the CORDIC algorithm of section 2 and 3. This extention is necessary to meet a true 16-bit accuracy which is minimally required in most of the applications we are referring to in this paper. *Section 5* gives the accuracy analyses of the CORDIC algorithm. In this section all parameters are determined that influence the accuracy and dynamic range of computations. In *section 6*, we present the specification of a prototype CORDIC which has been fabricated and used by the authors. Finally, in *sections 7*, conclusions about the novelty of the work are made.

2. The CORDIC Algorithm

In this section we briefly review the CORDIC concept in so far as our own needs go. Then we proceed with the the modifications we are proposing and the motivations for doing so.

The original CORDIC algorithm is a bit-recursive shift-and-add implementation of the forward and backward Givens rotation, that is, if we put

$$\begin{bmatrix} x' \\ y' \end{bmatrix} = \begin{bmatrix} \cos(\alpha) & \sin(\alpha) \\ -\sin(\alpha) & \cos(\alpha) \end{bmatrix} \begin{bmatrix} x \\ y \end{bmatrix}, \qquad (2.1)$$

then in the forward rotation, called *rotation*, the algorithm computes the rotation of a vector $[x\ y]^t$ over an angle α to a new vector $[x'\ y']^t$, and in the backward rotation, called *vectoring*, it computes the length x' and the inclination α of a vector $[x, y]^t$. The 2 procedures are eccentially the same. The function (2.1) can be easily generalized as follows[Wal71]:

$$\begin{bmatrix} x' \\ y' \end{bmatrix} = \begin{bmatrix} \cos(m^{1/2}\alpha) & -m^{1/2}\sin(m^{1/2}\alpha) \\ m^{-1/2}\sin(m^{1/2}\alpha) & \cos(m^{1/2}\alpha) \end{bmatrix} \begin{bmatrix} x \\ y \end{bmatrix}. \qquad (2.2)$$

which is the original Givens rotation for m=1, and includes, moreover, the hyperbolic rotation (m=-1) and the linear rotation (m=0). Figure 2.1 shows the three rotations. With appropriate definitions of m-norms (lengths) and m-arguments (angles) the 3 rotations (forward and backward) are vector norm preserving.

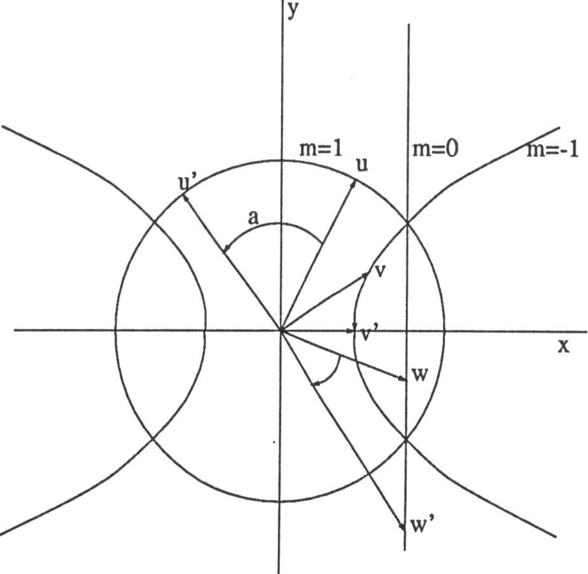

Figure 2.1. Rotation in the Circular, Hyperbolic and Linear Coordinate Systems. Shown is (1) rotation of a vector u to u′ over an angle α in the circular coordinate system, (2) vectorization of a vector v to v′ in the hyperbolic coordinate system, and (3) rotation of vector w to w′ in the linear coordinate system over an angle β.

The original CORDIC algorithm essentially decomposes the rotation in a minimal product of rotations, each one embedding an angle of fixed magnitude and variable sign such that the signed sum of angles can be any angle in the range of angles of resolution dependent on the number of elementary rotations. However the "minimal" number is not independent of the coordinate system (m) and the choice of elementary angle values is neither unique nor unifirm over the m-values. In traditional algorithms, the sets of fixed positive angles (called basic angles) are selected appropriately to each coordinate system and are denoted by $\alpha_{m,i}$, with $i = 0, 1, \cdots p(m)$, wher $p(m) + 1$ is the (minimal) number of elementary rotations. Hence the angle α, the number of basic rotations $p(m)$, and the set of basic angles $\alpha_{m,i}$ satisfy the following relation :

$$\sum_{i=0}^{p(m)} \sigma_i \alpha_{m,i} = \alpha, \quad \text{and} \quad \sigma_i \in [-1, 1]. \qquad (2.3)$$

So, equation 2.2 can be rewritten as :

$$\begin{bmatrix} x' \\ y' \end{bmatrix} = \prod_{i=0}^{p(m)} \begin{bmatrix} \cos(m^{\frac{1}{2}}\sigma_i\alpha_{m,i}) & -m^{\frac{1}{2}}\sin(m^{\frac{1}{2}}\sigma_i\alpha_{m,i}) \\ m^{-\frac{1}{2}}\sin(m^{\frac{1}{2}}\sigma_i\alpha_{m,i}) & \cos(m^{\frac{1}{2}}\sigma_i\alpha_{m,i}) \end{bmatrix} \begin{bmatrix} x \\ y \end{bmatrix}$$

$$= K \prod_{i=0}^{p(m)} \begin{bmatrix} 1 & -m^{-\frac{1}{2}}\tan(m^{\frac{1}{2}}\sigma_i\alpha_{m,i}) \\ m^{-\frac{1}{2}}\tan(m^{\frac{1}{2}}\sigma_i\alpha_{m,i}) & 1 \end{bmatrix} \begin{bmatrix} x \\ y \end{bmatrix}. \qquad (2.4)$$

where

$$K = \prod_{i=0}^{p(m)} \cos(m^{\frac{1}{2}}\alpha_{m,i}) .$$

The right hand side decomposition form in eqn. (2.4) is now inteerpreted as follows:

1. Choose the angles $\alpha_{m,i}$ in such a way that the term $m^{-\frac{1}{2}}\tan(m^{\frac{1}{2}}\alpha_{m,i})$ equals $2^{-S_{m,i}}$, where $S_{m,i}$ is a nonnegative integer.

2. Get the scaling factor $\prod_{i=0}^{p(m)}\cos(m^{\frac{1}{2}}\alpha_{m,i})$ included in pre-operations or post-operations, or distribute it over the elementary rotations, using (if possible) approximations $\cos(m^{\frac{1}{2}}\alpha_{m,i}) \approx (1-m\varepsilon_{m,i}2^{-2S_{m,i}})$, with either $\varepsilon_{m,i} = 0$, or $\varepsilon_{m,i} = 1$.

The CORDIC algorithms will then approximately implement the functions (2.2) in the following sense:

Let:

$$m^{\frac{1}{2}}\tan^{-1}(m^{\frac{1}{2}}\sigma_i\alpha_{m,i}) = 2^{-S_{m,i}}), \quad S_{m,i} \in N, \qquad (2.5)$$

$$\tfrac{1}{2}\alpha_{m,i} \le \alpha_{m,i+1} \le \alpha_{m,i}, \qquad (2.6)$$

$$|\alpha - \sum_{i=0}^{p(m)} \sigma_i \alpha_{m,i}| \le \alpha_{m,p(m)}, \qquad (2.7)$$

and:

$$\prod_{i=0}^{p(m)}(1-m\varepsilon_{m,i}2^{-S_{m,i}}) \approx \prod_{i=0}^{p(m)} \cos(m^{\frac{1}{2}}\alpha_{m,i}) \qquad (2.8)$$

Then eq. 2.2 can approximately be rewritten as :

for $i=0,1,...p(m)$

$$\begin{bmatrix} x_{i+1} \\ y_{i+1} \end{bmatrix} = (1-m\varepsilon_{m,i}2^{-S_{m,i}}) \begin{bmatrix} 1 & m\,\sigma_i\,2^{-S_{m,i}} \\ -\sigma_i\,2^{-S_{m,i}} & 1 \end{bmatrix} \begin{bmatrix} x_i \\ y_i \end{bmatrix}, \qquad (2.9)$$

with either $\varepsilon_{m,i} = 0$ or $\varepsilon_{m,i} = 1$, and either $\sigma_i = -1$ or $\sigma_i = 1^\dagger$.
In case of rotation, the σ_i $(i = 0,1,..,p(m))$ are computed by evaluating

$$\sigma_i = \text{sign}(\alpha - \sum_{j=0}^{i-1}\sigma_j\alpha_{m,j}). \qquad (2.10)$$

In case of vectoring, σ_i $(i = 0,1,..,p(m))$ are computed by inverting the signs of y_i $(i = 0,1,..,p(m))$:

$$\sigma_i = -\text{sign}(y_i). \qquad (2.11)$$

Due to the dependency of all coefficients of the parameter m, the set of algorithms as given above is much less unified as the function (2.2) is. For example, if we require an accuracy of M e.g 2^{-16} — and an angle range condition (2.7) — e.g. $-\pi \le \alpha \le \pi$ — and accuracy — e.g. 2^{-16} —, then the number of basic angles, hence of *micro rotations*, will be different for different values of m, since the convergence condition (2.6) depends on the parameter m, See[Dep84], and all shift exponents in the set m=1 will be different from those in the set m=-1, this is expensive, both in software and hardware implementations. The range $[-\pi, \pi]$ is very realistic; it is *the* range for the circular rotation and it is an *acceptable* range for the hyperbolicrotation (having an infinite range in theory). On the other hand; if the condition would be that execution times must be independent of the paprameter m — this is necessary in many applications that use rotations in both circular and hyperbolic coordinate systems —, then different conditions hold for the accuracy and range of angles α (eq. 2.7, accuracy is determined by $\alpha_{m,p(m)}$) in different coordinate system. This is unacceptable if both circular and hyperbolic rotations are used in the same algorithm (see e.g. speech coding[Jai86]). Another drawback of the original CORDIC algorithms as described above are the auxiliary angle *decomposer* (eq. 2.10) and a result angle *composer* : $\sum_{i=0}^{p(m)}\sigma_i\alpha_{m,i}$. These operations are neither necessary nor useful. For example, in many applications one has to rotate a set of vectors over an angle that is first to be computed from a given vector. In such cases the decomposer succeeds the composer creating a perfect identity that is not only useless but also prevents the sequence of rotations to be started almost concurrently with the vectorization. If, on the other hand encoded angle

---

$\dagger$ Some authors allow the σ_i's to become zero. This will speed up the determination of the sign of y_i and the sign of $\alpha-\sum_{j=0}^{i-1}\sigma_j\alpha_{m,j}$ if α and x/y_i have a binary redundant representation [Dup89], which will be explained in section 3.2. A serious drawback however is that the scaling factor $\prod_{i=0}^{p(m)}\cos(m^{\frac{1}{2}}\sigma_i\alpha_{m,i})$ (eq 2.10) will then depend on σ_i. This problem can be solved by either repeating the iteration serially[Tag87] (one for $\sigma=-1$ and one for $\sigma=+1$), or performing two iterations in parallel [Dup89]. See also page 11.

inputs/outpus are used, (σ_i), rotations can be initiated immediately after the first step in the vectorization has been completed.

Most of the drawbacks can be overcome by using a slightly more complex representation of the term "$m^{1/2}\tan(m^{1/2}\alpha_{i,m})$". Namely,

$$m^{1/2}\tan(m^{1/2}\alpha_{i,m}) = 2^{-S_{m,i}} - \eta_{m,i}2^{-S_{m,i}'}, \qquad (2.12)$$

where $\eta_{m,i}$ is either -1,0, or 1 for $i=0,1,..,p$. This modification allows us to impose the following constraints on the CORDIC algorithm :

1. The number of iterations $p+1$ is independent of the coordinate system.

2. Accuracy and range conditions are the same for each coordinate system.

3. The normalization (scaling) factors $K_m = \prod_{i=0}^{p(m)}\cos(m^{1/2}\alpha_{m,i})$ are simple powers of 2 :

$$K_m = \prod_{i=0}^{p(m)}\cos(m^{1/2}\alpha_{m,i}) \approx 2^{-S(m)}. \qquad (2.13)$$

A complete set of basic angles $\alpha_{m,i}$, η's, scaling constants $K_{m,i}$ and shift factors $S_{m,i}$, $S_{m,i}'$ has been deduced for the case $m=-1,+1$ (circular/hyperbolic coordinate systems) with an absolute error smaller than 2^{-15} and an angle range of $[-\pi,+\pi]$. This table is depicted in table 2.1.

TABLE 2.1. Optimal CORDIC parameters for 16-bits accuracy

index i	S,S',η_{-1}	S,S',η_1	α_{-1}	α_1
1	0 4 -1	0 3 +1	1.716994	0.844154
2	1 8 -1	1 8 +1	0.544111	0.466768
3	1 6 -1	1 6 +1	0.528685	0.476069
4	2 14 -1	2 14 +1	0.255348	0.245036
5 .	2 4 -1	2 4 -1	0.189745	0.185348
6	4 6 +1	4 6 +1	0.078285	0.077967
7	4 10 -1	4 10 -1	0.061601	0.061446
8	5	5	0.031260	0.031240
9	6	6	0.015626	0.015624
10	7	7	0.007813	0.007812
11	8	8	0.003906	0.003906
12	9	9	0.001953	0.001953
13	10	10	0.000977	0.000977
14	11	11	0.000488	0.000488
15	12	12	0.000244	0.000244
16	13	13	0.000122	0.000122
17	14	14	0.000061	0.000061
18	15	15	0.000031	0.000031
19	16	16	0.000015	0.000015

$$K_{-1} \approx 4.0000058891 \approx 2^2, \quad K_1 \approx 0.5000096618 \approx 2^{-1}$$

Note that for circular rotation, an additional initial rotation over $\alpha_{m,i} = \alpha_{1,0} = \pm\frac{1}{2}\pi$ is necessary, to cover the required angle range of $[-\pi,+\pi]$. This has not been depicted in the table.

The paprameters in the above table are slightly different than those buplished in [Bu86] where the number of micro-rotation

p(m)+1 was stilll different for circular and hyperbolic coordinate systems. The parameters given here have been given for the first time in [Lan88].

Notice that almost all shift exponents S and S' are independent of the parameter m. The problem of finding such set of exponents is not trivial. It is an NP complete problem and, moreover, the solution space is a very small intersection of the two solution spaces, one for m=1 and one for m=-1. The solution given here has been obtained by means of a simplified exhaustive search. It is not the only solution although the number of solutions quickly decreases with increasing accuracy. Only one solution was found for a 32 bit accuracy. It is not clear whether such a solution (such a set of constraints) is optimal or not. In fact, the *VLSI implementation of the* algorithm, based on the data of table 2.1, is not optimal in terms of silicon area used. Notice also that for most of the micro-rotations ($i=7,8,..,p(m)$) $\eta_{m,i}$ equals zero. This is not hard to understand, since for small basic angles the cordic becomes linear in α_m and

$$\cos(\alpha_{m,i}) = 1 - \frac{\alpha_{m,i}^2}{2} + O(\alpha_{m,i}^4) \qquad (2.14)$$
$$= 1 + O(2^{-24}), \qquad \text{for } \alpha_{m,i} < 2^{-12}.$$

This has several consequences, one being that the price to be payed for the "double rotations" is negligible, another being that increasing the accuracy (number of rotations) will only have a minor effect on the scaling factor. However, the price to be payed for one more bit of accuracy is a complete micro-rotation : the price is in the tail.

A detailed analysis of all different errors sources in our modified CORDIC algorithm is given in section 5. This will explain the need for additional micro-rotations to enhance the accuracy in hyperbolic rotations.

3. CORDIC Architectures

Many different architectures can be deduced for the CORDIC algorithm described in the previous section. These vary from bit-serial implementations to word parallel pipelined architectures. Which choice is made depends on the requirements for computing throughput and constraints that hold for area usage, latency and dissipation. At each level of abstraction of the CORDIC algorithm, such a tradeoff must be made. Two important levels of abstraction can be identified in the CORDIC algorithm :

1. *the micro-rotation level*
   This level describes the basic shift-add operation,

2. *the CORDIC top-level*
   This level describes the CORDIC operation as a sequence of micro-rotation operations.

3.1 Top Level CORDIC Architectures

Table 3.1 summarizes the main properties of some alternative top-level CORDIC architectures.

For any technology with a given T_c, table 3.1 can be used to

TABLE 3.1. Properties of Top Level CORDIC Architectures. T_c is the minimum clock period, T_μ is the delay of a micro-rotation, p+1 is the number of CORDIC iterations, L_μ is the latency of a micro-rotation (for pipelined CORDIC only) and area[+] indicates the additional amount of area due to multiplexers, rams, barrel-shifters, registers, wiring and control.

properties	Top Level CORDIC Implementations		
	sequential	ripple	pipeline
delay (sec)	$2(p{+}1)\max(T_c,T_\mu)$	$(p{+}1)T_\mu$	$L_\mu(p{+}1)\max(T_c,T_\mu)$
thru-put (rot/sec)	$\dfrac{1}{2(p{+}1)\max(T_c,T_\mu)}$	$\dfrac{1}{(p{+}1)T_\mu}$	$\dfrac{1}{\max(T_c,\frac{T_\mu}{L_\mu})}$
latency (# cycles)	0	0	$L_\mu(p{+}1)$
area (# μRot's)	½	$(p{+}1)$	$(p{+}1)$
area[+] (# μRot's)	16	$\dfrac{(p{+}1)}{4}$	$\dfrac{(p{+}1)}{2}$

determine the optimal top-level CORDIC architecture.

If we compare the performance figures for the different architectural alternatives (table 3.1), it is clear that the pipelined implementation has the highest throughput, however at the cost of more hardware. The amount of extra hardware is less than would be expected, because a number of simplifications in both top-level and micro-rotation architecture are possible.

1. The full-pipeline implementation does not need a barrel-shifter for shift-factors, because the shifts can be hard-wired.

2. No memory for shift-factors and η_i parameters is needed, because these are constant for each different micro-rotation.

3. The terms $m_i\sigma_i\eta_{m,i}$ and $\sigma_i\eta_{m,i}$ (eq. 3.5–3.12) can be simplified since $\eta_{m,i}$ only depends on m for a given i, hence reduces the complexity of the control-unit of each micro-rotation (identified by i).

4. A double shift factor 2^{-S_m} can be omitted for a large number of micro-rotations (see eq. 3.5–3.12) and table 2.1) when $\eta_{m,i}$ is zero. This reduces the micro-rotation complexity with a factor 2.

We conclude that a full pipeline implementation leads to a regular architecture with very few control. A compromise such as the usage of a limited number of micro-rotations, which are executed sequentially, looses all the advantages listed above.

3.2 Micro-Rotation Architectures

At the next level down we identify the micro-rotation operation. This operation can be implemented by add/subtract devices, control unit and shifter. Figure 3.1 shows two structures of a micro-rotation that implement a micro rotation operation given by eq. 3.5–3.12 : one with $\eta_{m,i} \neq 0$ and the other with $\eta_{m,i} = 0$. Both structures can be used in a ripple or pipeline CORDIC implementation.

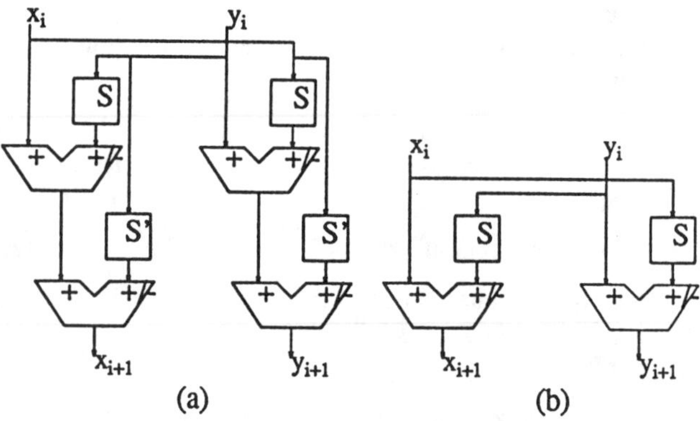

Figure 3.1. Micro-Rotation Structures for (a) $\eta_{m,i} \neq 0$ and (b) $\eta_{m,i} = 0$

High throughput with at the same time a limited latency and area occupation is a very important property of the CORDIC processor we want to design. We are therefore most interested in fast parallel add/subtract devices as far as area usage and dissipation do not create a problem.

Figure 3.2 shows the architecture of a parallel pipeline micro-rotation device for $\eta_{m,i} = 0$. The shift factors are hardwired and therefore not shown.

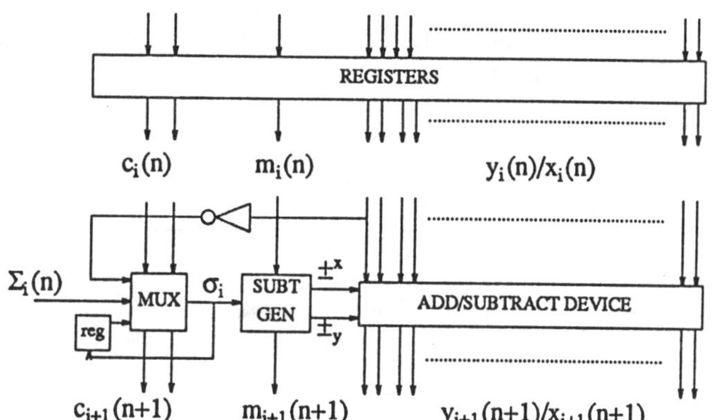

Figure 3.2. Architecture of a Word Parallel Pipeline Single Section Micro-Rotation. Σ_i is the encoded angle input signal, $c_i(n)$ the control signal that selects the input $\sigma_i : \Sigma_i(n)$, the previous $\sigma_i : \sigma_i(n-1)$, or the sign of y_i. The other signals are as defined in section 2

The best known fast devices are the carry-select adder/subtractor, the carry-look ahead adder/subtractor and its dynamic implementation in CMOS called the manchester chain adder/subtractor. See pages 322-325 of Weste and Eshraghian[Wes85]. Another very fast addition/subtraction device that is worth considering is the 'Binary Redundant' device See e.g.[Tak85]. A comparison of the transistor-count and gate-delay between the different types of add devices is depicted in table 3.2. The data for this table is obtained from Weste[Wes85] and Takagi[Tak85]. The numbers depicted are approximate figures for devices in static and dynamic CMOS. The extension to subtract devices involves only a small O(n) number of extra gates.

TABLE 3.2. Comparison of n-bit Adder Devices, consisting of k-bit digits

name	static		dynamic	
	# transistors	carry delay (# gates)	# transistors	carry delay (# gates)
carry ripple	28 n	2 n	22 n	2 n
carry save	34 n	2 k	28 n	2 k
carry select	28 (2 n - k)	2 k + 2 n / k	22 (2 n - k)	2 k + 2 n / k
static carry look ahead	$< 38 \cdot \frac{4}{3} n$ †	$2 \cdot {}^4\log(n)$	$< 32 \cdot \frac{4}{3} n$ †	$2 \cdot {}^4\log(n)$
manchester chain	-	-	30 n	$\frac{n}{2}$
binary redundant	116 n	3	103 n	3

Discussion.

Before going into the pros and cons of the different add/subtract devices, the following remark is in order. The CORDIC algorithm presented in this paper is nearly regular, meaning that it can be easily generated using repetition of (parametrized) bit-level as well micro-rotor level units. This means that recently proposed *systolic* or *Quasi-systolic* architecture design methodologies can be invoked to design such algorithms (and their implementations) as the CORDIC algorithm. See [Kun88, Dep91]. In this context, the implementation of the CORDIC algorithm using adders "such" or "so" is simply a matter of degree of systolization. See e.g., [McC90]. Therefore, the usage of a particular full-adder in the VLSI implementation of the algorithm is just a parameter in the design process. The implementation considered here uses carry-ripple adders only because such adders have been available in the cell library and not because of architecture-design level considerations. Of course, the performance and price of the ultimate implementation will depend on the particular adders used. We therefore include the following additional remarks.

- *carry-ripple*
  The slowest device is the carry-ripple adder, however its area consumption is low. Even more important is its simple and regular structure.

- *carry-save*
  The carry save adder has relatively low area consumption, paired with a high throughput. The main drawback of such a device, is its high latency. In the CORDIC algorithm, the latency will be increased with a factor n/k. This is undesirable in some filtering applications which require low computation latency[Dep89]. However, in case of a high computation latency, several filter applications can be executed in pipeline on the same CORDIC processor.

- *carry-select*
  The carry-select adder is a fast device, but occupies nearly

twice as much area. Application of this device in a micro-rotation can be considered in case small latency and high speed is paramount, while area consumption is of less importance.

- *carry look ahead*
  The area and delay of this device are logarithmically proportional to the number of bits. A drawback of this device is its less regular structure. This causes lower area efficiency of VLSI implementations, and complicates the task of constructing automatic module generators for this device (this is necessary if different accuracy requirements must be fulfilled). If only one level of carry-lookahead sections is used, the delay is equivalent to the delay of the manchester-chain, but the hardware of the device is reduced (30 % smaller for large devices : n > 32) and its structure is more regular.

- *manchester chain*
  This device is a more compact implementation of the carry-lookahead device. It is a dynamic implementation. This results in smaller area and its architecture is more regular, but the cascading of several sections cannot be done without clocking. However the applied pre/discharging scheme allows correct carry ripple propagation. It can be equipped with several levels of carry-lookahead, but this complicates the design of error free dynamic behavior. Like all dynamic devices, it must be designed carefully to ensure correct operation under all circuit conditions.

- *binary-redundant adder*
  Evidently, the fastest device is the binary-redundant adder. A binary redundant adder is a device that can perform additions in constant time, because there is no carry ripple. This is possible since two bits are used to encode one redundant bit, hence is sufficient to retain both sum and carry bit or an encoded version of these two.
  The area consumption of such a device is however very high. This is not only due to the larger basic add-cells, but also because of the additional wiring (twice as much, because each redundant bit consists of 2 standard bits). Finally, an area consuming conversion unit from binary redundant to standard two's complement encoding is necessary.

An additional problem arises in CORDIC vectoring operation. Here, the sign of the y-operand must be determined before the next addition/subtraction can take place. However, the sign of a binary redundant number cannot be determined in constant time, because there is no reserved sign bit. At best a $O(\log n)$ time complexity can be achieved. We can solve this problem for the CORDIC algorithm, by limiting the number of most significant bits (e.g. 1 to 4 bits) to be examined for sign determination. This means that the correct sign can not be determined if y_i has a too small magnitude. In this case however, vectorization (rotation over the next basic angle) is not necessary, because the accuracy of y_i will not be improved in this micro-rotation. In the following micro-rotation, a

---

† The number of bits that have carry look-ahead is 4 (k=4), because a larger number increases the complexity of the logic (and delay) dramatically. For 4 of such blocks (16-bits), a new block is added that performs carry lookahead for for the 16 th bit. This is repeated each time $^4\log(n)$ is integer. Hence the number of transistors = $38n (1 + 1/4 + 1/4^2 + ...) = 38$ $n (1 + 1/3 (1 - 1/4^m)) < 57$ n. Here m is the number of levels of carry-lookahead devices are added = ceiling($^4\log(n) - 1$).

few bits of y_i of lower weight are examined to determine the sign. If once again the magnitude of y_i is too small to allow correct sign determination, again no vectorizing micro-rotation is necessary. However, if the sign of y_i can be determined by examining these few bits, a micro-rotation must be performed. This way, in each micro-rotation, a few number of different bits are examined : the first micro-rotation examines the most significant bits, the last micro-rotation the least significant bits. A fast determination of the sign of y_i was proposed in [Tag87]. It can be shown that it is possible to combine Tagaki's approach with ours to obtain both fast determination of y_i signs and simple scaling factors.

The final problem that remains, is that if no micro-rotation is performed, this corresponds to $\sigma_i = 0$, hence eq. 3.8/12/13 are no longer valid. I.e., the scaling factor K_m is no longer constant. This can be solved by performing both a positive and negative micro-rotation (namely, $\sigma_i = 1$ and $\sigma_i = -1$). For this, the micro-rotation hardware must be doubled. We have now obtained, a very fast device. However, the maximum clock frequency will be smaller than the rate at which an adder device can execute. The only way to to retain performance, is to execute several microrotations within the same clock period. This is obtained by implementing the CORDIC as a rippling device.

Considering the previous discussion, most devices are either too costly in area, or too slow. A reasonable alternative that allows a trade-off between latency and throughput is the carry-save adder. This device has a very regular and simple structure, hence can easily be described as a parameterized module. Instead of using a static or dynamic fulladder in a carry-save device, the application of a manchester chain device seems attractive. This is because it is 4 times faster than the fulladder, hence less pipeline stages are needed in the carry save device.

4. Floating Point Extension of the CORDIC Processor

In many numerical applications, it is required that the CORDIC processor can operate on numbers with a large dynamic range. E.g. simulations of a speech coding application [Jai86] demonstrate the need for a CORDIC processor with an accuracy of 16 bits and a dynamic range of $\pm 2^{\pm 16}$. We therefore need to make a floating point version of the fixed-point CORDIC algorithm of section 2. Roughly speaking there are two options.

1. *Local floating point normalization*
   Each micro-rotation performs a floating point shift-add/subtract operation and normalizes the result.

2. *global floating point normalization*
   The CORDIC algorithm remains fixed-point. Only at the inputs and outputs of the algorithm floating to fixed point and fixed to floating point conversions are done.

The first approach is very costly in terms of hardware and speed in case several pipelined micro-rotations are used. Moreover, this approach is not very useful, because every micro rotation performs an addition or subtraction operation, which requires equal weights for the exponents of both operands. Hence, prior to each micro-rotation the floating point normalization operation (placing the binary point before the most significant bit of the mantissa and correcting the exponent accordingly) performed by the previous micro rotation has to be reversed.

The equalization of exponents before an add/subtract operation results in loss of accuracy because some bits of one of the operands will be shifted out of the data path. This problem also occurs in the global floating point normalization case. This is significant for the case of hyperbolic rotations/vectorizations where large input vectors (both x and y are large) become subsequently smaller during each micro rotation (this case is treated in section 5). The cases where the loss of accuracy leads to serious problems are very rare. Even then the relative error still remains acceptably small : ($e^{-\pi}$, see section 5).

The opposite case, where rotations can give rise to overflow, is easily prevented with local floating point normalizations. However, if we know the maximum possible overflow that can occur in hyperbolic and circular rotations, we can extend the data path to account for this. All these considerations are described in more detail in section 5.

Floating Point Input Processor
The floating point input processor adapts the mantissa whose associated exponent is the smallest exponent of the x/y vectors according to the difference between the x/y exponent values, such that the resulting x/y mantissas have equal exponents. The largest exponent is outputted. n and m are the sizes of respectively the mantissa and exponent data path. The architecture of the floating point preprocessor is shown in figure 4.1.

Floating Point Output Processor
The floating point output processor takes care of the output scaling (factor K_m) and the selection of significant bits. n denotes the size of the external datapath, F is the number of overflow bits, m the datapath size of the exponents and K_m the scaling factor in the CORDIC algorithm (see eq. 4.13). The generic architecture of the unit is given in figure 4.2.

---

† We would like to thank one of the reviewers for bringing this to our attention.

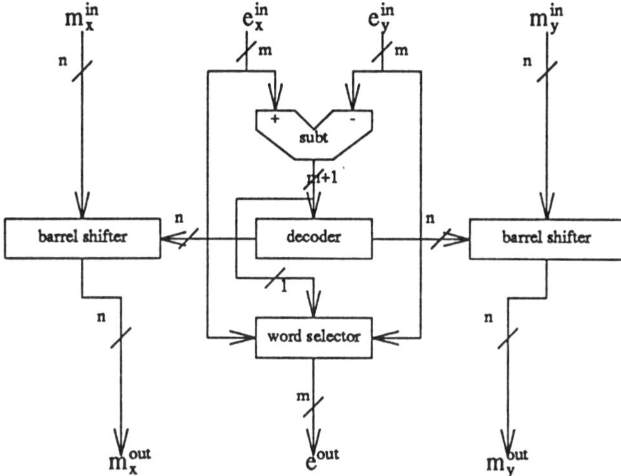

Figure 4.1. Floating Point Preprocessor Architecture. The parameter expressions n, m, m+1 indicate the number of bits of interconnections

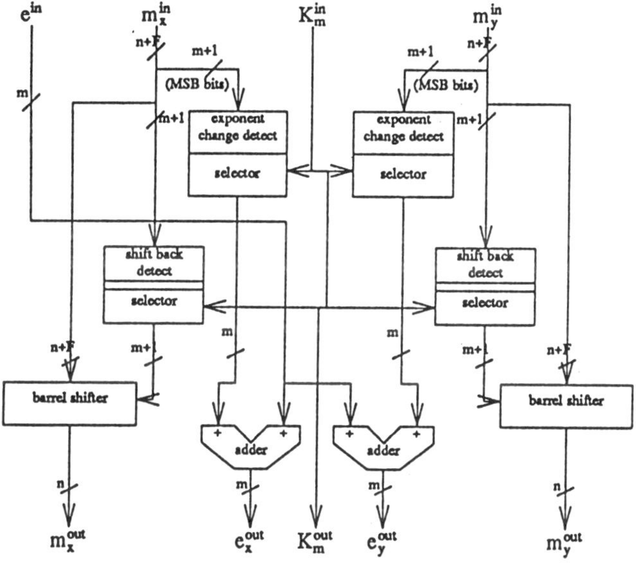

Figure 4.2. Generic Floating Point Postprocessor Structure

5. Accuracy Analysis of the Floating Point CORDIC Algorithm

In this section, the datapath widths of mantissas, exponents and control signals as present in floating point pre/post processors and CORDIC pipe are computed from the requirements for accuracy and dynamic range. The following items influence the computation accuracy.

5.1 Floating Point Input Normalization

The datapath size of the mantissa at the input and output of the floating point preprocessor are n-bits to obtain n bits (adder/subtractor) accuracy. In contrast to a floating point multiplier, no guaranteed accuracy improvement can be obtained by an extension of the internal datapath (micro-

rotations perform add/subtract operations). See[Lac88].

5.2 Floating Point Output Normalization

The floating point postprocessor performs a "back-shift" of x and y mantissas. The maximum shift size is determined by the number of overflow bits in the internal datapath in the CORDIC pipe and the scaling factor K_m that is to be performed at the output for circular ($K_1 = 2^{-1}$) and hyperbolic ($K_{-1} = 2^2$) rotations. Circular rotations are norm preserving according to $x^2 + y^2 = C$, with $C \geq 0$ is constant, hence x and y are always bounded by $\sqrt{C}$. On the other hand, hyperbolic rotations are norm preserving according to $x^2 - y^2 = C$, with $C \in Z$ and constant. This implies that the magnitude of x and y can be very large in hyperbolic rotations. Because the magnitude of x and y depends on the angle α, the number of overflow bits is determined by the maximum rotation angle α_{max} in hyperbolic rotations. For this, we write out equation 5.2 for x' with $m = -1$ (hyperbolic coordinate system) and $\alpha = \alpha_{max}$:

$$x' = \cos j\alpha_{max} \cdot x + \sin j\alpha_{max} \cdot y.$$
$$= \cosh\alpha_{max} \cdot x + \sinh\alpha_{max} \cdot x.$$
$$j = \sqrt{-1}.$$

Then

$$|x'| = |x \cosh(\alpha_{max}) + y_{in} \sinh(\alpha_{max})|$$
$$\leq \max(|x|, |y|) \, 2 \cosh(\alpha_{max}).$$

In the CORDIC algorithm that we have described in section 2, hyperbolic and circular rotations have the same domain for angles α. This is required, because a variety of algorithms (e.g. speech coding[Jai86]) apply both circular and hyperbolic and circular rotation/vectorization operations on the same (angle) data. In the circular case, an angle interval $[-\pi, +\pi]$ is sufficient to cover all possible rotations, hence we have chosen this angle range for both the circular and hyperbolic coordinate systems. The angle parameters of table 2.1 constitute an angle interval $|\alpha_{max}| = \pi + O(2^{-16})$, hence $2 \cosh(\alpha_{max}) < 2^5$, which means that a 5 bits overflow extension is sufficient. In case of hyperbolic vectoring, the angle α of an input vector (x, y) may be larger than $|\alpha_{max}|$. Then, this vector is rotated to the x-axes over an angle α_{max}, which introduces a relative error of $e^{-\pi}$ [Lan88].

The output scaling factor for hyperbolic rotations is : $K_{-1} \approx 2^2$ and is performed in the floating point postprocessor. Therefore, inside the fixed point CORDIC pipeline, only (5 - K_{-1} =) **3 MSB overflow bits** are needed, while **2 LSB extension bits** are needed to prevent the loss of accuracy due to the multiplication factor of $K_{-1} = 2^2$ in hyperbolic rotations.

5.3 Number of Micro Rotations

In the linear part of the CORDIC algorithm (see table 2.1) only single shift/add operations are performed. In this part of the pipe, the angles have become small, hence the micro-rotations can be approximated by the following equations (see eq. 5.9 and 5.8) :

$$x_{i+1} = x_i - m\alpha_i y_i, \quad y_{i+1} = y_i + m\alpha_i x_i$$

When $\alpha_i < 2^{-(n-1)}$ (n is the number of bits of the input/output mantissas), then the terms $m\alpha_i x_i$ and $m\alpha_i y_i$ can be neglected. This is because the selection of the n most significant bits from the mantissa x_i/y_i results the same string of bits as the selection of the n most significant bits from the mantissa x_{i+1}/y_{i+1}. If we denote the selection of the bits 0 to n-1 of a binary word z by $z[0..n-1]$, then

$x_{i+1}[0..n-1] = x_i[0..n-1]$, and
$y_{i+1}[0..n-1] = y_i[0..n-1]$, because
$2^{-(n-1)} x_i$ and $2^{-(n-1)} y_i$ are the LSB (and bits of lower weight) of x_i and y_i.

Therefore, the angle in the last useful micro-rotation satisfies : $\alpha_i = 2^{-(n-1)}$. However, the hyperbolic scaling factor $K_{-1} = 2^2$ propagates the error 2 bits towards the MSB, hence we require for the smallest angle α_i^{min} : $\alpha_i^{min} = 2^{-(n-1)-2} = 2^{-(n+1)}$. To reduce the cost of area we choose $\alpha_i^{min} = 2^{-n}$. This introduces a relative error of $2^{-(n-1)}$ for x_i/y_i numbers whose MSB equals one (negative number in Two's Complement), and an error of 2^{-n} otherwise.

5.4 Errors due to the Scaling Factors

In our CORDIC algorithm, the scaling factor K_m was determined for a total of $p+1 = 20$ micro rotors (see table 2.1), and an angle resolution of 2^{-n} ($S = n$). For hyperbolic scaling then holds :
$m = -1$, and $K_m = 4.0000058891 \approx 4 + 2^{-(n+1)}$, where $n = 16$
and for circular scaling :
$m = 1$, and $K_m = 0.5000096618 \approx 0.5 + 2^{-(n+1)}$
The normalization factor K_m (see eq. 5.13) hence introduces a relative error of 2^{-n}.

5.5 Accumulation of Truncation Errors

Every step in the CORDIC algorithm truncates the binary representation of the numbers at the least significant bit (LSB). The average error is equal to ½ the weight of the bit. For the total of $p + 1 = 20$ micro rotors, 26 shift/add operations are performed to achieve an angle resolution of 2^{-n} (Smax = n). Then the average error may have propagated over 4 bits towards the MSB, because $2^3 < \frac{1}{2} 26 < 2^4$. Therefore, another **4 bits must be added** to the data path. The total amount of bits in the CORDIC pipe now reaches $16 + 2 + 3 + 4 = 25$, 21 of which are accurate at the output. Figure 5.1 shows a simplified schematic for the floating point pipeline CORDIC architecture.

Detailed information considering the circuit and layout design of the CORDIC processor can be found in[Hoe88].

6. Implementation Results

A 21 bits floating point pipeline CORDIC processor has been designed and manufactured in a 1.5 μ CMOS process. It was originally designed in a 3 μ CMOS process[Phi82] and was later scaled down by a factor 0.72 to fit on a 1.2cm² chip. The CORDIC algorithm has been implemented as a pipeline of

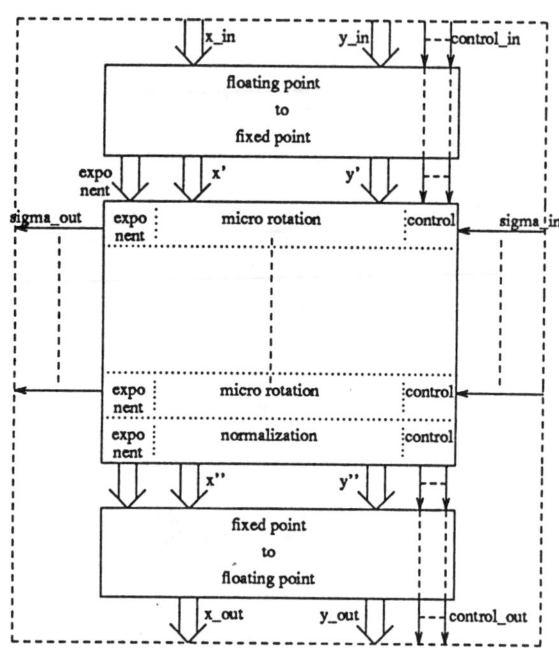

Figure 5.1. Floating Point Pipeline CORDIC Architecture

micro-rotations, while the micro-rotations consist of carry-ripple adders/subtractors (latency is zero), $2 \rightarrow 1$ multiplexers, hardwired shifts, word-pipeline registers and a small control unit that takes care of σ_i selection and coordinate system selection. See figure 3.2. Furthermore, floating point input and output convertors, a handshake module and clock buffers are integrated on the chip. The chip has a hold mode, in which the current state of pipeline registers is frozen, a scan-test mode, in which test vectors can be shifted in and out of the pipeline registers via the sigma-input/output pins, and a synchronous/asynchronous operation mode. The features of the chip are depicted in table 6.1.

TABLE 6.1. Implementation Results

accuracy	range	thru-put	latency	transistors
16-bits	$\pm 2^{16}$	5×10^6 rot/sec	4.4μs	70 000
# inputs	# outputs	dissipation	size	process
71	69	0.13 Watt	1.2cm²	CMOS(1.5μ)

7. Conclusions

In this paper we have described a novel CORDIC algorithm and processor. It is new with respect to its algorithm which is optimized to allow for both circular and hyperbolic rotations with low complexity, both in terms of software implementation and harware implementation. Moreover it is a very efficient algorithm when used in applications in which large amount of rotation and vectorizations are used, in any possible coordinate system and in any order. The required storage and/or silicon area is low and the execution time is independent of the particular operation performed. Another new feature of our CORDIC design is its pipelined architecture and floating point extension. It is angle pipelinable at the bit-level and has an execution time which is independent of any possible operation that can be executed. Its complexity is almost unaffected by the fact that a set of functions are implemented instead of just one. We are currently re-designing our floating point pipeline CORDIC to provide a professional very fast, small and high accuracy CORDIC core for signal processing applications. The novel design will be considerably smaller, mainly because the area consuming linear part of the cordic function can be implemented in a way which is much more efficient than has been done in our prototype CORDIC which is merely a straightforward map of the VLSI algorithm on silicon.

This architecture will enables the (real-time) application of CORDIC arithmetic in 2-dimensional high-speed systolic/wavefront arrays for parallel/pipelined signal processing algorithms and matrix computation applications. For an overview of applications see[Dep90]

References

Ahm85. H.M. Ahmed, "Alternative arithmetic unit architectures for VLSI digital signal processors", in *VLSI and modern signal processing*, ed. S.Y. Kung et.al., Prentice Hall, Inc., Englewood Cliffs, NJ (1985).

Bu86. J.C. Bu, E.F. Deprettere, and F. (A.A.J.) de Lange, "On the Optimization of a Pipelined Silicon Cordic Algorithm", *Proceedings European Signal Processing Conference*, (2) pp. 1227-1230 (September 1986).

Coc72. D.S. Cochran, "Algorithms and Accuracy in the HP-35", *Hewlett-Packard Journal* 10(11)(1972).

Dep84. E.F. Deprettere, P. Dewilde, and R. Udo, "Pipelined cordic architectures for fast VLSI filtering and array processing", *Proc. IEEE Int. Conf. Acoust., Speech, Signal Processing*, pp. 41A6.1-41A6.4 (March 1984).

Dep89. E.F. Deprettere and P.M. Dewilde, "Orthogonal Filter Design and VLSI Implementation", *Proc. International Conference on Computer Systems and Signal Processing*, pp. 779-790 (Dec. 1989).

Dep90. E.F. Deprettere and A.A.J. de Lange, "The Synthesis and Implementation of Signal Processing Application Specific VLSI CORDIC Arrays", *proc. International Symposium on Circuits and Signal Processing (ISCAS)*, (May 1-3. 1990).

Dep91. Ed F. Deprettere and P. Dewilde, "Architectural synthesis of large, nearly regular algorithms: design trajectory and environment", *Annales des telecommunications*, p. (to appear) (1991).

Dep89. J. Bu and E.F. Deprettere, "A VLSI Architecture for High Speed Radiative Transfer 3D Image Synthesis", *the VISUAL COMPUTER*, (5) pp. 121-133 (1989).

Dup89. Jean Duprat and Jean-Michel Muller, "The Cordic Algorithm : new results for fast VLSI implementation", Internal Report, CNRS, Laboratoire LIP-IMAG 69364 Lyon Cedex 07, France (1989).

Hav70. G.L. Haviland and A.A. Tuszynski, "A CORDIC Arithmetic Processor Chip", *IEEE trans. Computers Vol. C-29(2)* , pp. 68-79 (1970).

Hoe88. A.J. van der Hoeven and A.A.J. de Lange, "Synthesis and Verification of the Pipelined Floating Point Cordic Processor", pp. 109-125 in *Lecture Notes of the Nelsis Project*, ed. O.E. Hermann and B.J.F. van Beijnum, University Twente, Enschede (March 9-11 1988).

Hwa79. Kai Hwang, "Computer Arithmetic, Principles, Architecture and Design", *John Wiley & Sons*, (1979).

Jai86. K. Jainandunsing and E.Deprettere, "Design and VLSI Implementation of a Concurrent Solver for N Coupled Least-Squares Fitting Problems", *IEEE journal on SELECTED AREAS IN COMMUNICATIONS* , pp. 39-48 (Jan. 1986).

Jai89. K. Jainandunsing and E.F. Deprettere, "A New Class of Highly Structured Algorithms for Solving Systems of Linear Equations", *SIAM journal on Scient. and Stat. Computations*, pp. 880-912 (September 1989).

Kun88. S.Y. Kung, *VLSI Array Processors*, Prentice-Hall International, Englewood-Cliffs, NJ 07632 (1988).

Lac88. Arild Lacroix, "Floating-Point Signal Processing - Arithmetic, Roundoff-Noise, and Limit Cycles", *Proc. IEEE Int. Symp. on Circuits and Systems (ISCAS)*, pp. 2023-2030 (1988).

Lan88. A.A.J. de Lange, A.J. van der Hoeven, E.F Deprettere, and J. Bu, "An Optimal Floating Point Pipelined CMOS CORDIC Processor", *Proceedings International Symposium on Circuits and Systems (ISCAS)*, pp. 2043-2047 (June, 1988).

McC90. J.V. McCanny, J.G. McWhirter, and S.-Y. Kung, "The Use of Data Dependence Graphs in the Design of Bit-Level Systolic Arrays", *IEEE Transactions Acoustics, Speech, Signal Processing* ASSP-38(5) pp. 787-793 (May 1990).

Mul85. J.M. Muller, "Discrete Basis and Computation of Elementary Functions", *IEEE Transactions on Computers* C-34(9) pp. 857-862 (Sept. 1985).

Nav83. R. Nave, "Implementation of Transcendental Functions on a Numeric Processor", *Microprocessing and Microprogramming* 11 pp. 221-225 (1983).

Phi82. Philips, "Process Description - Philips C5th", USER's MANUAL, Philips Research Laboratries, Eindhoven (1982).

Sch83. C.W. Schelin, "Calculator Function Approximation", *American Mathematical Monthly*, pp. 317-325 (May 1983).

Tag87. N. Tagaki, T. Asada, and S. Yajima, "A hardware algorithm for computing sine and cosine using redundant binary representation", *Systems and Computers in Japan* 18(8) pp. 1-9 (Aug. 1987).

Tak85. Naofumi Takagi, Hiroto Yasuura, and Shuzo Yajima, "High-Speed VLSI Multiplication Algorithm with a Redundant Binary Addition Tree", *IEEE Transactions on Computers* c-34(9) pp. 789-796 (September 1985).

Vee88. Alle-Jan van der Veen and Ed F. Deprettere, "A parallel VLSI direction finding algorithm", *Proc. SPIE Conf. on Advanced Algorithms & Architecures* III(975) pp. 289-299 (1988).

Vol59. J.E. Volder, "The CORDIC trigonometric computing technique", *IRE Trans. Electronic Computers* EC-8 pp. 330-334 (Sep. 1959).

Wal71. J.S. Walther, "An Unified Algorithm for Elementary Functions", *Proc. Spring Joint Computer Conference Vol. 38* , p. 397 AFIPS press, (1971).

Wes85. Neil Weste and Kamran Eshraghian, *Principles of CMOS VLSI DESIGN, A Systems Perspective*, Addison Wesley Publishing Company (1985).

Wil89. F. Williams, "The CORDIC Algorithm - Cast in Silicon", *Electronic Engineering*, pp. 47-50 (Sept. 1989).

Author Index